The Limits of Logic

The International Research Library of Philosophy
Series Editor: John Skorupski

Metaphysics and Epistemology

Identity *Harold Noonan*

Personal Identity *Harold Noonan*

Scepticism *Michael Williams*

Infinity *A.W. Moore*

Theories of Truth *Paul Horwich*

Knowledge and Justification, Vols I & II
Ernest Sosa

Spacetime *Jeremy Butterfield,*
Mark Hogarth, Gordon Belot

Future Volumes: Substance and Causation; Necessity; The Existence and Knowability of God;
Faith, Reason and Religious Language.

The Philosophy of Mathematics and Science

The Ontology of Science *John Worrall*

Mathematical Objects and Mathematical Knowledge *Michael D. Resnik*

Theory, Evidence and Explanation *Peter Lipton*

Future Volumes: Proof and its Limits; Probability; The Philosophy of the Life Sciences; The
Philosophy of History and the Social Sciences; Rational Choice.

The Philosophy of Logic, Language and Mind

Understanding and Sense, Vols I and II *Christopher Peacocke*

Metaphysics of Mind *Peter Smith*

The Limits of Logic *Stewart Shapiro*

Events *Roberto Casati, Achille C. Varzi,*

Future Volumes: Truth and Consequence; Modality, Quantification, High-order Logic; Reference and Logical Form; The Nature of Meaning; Functionalism; Interpretation; Intentionality
and Representation; Reason, Action and Free Will.

The Philosophy of Value

Consequentialism *Philip Pettit*

Punishment *Antony Duff*

Meta-ethics *Michael Smith*

Duty and Virtue *Onora O'Neill*

The Ethics of the Environment *Andrew Brennan*

Medical Ethics *R.S. Downie*

Future Volumes: Aesthetics; The Foundations of the State; Justice; Liberty and Community.

The Limits of Logic
Higher-Order Logic and the Löwenheim-Skolem Theorem

Edited by

Stewart Shapiro

University of St Andrews and
The Ohio State University at Newark

Dartmouth
Aldershot • Brookfield USA • Singapore • Sydney

Published by
Dartmouth Publishing Company Limited
Gower House
Croft Road
Aldershot
Hants GU11 3HR
England

Dartmouth Publishing Company
Old Post Road
Brookfield
Vermont 05036
USA

British Library Cataloguing in Publication Data
The limits of logic : higher-order logic and the
    Löwenheim-Skolem theorem. – (The international research
    library of philosophy)
    1. Logic 2. Reasoning
    I. Shapiro, Stewart, 1951–
    160

Library of Congress Cataloging-in-Publication Data
The limits of logic : higher-order logic and the Löwenheim-Skolem
    theorem / edited by Stewart Shapiro.
                p. cm.— (The international research library of philosophy.
    The Philosophy of logic, language, and mind)
    Includes bibliographical references.
    ISBN 1-85521-731-7
    1. Logic, Symbolic and mathematical. I. Shapiro, Stewart, 1951–
    II. Series: International research library of philosophy.
    Philosophy of logic, language, and mind.
    BC135.L53 1996
    160—dc20   96–516
                                                                CIP

ISBN 1 85521 731 7

Printed in Great Britain at the University Press, Cambridge

Contents

Acknowledgements

The editor and publishers wish to thank the following for permission to use copyright material.

The Aristotelian Society for the essays: Paul Benacerraf (1985), 'Skolem and the Skeptic', *Proceedings of the Aristotelian Society, Supplementary Volume 59*, pp. 85–115; Crispin Wright (1985), 'Skolem and the Skeptic', *Proceedings of the Aristotelian Society, Supplementary Volume 59*, pp. 117–37. Reprinted by courtesy of the Editor of the Aristotelian Society. Copyright © 1985.

Association for Symbolic Logic for the essay: Stewart Shapiro (1985), 'Second-Order Languages and Mathematical Practice', *Journal of Symbolic Logic*, **50**, pp. 714–42. Copyright © 1985 by the Association for Symbolic Logic. All rights reserved.

Peter Clark (1993), 'Logicism, The Continuum and Anti-Realism', *Analysis*, **53**, pp. 129–41. Copyright © Peter Clark.

The Journal of Philosophy for the essays: George S. Boolos (1975), 'On Second-Order Logic', *Journal of Philosophy*, **LXXII**, pp. 509–27; Leslie H. Tharp (1971), 'Ontological Reduction', *Journal of Philosophy*, **68**, pp. 151–64; Stewart Shapiro (1990), 'Second-Order Logic, Foundations, and Rules', *Journal of Philosophy*, **87**, pp. 234–61; George Boolos (1984), 'To Be is To Be a Value of a Variable (or To Be Some Values of Some Variables)', *Journal of Philosophy*, **81**, pp. 430–49; Michael D. Resnik (1988), 'Second-Order Logic Still Wild', *Journal of Philosophy*, **85**, pp. 75–87.

Kluwer Academic Publishers for the essays: Leslie H. Tharp (1975), 'Which Logic is the Right Logic?', *Synthese*, **31**, pp. 1–21. All rights reserved. Copyright © 1975 by D. Reidel Publishing Company, Dordrecht-Holland; George Boolos (1987), 'A Curious Inference', *Journal of Philosophical Logic*, **16**, pp. 1–12. Copyright © 1987 by D. Reidel Publishing Company; Virginia Klenk (1976), 'Intended Models and the Löwenheim-Skolem Theorem', *Journal of Philosophical Logic*, **5**, pp. 475–89. All rights reserved. Copyright © 1976 by D. Reidel Publishing Company, Dordrecht-Holland; Charles McCarty and Neil Tennant (1987), 'Skolem's Paradox and Constructivism', *Journal of Philosophical Logic*, **16**, pp. 165–202. Copyright © 1987 by D. Reidel Publishing Company; Thomas Weston (1976), 'Kreisel, the Continuum Hypothesis and Second Order Set Theory', *Journal of Philosophical Logic*, **5**, pp. 281–98. All rights reserved. Copyright © 1976 by D. Reidel Publishing Company, Dordrecht-Holland; Nino B. Cocchiarella (1988), 'Predication Versus Membership in the Distinction Between Logic as Language and Logic as Calculus', *Synthese*, **77**, pp. 37–72. Copyright © 1988 by Kluwer Academic Publishers.

The Philosophical Review for the essay: George Boolos (1985), 'Nominalist Platonism', *Philosophical Review*, **XCIV**, pp. 327–44. Copyright 1985 Cornell University. Reprinted by permission of the publisher and the author.

Rodopi B.V. for the essay: Peter Simons (1993), 'Who's Afraid of Higher-Order Logic?', *Grazer Philosophische Studien*, **44**, pp. 253–64.

Taylor & Francis for the essays: Gregory H. Moore (1980), 'Beyond First-order Logic: The Historical Interplay between Mathematical Logic and Axiomatic Set Theory', *History and Philosophy of Logic*, **1**, pp. 95–137; Alfred Tarski (1986), 'What are Logical Notions?', *History and Philosophy of Logic*, **7**, pp. 143–54; Ignacio Jané (1993), 'A Critical Appraisal of Second-Order Logic', *History and Philosophy of Logic*, **14**, pp. 67–86; John Corcoran (1980), 'Categoricity', *History and Philosophy of Logic*, **1**, pp. 187–207; Alexander George (1985), 'Skolem and the Löwenheim-Skolem Theorem: A Case Study of the Philosophical Significance of Mathematical Results', *History and Philosophy of Logic*, **6**, pp. 75–89.

Every effort has been made to trace all the copyright holders, but if any have been inadvertently overlooked the publishers will be pleased to make the necessary arrangement at the first opportunity.

Series Preface

The International Research Library of Philosophy collects in book form a wide range of important and influential essays in philosophy, drawn predominantly from English-language journals. Each volume in the Library deals with a field of inquiry which has received significant attention in philosophy in the last 25 years, and is edited by a philosopher noted in that field.

No particular philosophical method or approach is favoured or excluded. The Library will constitute a representative sampling of the best work in contemporary English-language philosophy, providing researchers and scholars throughout the world with comprehensive coverage of currently important topics and approaches.

The Library is divided into four series of volumes which reflect the broad divisions of contemporary philosophical inquiry:

- Metaphysics and Epistemology
- The Philosophy of Mathematics and Science
- The Philosophy of Logic, Language and Mind
- The Philosophy of Value

I am most grateful to all the volume editors, who have unstintingly contributed scarce time and effort to this project. The authority and usefulness of the series rests firmly on their hard work and scholarly judgement. I must also express my thanks to John Irwin of the Dartmouth Publishing Company, from whom the idea of the Library originally came, and who brought it to fruition; and also to his colleagues in the Editorial Office, whose care and attention to detail are so vital in ensuring that the Library provides a handsome and reliable aid to philosophical inquirers.

John Skorupski
General Editor
University of St. Andrews
Scotland

Introduction

An interpreted formal language usually has *first-order variables* that range over a collection of objects, sometimes called a *domain-of-discourse* or simply a *domain*. A language may also contain *second-order variables* that range over properties, sets or relations on the domain-of-discourse or over functions from the domain to itself. For example, the statement 'there is a property that holds of all and only the prime numbers' invokes a second-order variable ranging over number-theoretic properties, and 'Clinton has several leadership qualities in common with Major' is naturally rendered with a second-order variable ranging over leadership qualities. *Third-order variables* range over items like properties of properties, relations over sets, and functions from relations to properties. *Fourth-order variables*, and beyond, are characterized similarly. A formal language is *first-order* if it does not have any variables other than first-order. A language is *second-order* if it has first-order and second-order variables and no others. A language is *third-order* if it has first-order, second-order and third-order variables and no others; and similarly for fourth-order languages and beyond. A language is *higher-order* if it is at least second-order.

Today, the study of first-order languages is the paradigm of mathematical logic. The field is called *first-order logic* or *elementary logic*. Most contemporary texts either ignore higher-order languages altogether or confine them to a short afterthought (e.g., Mendelson 1987, Enderton 1972). However, the systems developed by virtually all of the founders of modern logic, such as Frege (1879), Peano (1889) and Whitehead and Russell (1910), include higher-order languages. First-order logic only appeared as a separate study when some authors, like Löwenheim (1915) and Hilbert and Ackermann (1928), separated out first-order languages as *subsystems* for special investigation. Hilbert and Ackermann call first-order logic the '*restricted* functional calculus'. Moore (1980), reprinted here as Chapter 1, chronicles the historical emergence of first-order logic; see also Moore (1988) and Shapiro (1991, Chapter 7). Debates between the founding fathers of logic profoundly prefigure the issues treated in this volume.

Among contemporary authors, Quine is a longstanding and persistent critic of second-order logic. One broad attack, which dates to Quine (1941), targets traditional systems in which the second-order variables range over intensional items like properties, propositional functions or attributes. Even among friends of intensionality, there is no consensus on *which* properties, say, exist, nor on conditions under which two properties are identical or distinct – thus challenging the Quinean dictum 'no entity without identity'. Quine argues that these intractable metaphysical matters should not soil our pristine work in logic and the foundations of mathematics. Trying to be helpful, he suggests that variables ranging over properties be replaced with variables ranging over respectable *extensional* entities like sets. However, Quine (1986) later argues that, with this replacement, we cross the border out of logic and into mathematics. His conclusion is that second-order logic is not logic, but is 'set theory in disguise':

Set theory's staggering existential assumptions are cunningly hidden ... in the tacit shift from schematic predicate letter to quantifiable variable (Quine 1986, 68).

Historically, the shift went in the other direction, from quantifiable higher-order variables of second-order logic to schematic non-logical predicate letters of first-order logic.

In Chapter 5, Tarski explores the boundary of logic with characteristic rigour and insight. According to Tarski's conception, intensional items are not logical, but he leaves room for sets within logic. On the other hand, one might think that the border dispute between mathematics and logic is of little concern. Subject boundaries can be drawn or removed at will, can they not? The deep issues underlying the territorial dispute concern the purposes and goals of logical study and the metaphysical and epistemic status of sets and properties. The battles are joined in the essays reprinted in Part I of this volume, 'Is Second-Order Logic Logic?'

Most contemporary treatments of second-order logic, such as the development in my own 1991 book, skirt the issue of intensionality by not introducing an identity relation on the items in the range of the second-order variables. The reader is free to take them as ranging over sets, classes, properties or propositional functions at his or her pleasure. To be sure, the model-theory for second-order logic is done in a thoroughly extensional set theory. In each interpretation, the range of the second-order variables is a *set* of sets (of items in the domain-of-discourse). However, one can think of this set as the collection of extensions of whatever items are in the real range of the second-order variables. Since just about everything in mathematics is extensional, the set-theoretic model theory of second-order languages suffices for the foundations of mathematics, if not for the rest of discourse. In Chapters 9 and 22, Simons and Cocchiarella consider the nature of the items in the range of higher-order variables and the relation between such items and sets; see also Cocchiarella (1992).

There are two different model-theoretic systems for second-order logic. According to what is called *standard semantics*, an interpretation of a second-order language is the same thing as an interpretation of the corresponding first-order language; namely a domain-of-discourse and extensions for the non-logical terminology. The n-place second-order variables range over the collection of every set of n-tuples from the domain. That is, if D is the domain-of-discourse, then each n-place second-order variable ranges over the full powerset of D^n. See Shapiro (1991, Chapter 3) for details.

In the alternate model-theoretic semantics, due to Henkin (1950), each interpretation consists of a domain-of-discourse, extensions for the non-logical terminology, and separate ranges for the second-order variables. In a given interpretation, the range of the n-place second-order variables is a (possibly proper) subset of the powerset of D^n. This *Henkin semantics* might be attractive to an advocate of intensional items like properties. The specified range of the second-order variables would be the collection of sets that are the extensions of the relevant properties or propositional functions. On the other hand, if the friend of intensionality believes that for every arbitrary collection S of n-tuples of the domain, there is a property whose extension is S, then she will favour standard semantics.

In Chapters 15 and 16 in Part III ('Plural Quantification'), Boolos develops a way to finesse both the issue of intensionality and the issue concerning the status of sets. According to Boolos, an existential, monadic second-order quantifier is a *plural quantifier* of ordinary language. For example, let $\exists P(\Phi(P))$ be a second-order formula in the language of arithmetic. On the usual model-theoretic semantics, this would read 'there is a set (or property) P of numbers that such ...'. This is analogous to ordinary English locutions like 'there are some people' and 'there are some horses'. The idea is that the second-order quantifier is a different kind of quantifier over the same entities as the ordinary, first-order quantifier; and so second-order variables do not range over anything that the corresponding first-order variables do not range over. *Prima facie*, the plural construction does not presuppose abstract entities like sets or properties. However, Resnik (Chapter 17) argues that the plural construction does presuppose sets and so the Boolos semantics does not reduce the ontological burden of higher-order logic. See Lewis (1991, 62–71) for a clear and insightful discussion of plural quantification; see also Boolos (1985a).

There are several central results concerning first-order logic, sometimes called 'limitative theorems', which do not hold for second-order logic with standard (or Boolos) semantics. *Gödel's completeness theorem* is that there is a complete, sound and effective deductive system D for first-order logic: if Γ is a set of formulas in a first-order language and Φ is a single formula in that language, then Φ is deducible from Γ in D if and only if Φ is satisfied by every model of Γ. It follows that first-order logic is *compact*: for every set Γ of first-order formulas, if every finite subset of Γ has a model, then Γ itself has a model. The *downward Löwenheim-Skolem theorem* is that if Γ is a finite or denumerable set of first-order formulas that has a model whose domain is infinite, then Γ has a model whose domain is the natural numbers, i.e., a denumerable model. In general, if Γ is any satisfiable set of first-order formulas, then Γ has a model whose domain is at most denumerably infinite or the cardinality of Γ, whichever is larger. There is a stronger, *submodel* version of the downward Löwenheim-Skolem theorem: for any denumerable language L, if M is an interpretation with an infinite domain d, then there is a denumerably infinite subset d' of d such that the submodel M' of M generated by d' is elementarily equivalent to M. That is, if $\Phi(x_1 \ldots x_n)$ is a formula of L and $a_1 \ldots a_n$ are all in d', then M satisfies $\Phi(a_1 \ldots a_n)$ if and only if M' satisfies $\Phi(a_1 \ldots a_n)$. Finally, the *upward Löwenheim-Skolem theorem* is that if Γ is a set of first-order formulas such that for each natural number n, Γ has a model whose domain has at least n elements, then for every infinite cardinal κ, Γ has a model whose domain has cardinalty at least κ.

A set Γ of formulas is *categorical* if all of the models of Γ are isomorphic to each other, so that a categorical set of formulas characterizes a single structure 'up to isomorphism'. The Löwenheim-Skolem theorems entail that no set of first-order formulas that has an infinite model is categorical. In particular, there is no first-order characterization of the natural numbers, the real numbers, the Euclidean plane, or the set-theoretic hierarchy. Every satisfiable first-order theory of the natural numbers has uncountable models, and every satisfiable first-order theory of analysis, geometry and set theory has countable models. Such models are sometimes called *non-standard*.

Second-order languages with *standard* (or Boolos) *semantics* have none of these limitations. The upward Löwenheim-Skolem theorem fails because there is a sentence AR that is a categorical characterization of the natural numbers. The axiomatization in Dedekind (1888), equivalent to that in Peano (1889), is second-order and Dedekind established that it is categorical. The crucial second-order item is the induction principle:

$$\forall P[(P0 \ \& \ \forall x(Px \rightarrow Psx)) \rightarrow \forall xPx]$$

stating that every set (or property) of numbers that holds of 0 and is closed under the successor function holds of all natural numbers. In first-order arithmetic, this principle is replaced with an axiom scheme:

$$(\Phi(0) \ \& \ \forall x(\Phi(x){\rightarrow}\Phi(sx))){\rightarrow}\forall x\Phi(x),$$

one instance for each formula Φ of the respective first-order language. Thus, the difference between second-order arithmetic and its first-order counterpart is that, in the latter, one cannot directly state that the induction principle holds for *every* set or property of numbers. The best one can do is a separate induction principle for each set that is definable with a formula in the given first-order language.

Similarly, the downward Löwenheim-Skolem theorems fail in second-order logic with standard semantics because there is a sentence *AN* that is a categorical characterization of the real numbers. Every model of *AN* has the cardinality of the continuum and so is uncountable. The relevant second-order item is the completeness principle that every bounded non-empty set of real numbers has a least upper bound. In first-order analysis, a scheme replaces this principle and so, again, there is a separate completeness principle for each set of real numbers that is definable in the first-order language. Since some sets are not definable, first-order analysis is substantially weaker than its second-order counterpart.

It follows from the compactness theorem that many central mathematical notions, like finitude, countability and well-foundedness, cannot be characterized in a first-order language. There are, however, straightforward second-order characterizations of these notions. For example, Dedekind (1888) defined a set to be finite if there is no one-to-one function from the set to a proper subset of itself. This can be rendered as a second-order sentence with no non-logical terminology:

FIN (X): $\neg\exists f[\forall x(Xx{\rightarrow}Xfx) \ \& \ \forall x\forall y((Xx\&Xy\&fx{=}fy){\rightarrow}x{=}y) \ \& \ \exists z(Xz \ \& \ \forall x(Xx{\rightarrow}fx{\neq}z))]$

See Shapiro (1991, Chapters 3–5) for details.

A corollary of the categorical characterization of the natural numbers and Gödel's *in*completeness theorem is that second-order logic is not complete. There is no effective, sound and complete deductive system for it. The set of second-order logical truths is not recursively enumerable; the set is not even arithmetic. To see this, let Φ be a sentence in the language of arithmetic. Then Φ is true of the natural numbers if and only if $AR{\rightarrow}\Phi$ is logically true. Thus, if the set of valid second-order sentences were arithmetic, then arithmetic truth could be defined in arithmetic, contradicting Tarski's theorem on the definability of truth.

Another corollary is that second-order logic is not compact. Let Γ be the set

$$\{AR, \ a{>}0, \ a{>}s0, \ a{>}ss0, \ a{>}sss0, \ ...\}.$$

Notice that every finite subset of Γ has a model whose domain is the natural numbers. Since *AR* is categorical, any model of Γ itself would be isomorphic to the natural numbers. However, since there is no natural number larger than all of 0, *s*0, *ss*0, ..., there can be no denotation for *a* that makes every member of Γ true. Thus, Γ has no model. Moreover, since any given deduction has only finitely many premises and since every finite subset of Γ is consistent, no contradiction can be deduced from Γ. Thus Γ is consistent, but not satisfiable.

So, for second-order languages, consistency is not coextensive with satisfiability. See Shapiro (1991, Chapter 4) for details and Corcoran (1972) for a lucid discussion of the logical and meta-logical issues.

When it comes to *Henkin semantics*, things are different. The main result of Henkin (1950) is that there is a deductive system that is complete for higher-order languages with (a version of) Henkin semantics. With this semantics, second-order logic is compact and the various Löwenheim-Skolem theorems all hold. For example, there is a Henkin model R of second-order real analysis that is 'doubly countable' in the sense that the domain-of-discourse and the ranges of the second-order variables are all countable. That is, in R there are only countably many objects and countably many properties and relations (or sets). With Henkin semantics, there is no categorical characterization of any infinite structure. In effect, second-order logic with Henkin semantics is equivalent to a multi-sorted first-order language in which the 'predication' or 'membership' relation between objects and properties and sets is non-logical. See Shapiro (1991, Chapter 4).

A theorem of Lindström (1969) shows that limitative properties *characterize* first-order logic. Let L be any model-theoretic type of logic and assume that L has the property of the downward Löwenheim-Skolem theorem. If L is also compact (or if its set of logical truths is recursively enumerable, or if L has the property of the upward Löwenheim-Skolem theorem), then there is a clear sense in which L is equivalent to first-order logic: L cannot make any distinction among models that cannot be made with the corresponding first-order language.

For better or worse, then, standard (or Boolos) semantics is what makes second-order logic distinctive, and most of the contemporary discussion of second-order logic presupposes standard (or Boolos) semantics. The categoricity results and the concomitant failure of the limitative properties are the source of both the main strength and the main weakness of second-order logic.

To some extent, the issue comes down to what the purposes of logic are. If a logic is to be a *calculus* – an effective canon of inference – then second-order logic is not logic. Since, with standard semantics, the set of second-order logical truths is not effectively enumerable, there is no effective deductive system whose theorems are all and only the second-order logical truths. *A fortiori*, no effective deductive system captures the consequence relation of second-order languages.

There are, however, other purposes of logical study. The logician does not live by syntax alone. A logic should codify the standard of validity that arguments must meet, so that the logician can check if a deductive system is sound. An argument is valid if its conclusion is a logical consequence of its premises – even if the conclusion is not deducible from the premises in a favoured deductive system. The model-theoretic notion of logical consequence concerns the *semantics* of mathematical languages. To determine what follows from a given set of premises, we must know what the premises say. Which structures are described by which sentences? In short, the logician must attend to the expressive resources of the languages of mathematics.

Along these lines, it is widely agreed – at least implicitly – that mathematicians do succeed in describing various structures up to isomorphism and in communicating information about these structures to each other. When mathematicians refer to 'the natural numbers', 'the real numbers', etc., they are all talking about the same structures. Similarly, there is little doubt among mathematicians that notions like finitude and well-foundedness are clear and unequivocal. Among mathematicians, if not philosophers, there is a near consensus that the *informal* language of mathematics has the expressive resources sufficient for the ordinary description and communication of these basic structures and notions.

Thus, to capture the semantics of the informal languages of mathematics, a logical system should register the successful description and communication of mathematical structures and concepts. The desideratum is that the expressive resources of the formal languages of logic should match those of the mathematical discourse it models. As above, first-order logic falls short of this. Barwise (1985, 5) puts it well:

> As logicians, we do our subject a disservice by convincing others that logic is first-order and then convincing them that almost none of the concepts of modern mathematics can really be captured in first-order logic.

Chapter 12 by Corcoran is a study of the importance of categoricity and, thus, the inadequacies of first-order logic. Wang (1974, 154) takes a similar line:

> When we are interested in set theory or classical analysis, the Löwenheim-Skolem theorem is usually taken as a sort of defect (often thought to be inevitable) of the first-order logic ... . [W]hat is established (by Lindström's theorems) is not that first-order logic is the only possible logic but rather that it is the only possible logic when we in a sense deny reality to the concept of uncountable ... .

See also Montague (1965), Corcoran (1973) and Issacson (1985).

, In Chapter 6, Boolos presents a simple valid argument Γ in a first-order language such that (1) Γ has a relatively short derivation in a standard second-order deductive system, but (2) Γ has no first-order derivation with fewer lines than the number of particles in the galaxy. According to Boolos, this casts doubt on whether first-order deductive systems are the only way we learn about validity even in first-order languages. See Shapiro (1991, section 5.3.4) for other instances of this 'speed-up' phenomenon.

To be sure, the expressive resources of second-order logic bring a cost. Because important mathematical structures can be characterized up to isomorphism with second-order theories, some substantial mathematical statements can be formulated in second-order languages with no non-logical terminology. For example, there is a sentence that is a logical truth if and only if the continuum hypothesis is true, and there is a sentence that is a logical truth if and only if the axiom of choice holds. Some authors take this as a *reductio ad absurdum* against second-order logic. How can such powerful mathematical matters as the size of the continuum and choice be issues of pure logic? The underlying idea is that logic ought to be a weak discipline with very few substantial presuppositions. Tharp, Wagner and Jané all develop related themes (in Chapters 2, 7 and 8 respectively).

On yet another hand, a popular theme in contemporary philosophy is that there are no sharp borders between disciplines. Ironically, Quine himself is a champion of this idea, arguing that in the seamless web of belief, there is no sharp distinction – no difference in kind – between mathematics and, say, zoology. Why should logic, especially the logic of mathematics, be different? Why expect logic to be free of ontology and substantial expressive resources?

Kreisel (1967) lays out several epistemological advantages of the second-order axiomatizations of mathematical theories. He argues that relying on infinitely many axioms presented via an axiom scheme is unnatural when a single second-order axiom is available.

Suppose, for example, that someone is asked why he believes each instance of the completeness scheme of first-order real analysis. A separate justification for each axiom is out of the question, since there are infinitely many of them. The respondent cannot claim that the scheme characterizes the real numbers since, as we have seen, no first-order axiomatization characterizes this structure. Kreisel suggests that the reason that mathematicians believe the instances of the axiom scheme is that each instance follows from the second-order completeness *axiom*. The respondent might reply that he can see from the scheme itself that all of its instances are true of the real numbers. Notice, however, that this meta-linguistic manoeuvre invokes a generalization over all instances of the scheme. It is not clear that the generalization over the instances of a scheme is any less problematic than the generalization over properties or sets. Also, the scheme itself is an infinite structure and cannot be characterized in a first-order (meta-)language.

A related problem is that each first-order scheme is tied to the ingredients of the particular first-order language in use at the time. Mathematicians, however, are quick to apply the induction or completeness principles to sets regardless of whether they are definable in the given first-order language. Indeed, they usually do not check for definability. This is manifest in the practice of embedding structures in each other. For example, when one sees that there is a structure isomorphic to the real numbers in the set-theoretic hierarchy, one can use set theory to shed light on the real numbers. This works by applying the completeness principle to sets of real numbers definable in set theory. One cannot tell in advance what resources are needed to shed light on a mathematical structure – even after the structure has been adequately characterized.

Chapters 3 and 4 by Boolos and Shapiro pursue the case begun by Kreisel (1967). Weston's Chapter 18 is a reply to some of Kreisel's points, and Hellman (1990) is a reply to the reply. See also Shapiro (1991, Chapters 5, 8) and the discussion remarks on second-order logic in Lakatos (1967).

To sum up, then, advocates of second-order logic focus on the thesis that 'the natural numbers', 'the real numbers', 'finitude', 'countability', etc. all have more or less unequivocal informal interpretations that outrun what is captured in first-order languages, and these informal concepts play a central role in actual mathematical practice. In the section on second-order logic of his classic textbook, Church (1956, 326n) wrote:

> ... our definition of the consequences of a system of postulates. ... can be seen to be not essentially different from [that] required for the ... treatment of classical mathematics. ... It is true that the non-effective notion of consequence, as we have introduced it ... presupposes a certain absolute notion of ALL propositional functions of individuals. But this is presupposed also in classical mathematics, especially classical analysis. ...

As noted, many of those who reject second-order logic (as logic) agree that the informal language of real mathematics is somehow adequate to describe and communicate the various structures and concepts of mathematics. As might be expected, however, some philosophers reject the presupposition that the mathematical structures and notions are unequivocal. The Löwenheim-Skolem theorems are sometimes used to support this sceptical position, and here we turn to the other main theme of this volume, the Skolem paradox.

The technical situation is as follows: let T be any first-order axiomatization of set theory, such as the Zermelo-Fraenkel system. Assume that the language of T has a symbol 'ω' for the finite ordinals (or the natural numbers) and a symbol 'ρ' for the powerset operator. It is a well-known theorem, due to Cantor, that there are uncountable sets and, in particular, that $\rho(\omega)$ is uncountable – there is no function from ω onto $\rho(\omega)$. However, it follows from the downward Löwenheim-Skolem theorem that if T has an infinite model, then it has a countable model C, say a model whose domain is the natural numbers. Let c be the number that is the denotation of '$\rho(\omega)$' in C. Since C is a model of T, it must satisfy Cantor's theorem. Thus, c is uncountable in C. However, since the entire domain of C is countable, c has only countably many 'elements' in the model: only countably many numbers are 'members' of c. How can the very same thing – the denotation of '$\rho(\omega)$' – be both uncountable and yet have only countably many 'elements'? Let us dub this the 'numerical version' of the Skolem paradox.

In C the symbol for the membership relation is not interpreted to be the membership relation. Instead, the membership symbol denotes a complicated number-theoretic relation. For this reason, some authors prefer to consider the stronger submodel version of the Löwenheim-Skolem theorem. Let M be an intended model of T. That is, the domain of M is a collection of sets and the membership relation of M is real membership. The symbol 'ω' is interpreted as the set of finite ordinals (i.e., as ω) and 'ρ' is interpreted as the powerset operation (i.e., as ρ). In Quinean terms, M is homophonic. It follows from the submodel theorem that there is a countable substructure M' of M that is elementarily equivalent to M. That is, the domain of M' is a countable set of sets, with the membership symbol still interpreted as membership – M' is still homophonic. Since both ω and $\rho(\omega)$ are in the domain of the submodel M', there is not even a *prima facie* problem thus far. The countable model M' satisfies the assertion that $\rho(\omega)$ is uncountable and $\rho(\omega)$ *is* uncountable. Of course, not every member of $\rho(\omega)$ is in the domain of M'. It follows from another theorem, sometimes called the 'Mostowski collapse', that there is a model M'' isomorphic to M' whose domain is a countable transitive set with the membership symbol interpreted as membership. The domain of M'' is a countable set of sets and every member of every member of this domain is itself in the domain. In particular, every member of the domain of M'' is countable. (Some writers claim that M'' is a submodel of the given model M, but as far as I can tell, this does not hold in general; M'' is a submodel of M if the domain of M contains every subset of every member of the domain.)

Now we have at least a passing appearance of trouble. The model M'' is isomorphic to M' which is equivalent to M. Thus, M'' satisfies the statement that $\rho(\omega)$ is uncountable. However, since every member of the domain of M'' is countable, $\rho(\omega)$ is not in this domain. Instead, the set donated by '$\rho(\omega)$' in M'' is a countable set d. Thus, M'' is homophonic on membership, but not powerset. The model M'' satisfies the statement that d is uncountable, but manifestly d is a countable set of sets. The very same set d appears to be both countable and uncountable. This is sometimes called the 'submodel version' of the Skolem paradox.

To be sure, virtually all theorists agree that neither the numerical version nor the submodel version of the Skolem paradox represents a contradiction within set theory – within T say. Let COUNT(x, y) be a formula that states that x is a function whose domain is ω and whose range is y. That is, COUNT (x, y) is the statement that y is counted by x. Cantor's theorem is $\neg\exists x(\text{COUNT}(x, \rho(\omega)))$. This sentence holds in C and in M'' (and in M and M'). In the numerical version, it follows that there is no natural number interpreted in C to be an enumeration of the 'elements in C' of c. Since c has only countably many 'elements in C', there is a function from ω on to these 'elements', but no such function is represented in C.

The same goes for the submodel version. The denotation of '$\rho(\omega)$' in M'' is d, and so we have that M'' satisfies $\neg\exists x(\text{COUNT}(x,d))$. That is, there is no function *in the domain of M''* that enumerates the members of d. However, since d is in fact countable, there is a function from ω onto d. All we can conclude is that no such function is in the domain of M''. In sum, when we say that d is countable (really or in M), we mean that there is a function (really or in M) that enumerates the members of d. When we say that d is not countable 'in M'''', we mean that there is no function in M'' that enumerates its members. Thus no contradiction.

Still, some philosophers and logicians have drawn conclusions about the nature of mathematical domains like that of the natural and real numbers, as well as the nature of mathematical concepts like finitude, countability and powerset. One natural 'application' of the theorems is to the philosophical problem of reduction. Does the downward Löwenheim-Skolem theorem entail that any infinite domain can be 'reduced' to the natural numbers? Can the ontologist safely ignore Cantor's theorem and claim that the only mathematical objects we need or, indeed, the only objects at all, are natural numbers? Quine (1966) raises this possibility but goes on to reject it, arguing that a true reduction requires 'proxy functions' not delivered by the Löwenheim-Skolem theorems. Quine's thesis on reduction generated a spirited discussion, including the lucid accounts in Grandy (1969) and Gottlieb (1976). Chapter 10 by Tharp is a clear and useful critique of Quine's views on reduction and the relevance of the Löwenheim-Skolem theorem to this issue.

Skolem himself concluded from the Skolem paradox that, in axiomatic set theory, notions like finitude and countability are inherently relative. They do not have an absolute meaning or extension (see, for example, Skolem 1922, 1941, 1958, 1961). It is not clear what Skolem meant by this relativity. One interpretation is that in axiomatic set theory there is no saying that a set is uncountable *simpliciter*, but only uncountable relative to a given background structure, such as a model of the theory. Accordingly, the statement that a set s is uncountable makes (implicit or explicit) reference to a background like M or M''. On the other hand, Skolem may have meant that the notion of countability is relative to a *theory*, such as one or another axiomatic presentation of set theory. A given set may be uncountable in one theory but countable in another.

Skolem's conclusions concerning this relativity changed during his career. The thesis of the early 1922 classic is that there may well be absolute notions of countability, finitude, etc., but these notions are not registered in axiomatic set theory, or any other axiomatic theory for that matter. The argument leading to the Skolem paradox is a *reductio ad absurdum* against the idea that any axiomatic theory can be an adequate foundation for mathematics. Skolem's early orientation is thus consonant with the view, noted above, that the informal language of mathematics does succeed in describing and communicating unequivocal mathematical concepts and structures, and that formal logic – here axiomatics – must fail where informal language succeeds. Myhill (1951) is an explicit and clear exposition of this 'lesson' of the Löwenheim-Skolem theorems. Of course, this Skolem-Myhill thesis presupposes that the languages and logic are all first-order – an assumption Skolem explicitly embraced and defended throughout his career. Later, Skolem took the Skolem paradox to show that even the *informal* mathematical structures and notions in question are inherently relative. Apparently, the mature Skolem held that the model-theoretic semantics of first-order languages accurately reflects the situation of the mathematician. Since the extensions of countability, finitude, natural number, etc. vary from model to model (of the respective first-order theories), the notions of countability, finitude, natural number, etc. themselves are not fixed once and for all. Skolem's views on the Skolem paradox are treated here in Chapters 1, 19 and 20 by Moore, George and Benacerraf respectively and elsewhere by Hart (1970), McIntosh (1979), Moore (1988) and Shapiro (1991, Chapter 7).

Let us say that s is 'absolutely countable' if s is countable relative to some structure or theory. Resnik (1966) attributes the view that all sets are absolutely countable to the 'Skolemite' (and to Skolem). It is crucial for the Skolemite to be able to say that *one and the same s* is uncountable in one context and countable in another. What is this s? If s is a *linguistic* item, such as a term or definite description, then the Skolemite thesis is only that s is ambiguous. A linguistic item like '$\rho(\omega)$' denotes an uncountable set ($\rho(\omega)$) in one context and a countable set (c or d) in another. This is not a surprising situation, since virtually any language is up for radical (mis)interpretation. More important, this reading of the Skolemite notion does not support the far-reaching relativity. The argument leaves open, and presupposes, that sets themselves are countable, or not, in a non-relative sense.

Suppose, then, that the Skolemite claims that it is *the same set* that is countable in one context and uncountable in another. It is $\rho(\omega)$ itself that is uncountable in some models (or theories) and countable in others. Notice, however, that the argument for the Löwenheim-Skolem theorem does not support this. There is no model delivered in which $\rho(\omega)$ is countable – we only get a model in which '$\rho(\omega)$' denotes a set (other than $\rho(\omega)$) which is countable. To be sure, there are cases in which *some* set d or number c is uncountable in a model (M'' or C), but is absolutely countable. However, returning to Resnik (1966), even this requires a distinction between the 'internal' perspective of the model and the 'external' perspective in which the sets are countable. Resnik argues that the external perspective is not relative, and thus that the Skolemite position is not coherent.

Thomas (1968) and Fine (1968) both claim that Resnik's response to the Skolemite presupposes 'platonism', a thesis that there is an unequivocal set of real numbers, powerset of the natural numbers, etc. Resnik (1969) accepts this point, conceding that some of his arguments are 'predicated upon the existence of the intended (*prima facie*) uncountable models for the standard formal systems for set theory'. He agrees that the term '$\rho(\omega)$' does denote something – really – not just in this or that model.

We have a standoff. Sceptics and anti-realists take the Löwenheim-Skolem theorem to support their position since they see nothing beyond first-order formal languages and their model-theoretic semantics. Realists regard the theorem as irrelevant either because they hold that the underlying language of mathematics is best construed as second-order, or that the informal language has resources beyond what is captured in first-order model theory. Among the papers reprinted here, Klenk and George (Chapters 11 and 19) note the stalemate, the former citing the Resnik/Fine/Thomas exchange, the latter dealing with Skolem's own shifting views. Klenk takes the standoff to support formalism. Chapter 20 by Benacerraf and Chapter 14 by myself show how the Skolem paradox only convinces the converted. Benacerraf suggests the presence of ill-understood informal semantic resources, and my paper supports second-order logic (see also my 1991). Chapter 21 by Wright is a response to Benacerraf, and Chapter 23 by Clark a response to Wright.

There is yet another avenue. In Chapter 13, McCarty and Tennant point out that the normal proofs of all of the Löwenheim-Skolem theorems invoke the law of excluded middle, and it seems that these uses are essential. None of the results holds in intuitionistic or constructive mathematics. Thus, like the advocate of second-order logic, the intuitionist is not saddled with unintended models (see also Chapter 21 by Wright).

Putnam's 'Models and reality' (1980) is not included here since it is reprinted elsewhere (e.g., in Benacerraf and Putnam 1983) and is readily available. Putnam defends a version of Skolemism and claims that it is conclusive against any moderate sort of realism. The only realism that escapes the argument is one with a mystical epistemology. Chapter 4 in this volume contains a criticism by me of Putnam's view.

The essays reprinted here represent a major part of the philosophical literature on higher-order logic and the Skolem paradox. Any omissions are the result of my own ignorance, with apologies to the authors and defenders of any slighted views.

References

Barwise, J. (1985), 'Model-theoretic logics: Background and aims', in J. Barwise and S. Feferman (eds), *Model-Theoretic Logics*, New York: Springer-Verlag, 3–23.

Benacerraf, P. and H. Putnam (eds) (1983), *Philosophy of Mathematics*, 2nd edition, Cambridge: Cambridge University Press.

Boolos, G. (1985a), 'Reading the *Begriffsschrift*', *Mind*, **94**, 331–44.

Church, A. (1956), *Introduction to Mathematical Logic*, Princeton: Princeton University Press.

Cocchiarella, N.B. (1992), 'Conceptual realism versus Quine on classes and higher-order logic', *Synthese,* **90**, 379–436.

Corcoran, J. (1972), 'Conceptual structure of classical logic', *Philosophy and Phenomenological Research*, **33**, 25–47.

Corcoran, J. (1973), 'Gaps between logical theory and mathematical practice', in M. Bunge (ed.), *The Methodological Unity of Science*, Dordrecht: D. Reidel, 23–50.

Dedekind, R. (1888), 'The nature and meaning of numbers', in W.W. Beman (ed.) (1963), *Essays on the Theory of Numbers*, New York: Dover Press, 31–115.

Enderton, H. (1972), *A Mathematical Introduction to Logic*, New York: Academic Press.

Fine, A. (1968), 'Quantification over the real numbers', *Philosophical Studies*, **19**, 27–32.

Frege, G. (1879), *Begriffsschrift, eine der arithmetischen nachgebildete Formelsprache des reinen Denkens*, Halle: Louis Nebert; trans. in van Heijenoort (1967), 1–82.

Gottlieb, D. (1976), 'Ontological reduction', *Journal of Philosophy*, **73**, 57–76.

Grandy, Richard E. (1969), 'On what there need not be', *Journal of Philosophy*, **66**, 806–12.

Hart, W.D. (1970), 'Skolem's promises and paradoxes', *Journal of Philosophy*, **67**, 98–109.

Hellman, G. (1990), 'Towards a modal-structural interpretation of set theory', *Synthese*, **84**, 409–43.

Henkin, L. (1950), 'Completeness in the theory of types', *Journal of Symbolic Logic*, **15**, 81–91.

Hilbert, D. and W. Ackermann (1928), *Grundzüge der theoretischen Logik*, Berlin: Springer.

Isaacson, D. (1985), 'Arithmetical truth and hidden higher-order concepts' in *Logic Colloquium 85*, edited in 1987 by The Paris Logic Group, Amsterdam: North Holland, 147–69.

Kreisel, G. (1967), 'Informal rigour and completeness proofs', in Lakatos (1967), 138–86.

Lakatos, I. (ed.) (1967), *Problems in the Philosophy of Mathematics*, Amsterdam: North Holland.

Lewis, D. (1991), *Parts of Classes*, Oxford: Blackwell.

Lindström, P. (1969), 'On extensions of elementary logic', *Theoria*, **35**, 1–11.

Löwenheim, L. (1915), 'Uber Möglichkeiten im Relativkalkül', *Mathematische Annalen*, **76**, 447–79; trans. in van Heijenoort (1967), 228–51.

McIntosh, C. (1979), 'Skolem's criticisms of set theory', *Noûs*, **13**, 313–34.

Mendelson, E. (1987), *Introduction to Mathematical Logic*, 3rd edition, Princeton, NJ: van Nostrand.

Montague, R. (1965), 'Set theory and higher-order logic', in J. Crossley and M. Dummett (eds), *Formal Systems and Recursive Functions*, Amsterdam: North Holland, 131–48.

Moore, G.H. (1988), 'The emergence of first-order logic', in W. Aspray and P. Kitcher (eds), *History and Philosophy of Modern Mathematics*, Minneapolis: Minnesota Studies in the Philosophy of Science, Volume 11, University of Minnesota Press, 95–135.

Myhill, J. (1951), 'On the ontological significance of the Löwenheim-Skolem theorem', in M. White (ed.), *Academic Freedom, Logic and Religion*, Philadelphia: American Philosophical Society, 57–70; reprinted in 1967 in I. Copi and J. Gould (eds), *Contemporary Readings in Logical Theory*, Macmillan, 40–54.

Peano, G. (1889), *Arithmetices principia, nova methodo exposita*, Turin; trans. as 'The principles of arithmetic, presented by a new method' in van Heijenoort (1967), 85–97.

Putnam, H. (1980), 'Models and reality', *Journal of Symbolic Logic*, **45**, 464–82.

Quine, W.V.O. (1941), 'Whitehead and the rise of modern logic', in P.A. Schilpp (ed.), *The Philosophy of Alfred North Whitehead*, New York: Tudor Publishing Company, 127–63.

Quine, W.V.O. (1966), 'Ontological reduction and the world of numbers', in his *The Ways of Paradox and Other Essays*, New York: Random House, 199–207.

Quine, W.V.O. (1986), *Philosophy of Logic*, 2nd edition, Englewood Cliffs, NJ: Prentice-Hall.

Resnik, M. (1966), 'On Skolem's paradox', *Journal of Philosophy*, **63**, 425–38.

Shapiro, S. (1991), *Foundations without Foundationalism: A Case for Second-Order Logic*, Oxford: Oxford University Press.

Skolem, T. (1922), 'Einige Bemerkungen zur axiomatischen Begründung der Mengenlehre', *Mathematikerkongressen I Helsingfors den 4–7 Juli 1922*, Helsinki: Akademiska Bokhandeln, 217–32; trans. as 'Some remarks on axiomatized set theory' in van Heijenoort (1967), 291–301.

Skolem, T. (1941), 'Sur la porté du théorème de Löwenheim-Skolem', in F. Gonseth (ed.), *Les entretiens de Zurich, 6–9 décembre 1938*, Zurich: Leeman, 25–52.

Skolem, T. (1958), 'Une relativisation des notions mathématiques fondamentales', *Colloques internationaux du Centre National de la Recherche Scientifique*, Paris, 13–18.

Skolem, T. (1961), 'Interpretation of mathematical theories in the first-order predicate calculus', in Y. Bar-Hillel et al. (eds), *Essays on the Foundations of Mathematics, Dedicated to A.A. Fraenkel*, Jerusalem: Magnes Press, 218–25.

Thomas, W.J. (1968), 'Platonism and the Skolem paradox', *Analysis*, **28**, 193–6.

Van Heijenoort, J. (ed.), (1967) *From Frege to Gödel*, Cambridge, MA: Harvard University Press.

Wang, H. (1974), *From Mathematics to Philosophy*, London: Routledge and Kegan Paul.

Whitehead, A.N. and B. Russell (1910), *Principia Mathematica 1*, Cambridge: Cambridge University Press.

Part I
Is Second-Order Logic Logic?

[1]

History and Philosophy of Logic, 1 (1980), 95–137

Beyond First-order Logic: The Historical Interplay between Mathematical Logic and Axiomatic Set Theory

Dedicated to Kurt Gödel and Ernst Zermelo

GREGORY H. MOORE

Department of Mathematics, University of Toronto, Toronto, Ontario, Canada, M5S 1A1

Received 31 December 1979

What has been the historical relationship between set theory and logic? On the one hand, Zermelo and other mathematicians developed set theory as a Hilbert-style axiomatic system. On the other hand, set theory influenced logic by suggesting to Schröder, Löwenheim and others the use of infinitely long expressions. The question of which logic was appropriate for set theory – first-order logic, second-order logic, or an infinitary logic – culminated in a vigorous exchange between Zermelo and Gödel around 1930.

1. INTRODUCTION

How has the historical boundary developed between logic and set theory? Prior to Georg Cantor's researches, the notion of class belonged to logic and hence constituted a part of philosophy. In particular George Boole, whose logic continued to be influenced by the Aristotelian tradition, employed the notion of class in this fashion.[1] Yet during the 1880s, when Cantor generalized his fundamental notion from that of a point-set (a set of *n*-tuples of real numbers) to that of a set with elements of arbitrary nature, the time was ripe for an interplay between set theory as a part of mathematics and logic as a part of philosophy.[2]

---

1. Boole wrote in *An investigation of the laws of thought* (1854, London), 28: 'By a class is usually meant a collection of individuals, to each of which a particular name or description may be applied; but in this work the meaning of the term will be extended so as to include the case in which but a single individual exists, answering to the required name or description, as well as the cases denoted by the terms "nothing" and "universe", which as "classes" should be understood to comprise respectively "no beings", and "all beings"'.

2. G. Cantor, 'Über unendliche lineare Punktmannigfaltigkeiten, III', *Math. Annalen*, **20** (1882), 113–121 (pp. 114–115); also in G. Cantor, *Gesammelte Abhandlungen mathematischen und philosophischen Inhalts* (1932, Berlin), 149–150. This edition is hereafter cited as '*Abhandlungen*'.

Thus it was only natural that both mathematicians and philosophers began to inquire what relationship existed, or ought to exist, between set theory and logic.

If set theory were to form a subdivision of logic, then either one pre-empted its concepts, as both Gottlob Frege and Bertrand Russell were to do, or else one had to formulate set theory as a deductive system with explicit postulates. In 1899 David Hilbert had provided a paradigm for such a deductive system by axiomatizing geometry in a way that did not require the postulates to be supplemented by intuition. Influenced by Hilbert while at Göttingen, Ernst Zermelo axiomatized set theory – not in order to reduce it to logic but so that set theory could provide a foundation for all of mathematics. Nevertheless, the criticisms directed against his axiomatization urgently required that he specify the underlying logic as well.

At the time mathematical logic was very much in flux. As the nineteenth century ended, the distinction between syntax and semantics was not uniformly observed nor even clearly understood (with the exception of Frege and to a lesser extent Hilbert). This partial conflation of syntax and semantics occurred frequently within the Boolean tradition of logic, as developed by C. S. Peirce and Ernst Schröder. Consequently, the door was opened to an infinitary logic – one employing either infinitely long expressions or rules of inference with infinitely many premises. In particular, Peirce treated the notions of universal and existential quantifier as all but identical with infinite conjunctions and disjunctions. Thus the actual infinite entered logic not only in the semantic guise of infinite sets but also in the syntactic guise of infinitely long expressions – a fact that has long been neglected by both historians and philosophers. The situation was further complicated because first-order logic, which permitted quantification only over individuals, had not yet been separated from second-order logic, whose quantifiers could range over predicates as well as over individuals.

Indeed, the question of how to formulate set theory became entangled with the distinction between first-order and second-order logic, as well as with the related problem of categoricity. Certain mathematicians, such as Abraham Fraenkel, sought to modify Zermelo's axiomatization in order to render it categorical. On the other hand, Thoralf Skolem insisted that *no* system of axiomatic set theory could be categorical since every such system had a denumerable model within first-order logic. This so-called 'Skolem paradox' caused Zermelo to reject first-order logic as inadequate to the needs of mathematics – a conclusion which, he believed, was corroborated by Kurt Gödel's incompleteness theorem. In response, Zermelo proposed a very strong infinitary logic, with arbitrarily long expressions and even with infinitely long proofs, to serve as the underlying logic for mathematics.

Yet Zermelo's proposal to strengthen logic attracted no interest at the time. In fact, Skolem's desire to base set theory *solely* on first-order logic became increasingly the standard approach. This article investigates how the restrictive logic of Skolem triumphed over the rich logic of Zermelo, and how this triumph affected the development of set theory.

2. MATHEMATICAL LOGIC AROUND 1900

Historians óf logic have often distinguished between two logical traditions which arose during the latter half of the nineteenth century.[3] The first of these, originating with Boole, grew through the researches of his English followers as well as through those of Peirce in the United States and of Schröder in Germany. Frequently these logicians treated the analogy between the laws of algebra and the laws of logic as a guide in their logical investigations. For Schröder and often for Peirce, the same symbols represented both classes and propositions – an ambiguity important in what follows.

The second tradition, often designated the logistic method, had two sources: Frege and Peano. Yet despite his many insights into the nature of logic, Frege had little influence upon his contemporaries. Even Russell was most affected not by Frege but by Peano, whose symbolism Russell adopted.[4] Above all, it was through Russell that the logistic method influenced the development of mathematical logic.[5]

Within these two logical traditions there arose quite different notions of quantifier. In his *Begriffsschrift* of 1879, Frege introduced a universal quantifier that resembles its modern counterpart, except that he placed no restriction on what may be quantified. At the same time he treated the existential quantifier, for which he did not propose any special symbol, as 'not for every x not'. A decade later Peano introduced a universal quantifier as a subscript to equivalence and implication, and an existential quantifier as 'not for every x, $A(x)$ is the false'.[6] Thus Peano understood quantifiers, in effect,

---

3. J. van Heijenoort, 'Set-theoretic semantics', in *Logic colloquium 76* (1977, Amsterdam), 183–190 (p. 183); R. Dipert, 'Development and crisis in late Boolean logic: the deductive logics of Peirce, Jevons, and Schröder' (1978, Ph.D dissertation, Indiana), .1–5; and I. Grattan-Guinness, 'Wiener on the logics of Russell and Schröder', *Ann. sci.*, 32 (1975), 103–132.

4. For an informed discussion of Frege's limited influence, see C. I. Lewis, *A survey of symbolic logic* (1918, Berkeley), 114–115; and M. Dummett, *Frege: philosophy of language* (1973, London), xxiv–xxv. Concerning Peano's approach, Russell wrote in a letter of 15 April 1910 to Jourdain: 'Until I got hold of Peano, it had never struck me that Symbolic Logic would be of any use for the Principles of mathematics, because I knew the Boolian stuff and found it useless. It was Peano's ε, together with the discovery that relations could be fitted into his system, that led me to adopt symbolic logic. . . . The only part played by Cantor's work was that I tested my logic of relations by its applicability to Cantor' (quoted from I. Grattan-Guinness, *Dear Russell – dear Jourdain* (1977, London), 133–134).

5. For a discussion of how Frege's and Russell's logics differ from modern mathematical logic, see W. Goldfarb, 'Logic in the twenties: the nature of the quantifier', *Journal of symbolic logic*, 44 (1979), 351–368 (pp. 351–354).

6. G. Frege, *Begriffsschrift* (1879, Halle), section 11; G. Peano, *Arithmetices principia* (1889, Turin), sections 2–3. Both are translated in J. van Heijenoort, *From Frege to Gödel: A source book of mathematical logic, 1879–1931* (1967, Cambridge, Mass.), 24, 27, 87–89.

similarly to Frege. Nevertheless, at the time Peano was not acquainted with Frege's work, but cited instead an article of 1885 in which Peirce had introduced quantifiers quite differently. In fact, Peirce defined universal and existential quantifiers by analogy to infinite algebraic sums and products.

Peirce's notion of quantifier, a term that he suggested, merits closer examination. His symbols Σ and Π, originally representing algebraic sum and product, stood respectively for set-theoretic union and intersection on the one hand and for existential and universal quantifiers on the other. Since he used this notation for relations (or classes) and for propositions, the reader might wonder at times which meaning was intended. Yet this very confusion was to prove fruitful, since it encouraged the introduction of infinitely long expressions into logic.

Crediting the invention of quantifiers to his student O. H. Mitchell, Peirce wrote:

Here, in order to render the notation as iconical as possible we may use Σ for *some*, suggesting a sum, and Π for *all*, suggesting a product. Thus $\Sigma_i x_i$ means that x is true for some one of the individuals denoted by i, or $\Sigma_i x_i = x_i + x_j + x_k$ etc. In the same way $\Pi_i x_i$ means that x is true of all these individuals, or $\Pi_i x_i = x_i x_j x_k$ etc. .... It is to be remarked that $\Sigma_i x_i$ and $\Pi_i x_i$ are only *similar* to a sum and product; they are not strictly of that nature, because the individuals of the universe may be innumerable.[7]

Peirce's last sentence suggests two interpretations. On the one hand, it may be taken to assert that $\Sigma_i x_i$ and $\Pi_i x_i$ are not to be considered as infinitely long propositions. On the other hand, it may mean that $\Sigma_i x_i$ and $\Pi_i x_i$ stand for such infinitely long propositions, while infinite sums and products are not strictly permitted in algebra (where one replaces them by limits of finite sums and products). The latter interpretation is more reasonable in the given context, and was the one evidently taken by Schröder.

Thus Frege and Peano, the founders of the logistic method, introduced quantifiers in the modern sense (but without distinguishing between first-order and second-order logic), while Peirce and those who followed him in developing the algebraic tradition of logic had something rather different in mind. To identify a universal quantifier with an infinite conjunction, one must specify a particular domain. If this domain is infinite and if there is an individual constant naming each individual of the domain, it becomes possible to treat each universal quantifier as an infinite conjunction, and likewise each existential quantifier as an infinite disjunction, in the manner that Peirce did. Of course, if the given domain is finite, then finite conjunctions and disjunctions suffice to represent quantifiers.

It is intriguing to find that Hilbert, who was later much influenced by

---

7. C. S. Peirce, 'On the algebra of logic: a contribution to the philosophy of notation', *Amer. jour. math.*, 7 (1885), 180–202 (pp. 194–195); also in his *Collected papers*, vol. 3 (1933, Cambridge, Mass.), 228.

Russell's *Principia mathematica*, represented quantifiers in 1904 by essentially the same means as Peirce. At the Third International Congress of Mathematicians, held at Heidelberg, Hilbert analyzed the foundations of logic and of the real numbers. To secure these foundations properly and to circumvent the set-theoretic paradoxes, he insisted that the laws of logic and some of those for arithmetic must be developed simultaneously. Above all, he considered such paradoxes to indicate that traditional logic had failed to fulfill the rigorous demands which set theory now imposed on it. In the course of outlining a logical theory for the positive integers, Hilbert employed both infinite conjunctions $A(1)$ & $A(2)$ & ... and infinite disjunctions $A(1) \vee A(2) \vee ...$, where $A(x)$ was a number-theoretic proposition. Moreover, he introduced $A(x^{(u)})$ and $A(x^{(o)})$ – meaning 'for every x, $A(x)$' and 'there exists an x, $A(x)$' respectively, where x was a positive integer – as abbreviations for $A(1)$ & $A(2)$ & ... and $A(1) \vee A(2) \vee ...$.[8] Since he did not cite either Peirce or Schröder, it appears that Hilbert independently formulated this method of defining quantifiers with a fixed domain. Nevertheless, he did not further elaborate a syntax for his infinitely long expressions, nor did he investigate the relationship between this and any other form of logic. Two decades later he came to employ a version of the Axiom of Choice, rather than infinite expressions, in order to define quantifiers (see section 6).

Since Schröder did not always distinguish clearly between the calculus of domains and the calculus of propositions (with quantifiers), it remains uncertain whether he intended his infinite expressions to be understood as referring only to classes or also to propositions.[9] However, his successors understood him to use quantifiers, in effect, as infinite conjunctions and disjunctions. After Schröder's death in 1902, Eugen Müller reconstructed from his *Nachlass* a book which Schröder had intended to write but had left unfinished – a synopsis of his extensive investigations in the algebra of logic. The second volume of this *Abriss der Algebra der Logik*, which Müller published in 1910, contained a brief treatment of infinite conjunctions and disjunctions reminiscent of Hilbert's. If $P(x)$ was a proposition in a denumerable domain of individuals (with names $x_1, x_2, ...$ respectively), then the universal proposition $\Pi_x P(x)$ held if and only if $P(x_1) \cdot P(x_2) \cdot ...$ was satisfied, while the existential proposition $\Sigma_x P(x)$ held if and only if $P(x_1) + P(x_2) + ...$ was satisfied.[10] Furthermore, he went beyond Hilbert by discussing repeatedly quantified

---

8. D. Hilbert, 'Über die Grundlagen der Logik und der Arithmetik', *Verhandlungen des Dritten Internat. Math.-Kongresses in Heidelberg* (1905, Leipzig), 174–185 (p. 178). Strictly speaking, he used $A(1)$ u. $A(2)$ u. ... (rather than his later notation $A(1)$ & $A(2)$ & ...) and $A(1)$ o. $A(2)$ o. ... (rather than $A(1) \vee A(2) \vee ...$), where 'u.' abbreviated *und* (and) and 'o.' abbreviated *oder* (or). In the English translation in van Heijenoort (footnote 6), 'u.' is rendered as 'a.'.

9. See, for example, E. Schröder, *Vorlesungen über die Algebra der Logik (exacte Logik)* (3 vols., 1895–1903), vol. 3, 514–516.

10. Note that Müller followed Schröder in using · for conjunction and + for disjunction.

100
                                    G. H. MOORE

propositions, such as $\Pi_x \Sigma_y P(x, y)$, analogously. All the same, Müller did not
extend the analogy by considering infinite strings of quantifiers.[11]

In 1915 Leopold Löwenheim did make this analogy. At first, while reviewing
volume two of Müller's *Abriss* in 1911, Löwenheim was content to emphasize
the role of infinite sums (disjunctions) and products (conjunctions) in
Schröderian logic. He criticized Müller for not explicitly defining such infinite
sums and products:

> Admittedly the reader can easily extend the definition in §91 [for finite disjunctions] himself.
> However, the rules for calculating infinite sums of propositions remain unproved. In each case
> the extension of the axioms and proofs to infinite products and sums [of propositions] is
> unconditionally required for the general calculus of domains, since otherwise the most important
> applications become impossible....[12]

Here, as elsewhere, Löwenheim worked fully within the Schröderian algebraic
logic – including infinitely long expressions.

With such a wealth of expressive power, it may seem odd that in 1915
Löwenheim also considered a possible restriction on logic as it then existed. At
that time he distinguished *Relativausdrücke*, expressions in which the
quantifiers ranged over relations, from *Zählausdrücke*, expressions whose
quantifiers acted only on individuals. Moreover, his term *Zählausdruck*
suggests strongly that he conceived of those individuals as represented by
natural numbers. Here he became the first logician to separate first-order logic
clearly from second-order logic, and to acknowledge that first-order logic
deserved to be studied in its own right. Like Schröder, on the other hand,
Löwenheim's expressions and quantifiers involved semantics as well as syntax.
That is, he considered a first-order expression to require that one specify a
domain of individuals, where the quantifier ranged over the name of each
individual in the domain. Thus the domain (a semantic concept) was built into
the syntactic expression. Likewise he did not explicitly distinguish the names of
individuals from the individuals themselves, nor did he introduce relation-
symbols in a fashion distinct from relations.[13] Once again the partial conflation
of syntax and semantics encouraged the use of infinitely long expressions.

At this juncture Löwenheim advanced beyond Hilbert and Müller by combining first-order expressions not only with infinite conjunctions and disjunctions but with transfinitely many quantifiers as well. These infinitary propositions occurred in his proof of the 'Löwenheim–Skolem theorem', which he
stated in the following fashion: if a first-order expression is satisfied in every

11. E. Müller (ed.), *Abriss der Algebra der Logik* (2 vols., 1909–1910, Leipzig), vol. 2, 35–37.
It is not known how much of the *Abriss* was written by Schröder, since his *Nachlass* disappeared
during the Second World War.

12. L. Löwenheim, review of Müller (footnote 11), vol. 2, in *Archiv der Math. und Physik*, **17**
(1911), 71–73 (p. 72).

13. L. Löwenheim, Über Möglichkeiten im Relativkalkül', *Math. Annalen*, **76** (1915),
447–470 (pp. 450–459); translated in van Heijenoort (footnote 6), 228–251.

MATHEMATICAL LOGIC AND AXIOMATIC SET THEORY 101

finite domain but not in every domain, then it is not satisfied in some denumerable domain. To begin his demonstration, he showed that every first-order expression can be transformed into an equivalent one having a normal form such that all the existential quantifiers precede the universal ones. However this string of quantifiers might be infinite – with one such quantifier for each element in the domain under consideration. As an example, he examined the expression $\Pi_i \Sigma_k A_{ik}$ where the domain consisted of the positive integers. This expression was equivalent to

$$\prod_i (A_{i1} + A_{i2} + A_{i3} + \cdots)$$

and hence to

$$\sum_{k_1, k_2, k_3, \ldots} (A_{1k_1} \cdot A_{2k_2} \cdot A_{3k_3} \cdots).$$

and finally to

$$\sum_{k_1 = 1,2,3,\ldots} \sum_{k_2 = 1,2,3,\ldots} \sum_{k_3 = 1,2,3,\ldots} (\ldots A_{1k_1} \cdot A_{2k_2} \cdot A_{3k_3} \cdots).^{14}$$

Five years later Skolem was particularly careful to recast this portion of Löwenheim's proof by introducing Skolem functions (see section 5).

To sum up, Löwenheim undertook to separate first-order logic from second-order logic and to establish first-order logic as a subject worthy of study within the Schröderian tradition. Yet at the very same time he developed an infinitary logic as an adjunct to first-order logic and utilized this infinitary logic to demonstrate the Löwenheim–Skolem Theorem for first-order logic. By contrast Skolem, who initially extended Löwenheim's research in infinitary logic, soon abandoned the use of infinitely long expressions and their set-theoretic emphasis. Instead Skolem argued that mathematical logic ought to be identified *solely* with first-order logic. In this way he motivated Zermelo to counter such a restriction by introducing an infinitary logic considerably more powerful than even Löwenheim's (see section 8).

3. FROM CANTOR TO ZERMELO: THE FOUNDATIONS OF SET THEORY

The path leading from Cantor's researches to Zermelo's axiomatization of set theory was circuitous. Little concerned with axiomatic systems or with logic in general, Cantor did not rely on Boole's investigations or those of Boole's successors. Nor did he conceive of his results within a formal system, such as the one that Frege proposed in his *Begriffsschrift* of 1879. In fact, while reviewing each other's publications, Cantor and Frege disagreed in print over various matters, including the role of definitions in set theory.[15]

14. *Ibid.*
15. For a discussion of the controversy between Cantor and Frege, see J. Dauben, *Georg Cantor: his mathematics and philosophy of the infinite* (1979, Cambridge, Mass.), 220–228.

102   G. H. MOORE

Although Cantor made no attempt to axiomatize set theory, he did propose
for it a metamathematical axiom (to be discussed below) as well as what he
termed 'a law of thought'. The latter arose in 1883 from his concept of a well-
ordered set: 'It is always possible to bring any *well-defined* set into the *form* of
a *well-ordered* set. Since to me this law of thought appears to be fundamental,
rich in consequences, and particularly marvelous for its general validity. I shall
return to it in a later article'.[16] Indeed this law of thought, hereafter termed 'the
well-ordering principle', proved vital to Cantor's theory of infinite cardinals.
After repeatedly relying on the well-ordering principle in articles and
correspondence over the next decade, he revised its logical status in 1895. At
that time he published the first article of his *Beiträge*, which presented his
mature theory of ordinals and cardinals to a mathematical audience, and there
he conceded that the well-ordering principle required a proof. More precisely,
while he did not directly refer to this principle, he admitted that its consequence
the trichotomy law of cardinals ($m < n$ or $m = n$ or $n < m$ for all cardinals m
and n) remained unproven, and he intentionally refrained from using
trichotomy to deduce other theorems.[17]

While the well-ordering principle had found no supporters by 1895, Cantor's
problem of establishing the trichotomy of cardinals attracted the interest of
various mathematicians soon thereafter. In Germany, Schröder asserted that a
demonstration for trichotomy could only be obtained from algebraic logic and
even then would be quite difficult.[18] However, he made no real progress toward
such a goal.

By contrast the Italian mathematician Cesare Burali-Forti, who was
collaborating with Peano on the *Formulaire de mathematiques*, approached
the problem from an axiomatic perspective. In 1896, when adopting Richard
Dedekind's definition of finite set, Burali-Forti had expressed a firm opinion on
the axiomatization of set theory within a Peanoesque framework: 'We ought to
consider the concepts of *class* and of *correspondence* as primitive (or
irreducible) and to assign to them a system of properties (*postulates*) from
which it is possible to *deduce* logically all the properties which are usually
attributed to those concepts. At present, such a system of postulates is
unknown'.[19] Burali-Forti did not attempt to formulate such a system.

Yet in order to deduce the properties of finite sets from Dedekind's defini-

16. G. Cantor, 'Über unendliche, lineare Punktmannigfaltigkeiten. V', *Math. Annalen*, **21**
(1883), 545–591 (p. 550); *Abhandlungen*, 169.
17. G. Cantor, 'Beiträge zur Begrüdung der transfiniterr Mengenlehre. I', *Math. Annalen*, **46**
(1895), 481–512 (p. 484); *Abhandlungen*, 285.
18. E. Schröder, 'Ueber zwei Definitionen der Endlichkeit und G. Cantor'sche Sätze', *Nova
acta Leopoldina*, **71** (1898), 303–376 (p. 349). Unfortunately, Schröder's proof of what has
become known as 'the Schröder–Bernstein theorem' was erroneous; see A. Korselt, 'Über einen
Beweis des Äquivalenzsatzes', *Math. Annalen*, **70** (1911), 294–296.
19. C. Burali-Forti, 'Le classi finite', *Accad. delle Sci. di Torino, atti*, **32** (1896), 34–52
(p. 36).

tion, Burali-Forti proposed the following postulate (I): if A is a family of non-empty classes, then there exists a one–one function from A into the union of A, i.e., $A \leqslant \overline{\overline{\cup A}}$.[20] As Russell pointed out a few years later, (I) was false unless one also required A to be disjoint.[21] With that stipulation, (I) could well have formed part of an axiomatization for set theory. While investigating the problem of trichotomy, Burali-Forti proposed an additional postulate (II), which he described as a truth possessing 'the degree of simplicity and clarity appropriate to primitive propositions': for any classes A and B, there exists a function $f : A \rightarrow B$ which is either one–one or onto.[22] From (I) and (II) he correctly deduced the trichotomy of cardinals, but neither his proof nor his postulates attracted any interest at the time. Thus the first attempt to begin axiomatizing set theory – one, it is important to note, that occurred *before* any set-theoretic paradoxes were known – ended inconclusively.

Meanwhile Cantor had become convinced that he could demonstrate the well-ordering principle and hence the trichotomy of cardinals as well. While the letter from Cantor to Hilbert which contained this proof has evidently not survived, similar letters written to Dedekind in the summer of 1899 are extant.[23] What is intriguing about this demonstration, in terms of the relationship between logic and set theory, is Cantor's recognition that his concept of set cannot be identified with the most general concept of class or collection. For Cantor, a set (*Menge*) was a collection (*Vielheit*) that could be considered as a unity (*Einheit*), a single object, without contradiction. On the other hand, he insisted, there were collections which could not be considered as completed or finished without generating a contradiction, and these he termed 'absolutely infinite' (*absolut unendliche*) or 'inconsistent' (*inkonsistente*) collections. In order to show that the collection Ω of all ordinals was absolutely infinite, he employed an argument which closely resembles what was later named Burali-Forti's paradox. To establish that every set can be well-ordered, it then sufficed to demonstrate that every transfinite cardinal is an aleph. Suppose, he argued, that some infinite collection V cannot be well-ordered. Then Ω can be mapped one–one onto some subcollection V' of V. Since Ω is inconsistent, so is V' and hence V as well. But as only consistent collections (that is, sets) have cardinal

---

20. Ibid., *46*. This postulate, which I have termed 'the partition principle', played an important role in the early history of the axiom of choice; see my 1979 (footnote 40 below), 14–15, 39–41, 130–135, 264–265.

21. B. Russell, 'On some difficulties in the theory of transfinite numbers and order types', *Proc. London Math. Soc.*, (2) **4** (1906), 29–53 (p. 49). Consider, for example, the class $A = \{\{1\}, \{2\}, \{1,2\}\}$.

22. C. Burali-Forti, 'Sopra un teorema del sig. G. Cantor', *Accad. delle Sci. di Torino, atti*, **32** (1896), 229–237 (p. 236). Strictly speaking, Burali-Forti assumed (II) only if A and B were uncountable, since he believed that he could prove (II) in the countable case.

23. G. Cantor, *Abhandlungen*, 443–451. For a discussion of how the letters of 28 July and 3 August 1899 became conflated, see I. Grattan-Guinness, 'The rediscovery of the Cantor–Dedekind correspondence', *Jahresbericht Deut. Math.-Verein.*, **76** (1974), 104–139 (pp. 126–131, 134–135).

104   G. H. MOORE

numbers, then every transfinite cardinal is an aleph and the well-ordering principle is shown.[24]

The question of absolutely infinite collections, such as Ω and that of all alephs, prompted Cantor to examine the consistency of each of his alephs. A few weeks later he wrote to Dedekind that it was impossible to demonstrate the consistency even of each finite set. Such consistency was 'a simple indemonstrable truth', which he named 'the axiom of arithmetic'.[25] Likewise he considered the consistency of each aleph to be an indemonstrable truth, which he termed 'the axiom of Extended transfinite arithmetic'. The metamathematical axiom that each aleph cannot lead to a contradiction appears to be the only postulate, other than the well-ordering principle, that Cantor ever proposed for set theory. How to establish the consistency of each aleph, and of set theory in general, was a question that greatly interested Hilbert (see section 6).

It is important to realize that Cantor did not view his argument, which established Ω to be absolutely infinite, as paradoxical. Upon discovering this argument, Cantor exhibited no alarm over the health of set theory – in decisive contrast to Frege's dismay in 1902 when he learned of Russell's paradox. What Cantor believed himself to have discovered was that certain collections are too large to be handled consistently as unities, and so he designated them as absolutely infinite (see section 5). Instead of regarding these absolutely infinite or inconsistent collections as a source of paradoxes, he treated them as tools with which to obtain new mathematical results – an approach adopted later by Zermelo in his axiomatization of set theory, and by Gödel in his incompleteness theorem.

Similarly, the mathematical community was little occupied with set-theoretic or logical paradoxes before Russell's *The principles of mathematics* appeared in 1903. Despite received tradition, Burali-Forti did *not* believe that he had found a paradox in 1897.[26] Neither the mathematician who reviewed Burali-Forti's article for the abstracting journal *Jahrbuch über die Fortschritte der Mathematik*,[27] nor anyone else at the time, considered the article to contain a paradox. What Burali-Forti demonstrated was rather that there exist order-types α and β of perfectly ordered sets such that neither $\alpha \leqslant \beta$ nor $\beta \leqslant \alpha$. On the other hand, Cantor established later the same year that $\alpha \leqslant \beta$ or $\beta \leqslant \alpha$ for all order-types α and β of well-ordered sets. Shortly afterward Burali-Forti published an addendum noting that his perfectly ordered sets did not coincide

24. G. Cantor, *Abhandlungen*, 443–445.

25. *Ibid.*, 447–448.

26. C. Burali-Forti, 'Una questione sui numeri transfiniti', *Rend. Circ. Mat. Palermo*, **11** (1897), 154–164; translated in van Heijenoort (footnote 6), 104–111.

27. G. Vivanti, review of Burali-Forti (footnote 26), in *Jahrbuch Fortschritte Math.*, **28** (1897), 62–63.

with Cantor's well-ordered ones.[28] In this he was correct. Yet, while these two notions were distinct, some properties of the first satisfied the second. Recognizing this fact in 1902, Russell transferred Burali-Forti's argument to Cantor's well-ordered sets. Since he detected no error in Cantor's proof of 1897, Russell concluded that the class of all ordinals was not well-ordered by its usual ordering.[29] While he did not refer to any paradox or contradiction at the time, he did so the following year in the *Principles*.[30] Thus Russell, predisposed to find paradoxes by discovering his own in 1901, fused Burali-Forti's and Cantor's arguments in order to create what became known as 'Burali-Forti's paradox'.[31]

Zermelo's attitude toward the paradoxes differed appreciably from Russell's. Coming from Berlin, where he had worked in mathematical physics, Zermelo obtained his *Habilitationsschrift* in the same field at Göttingen during 1899. Under Hilbert's influence, which Zermelo later described as the most important of his mathematical career, his interests soon turned to set theory and the foundations of mathematics.[32] Hilbert had just completed his researches on the foundations of geometry and was now occupied with problems concerning the real numbers: well-ordering them, establishing the consistency of his axiomatization for them, and proving the Continuum Hypothesis. By 1900 Zermelo had discovered Russell's paradox, a year before Russell himself, and had communicated it to Hilbert.[33] Nevertheless, Zermelo did not publish this paradox and continued to demonstrate results in set theory.[34] Thus he acted in accord with his belief that this paradox did not seriously endanger the Cantorian theory.

What did concern Zermelo, however, was the problem of establishing that every set (and particularly the set of all real numbers) can be well-ordered. At the same Heidelberg Congress where Hilbert spoke in August 1904, the Hungarian mathematician Julius König had attempted to show that the real numbers cannot be well-ordered. Within a day Zermelo found the gap in

28. G. Cantor, 'Beiträge zur Begrüdung der transfiniten Mengenlehre. II', *Math. Annalen*, 49 (1897), 207–246 (p. 216); *Abhandlungen*, 321. C. Burali-Forti, 'Sulle classe ben ordinate', *Rend. Circolo Mat. Palermo*, 11 (1897), 260; translated in van Heijenoort (footnote 6), 111–112.

29. B. Russell, 'Théorie générale des séries bien ordonnées', *Rivista di mat.*, 8 (1902), 12–43 (pp. 33, 43).

30. B. Russell, *The principles of mathematics* (1903, Cambridge, England), 323.

31. A detailed reinterpretation of the early history of the set-theoretic paradoxes, by Alejandro Garciadiego and myself, is forthcoming in *Historia mathematica*.

32. See sub-section 10.1, p. 14 of Zermelo's report.

33. See Hilbert's letter of 1903 to Frege in the latter's *Wissenschaftliche Briefwechsel* (1976, Hamburg), 79–80. In 1902 Zermelo discussed this paradox with E. Husserl, who made notes of the conversation; see *Husserliana*, vol. 22 (1980, Den Haag), 399; and B. Rang and W. Thomas, 'Zermelo's discovery of the "Russell paradox" ', *Hist. math.* (to appear).

34. E. Zermelo, 'Addition transfiniter Cardinalzahlen', *Göttinger Nachrichten*, (1901), 34–38.

König's argument.[35] Furthermore, by late September Zermelo had succeeded in establishing the well-ordering theorem (every set can be well-ordered) and hence the trichotomy of cardinals as well. Sent to Hilbert as editor of *Mathematische Annalen*, this demonstration was quickly published.[36]

Almost at once an intense controversy arose. Although Cantor's argument for the well-ordering theorem remained unpublished at his own request, it circulated among mathematicians. Yet it met with little approval and little comment in print. Only Philip Jourdain, who had independently discovered Cantor's argument, supported it publicly.[37] On the other hand, Zermelo's demonstration provoked critics to object strongly in England, France, Germany, Holland, Hungary, Italy and the United States. Above all, the axiom of choice (which provided the basis for the proof) encountered fierce resistance. During July 1907 Zermelo responded by composing two articles, completed within sixteen days of each other and connected by numerous internal links. The first included another demonstration for the well-ordering theorem – based explicitly on his new axiomatization for set theory – and vigorously defended his earlier proof against his opponents.[38] The second article detailed the axiomatization itself and referred to the first for a justification of the axiom of choice.[39] Thus, as I have documented in detail elsewhere, Zermelo was primarily motivated to axiomatize set theory not by the paradoxes but by the controversy surrounding his proof that every set can be well-ordered and especially by a desire to secure his axiom of choice against its numerous critics.[40]

Zermelo's system of 1908 was fully in the tradition of Hilbert's axiomatic method. In a manner analogous to Hilbert's *Grundlagen der Geometrie*, he began with a domain (*Bereich*) B of objects – containing the sets and possibly certain non-sets or urelements – and with the primitive relation $\in$ of membership holding between certain objects in B. While he adopted the symbol $\in$ for the membership relation from Peano, he used Schröder's symbol $\in$ for the subset relation rather than Peano's $\supset$. His axioms were seven in number. The axiom of extensionality stated that two sets with the same elements are identical, while his axiom of infinity asserted the existence of a denumerable

35. G. Kowalewski, *Bestand und Wandel*(1950, Munich), 202.

36. E. Zermelo, 'Beweis, dass jede Menge wohlgeordnet werden kann', *Math. Annalen*, 59 (1904), 514–516; translated in van Heijenoort (footnote 6), 139–141.

37. P. Jourdain, 'On the transfinite cardinal numbers of well-ordered aggregates', *Phil. mag.*, (6) 7 (1904), 61–75.

38. E. Zermelo, 'Neuer Beweis für die Möglichkeit einer Wohlordnung', *Math. Annalen*, 65 (1908), 107–128; translated in van Heijenoort (footnote 6), 183–198.

39. E. Zermelo, 'Untersuchungen über die Grundlagen der Mengenlehre. I', *Math. Annalen*, 65 (1908), 261–281; translated in van Heijenoort (footnote 6), 199–215.

40. G. Moore, 'The origins of Zermelo's axiomatization of set theory', *Journal of philosophical logic*, 7 (1978), 307–329. For a detailed discussion of the Axiom of Choice, see my 'Zermelo's axiom of choice: its origins and role in the development of mathematics' (1979, Toronto University Ph.D. dissertation); to be published as *Zermelo's axiom of choice; its origins, development, and influence* (1980, Berlin).

set. The remaining five axioms of conditional existence – elementary sets (singletons and unordered pairs, as well as the empty set), power set, union, choice and separation – permitted the existence of one set to yield that of another. By means of his system, Zermelo intended axiomatic set theory to serve as a foundation for mathematics as a whole.[41]

How did Zermelo come to formulate these particular axioms? Since he explicitly refused to discuss their origins, any reconstruction is necessarily tentative.[42] Certainly his axioms were consonant with set theory as propounded by Cantor and Dedekind, neither of whom, it must be stressed, took an axiomatic approach to set theory. In 1888 Dedekind had stated as a fact, rather than as an axiom, that if two sets S and T have the same elements, then $S = T$.[43] By proposing his axiom of extensionality, Zermelo adopted an extensional view of sets similar to Dedekind's and Peano's. Furthermore, in a letter sent to Dedekind in 1899, Cantor had regarded as true two propositions which resembled Zermelo's later axioms of union and of separation. Yet there is no reason to suspect that Zermelo had access to this letter or Cantor's views, by 1908.[44] The axiom of elementary sets may have been motivated by a desire to acknowledge the empty set to be a legitimate set, as did Schröder but not Frege, and to distinguish a set A from the singleton $\{A\}$, as did Peano but not Dedekind or Schröder. Zermelo did not know of the two postulates proposed by Burali-Forti a decade earlier.

The prehistory of the axiom of infinity merits a more thorough discussion. In 1888 Dedekind had contended on psychological grounds that an infinite set exists: the totality of all objects of one's thought is infinite, since for any thought A there is the thought of A, which is distinct from A.[45] While investigating finite classes in 1896, Burali-Forti expressed his dissatisfaction with this alleged proof, and took the existence of an infinite class as a hypothesis wherever necessary.[46] The following year he again mentioned that the existence of some infinite class must be assumed, although he did not elevate this presupposition to the status of a postulate, as he had done in the two other cases discussed above.[47] In 1904 the American mathematician Cassius Keyser independently criticized Dedekind's argument, as well as a similar one offered by Russell in his *Principles*. In fact, Keyser expressly assumed what he termed 'the axiom of infinity': The set of possible repetitions of an act, mentally performable once, is infinite.[48] Russell, who believed that no

41. E. Zermelo (footnote 39), 261–267.

42. *Ibid.*, 261.

43. R. Dedekind, *Was sind und was sollen die Zahlen?* (1888, Braunschweig), 2–4.

44. G. Cantor, *Abhandlungen*, 444.

45. R. Dedekind (footnote 43), theorem 66.

46. C. Burali-Forti (footnote 19), 38.

47. C. Burali-Forti (footnote 26), footnote 12.

48. C. Keyser, 'The axiom of infinity: a new presupposition of thought', *Hibbert journal*, 2 (1904), 532–552 (p. 551).

such non-logical assumption was required in pure mathematics, argued against the need for this axiom and especially against its psychological character.[49] When Keyser remained unpersuaded, the discussion came to an end.[50]

Zermelo, it should be noted, was not acquainted with the articles of either Burali-Forti or Keyser, and he responded instead to Dedekind's alleged proof. Since Zermelo's domain B could be proved not to be a set in his system, he rejected Dedekind's argument for the existence of an infinite set. In its place he postulated the existence of a set Z, containing the empty set, such that $A \in Z$ implies $\{A\} \in Z$. Such a set Z was infinite, and the least such Z acted as the set of natural numbers. Thus Zermelo's axiom of infinity was considerably more precise than the earlier assumption of Burali-Forti and lacked the psychologism apparent in Keyser's postulate.

Significantly, three of the most fundamental of Zermelo's axioms – those of choice, power set and separation – originated in his first proof of the well-ordering theorem. As initially formulated in 1904, the axiom of choice asserted that for every set M there exists a function assigning to each non-empty subset M' of M an element of M'.[52] In effect he relied on the assumption that the set of all subsets of a given set M exists, embodied in the power set axiom which he first stated explicitly within his second proof of the well-ordering theorem.[53] While criticizing Zermelo's first proof in 1905, the French mathematician René Baire had claimed that, if an infinite set M is given, it is not necessarily true that all the subsets of M are given.[54] It may well have been Baire's claim, well known to Zermelo, which prompted him to postulate the opposite in his power set axiom. Lastly, the axiom of separation acted in large part as a substitute for Cantor's definition of set, a definition which contained the essence of the principle of comprehension. In his proof of 1904, Zermelo already envisioned a form of this axiom as a suitable restriction on Cantor's concept of set.[55] Yet it was precisely this axiom which provoked the most intense criticism against his axiomatization and which made it imperative to specify how his system was related to logic.

4. ZERMELO'S EARLY CRITICS

Zermelo's axiom of separation asserted that for every propositional function $P(x)$ (*Klassenaussage*), if $P(x)$ is *definit* for a set S, then there exists a set containing exactly those elements x of S for which $P(x)$ is true. He defined a given

49. B. Russell, 'The axiom of infinity', *Hibbert journal*, **2** (1904), 809–812.

50. C. Keyser, 'The axiom of infinity', *Hibbert journal*, **3** (1905), 380–383.

51. E. Zermelo (footnote 39), 266.

52. E. Zermelo (footnote 36), 514.

53. E. Zermelo (footnote 38), 107.

54. R. Baire *et alii*, 'Cinq lettres sur la théorie des ensembles', *Bull. Soc. Math. de France*, **33** (1905), 261–273 (p. 264).

55. E. Zermelo (footnote 38), 118–119.

$P(x)$ to be *definit* for a set S if the relation $\in$ on the domain B and 'the universally valid laws of logic' determine whether or not $P(x)$ holds for each x in S.[56] Nevertheless, it remains uncertain what laws of logic he had in mind. Certainly such laws were part of a two-valued classical logic, including the principle of the excluded middle. On the other hand, Zermelo by no means conceived of his axiomatization as restricted to first-order logic, which in any case Löwenheim had not yet separated from second-order logic. Moreover, Zermelo formulated the axiom of separation as an axiom rather than as an axiom schema. Later, indeed, Zermelo inveighed against Skolem's attempts to confine set theory to first-order logic (see section 7).

While the axiom of separation absorbed most of the early criticism directed against Zermelo's system, outside of Germany the public reaction was an almost unbroken silence. In England, Russell published nothing on the subject but wrote privately to Jourdain:

> I have only read Zermelo's article once as yet, and not carefully, except his new proof of [the] Schröder–Bernstein [Theorem] which delighted me. I agree with your criticisms of him entirely. I thought his axiom [of Separation] for avoiding illegitimate classes so vague as to be useless; also, since he does not recognize the theory of types, I suspect that his axioms will not really avoid contradictions, i.e., I suspect new contradictions could be manufactured specially designed to be consistent with his axioms.[57]

In France, Henri Poincaré was the only non-German to discuss Zermelo's system in print. After attacking Russell's theory of types in 1909, Poincaré criticized Zermelo. Poincaré found the axiom of separation objectionable because a proposition $P(x)$ that was *definit* might not be well-defined – an argument that depended heavily on his rejection of the actual infinite.[58] Yet his objections were more complex since they concerned both the self-evidence and the consistency of Zermelo's system:

> Zermelo wanted to construct an impeccable system of axioms; but these axioms cannot be regarded as arbitrary decrees [in the manner of Hilbert's *Grundlagen*], since one would need to prove that these decrees are consistent; and since Zermelo made a *tabula rasa* there is nothing upon which one could base such a proof. Hence these axioms must be self-evident. But what is the mechanism by which they were constructed? One took axioms which are true for finite collections; since one could not extend all of them to infinite collections, one made this extension only for a certain number of them, chosen more or less arbitrarily.[59]

This objection to Zermelo's choice of axioms would not be answered cogently

56. E. Zermelo (footnote 39), 262–263. In a letter to Dedekind in 1899, Cantor had stated – as a fact about sets rather than as an axiom – that every subcollection (Teilvielheit) of a set is a set (see Cantor's *Abhandlungen*, 444).

57. Quoted in I. Grattan-Guinness (footnote 5), 109.

58. H. Poincaré, 'La logique de l'infini', *Revue de métaphysique et morale*, **17** (1909), 461–482 (pp. 476–477).

59. *Ibid.*, 481–482.

until he formulated his cumulative type hierarchy for set theory two decades later (see section 7).

Even within Germany, Zermelo's system found few adherents at first. Only two mathematicians, who were investigating ordinals, applied his axiomatization to obtain new results.[60] Meanwhile Arthur Schoenflies contributed to the public criticism directed against the axiom of separation because he misunderstood the term *definit*.[61] Zermelo might have hoped for support from Felix Hausdorff, who at the time was developing set theory more deeply than any other mathematician. Yet Hausdorff, who had little patience with foundational discussions, refrained from employing the axiomatization.[62] Not until 1915 did Zermelo's system serve as the basis for a major mathematical discovery – Friedrich Hartogs's demonstration that the trichotomy of cardinals implies the well-ordering theorem.[63]

The first critic of Zermelo's system who contributed positively to its development was Hermann Weyl. In 1910, convinced that the system was fundamentally correct, Weyl could not state a definition of *definit* property which fully satisfied him. Nevertheless, he described a property as *definit* if it was built up from the relations $\in$ and $=$ by a finite number of unspecified definition-principles. Moreover he believed that, in the absence of a precise statement for such principles, one could not hope to resolve Cantor's continuum problem. Yet, influenced by Dedekind and Zermelo, Weyl regretted that his reformulation of *definit* property presupposed the natural numbers.[64]

Over the next few years Weyl came to believe that he had successfully defined the notion of *definit* property. A *definit* property was either a proposition of the form $x \in y$, or $x = y$, or was obtained from such propositions by finitely many uses of negation, conjunction, disjunction, existential quantification, or substitution of a constant for a variable. However, as he sought to express this definition without the notion of natural number, his formulation became forbiddingly convoluted. By 1917 he concluded that such a requirement was merely a scholastic pseudoproblem and that set theory must be founded on the natural numbers rather than conversely. Hence he rejected the efforts of Dedekind and Zermelo to build the natural numbers within set

60. G. Hessenberg, 'Kettentheorie und Wohlordnung', *J. reine angew. Math.*, **135** (1909), 81–133; E. Jacobsthal, 'Über den Aufbau der transfiniten Arithmetik', *Math. Annalen*, **66** (1909), 145–194.

61. A. Schoenflies, 'Über die Stellung der Definition in der Axiomatik', *Jahresbericht Deut. Math.-Verein.*, **20** (1911), 222–255 (pp. 231–232, 251).

62. F. Hausdorff, 'Grundzüge einer Theorie der geordneten Mengen', *Math. Annalen*, **65** (1908), 435–505 (pp. 436–437); *Grundzüge der Mengenlehre* (1914, Leipzig), 2.

63. F. Hartogs, 'Über das Problem der Wohlordnung', *Math. Annalen*, **76** (1915), 436–443.

64. H. Weyl, 'Über die Definitionen der mathematischen Grundbegriffe', *Math. naturwiss. Blatter*, **7** (1910), 93–95, 109–113; reprinted in his *Gesammelte Abhandlungen* (4 vols., 1968, Berlin), vol. 1. 298–304 (pp. 303–304).

theory.[65] After accepting such a constructivistic perspective on mathematics, Weyl soon embraced Brouwer's intuitionism.

A number of other mathematicians considered the possibility that sets which were not well-founded occurred in Zermelo's system. Thus in 1911 Schoenflies disallowed sets A such that $A \in A$ by postulating that for all objects A and B, $A \in B$ and $B \in A$ cannot both hold. Moreover, Schoenflies blamed Russell and his paradox for the possibility that $A \in A$ for some set A in Zermelo's system, and then upbraided Zermelo for allowing a philosopher like Russell to influence him.[66] On the other hand, Schoenflies made no attempt to exclude from his own system sets A, B, C such that $A \in B \in C \in A$, much less to rule out descending $\in$-sequences such as $\ldots \in A_3 \in A_2 \in A_1$.

Such descending $\in$-sequences greatly interested Dimitry Mirimanoff of Geneva. In 1917 he endowed them with a special status by designating sets A_1 at the beginning of some such sequence as extraordinary sets, while all others were termed ordinary sets. His chief aim was to resolve what he considered the fundamental problem of set theory: to find necessary and sufficient conditions for a set of objects to exist. In the particular case of ordinary sets he believed himself to possess a solution, but *not* for the general case of all arbitrary sets. In order to carry out this partial solution, he introduced the notion of rank. By definition the empty set and each urelement had a rank of zero, while the rank of any other set was the least ordinal greater than the ranks of all its elements. Consequently every ordinary set had a rank, and the collection O_α of all ordinary sets of rank α was a set. Then his partial solution to the fundamental problem asserted that a collection B of ordinary sets is a set if and only if there exists an ordinal greater than the rank of every set in B.[67]

Two aspects of Mirimanoff's contribution deserve emphasis. Although he did not formulate an axiom which prohibited descending $\in$-sequences, he did investigate the structure of sets which do not form part of such sequences. Hidden within his system but unnoticed at the time were rudiments of Zermelo's later cumulative type hierarchy. The second aspect concerns the axiom of replacement, usually ascribed to Abraham Fraenkel, which Mirimanoff formulated in 1917 as follows: if a collection B of ordinary sets is equipollent to a set, then B is also an ordinary set.[68] While Mirimanoff first introduced the axiom of replacement as a postulate, its ascription to Fraenkel remains accurate in part. It was through Fraenkel rather than Mirimanoff that this postulate entered Zermelo's axiomatization – after a second round of criticism directed at that system.

65. H. Weyl, *Das Kontinuum* (1917, Leipzig), 4–6, 36–37.

66. Schoenflies (footnote 61), 242–244, 229.

67. D. Mirimanoff, 'Les antinomies de Russell et de Burali-Forti et le problème fondamental de la theorie des ensembles', *L'enseignement math.*, **19** (1917), 37–52 (pp. 42–45, 48–52).

68. *Ibid.*, 49. In the letter mentioned in footnote 56, Cantor stated as a fact (but not as an axiom) that every collection equipollent to a set is a set.

112   G. H. MOORE

5. NEW CRITICS AND NEW SYSTEMS

The observations of Poincaré, Schoenflies, Weyl and Mirimanoff did not directly influence the further development of Zermelo's axiomatization. By contrast, the researches of Fraenkel and Skolem during 1921–1922 instigated a second round of criticism which enriched Zermelo's system and which placed it firmly within the framework of mathematical logic. Yet the perspectives of these two critics diverged substantially. While Skolem viewed Zermelo's system from the standpoint of first-order logic, Fraenkel remained suspicious of mathematical logic in general. Indeed, Fraenkel believed that Zermelo's notion of *definit* property was inadequate precisely because it relied on 'general logic', which he considered to be a very insecure base.[69] In order to avoid general logic, Fraenkel introduced the notion of *definitorisch* function as an alternative to that of *definit* property. A function $f(x)$ was *definitorisch* if it could be obtained from the empty set and the set of natural numbers by the finite iteration of the operations of power set, union, and unordered pair.[70]

This fundamental divergence between Fraenkel and Skolem also appeared in their independent formulations of the axiom of replacement. Both arrived at such an axiom through discovering that Zermelo's system did not yield the denumerable set $\{N, \mathscr{P}(N), \mathscr{P}(\mathscr{P}(N)), \ldots\}$, where $\mathscr{P}(N)$ was the power set of the set N of all natural numbers. As proposed by Skolem, the axiom stated that if $A(x, y)$ is a *definit* property (a notion which he revised) and if for every x there is at most one y such that $A(x, y)$ holds, then for each set M there exists a set M_A consisting of every y for which there is some x in M such that $A(x, y)$ is true.[71] By contrast, Fraenkel formulated this axiom rather vaguely at first: If M is a set and if M' is obtained from M by replacing each member of M with some object, then M' is also a set.[72] When he revised the axiom to accord with his more precise notion of *definitorisch* function, it no longer yielded $\{N, \mathscr{P}(N), \mathscr{P}(\mathscr{P}(N)), \ldots\}$. In fact, as John von Neumann demonstrated in 1928, Fraenkel's revised version of replacement could be deduced in Zermelo's system. Consequently von Neumann insisted that, in order to develop the theory of ordinals and cardinals properly, one had to strengthen the axiom

69. A. Fraenkel, 'Zu den Grundlagen der Mengenlehre', *Jahresbericht Deut. Math.-Verein. (Angelegenheiten)*, **31** (1922), 101–102 (p. 101).

70. A. Fraenkel, 'Untersuchungen über die Grundlagen der Mengenlehre', *Math. Zeitschrift*. **22** (1925), 250–273 (p. 254). See also his 'Über den Begriff "definit" und die Unabhängigkeit des Auswahlaxioms', *Sitzungber. Preuss. Akad. Wiss., physik.-math. Klasse*, (1922), 253–257 (pp. 253–254); translated in van Heijenoort (footnote 6), 286.

71. T. Skolem, 'Einige Bemerkungen zur axiomatischen Begründung der Mengenlehre', *Den femte skandinav. mat.-kongressen, Redogörelse* (1922, Helsinki), 217–232 (pp. 225–226); translated in van Heijenoort (footnote 6), 297.

72. A. Fraenkel, 'Zu den Grundlagen der Cantor–Zermeloschen Mengenlehre', *Math. Annalen*, **86** (1922), 230–237 (pp. 230–231).

MATHEMATICAL LOGIC AND AXIOMATIC SET THEORY   113

of replacement by introducing a notion of function which was more general than Fraenkel's.[73] Von Neumann did so.

On the other hand, the sole but substantial difference between Weyl's revision of the property *definit* in 1917, and that introduced independently by Skolem in 1922, concerned first-order logic. Skolem restricted quantification to individuals, whereas Weyl had mentioned no such restriction. In the same vein Skolem replaced Zermelo's axiom of separation by an axiom schema — with one axiom for each of the denumerably many *definit* properties expressible in first-order logic — as is the usual procedure today. Furthermore, Skolem emphasized, the Löwenheim–Skolem theorem implies that Zermelo's system has a countable model even though the system yields the existence of uncountable sets.[74] In this way there arose the justly famous Skolem paradox, which encouraged Skolem to found mathematics on the natural numbers rather than on set theory.[75] Zermelo disagreed vehemently.

Despite Skolem's increasingly constructivistic tendencies, at first he expanded the compass of logic. Like Löwenheim, he based his mathematical logic on Schröder's system, but he emphasized even more strongly than had Löwenheim the pre-eminence of first-order logic. During 1920 he continued to use propositions with countably many quantifiers and even went beyond Löwenheim by permitting uncountable conjunctions and disjunctions. Thus, after introducing Skolem functions to eliminate Löwenheim's arbitrarily large transfinite quantifiers, he deduced the Löwenheim–Skolem theorem for propositions with countably many quantifiers and for denumerable sets of such propositions.[76] It may well have been his distrust of axiomatic set theory and of the uncountable that led him in 1922 to confine first-order logic to expressions of finite length and hence to abandon infinite conjunctions, disjunctions and strings of quantifiers. Thereby Skolem brought to an end the most profound early developments in infinitary logic.

Although Skolem shared with Fraenkel and von Neumann a common concern about the non-categoricity of Zermelo's system, their responses diverged greatly. Skolem viewed non-categoricity from two perspectives. On the one hand, the Löwenheim–Skolem theorem yielded the relativism of set-theoretic notions — regardless of which axiomatization one used for set theory.

73. J. von Neumann, 'Über die Definition durch transfinite Induktion und verwandte Fragen der allgemeinen Mengenlehre', *Math. Annalen*, 99 (1928), 373–391 (pp. 375–377); also in *Collected works*, vol. 1 (1961, New York), 320–338 (pp. 322–324).

74. T. Skolem (footnote 71), 218–220.

75. *Ibid.*, 229–230. For a detailed discussion of this paradox, see P. Bernays, 'Betrachtungen zum Paradoxon von Thoralf Skolem', *Avhandlinger utgitt av Det Norske Videnskaps-Akademi i Oslo, I. Mat.-Nat. Klasse*, (1957), No. 5, 1–9.

76. T. Skolem, 'Logisch-kombinatorische Untersuchungen über die Erfüllbarkeit oder Beweisbarkeit mathematischer Sätze...', *Videnskapsselskapets skrifter. I, Mat.-nat. Klasse*, 4 (1920), 1–36 (pp. 4–13).

On the other hand, even if one strengthened Zermelo's system by adjoining the axiom of replacement, there would remain other sources of non-categoricity. In particular, Skolem continued, if B was a domain which satisfied Zermelo's axioms and which contained a descending $\in$-sequence, then the subdomain B' of B consisting of all those members of B which did not begin any such sequence also satisfied the axioms. Moreover the subdomain B'' of B', consisting of all the elements of B' whose finite descending $\in$-sequences terminated with the empty set rather than with an urelement, likewise satisfied the axioms. Conversely, one could extend a domain B to a larger domain by adjoining a new urelement. Whether it was possible to adjoin some set of natural numbers not contained in a particular countable domain was a difficult problem which Skolem left open. Furthermore, he suspected that Zermelo's system did not settle every question concerning cardinality – especially the continuum problem.[77] Four decades later Paul Cohen would resolve both of these problems decisively in first-order logic.[78]

However, in 1922 Fraenkel had already investigated the related question whether the axiom of choice was independent of Zermelo's other axioms. To establish its independence, he required infinitely many urelements – an assumption which he would have preferred to avoid. His essential idea was to specify a denumerable set A, each of whose members contained a pair of urelements, and to show that from the rest of Zermelo's axioms one could not derive a choice function for A. In order to generate a model of set theory from the set A, Fraenkel relied on a new postulate to be discussed below.[79] A few years later Heinrich Vieler, a doctoral student of Fraenkel's, studied the independence of Zermelo's other axioms as well.[80]

In contrast to Skolem, Fraenkel sought to render set theory categorical by introducing his new *Beschränktheitsaxiom*, or axiom of restriction, inverse to the completeness axiom which Hilbert had contributed in his *Grundlagen der Geometrie*.[81] Whereas Hilbert had postulated the existence of a maximal model of Euclidean geometry satisfying his other axioms (an act possible because these included the archimedean axiom), Fraenkel's axiom of restriction asserted that the only sets that exist are those whose existence is implied by Zermelo's axioms and replacement. In particular, Fraenkel's system contained no urelements, except when he had to adjoin them in order to obtain an independence proof. Neither Hilbert nor Fraenkel remarked at first that their axioms, as proposed, conflated the object language with the meta-language.

77. T. Skolem (footnote 71), 229.

78. P. Cohen, 'The independence of the continuum hypothesis', *Proc. Nat. Acad. Sci. (U.S.A.)*, **50** (1963), 1143–1148; **51** (1964), 105–110.

79. A. Fraenkel, 'definit' (footnote 70), 253–257.

80. H. Vieler, *Untersuchungen über Unabhängigkeit und Tragweite der Axiome der Mengenlehre in der Axiomatik Zermelos und Fraenkels* (1926, Göttingen).

81. A. Fraenkel, 'Über die Zermelosche Begründung der Mengenlehre', *Jahresbericht Deut. Math.- Verein.*, **30** (1921), 97–98 (p. 98).

MATHEMATICAL LOGIC AND AXIOMATIC SET THEORY   115

Fraenkel believed that the intersection of all domains satisfying the rest of his axioms provided a minimal domain in which the axiom of restriction held.[82] Von Neumann would not agree.

Inspired by Zermelo's axiomatization, von Neumann formulated his own system which he characterized in a letter sent to Zermelo during 1923. Sets were built out of the primitive notions of function and argument; Fraenkel's axiom of replacement was adopted in a strengthened form; and classes that were 'too large' were permitted in the system but not allowed to be members of any set.[83] When von Neumann published an exposition of his system two years later, he emphasized the third characteristic. Embodied as an axiom, it provided a precise criterion that a class was too large to be a set or was, in Gödel's later terminology, a proper class: a class S is a proper class if and only if S is equipollent to the class of all sets. Thus any class cardinally smaller than the class of all sets was a set. Implying both the axiom of separation and the axiom of replacement, von Neumann's axiom also yielded the well-ordering theorem (as well as a strong form of the axiom of choice) by modifying Burali-Forti's 'paradox'.[84] Here von Neumann, as Zermelo had done before him, turned a paradox into a mathematical theorem. Von Neumann could do so precisely because of his criterion for Cantor's 'absolutely infinite collections' (proper classes), a criterion that had been missing from both Cantor's and Jourdain's attempted proofs for the well-ordering theorem (see section 3). Thus von Neumann established that Canor was mistaken in his belief that absolutely infinite collections, such as the collection Ω of all ordinals, cannot be consistently treated as unities.[85]

Next von Neumann turned to the question of categoricity. To render his system as likely as possible to be categorical, he went beyond Mirimanoff by appending an axiom which prohibited those descending $\in$-sequences expressible in the system. Moreover, he recognized that his system would lack categoricity unless he excluded those large cardinals, later termed weakly inaccessible, which were regular alephs $\aleph_\alpha$ having α as a limit ordinal.

82. A. Fraenkel (footnote 72), 234; 'Axiomatische Begründung der transfiniten Kardinalzahlen. I', *Math. Zeitschrift*, **13** (1922), 153–188 (p. 163). For a recent discussion of such axioms of restriction in set theory, see A. Levy in A. Fraenkel *et alii, Foundations of set theory* (second edition 1973, Amsterdam), 113–119.

83. See von Neumann's letter in H. Meschkowski, *Probleme des Unendlichen: Werk und Leben Georg Cantors* (1967, Braunschweig), 271–273.

84. J. von Neumann, 'Eine Axiomatisierung der Mengenlehre', *J. reine angew. Math.*, **154** (1925), 219–240 (p. 223); translated in van Heijenoort (footnote 6), 398. See also his 'Die Axiomatisierung der Mengenlehre', *Math. Zeitschrift*, **27** (1928), 669–752 (pp. 726–727). These papers are also published in von Neumann *Works* (footnote 73), 34–56 and 339–422 respectively.

85. Jourdain's criterion for a collection to be absolutely infinite, or inconsistent, was that it have a subcollection equipollent to Ω. This criterion did not even provide a well-ordering for the set of all real numbers, a weakness that Jourdain himself soon recognized; see his 'On transfinite cardinal numbers of the exponential form', *Phil. mag.*, (6) **9** (1905), 42–56 (p. 42).

Even then von Neumann believed that categoricity remained beyond reach. Firstly, he dismissed Fraenkel's axiom of restriction as untenable since it relied on the concept of subdomain and hence on inconsistent 'naive' set theory. One could overcome this objection, he added, by presupposing a model M of Fraenkel's system and considering its subdomains to be those subclasses of M satisfying the axioms. Nevertheless, the system remained non-categorical since a model M', extending M, might alter which sets were contained in the intersection and hence might exclude new extraordinary sets whose descending $\in$-sequences were 'outside' of M. Secondly, von Neumann analyzed the effect of the Löwenheim–Skolem theorem on categoricity. On the one hand he noted that Euclidean geometry was categorical but, on the other, set theory as presently formulated was not. Yet since geometry was based on set theory he concluded that probably no categorical axiomatization could exist for set theory or, indeed, for any infinite system. Moreover, he insisted that the same critique applied to the axiomatic treatment of the notion of finite set, since a set infinite in one model could be finite in an extension. Consequently the intuitionists were vulnerable as well. Even Hilbert's approach was powerless in these matters, von Neumann added, since they concerned categoricity rather than consistency.[86]

What von Neumann's analysis lacked was a clear understanding of the divergent effects of first-order and second-order logic on categoricity. Today it is well known that the Dedekind–Peano postulates for the natural numbers, like Hilbert's axioms for Euclidean geometry and for the real numbers, are categorical in second-order logic. On the other hand, by 1922 it was clear to Skolem that any theory with an uncountable model, especially set theory, cannot be categorical in first-order logic.[87] Yet, as von Neumann's discussion illustrates, logicians other than Skolem had not yet digested fully the differences between first-order and second-order logic. Once the effect of first-order logic on categoricity had been grasped, one could inquire whether first-order logic truly met the needs of mathematics. Both Hilbert and Zermelo were soon to ask this question.

6. HILBERT: QUANTIFIERS, INFINITARY PROPOSITIONS AND THE AXIOM OF CHOICE

To understand the context in which Zermelo argued for a stronger logic, in fact an infinitary one, we must first see how set theory and logic were related in the researches of Hilbert, the most influential logician of the 1920s. Hilbert had

86. J. von Neumann, 'Eine Axiomatisierung' (footnote 84), 230–232, 238–240.
87. That the theory of the natural numbers lacks categoricity in first-order logic did not emerge clearly in print until Skolem exhibited a non-standard model for arithmetic in 1934; see his 'Über die Nicht-charakterisierbarkeit der Zahlenreihe...', *Fundamenta math.*, 23 (1934), 150–161, and Tarski's comment at the end of that article.

MATHEMATICAL LOGIC AND AXIOMATIC SET THEORY   117

already pondered this relationship when, in his Paris address of 1900, he claimed that each of Cantor's alephs existed but that no class containing all of them existed. What was involved in this assertion was his fervent belief, at that time a philosophical dogma rather than a proposition capable of proof, that the consistency of the axioms for a concept implies the existence of the concept.[88] As we would say now, the consistency of an axiom system implies the existence of a model. When he returned to mathematical logic in 1917, he stressed that both the natural numbers and set theory constituted parts of logic. Then, after he had adopted in effect the logicist thesis of Frege and Russell, he praised *Principia mathematica* for fulfilling the axiomatic method by its axiomatization of logic proper. The present task of foundational research, he concluded, was to demonstrate the consistency of both number theory and set theory, as well as to establish the decidability of every well-formulated mathematical problem in a finite number of steps.[89]

Five years later, responding to the intuitionistic criticisms by which Brouwer and Weyl had buffeted classical analysis, Hilbert sought to formulate each foundational question so clearly that its answer would be unique. He especially desired to restate Zermelo's axiom of choice so as to render it completely self-evident. Against the claims of Brouwer, he reaffirmed the truth of the principle of the excluded middle and hoped to deduce this principle, applied to arbitrary sets of real numbers, via an unspecified function τ. These tasks soon fused into one.[90]

In 1923, beginning a long collaboration with Paul Bernays, Hilbert proposed his *Beweistheorie* (or proof theory) as a consistent foundation for analysis and set theory. Hilbert treated both of them, together with the underlying logic, as syntactic strings of uninterpreted symbols. In this way proofs became strings having a certain form and obtained mechanically from previous strings. Meaning entered only at the metamathematical level, where the permissible methods were finitistic and combinatorial.[91]

As the cornerstone for his *Beweistheorie*, Hilbert turned to an unexpected source. Dismissing the critics of Zermelo's Axiom of Choice, he wrote grandly: 'The essential idea on which the Axiom of Choice is based constitutes a general logical principle which, even for the first elements of mathematical inference, is necessary and indispensable. When we secure these first elements, we obtain at

---

88. D. Hilbert, 'Mathematische Probleme', *Göttinger Nachrichten*, (1900), 253–297 (pp. 264–266); also in *Gesammelte Abhandlungen*, vol. 3 (1935, Berlin), 290–329.

89. D. Hilbert, 'Axiomatisches Denken', *Math. Annalen*, **78** (1917), 405–415 (pp. 411–413); *Abhandlungen*, 146–156.

90. D. Hilbert, 'Neubegründung der Mathematik (Erste Mitteilung)', *Abhand. math. Seminar Hamburg. Univ.*, **1** (1922), 157–177 (pp. 157, 176–177); *Abhandlungen*, 157–177.

91. D. Hilbert, 'Die logischen Grundlagen der Mathematik', *Math. Annalen*, **88** (1923), 151–165 (pp. 151–155); *Abhandlungen*, 178–191. This article was originally titled 'Das Auswahlaxiom in der mathematischen Logik'; see *Jahresbericht Deut. Math.-Verein. (Angelegenheiten)*, **31** (1922), 101.

the same time the foundation for the Axiom of Choice; both are done by means of my proof theory'.[92] What Hilbert had in mind was a new postulate for logic. Since he considered this postulate to go beyond finitistic logic, he named it 'the transfinite axiom':

$$A(\tau_x(A(x))) \rightarrow A(y).$$

In effect it asserted that if a proposition $A(c)$ is provable when c is the value given by the function τ operating on the formula $A(x)$, then $A(y)$ is provable for every value of y. Thus τ selected a value x for which $A(x)$ would be unprovable if this could occur at all.

In what sense did Hilbert consider this axiom to be transfinite? He argued that the transfinite first entered logic with the notions of 'all' and 'there exists'. The proposition that all the objects in a finite set have a certain property $A(x)$ was equivalent, he continued, to a finite conjunction

$$A(x_1) \& A(x_2) \& \ldots \& A(x_n),$$

where $x_1, x_2, \ldots, x_n$ were names of all the members of the set. An analogous result held between 'there exists', applied to the members of a finite set, and a finite disjunction. Here the principle of the excluded middle for finite sets took the form of two equivalences:

$$\sim(\forall x)A(x) \leftrightarrow (\exists x) \sim A(x) \qquad \text{and} \qquad \sim(\exists x)A(x) \leftrightarrow (\forall x) \sim A(x).$$

'These equivalences', he added,

are commonly assumed in mathematics to be valid for infinitely many individuals as well. However, we thereby abandon the ground of the finite and enter the domain of transfinite inferences. If we were always to apply to infinite sets the processes admissible for finite sets, then we would open the gates to error. This is the same source of error which we know well in analysis. There the theorems valid for finite sums and products can be transformed into theorems valid for infinite sums and products only if the inference is secured, in a particular case, by convergence. Analogously, we must not treat the infinite logical sums and products, $A(1) \& A(2) \& A(3) \& \ldots$ and $A(1) \lor A(2) \lor A(3) \lor \ldots$, in the same way as finite ones; my *Beweistheorie*, to be discussed now, provides such a treatment.[93]

This subtle passage deserves a careful reading. At this juncture Hilbert, answering the attacks of the intuitionists, began to fall back to the secure ground of the finite. Hence he needed a new method for handling the content of infinite conjunctions and disjunctions, by which he had defined the universal and existential quantifiers in 1904. His transfinite axiom provided such a method, for he now defined $\forall y A(y)$ as $A(\tau_x(A(x)))$ and $\exists y A(y)$ as $A(\tau_x(\sim A(x)))$.[94] From 1923 onward, quantifiers (based on the transfinite

---

92. D. Hilbert (footnote 91), 152.
93. *Ibid.*, 155.
94. *Ibid.*, 157.

axiom) superseded infinite conjunctions and disjunctions in Hilbert's writings. Thus, as with Skolem, constructivistic objections caused an infinitary logic to be replaced by a finitary one.

Through his *Beweistheorie* Hilbert next endeavored to establish the consistency of his formal system for logic, which included axioms for equality as well as for the natural numbers, and above all to obtain the consistency of the transfinite axiom. From this, he insisted, the consistency of the principle of the excluded middle would follow.[95]

When Hilbert returned to the transfinite axiom in 1926, he reformulated it in a way that clearly paralleled the axiom of choice:

$$A(y) \rightarrow A(\varepsilon_x(A(x))),$$

where ε was a universal choice function.[96] Immediately he applied this ε-axiom in his unsuccessful attempt to establish the continuum hypothesis via the definability of number-theoretic functions. Moreover, he now wished to extend the process, begun in his treatment of quantifiers, by which the infinite would be reduced to the finite. Karl Weierstrass's reduction of infinitesimal analysis to finite sets provided Hilbert with a paradigm. In this spirit he treated propositions about infinite sets as a type of ideal element, analogous to the points at infinity which mathematicians adjoined to the Euclidean plane in order to obtain projective geometry.[97] Two years later he cited the ε-axiom as the source of all such ideal propositions and relied on it to parry Brouwer's continuing thrusts against the principle of the excluded middle.[98] Indeed, Hilbert argued in 1930, the most important questions of his *Beweistheorie* centered around the ε-axiom.[99]

At the same time Hilbert took a public stance toward Zermelo's axiomatization. Expressing his full approval, he praised Zermelo for making precise what Cantor and Dedekind had left vague or partly unconscious. Zermelo's axioms, he insisted, met the demands of the axiomatic method and were such that serious doubts about their correctness could not arise. Regretting that the

---

95. *Ibid.*, 157–165. As stated, both the transfinite axiom and an axiom of equality were second-order.

96. In the recent literature there is some confusion concerning how Hilbert's ε-axiom is related to the axiom of choice. If one begins with a theory in first-order logic, formulated without the ε-axiom, then by adjoining the ε-axiom one obtains a conservative extension according to the second ε-theorem of Hilbert and Bernays. In particular, the ε-axiom is a conservative extension of ZF and hence does not imply the axiom of choice in the language of ZF. On the other hand, if one permits terms of the form $\varepsilon_x(A(x))$ to appear in the axiom schema of replacement, then the ε-axiom yields a strong form of the axiom of choice. See A. Levy (footnote 82), 72–73.

97. D. Hilbert, 'Über, das Unendliche', *Math. Annalen*, **95** (1926), 161–190; translated in van Heijenoort (footnote 6), 367–392.

98. D. Hilbert, 'Die Grundlagen der Mathematik', *Abhand. math. Seminar Hamburg. Univ.*, **6** (1928), 65–92 (pp. 67–68, 81).

99. D. Hilbert, 'Probleme der Grundlegung der Mathematik', *Math. Annalen*, **102** (1930), 1–9 (p. 3).

120   G. H. MOORE

objections of eminent mathematicians, such as Poincaré, had caused Zermelo's
system to be ignored at first, Hilbert praised the young mathematicians
(presumably Fraenkel and von Neumann) who in recent years developed and
completed that system. The chief task of foundational research, he concluded,
was to discover a proof for the substantive, non-logical (*inhaltliche*) assumptions underlying Zermelo's system and, as the first step, for the consistency of
the theory of real numbers.[100]

Yet the question lying just beneath the surface was what sort of logic was
appropriate for set theory and hence for mathematics as a whole. In 1928
Hilbert had argued that first-order logic did *not* suffice. Thus he formulated the
principle of mathematical induction (for the natural numbers) in second-order
logic. While he granted that first-order logic allowed one to deduce the consequences of theories taken individually, he added: 'As soon as the object of
investigation becomes the foundation of . . . mathematical theories, as soon as
one wants to determine in what relation the theory stands to logic and to what
extent it can be obtained from purely logical operations and concepts, then
second-order logic is essential'.[101] In particular, he defined set-theoretic concepts, such as well-ordering, by means of second-order rather than first-order
logic. It was precisely over the role of second-order logic in set theory that
Skolem and Zermelo soon clashed. For Zermelo was not about to abandon an
infinitary logic, as Hilbert had done, but would come to insist that such an
infinitary logic provided exactly what mathematics required.

7. ZERMELO AND SKOLEM: A SECOND-ORDER LOGIC FOR SET THEORY?

In 1929, after more than a decade of ill health had curtailed his mathematical
activity, Zermelo returned to foundational questions. During May and June he
delivered a series of lectures on such questions at the University of Warsaw.
The topics ranged from 'What is mathematics? Mathematics as the logic of the
infinite' (first lecture) to 'On the possibility of an independent definition of set'
(seventh lecture).[102] In this way he established close relations with the Warsaw
school of mathematicians, which Wacław Sierpiński began when Poland was
reunited at the end of the First World War and whose finest student was Alfred
Tarski.[103]

Zermelo's lectures marked a new stage in the development of his ideas on
how mathematics and logic are interrelated. Indeed, his first lecture opened:

Mathematics is *not* to be characterized by its objects (such as: space and time, forms of inner
intuition, theories of number and measurement, and the like) but only, if one wishes to circums-

100. *Ibid.*, 1–3.
101. D. Hilbert and W. Ackermann, *Grundzüge der theoretischen Logik* (1928, Berlin), 86;
compare pp. 83–92.
102. See the lectures in sub-section 10.2 below.
103. For a discussion of the Warsaw school, see G. Kuzawa. *Modern mathematics: the
genesis of a school in Poland* (1968, New Haven).

MATHEMATICAL LOGIC AND AXIOMATIC SET THEORY   121

cribe it completely, by its peculiar process: the proof. Mathematics is a systematization of the provable and, as such, an applied logic; its task is the systematic development of 'logical systems', whereas 'pure logic' only investigates the general theory of logical systems. Now what does 'prove' mean? A 'proof' is the derivation of a new proposition from other previously given propositions, by whose truth its own is established through general logical rules or laws.[104]

While he adopted a version of Russell's logicist thesis, Zermelo's notion of proof was in flux. As stated, it lacked the usual stipulation of finite length. But he had not yet replaced this stipulation by the weaker one of well-foundedness, as he was to do later. In his fourth lecture he insisted that mathematics was concerned primarily with infinite structures: 'The true mathematics is infinitistic in its essence and is founded on the assumption of infinite domains; it can be designated precisely as the "logic of the infinite"'.[105] Increasingly the question of what constituted such a logic occupied the forefront of his researches.

That same year Zermelo published in the Warsaw journal *Fundamenta mathematicae* his first contribution to set theory and logic in nearly two decades, a response to the objections directed against his original notion of *definit* property. His reply rested squarely on second-order logic. Although he did not refer explicitly to Skolem or Weyl, his article seems to have been aimed in part at their critiques. Thus he noted that some authors had treated his notion of *definit* property as superfluous, since they reduced it to general logic. On the other hand, he insisted, 'At the time [1908] there did not exist a generally recognized "mathematical logic", to which I could appeal, any more than today when every foundational researcher has his own logistical system'.[106] Moreover, in 1908 many mathematicians had distrusted logistic because of the paradoxes, and he himself preferred to avoid the unrestricted use of propositional functions.

Yet he was not completely satisfied with either Fraenkel's or von Neumann's revisions of *definit* property. He rejected Fraenkel's proposal to bypass the general notion of *definit* property by admitting only propositional functions of a restricted form, since Fraenkel had characterized those functions by a construction. Such a procedure, Zermelo believed, contradicted the essence of the axiomatic method. As an alternative he proposed to axiomatize the notion of *definit* property – a step which he had contemplated in 1908 without executing it in detail. Indeed, he had kind words for von Neumann's attempt to do so by axiomatizing the general notion of function. Nevertheless, he regarded von Neumann's approach as unnecessarily convoluted since it replaced simple set-theoretic concepts with complicated functions.

Zermelo axiomatized the notion of *definit* property, within what was

104. See sub-section 10.2, p. 1 of lecture 1.
105. *Ibid.*, p. 1 of lecture 4.
106. E. Zermelo, 'Über den Begriff der Definitheit in der Axiomatik', *Fundamenta math.*, **14** (1929), 339–344 (p. 340).

G. H. MOORE

essentially second-order logic, for an arbitrary axiom system rather than for set theory proper. In any such system the primitive relations were to be *definit* for any values of the variables. If the propositions P and Q were *definite*, then so were $\sim P$, $P \& Q$, and $P \vee Q$. If $P(x)$ was *definit* for every value of its free individual variable x, then $\forall x P(x)$ and $\exists x P(x)$ were also *definit*. Furthermore, if $P(f)$ was *definit* for each propositional function f with only individual variables, then so were $\forall f P(f)$ and $\exists f P(f)$. Lastly, the class of *definit* propositions was the intersection of the classes of propositions closed under the logical operations mentioned above. Zermelo was especially pleased with this axiomatization of *definit* property because it did not rely on the notion of natural number – a remark perhaps directed at Skolem's finitism.[107]

Soon Skolem published a critique of this new characterization of *definit* property. After stressing that his article of 1922 contained a similar characterization, he underlined a vital point of difference. While Zermelo permitted quantification over predicates, he did not. Thus he refused to allow the clause whereby $\forall f P(f)$ and $\exists f P(f)$ could be *definit* properties. Moreover, Skolem believed this clause to be obscure and possibly to engender Russell's paradox. Did Zermelo, he wondered, intend to characterize such functions of propositional functions by additional axioms? Most importantly, he insisted, if one considered functions of propositional functions to remain within first-order logic, then no new sets were obtained beyond those generated by his own version of *definit* property. Lastly, he argued that one could not characterize the notion of *definit* property by a finite number of axioms, because of the Löwenheim–Skolem theorem.[108] In this way Skolem strongly attacked Zermelo's attempt to base set theory explicitly on a second-order logic.

In 1930, acknowledging Skolem's critique but maintaining his own position, Zermelo proposed Zermelo–Fraenkel set theory (ZF) almost as we know it today. In particular, he included the original form of Fraenkel's axiom of replacement as well as his own new axiom of foundation. This latter axiom, which seems to have been formulated independently of von Neumann, stated that every non-empty set S was well-founded with respect to $\in$, i.e., S contained an $\in$-minimal element. Thus there existed no infinite descending $\in$-sequence $\ldots \in A_3 \in A_2 \in A_1$. On the other hand, Zermelo did not explicitly include the axiom of choice because its character differed from the other axioms and because it could not serve to delimit models. Nevertheless, he assumed the axiom of choice as a general logical principle within his metamathematics. Although he omitted the axiom of infinity from the system which he called ZF', he included it in ZF.[109]

107. *Ibid.*, 341–344.

108. T. Skolem, 'Einige Bemerkungen zu der Abhandlung von E. Zermelo: |"Über die Definitheit in der Axiomatik" ', *Fundamenta math.*, 15 (1930), 337–341.

109. E. Zermelo, 'Über Grenzzahlen und Mengenbereiche: neue Untersuchungen über die Grundlagen der Mengenlehre', *Fundamenta math.*, 16 (1930), 29–47 (pp. 29–31). Unlike Fraenkel and many later set-theoreticians, Zermelo insisted on retaining urelements.

MATHEMATICAL LOGIC AND AXIOMATIC SET THEORY   123

Zermelo's principal result was that any normal domain D (or, as we would say now, any standard complete model) of ZF' can be characterized up to isomorphism by two cardinals, chosen independently of each other. The first was the cardinal of the basis for D (the set of all urelements in D) while the second was the characteristic of D (in effect, the least ordinal greater than all the ordinals in D). In particular, he established that the characteristics of a normal domain D must be either aleph-zero or a strongly inaccessible cardinal.[110]

To obtain these theorems, Zermelo introduced what is now called the cumulative type hierarchy for set theory. Suppose Q_0 is the set of urelements in a particular domain D with characteristic κ. Zermelo proved that D could be decomposed into κ disjoint levels Q_α, where Q_α consisted of all those sets S which did not occur at earlier levels but all of whose members did. The cumulative levels P_α were defined by letting

$$P_1 = Q_0, \qquad P_{\alpha+1} = P_\alpha \cup Q_\alpha, \qquad \text{and} \qquad P_\beta = \bigcup_{\alpha < \beta} P_\alpha,$$

for every limit ordinal β less than κ. By means of his cumulative type hierarchy, Zermelo deduced that any two normal domains with the same characteristic and equipollent bases were isomorphic and that the isomorphism was completely determined by that between their bases. Furthermore, if two normal domains had equipollent bases but unequal characteristics, then one was isomorphic to a subdomain of the other.[111]

Thus Zermelo offered a vision of the models of set theory, formulated within second-order logic, that differed fundamentally from Skolem's. At the same time Zermelo provided a rationale for the choice of axioms that had been lacking in their initial presentation in 1908 and that Mirimanoff had seen dimly in 1917. In effect Zermelo's system ZF determined initial segments of the cumulative type hierarchy that were closed under certain operations such as union and power set.

By way of conclusion, Zermelo turned to questions of consistency and categoricity. He noted that his researches on models of set theory presupposed the consistency of ZF, which he made no attempt to establish at that time. While the existence of P_ω insured the consistency of ZF', it did not contain an infinite set and hence did not constitute an adequate model for Cantorian set theory. He argued that the smallest adequate model was P_{π_1}, whose subscript was the least strongly inaccessible ordinal. Since any other strongly inaccessible ordinal also provided a model of ZF, he added, then ZF was far from being categorical. While one could enforce categoricity as Fraenkel and von Neumann had wished to do, he argued that its absence was a virtue of ZF. By insisting on categoricity they confined set theory to a single model, whereas

---

110. *Ibid.*, 29–34.
111. *Ibid.*, 36–44.

124   G. H. MOORE

he preferred to develop set theory in full generality and thus to preserve its unlimited applicability. Indeed, he assumed in the metalanguage the large cardinal axiom that the class of all strongly inaccessible cardinals is equipollent to the class of all ordinals.[112]

Sometime between 1930 and 1933, and probably at the beginning of that period, Zermelo wrote a report which he intended to submit to the Emergency Society for German Science, but for unknown reasons it was never published. It contained an overview of his investigations on set theory together with his plans for future research. In this report he consciously staked out a position different from both the intuitionists and the formalists, but emphasized that his own contribution would be primarily mathematical rather than philosophical. Using his investigations of 1930 on normal domains, he developed a type of set-theoretic relativism which he carefully distinguished from that of Skolem, who treated the notions of power set and cardinal number as relative to a model. Moreover, he believed that for Skolem, who insisted that all of set theory could be represented by a denumerable model, the Continuum Problem lost its true meaning.[113]

To the end of this unpublished report Zermelo appended a description of his work in progress. He intended to study the construction of models for set theory and particularly the consistency of set theory. Further he underlined the distinction (introduced in his report) between a closed domain, which became a set in any larger model, and an open domain, such as that of all normal domains, which corresponded to the whole of Cantorian set theory. At a more philosophical level, he argued that mathematics originates from intuitions but cannot be founded on them. Finally, his most intriguing proposal was to investigate 'mathematics as the "logic of the infinite" and the imposibility of a "finitistic mathematics"'.[114] It was this proposal, echoing his Warsaw lectures, which soon brought him into direct conflict with Gödel.

8. ZERMELO AND GÖDEL: FINITARY VERSUS INFINITARY LOGIC

In 1928 Hilbert and Ackermann published a slim volume, *Grundzüge der theoretischen Logik*. Based on courses which Hilbert had given at Göttingen between 1917 and 1922, it discussed both first-order and higher-order logic. This treatise, much concerned with syntax but little with semantics, soon prompted mathematicians to try their ingenuity on two fundamental problems. One was to establish the completeness of first-order logic: if a first-order sentence is valid in every domain, then it can be deduced from the axioms of

112. *Ibid.*, 44–47. This occurred eight years before Tarski proposed the equivalent axiom of inaccessible cardinals in the object language; see A. Tarski, 'Über unerreichbare Kardinalzahlen', *Fundamenta math.*, **30** (1938), 68–89.

113. See pp. 1–5 of the notes in sub-section 10.1 below.

114. *Ibid.*, p. 7.

first-order logic. The other was the *Entscheidungsproblem*, or 'decision problem', which asked for an algorithmic procedure determining the provability or refutability of every first-order (or higher-order) sentence.[115] It remained for a doctoral student of Hans Hahn at Vienna, Kurt Gödel, to resolve the first problem positively and to make a negative solution highly probable for the second.

In his doctoral dissertation, which established the completeness theorem for first-order logic, Gödel exhibited a more profound understanding of the distinction between syntax and semantics – as well as their interrelationship – than had his predecessors. Skolem had failed to observe this distinction, especially as it concerned consistency and satisfiability, by expressing Löwenheim's Theorem in the following form: A first-order sentence is either inconsistent or else satisfiable in a countable domain. However, Skolem demonstrated only that if a first-order sentence is satisfiable in a set M, it is satisfiable in a countable subset of M.[116] What Gödel later established was essentially Skolem's stated theorem. Thus the completeness theorem for first-order logic arose in 1930 rather than a decade earlier.[117]

Shortly afterward Gödel published an abstract of his incompleteness theorem, which made a positive solution to the *Entscheidungsproblem* highly unlikely. The theorem stated that Peano arithmetic contains undecidable propositions if formulated in the (first-order) logic of *Principia mathematica* (*PM*). Moreover, no consistency proof for Peano arithmetic – or, *a fortiori*, for Zermelo–Fraenkel set theory – could be given within *PM*.[118] The following year Gödel published the details.[119]

On 15 September 1931 the annual meeting of the *Deutsche Mathematiker-Vereinigung* took place at Bad Elster, where Fraenkel, Gödel and Zermelo all spoke on the same afternoon. Fraenkel reported on his recent studies of models for set theory, such as his attempt to establish that the countable axiom of choice does not imply the full axiom of choice within the rest of Zermelo's system.[120] Gödel elaborated his incompleteness theorem and its consequences. By contrast, Zermelo's lecture fused polemic with a radically new perspective on the relationship between mathematics and logic.

115. D. Hilbert and W. Ackermann (footnote 101), 65–68, 72–81.

116. T. Skolem (footnote 76), 4–13.

117. K. Gödel, 'Die Vollständigkeit der Axiome des logischen Funktionenkalkuls', *Monatshefte für Math. und Physik*, 37 (1930), 349–360; translated in van Heijenoort (footnote 6), 582–591.

118. K. Gödel, 'Einige metamathematische Resultate über Entscheidungsdefinitheit und Widerspruchsfreiheit', *Anzeiger Akad. Wiss. Wien. Math.-nat. Klasse*, 67 (1930), 214–215; translated in van Heijenoort (footnote 6), 595–596.

119. K. Gödel, 'Über formal unentscheidbare Sätze der Principia Mathematica und verwandter Systeme. I', *Monatshefte für Math. und Physik*, 38 (1931), 173–198; translated in van Heijenoort (footnote 6), 596–616.

120. A. Fraenkel, 'Über eine wesentliche Spezialisierung des Auswahlaxioms', *Jahresbericht Deut. Math.-Verein. (Angelegenheiten)*, 31 (1931), 88.

126 G. H. MOORE

Entitled 'On the levels of quantification and the logic of the infinite', Zermelo's lecture was motivated by his fear that Skolem's views on the logic appropriate for mathematics would prevail. He began forcefully:

From the assumption that all mathematical concepts and theorems must be representable by a *fixed finite system of signs*, one falls inevitably into '*Richard's Paradox*'. Recently this paradox, long after it seemed dead and buried, found a happy resurrection in *Skolemism*, the doctrine that *every* mathematical theory, in particular set theory, is satisfiable in a *countable model*. As is well known, from contradictory premises one may deduce whatever one wishes. Even the strangest consequences which Skolem and others have drawn from their basic assumptions, e.g. the 'relativity' of the concepts of power set and equipollence, have not sufficed to turn them away from a doctrine which for many has seemed to degenerate into a dogma beyond criticism. However, a healthy 'metamathematics', a true 'logic of the infinite', first becomes possible through a *fundamental renunciation* of the assumption characterized above, which I term the '*finitistic prejudice*'. In any case the true subject matter of mathematics is *not*, as many would have it, 'combinations of signs' but *conceptually-ideal relations* between the elements of a conceptually determined *infinite manifold*. Thus our system of signs is always an *incomplete* device, shifting from case to case. It reflects our *finite* understanding of the infinite, which we cannot *immediately* and *intuitively* 'survey' or comprehend, though at least we can approach mastery step by step. In the following an attempt is made to develop the foundations of a 'mathematical logic' which, free from the 'finitistic prejudice' and from inner contradictions, offers enough room for the whole of mathematics as it exists at present (and for its fruitful future development) while *abandoning* all *arbitrary prohibitions* and *restrictions*.[121]

Zermelo modeled his new mathematical logic on the cumulative type hierarchy which he had introduced for ZF set theory the previous year. In fact, it constituted a richer logic than anyone had proposed before. Yet like Schröder, Löwenheim and Skolem, he did not distinguish clearly enough between his syntax and his semantics. Zermelo began by positing that any mathematical theory presupposed a fundamental domain *D* (*Urbereich*) of elements among which certain primitive relations held, or failed to hold, for each combination of elements.[122] From these primitive relations one obtained compound relations by the operations of negation, (finite or infinite) conjunction, and (finite or infinite) disjunction. Here he considered conjunction and disjunction to operate on any ordinal number of propositions. Indeed, he termed these two operations as 'quantifiers' (in the spirit of Schröder) since he did not introduce symbols for existential or universal quantification as such. Through the three operations which he included there arose a system *S* of theorems, which he required to be well-founded with respect to the binary relation of implication. Moreover, the class *Q* of all primitive relations was well-founded with respect to the compound relations. Thus, like the cumulative type hierarchy within set theory, the system *S* could be uniquely decomposed into a

121. E. Zermelo, 'Über Stufen der Quantifikation und die Logik des Unendlichen', *ibid.*, 85–88 (p. 85).
122. To translate this semantic assumption into syntactic terms, one must presuppose that the language contained a name for each individual of *D*.

MATHEMATICAL LOGIC AND AXIOMATIC SET THEORY　　127

well-ordered transfinite sequence of levels Q_α such that the elements of Q_α depended only on previous levels and belonged to the least level with this property.[123]

Next Zermelo turned to the notion of proof, which he defined to be, in effect, the (semantic) notion of logical consequence. For him it was vital that every true sentence be provable. Consequently he found distasteful Gödel's incompleteness theory, whose limitations he believed to lie in its restriction on quantification. Since in Gödel's *PM*-system containing arithmetic there were only a countable number of provable propositions, while the system contained uncountably many propositions, there had to exist undecidable propositions. In fact, he insisted. Gödel's reasoning showed the inadequacies of any finitistic theory of proof. The question whether mathematics possessed any absolutely undecidable propositions would not be resolved, Zermelo concluded, by such relativistic considerations.[124]

On 21 September, still dissatisfied with Gödel's theorem but wishing to dispel any misconceptions, Zermelo wrote to Gödel. While this letter has apparently not survived, Gödel's reply of 12 October is extant. In his letter Zermelo had introduced the class K^* of all formulas valid in Gödel's system. Gödel responded that K^*, unlike his own class K of all provable formulas, was not expressible within the system. At the same time he agreed with Zermelo that 'the existence of such classes [undefinable in the system] naturally follows even more simply from diagonalization or from cardinality considerations'.[125] However, Gödel added:

> I would like to remark that I do not see the essential point of my result to be that one can somehow go beyond the limits of a formal system ... but that for any formal system in mathematics there exist propositions which are *expressible* within this system but which *cannot be decided* by the axioms of this system. ... That one cannot capture the whole of mathematics in a formal system follows already from Cantor's process of diagonalization. However, it was still conceivable that one could formalize certain subsystems of mathematics completely (i.e., in a decidable way). My demonstration shows that this too is impossible, if the subsystem includes the concepts of the addition and multiplication of whole numbers. (Here one should understand by formalization: reduction to finitely many axioms and rules of inference.)[126]

In conclusion, Gödel granted that relatively undecidable propositions were always decidable in higher systems, which nevertheless included other undecidable propositions.

---

123. E. Zermelo (footnote 114), 86. He could have expressed this decomposition in terms of the number of conjunctions and disjunctions in a proposition.

124. *Ibid.*, 87.

125. See the letters, in German, of Gödel and Zermelo in I. Grattan-Guinness, 'In memoriam Kurt Gödel: his 1931 correspondence with Zermelo on his incompletability theorem', *Historia mathematica*, 6 (1979), 294–304 (pp. 298–299).

126. *Ibid.*, 301. By restricting formalization to finitely many axioms, Gödel excluded ZF from consideration within first-order logic.

128   G. H. MOORE

On 29 October Zermelo responded by discussing such higher systems, which he understood quite differently from Gödel:

But such a 'higher' system differs from the original one not through any assumption of *new propositions*, as one could think from your formulation, but solely through new *methods of proof*. And all that your article proves, which *I* always ·stress, comes down to the fact that a 'finitistically restricted' proof-schema does *not suffice* to 'decide' the propositions of an uncountable mathematical system. . . . For what a 'proof' actually is cannot be 'proved' but must be *presupposed* or *assumed* in some form. . . . What does one understand by a 'proof'? In complete generality one understands thereby *a system of propositions such that from the assumption of the premises the validity of the conclusion can reasonably [einsichtig] be asserted.* And still the question remains, what is 'reasonable'? In any case *not merely* (and this you yourself have shown) the propositions of a finitistic schema which in your case can always be 'extended'. Thereby we would actually be *in agreement*: only that from the beginning I have been working on the basis of a *more general* schema, which does not *need* to be extended. And in such a system *all* propositions are actually decidable.[127]

Thus a potentially fruitful dialogue came to an end. While Gödel used Zermelo's cumulative type hierarchy in his later researches that established the relative consistency of the Axiom of Choice and the Generalized Continuum Hypothesis, he did not pursue Zermelo's infinitistic conception of logic.[128] Zermelo himself published only two further articles, which did not go appreciably beyond his lecture of 1931, although the second article explicitly allowed proofs of infinite length.[129] Yet what the situation demanded was a comprehensive and detailed investigation of infinitary logic. What could each of the levels Q_α express that first-order logic could not? After the University of Freiburg im Breisgau fired Zermelo in 1935 for failing to give the Hitler salute, the doors to such an investigation were shut.

9. CONCLUSION

When Zermelo published Cantor's collected works in 1932, he regretted that Cantor the mathematician and Frege the logician had so little understood and appreciated each other.[130] Yet similar misunderstandings marred the relationship between Zermelo and Gödel. Dissatisfied with finitistic restrictions on logic, and with the concomitant Skolem paradox, Zermelo argued for the

127. *Ibid.*, 302–303.
128. K. Gödel, 'The consistency of the axiom of choice and of the generalized continuum-hypothesis', *Proc. Nat. Acad. Sci. (U.S.A.)*, **24** (1938), 556–557; 'Consistency-proof for the generalized continuum-hypothesis', *ibid.*, **25** (1939), 220–224; *The consistency of the continuum hypothesis* (1940, Princeton).
129. E. Zermelo, 'Über mathematische Systeme und die Logik des Unendlichen', *Forschungen und Fortschritte*, **8** (1932), 6–7; 'Grundlagen einer allgemeinen Theorie der mathematischen Satzsysteme (erste Mitteilung)', *Fundamenta math.*, **25** (1935), 136–146 (p. 144).
130. See Zermelo's remarks in Cantor's *Abhandlungen*, 442.

study of infinitary languages and their semantics in order to express fully the content of mathematics. This bold proposal ran counter to the finitism which increasingly dominated mathematical logic by 1930. Influenced by Hilbert and Skolem, Gödel operated within such a finitistic tradition of logic. Thus he confined his researches to first-order logic, where he investigated the role of recursive arithmetic in questions of decidability, and was little inclined to follow the sweeping expansion of logic which Zermelo urged.

Nevertheless, as we have seen, this expansion of logic had its roots in the Peirce–Schröder tradition of algebraic logic and had been developed in part by Hilbert, Löwenheim and Skolem. For philosophical reasons both Hilbert and Skolem retreated to a constructivistic perspective. Thus when Hilbert formulated his transfinite axiom, he borrowed from set theory largely in order to tame infinite logical expressions.

During the decades that followed, logicians have tended to forget that the infinite entered logic from set theory in syntactic as well as semantic guises. Always there was an essential tension between logic and set theory, for while one could base set theory on logic, as is usually the case today, one could also use set theory as a *paradigm* for logic. In a sense the early appearance of infinite conjunctions and disjunctions gave evidence of such a paradigm, which Zermelo vigorously attempted to extend. All the same, the nature of the paradigm changed as pre-axiomatic set theory grew into Zermelo's system of 1908 and then into his cumulative type hierarchy. Above all, infinitely long propositions acquired a clearer structure.

From the perspective of many mathematicians, first-order logic could not be fully adequate since it could not characterize the real numbers, or even the natural numbers, up to isomorphism.[131] Nevertheless, after 1930 mathematical logic became increasingly identified with first-order logic. The logicians (such as Gödel and Skolem) who argued for a restrictive logic had triumphed. Set theory was formulated more and more frequently within first-order logic, rather than within the second-order logic preferred by Zermelo. Even the use of first-order languages with uncountably many symbols, a legacy of researches by the Russian logician Anatol Malcev in 1936, found little application until the 1950s.[132]

Yet what has ebbed can also flow. In 1956 the study of infinitary languages began again, though with little knowledge of the previous labors in this field. Writing a doctoral dissertation under Leon Henkin at the University of Southern California, Carol Karp began to investigate propositions of the form $x = 1 \lor x = 2 \lor x = 3 \lor \ldots$.[133] Later the same year Henkin and Tarski led a

131. Bernays, for example, spoke of the inadequacy of first-order logic in 1939; see his comments in F. Gonseth (ed.), *Les entretiens de Zurich* (1941, Zurich).
132. A. Malcev, 'Untersuchungen aus dem Gebiete der mathematischen Logik', *Mat. sbornik*, **43** (1963), 323–336.
133. C. Karp, *Languages with expressions of infinite length* (1964, Amsterdam), v–vi.

130   G. H. MOORE

seminar on infinitary logic at Berkeley, where their semantic results were presented together with Karp's syntactic ones.[134] During the two decades that
followed, the study of infinitary languages has flourished as a branch of
mathematical logic.[135] On the other hand, the philosophical significance of
infinitary logic has yet to receive serious analysis.

Why did Zermelo's ideas on infinitary logic not take root in the 1930s?
Perhaps they would have if Zermelo had been near the beginning of his career
rather than close to the end. Indeed, his controversial axiom of choice, as well
as his axiomatization for set theory, required considerable time to gain
widespread acceptance. Furthermore, by 1931 first-order logic had not yet
been investigated in depth, and consequently it was difficult to persuade
logicians such as Gödel that a radical extension was needed. Zermelo's ideas
were suggestive but perhaps too vague – and too powerful; when infinitary
logic took root after 1956, it was the weakest infinitary logic $L_{\omega_1,\omega}$ that
attracted the most interest. Perhaps also the difficulty resided in the connection
between higher-order logic and set theory, for if one stepped beyond first-order
logic, one necessarily depended on set-theoretic assumptions such as the
meaning of the power set operation. Finally, but most importantly, what
Zermelo denigrated as the 'finitistic prejudice' was a strong historical force
in the work of Brouwer, Hilbert, Skolem and Gödel. What remains to be
discovered is why, two decades later, infinitary logic took root.

10. APPENDICES: DOCUMENTS FROM ZERMELO'S *NACHLASS*

I have preserved Zermelo's emphases, but expanded some contracted words
and corrected a few errors in spelling. The pagination is indicated by two
vertical lines in the text, while the page number is given in the margin.

10.1. *Zermelo's report to the Emergency Society of German Science*

1 *Bericht* an die Notgemeinschaft der Deutschen Wissenschaft über meine
Forschungen betreffend die *Grundlagen der Mathematik*

Schon vor 30 Jahren, als ich Privatdozent in Göttingen war, begann ich
unter dem Einflusse D. HILBERTs, dem ich überhaupt das meiste in meiner
wissenschaftlichen Entwickelung zu verdanken habe, mich mit den
Grundlagenfragen der Mathematik zu beschäftigen, insbesondere aber mit
den grundlegenden Problemen der CANTORschen *Mengenlehre*, die mir in
der damals so fruchtbaren Zusammenarbeit der Göttinger Mathematiker
erst in ihrer vollen Bedeutung zum Bewusstsein kamen. Es war damals die

134. These were soon published. See A. Tarski, 'Remarks on predicate logic with infinitely
long expressions', *Colloquium math.*, 6 (1958), 171–176; D. Scott and A. Tarski, 'The sentential
calculus with infinitely long expressions', *ibid.*, 165–170; and C. Karp (footnote 133).
135. See, for example, H. Keisler, *Model theory for infinitary logic* (1971, Amsterdam); M.
Dickman, *Large infinitary languages – model theory* (1974, Amsterdam); and J. Barwise,
Admissible sets and structures (1975, Berlin).

Zeit, wo die 'Antinomien', die scheinbaren 'Widersprüche' in der Mengenlehre, die allgemeinste Aufmerksamkeit auf sich zogen und berufene wie unberufene Federn zu den kühnsten wie zu den ängstlichsten Lösungsversuchen veranlassten. In der Überzeugung, dass in diesem Komplexe von Fragen auch die tiefsten Einblicke in das Wesen der Mathematik überhaupt zu gewinnen seien, wandte ich mich diesen Problemen zu in einer Reihe von Arbeiten, die u.a. die damals noch sehr umstrittene Möglichkeit der 'Wohlordnung' betrafen und im Jahre 1907–8 in der Abhandlung 'Über die Grundlagen der Mengenlehre' in den 'Mathematischen Annalen' Bd. 65 zum vorläufigen Abschlusse kamen. Das von mir damals eingeführte Axiomen-System ist seitdem für die axiomatische Forschung auf diesem Gebiete im Wesentlichen massgebend geblieben und hat inzwischen nur einmal durch das von FRAENKEL vorgeschlagene 'Ersetzungs-Axiom' eine wertvolle Erweiterung erfahren, während andererseits der von HAUSDORFF eingeführte Begriff der 'Konfinalität' eine fruchtbare Anwendung des neuen Axioms erst ermöglichte. Mittlerweile war aber auch die Frage nach den 'Grundlagen' aufs Neue wieder in Fluss gekommen durch das etwas geräuschvolle Auftreten der 'Intuitionisten', die in temperamentvollen
2 Streitschriften eine ‖ 'Grundlagen-Krisis' der Mathematik verkündeten und so ziemlich der ganzen modernen Wissenschaft den Krieg erkläten – ohne selbst etwas Besseres an ihre Stelle setzen zu können. 'Eine Mengenlehre als besondere mathematische Disziplin wird es nicht mehr geben.' dekretierte einer ihrer eifrigsten Adepten – während gleichzeitig die neuen Lehrbücher der Mengenlehre nur so ins Kraut schossen. Diese Sachlage veranlasste auch mich damals, den Grundlagen-Problemen wieder meine forschende Tätigkeit zuzuwenden, nachdem ich durch langwierige Krankheit und geistige Isolierung im Auslande der wissenschaftlichen Produktion schon fast entfremdet war. Ohne in dem proklamierten Streite zwischen 'Intuitionismus' and 'Formalismus' Parteigänger zu werden – ich halte diese Alternative überhaupt für eine logisch unzulässige Anwendung des 'Tertium non datur' – glaubte ich doch zu einer Klärung der einschlägigen Fragen beitragen zu können: nicht als 'Philosoph' durch Verkündung 'apodiktischer' Prinzipien, welche durch Vermehrung der bestehenden Meinungen die Verwirrung nur noch zu steigern pflegen, sondern als Mathematiker durch Aufweisung objektiver mathematischer Zusammenhänge, die erst eine gesicherte Grundlage für alle philosophische Theorien abgeben können. In der besonderen Frage der Mengenlehre nun, wo es sich vor allem um die Aufklärung der 'Antinomien' handelt, stellte ich mir jetzt, dem aufgestellten Grundsatze entsprechend, die entscheidende Vorfrage: Wie muss ein 'Bereich' von 'Mengen' und 'Urelementen' beschaffen sein, um den 'allgemeinen' Axiomen der Mengenlehre zu genügen? Ist unser Axiomen-System 'kategorisch' oder gibt es eine Vielheit wesentlich verschiedener 'mengentheoretischer Modelle'? Ist der Begriff einer 'Menge' im Gegensatz zu einer blossen 'Klasse' ein absoluter, durch logische Merkmale bestimmbarer oder nur ein relativer, abhängig von dem jeweils zugrunde

3 gelegten mengentheoretischen Modell? ‖ Dieses Problem ist es, das ich mir in meiner 1930 erschienenen Fundamenta-Arbeit 'Über Grenzzahlen und Mengenbereiche' gestellt und gelöst habe. Um es aber erfolgreich in Angriff nehmen zu können, ergab sich zunächst die Notwendigkeit, das 'Zermelo–Fraenkel'sche Axiomensystem' durch ein weiteres, das 'Fundierungs-Axiom' zu ergänzen, das u.a. 'sich selbst enthaltende' und 'zirkelhafte Mengen' ausschliesst und in allen praktisch wichtigen Fällen tatsächlich erfüllt ist. Mit Hilfe des neuen Axioms konnte eine schichtenförmige Zerlegung, die 'Entwickelung' eines 'Normalbereiches' (d.h. eines dem Axiomensystem genügenden Mengenbereiches) durchgeführt und in den 'Isomorphie-Sätzen' die entscheidende Hauptfrage beantwortet werden. Zwei Normalbereiche sind dann und nur dann 'isomorph', wenn 1) ihre 'Basen' (d.h. die Gesamtheit ihrer Urelemente) einander äquivalent und 2) ihre 'Charakteristiken' (d.h. die oberen Grenzen der vorkommenden Alefs) einander gleich sind, wenn also auch jeder Menge des einen Bereiches mindestens eine Äquivalente im anderen entspricht. Von zwei Bereichen mit äquivalenten Basen (aber verschiedenen Charakteristiken) ist immer der eine isomorph einem 'kanonischen' Entwickelungs-Abschnitte des anderen. Hieraus folgt u.a. bereits, dass die (verallgemeinerte) CANTORsche Vermutung (wonach die Potenzmenge jeder Menge immer gerade die nächstfolgende Mächtigkeit haben soll) *nicht* von der Wahl des Modells abhängt, sondern durch unser Axiomensystem ein für allemal (als wahr oder als falsch) entschieden ist. Ein 'Normalbereich' ist (bis auf isomorphe Abbildungen) bestimmt durch zwei Alefs, durch seine 'Breite' (d.h. die Mächtigkeit seiner Basis) und durch seine 'Höhe' (d.h. seine Charakteristik), und die Gesamtheit aller möglichen 'Modell-Typen' wird also dargestellt durch eine (zweifach wohlgeordnete) Doppel-Reihe von Alefs, in welcher die 'Breite' sämtliche Alefs durchläuft, die 'Höhe' aber auf die Reihe der 'Grenzzahlen' beschränkt ist. Um 'Grenzzahl' oder 'Charakteristik eines

4 ‖ Normalbereiches' zu sein muss eine (transfinite) 'Anfangszahl' zwei charakteristische Eigenschaften haben: sie muss 'Eigenwert' oder 'kritische Zahl' einer gewissen 'Normalfunktion' sein, darf aber keiner kleineren Ordnungszahl 'konfinal', muss vielmehr immer eine 'reguläre Anfangszahl zweiter Art' im Sinne HAUSDORFFs, eine 'exorbitante' Zahl sein. Die *Existenz* solcher 'Grenzzahlen' kann nun freilich nicht aus dem Axiomensystem erwiesen sondern, weil sie eben nicht für *alle* Normalbereiche gilt, nur (für die höheren Bereiche metamathematisch) postuliert werden. Die (absolut) kleinste Grenzzahl ω begrenzt den 'finitistischen Bereich', der nur endliche Mengen enthält, die nächstfolgende π_1 den 'Cantorschen Bereich', in welchem bereits die ganze Cantorsche Mengenlehre einschliesslich des Unendlichkeits-Axioms zur Darstellung gelangt. Aber auch dieser Bereich ist (wie jeder andere Normalbereich) noch erweiterungsfähig zu einem 'höheren' Bereiche, welcher u.a. auch 'Mengen' von der Mächtigkeit π_1 enthält. Ganz allgemein gesprochen lösen sich die 'ultrafiniten Antinomien' bei dieser Betrachtungsweise dadurch, dass jeder

Normalbereich *N* zwar Teilbereiche *M* besitzt, die 'zu gross' sind, um in ihm 'Mengen' zu sein, dass aber alle solchen Teilbereiche *M* wie auch *N* selbst bereits im nächstfolgenden Normalbereiche *N'* durch vollgültige 'Mengen' vertreten sind. Jeder Normalbereich selbst ist 'Menge' in allen höheren Bereichen, aber es gibt keinen höchsten Normalbereich, welcher *alle* Normalbereiche als Mengen enthielte. Charakterisiert wird der einzelne Normalbereich (abgesehen von seiner Basis) durch die in ihm vorkommenden 'Mengen von Grenzzahl-Mächtigkeit', der 'Cantorsche' z.B. durch die Eigenschaft, ausser den abzählbaren *keine* weiteren Mengen von Grenzzahl-Mächtigkeit zu enthalten. So, gelangen wir also zu einer Art von 'mengentheoretischen Relativismus', der sich aber von dem Skolemschen

5 'Relativismus', in welchem sogar die Begriffe von ‖ 'Teilmenge' und 'Mächtigkeit' relativiert werden, grundsätzlich unterscheidet: SKOLEM will die Bildung der Untermengen auf besondere Klassen definierender Funktionen einschränken, während bei mir, dem wahren Geiste der Mengenlehre entsprechend, die freie Teilung zugelassen und die Existenz aller irgendwie gebildeten Teilmengen postuliert wird. Bei SKOLEM soll schon in einem *abzählbaren* Modell die *ganze* Mengenlehre dargestellt werden können, und für ihn verliert z.B. auch schon das Problem von der Mächtigkeit des Kontinuums seine eigentliche Bedeutung. Die Frage nach der 'Existenz' oder 'Widerspruchslosigkeit' der 'höheren' Normalbereiche sind freilich in der gedruckt vorliegenden Arbeit noch nicht völlig erledigt; aber ich glaube ein Verfahren gefunden zu haben, durch systematischen Aufbau eines 'mengentheoretischen Modelles' und mit Hilfe der 'unbegrenzten Zahlenreihe' diese Widerspruchslosigkeit einsichtig machen zu können. Ich brauche dazu den 'metamathematischen' Begriff eines 'geschlossenen Bereiches', der etwa dem CANTORschen 'Mengenbegriffe' entspricht und auf den eines 'kategorischen Postulatsystems' zurückgeführt werden kann. Jeder 'Normalbereich' ist ein 'geschlossener Bereich' und kann daher in einem 'höheren' auch als 'Menge' aufgefasst werden. Jeder *Abschnitt* der (transfiniten) Zahlenreihe ist ein geschlossener Bereich, aber auf jede 'Menge', jeden 'geschlossenen' Bereich von Ordnungszahlen folgen immer noch weitere Ordnungszahlen: die 'unbegrenzte' Zahlenreihe selbst ist ein 'offener Bereich'. Kein (geschlossener) Normalbereich kann die ganze Mengenlehre darstellen, da jede 'Grenzzahl' einem Abschnitt der Zahlenreihe entspricht und daher kein Normalbereich alle Grenzzahlen enthält. Die ganze Mengenlehre ist allein darstellbar in der 'offenen' Gesamtheit aller Normalbereiche.

6 Anhang I: Zusammenstellung meiner bisherigen Fortschritte und Resultate zum Neuaufbau der Mengenlehre

  1) Unterscheidung der allgemeinen Mengenlehre von den verschiedenen, sie darstellenden Modellen, den 'Normalbereichen'.
  2) Ausscheidung 'zirkelhafter' und ähnlicher Mengen durch Hinzufügung des 'Fundierungs-Axioms'.

134   G. H. MOORE

3) Einführung einer 'Basis' von 'Urelementen' anstelle der 'Nullmenge'.
4) Vermeidung des SKOLEMschen 'Relativismus' durch freie Bildung von Untermengen *ohne* 'Definitheits-Beschränkung'.
5) Schichtenförmige 'Entwickelung' eines 'Normalbereiches' aus gegebener 'Basis' mit Hilfe der Wohlordnung.
6) Begriff der 'Grenzzahl' als 'Charakteristik' eines Normalbereiches. 'Finitistische' und 'infinitistische' Normalbereiche.
7) Eigenschaften der 'Grenzzahlen': Sie sind 'Eigenwerte' oder 'kritische Zahlen' einer gewissen 'Normalfunktion', die gleichzeitig 'Kernzahlen' d.h. 'reguläre Anfangszahlen zweiter Art' sind.
8) 'Isomorphie-Sätze' über Normalbereiche: jeder Modell-Typus ist bestimmt durch zwei Alefs, die Mächtigkeit seiner Basis und durch seine 'Charakteristik' d.h. die obere Grenze der als Mengen vorkommenden Mächtigkeiten. Die Modelltypen bilden eine zweifach wohlgeordnete Doppelreihe. Die Gültigkeit der 'CANTORschen Vermutung' ist *unabhängig* vom gewählten Modell.
9) Aufklärung der 'ultrafiniten Antinomien' durch Unterscheidung der Modelltypen: Relativismus des Mengenbegriffes – in einem anderen als dem SKOLEMschen Sinne.

7 Anhang II: Zusammenstellung weiterer in Vorbereitung begriffener Untersuchungen

1) Über die Konstruktion eines mengentheoretischen Modells und die Widerspruchslosigkeit der Mengenlehre.
2) Über 'geschlossene' und 'offene Bereiche' und den CANTORschen absoluten Mengenbegriff.
3) Über den mengentheoretischen Relativismus bei SKOLEM und mir und seine Bedeutung für das Kontinuum-Problem.
4) Über die Mathematik als die 'Logik des Unendlichen' und die Unmöglichkeit einer 'finitistischen Mathematik'.
5) Über das Verhältnis der Mathematik zur *Anschauung*: erst mit der infinitistisch-logischen Verarbeitung eines anschaulich gegebenen Materiales beginnt die mathematische Wissenschaft und kann daher selbst *nicht* auf 'Anschauung' gegründet werden. Auch in der Geometrie beruht der Vorzug der 'Euklidischen Geometrie' *nicht* auf ihrer 'anschaulichen Gegebenheit' sondern lediglich auf ihrer logisch-mathematischen Einfachheit.

10.2. *Zermelo's 1929 Warsaw lectures*

Neun Vorträge über die Grundlagen der Mathematik (Universität Warschau, 27 Mai–8 Juni 1929)

1 Vortrag Nr. 1: Was ist Mathematik?

Die Mathematik ist *nicht* nach ihrem Gegenstande (etwa: Raum und Zeit,

Formen der inneren Anschauung, Lehre vom Zählen und Messen u. dergl.) zu charakterisieren sondern, wenn mann ihren ganzen Umfang erschöpfen will, allein durch ihr eigentümliches Verfahren, den Beweis. Die Mathematik ist eine Systematik des Beweisbaren und als solche eine angewandte Logik; sie hat zur Aufgabe die systematische Entwickelung der 'logischen Systeme', während die 'reine Logik' nur die allgemeine Theorie der logischen Systeme untersucht. Was heisst nun 'beweisen'? Ein 'Beweis' ist die nach allgemeinen logischen Regeln oder Gesetzen erfolgende Ableitung eines neuen Satzes aus anderen, vorgegebenen Sätzen, durch deren Wahrheit seine eigene gesichert ist. Das Ideal einer mathematischen Disziplin wäre demnach ein System von Sätzen, welches alle aus ihm rein logisch ableitbaren Sätze bereits in sich enthält d.h. ein 'logisch vollständiges System'. Ein 'vollständiges' System ist z.B. die Gesamtheit aller logischen Folgerungen, die aus einem vorgegebenem System von Grund-Annahmen, einem 'Axiomen-System' ableitbar sind. Aber nicht jedes 'vollständige System' ist notwendig durch eine endliche Anzahl von Axiomen bestimmt. Ein und dasselbe vollständige System kann durch mehrere, ja durch unendlich viele verschiedene Axiomen-Systeme gegeben sein, z.B. die Euklidische Geometrie oder die Arithmetik der reellen bzw. der komplexen Zahlen. Ein vollständiges System ist also gleichsam die 'Invariante' aller äquivalenten Axiomen-Systeme, und die Frage nach der 'Unabhängigkeit' der Axiome geht es nichts an. Ein vollständiges System verhält sich zu jedem es bestimmenden Axiomensystem wie ein 'Körper' zu seiner 'Basis', und die Untersuchung solcher 'vollständigen Systeme' verspricht vielleicht ähnliche Vorteile der grösseren Allgemeinheit und Übersichtlichkeit wie der Übergang von den algebraischen Gleichungen zu den algebraischen Körpern.

Die bisher entwickelten mathematischen Disziplinen beziehen sich immer auf einen Bereich von 'Dingen', Objekten, Gegenständen, zwischen denen gewisse 'Grund-Relationen' bestehen – oder nicht bestehen, z.B. die Relation $x + 1 = y$ im System der 'Peano'schen Axiome'. Ein Axiomensystem oder ein vollständiges System wird dann 'realisiert' durch ein 'Modell' d.h. durch eine volle 'Matrix' eines speziellen Bereiches, durch welche das Bestehen oder Nicht-bestehen der Grundrelation zwischen je zwei (oder mehr) Dingen des Bereiches eindeutig entschieden ist, und die gleichzeitig alle Sätze unseres Systems erfüllt. Nur wenn eine solches 'Modell' existiert oder wenigstens denkbar ist, gilt unser System als 'konsistent' d.h. realisierbar; ist die Existenz eines Modelles als logisch unmöglich nachweisbar, so ist damit das System selber 'inkonsistent' und 'widerspruchsvoll'. Sind je zwei Modelle,

2 ‖ welche ein und dasselbe System realisieren, unter einander 'isomorph' d.h. ihre Bereiche derart ein-eindeutig auf einander abbildbar, dass die Grund-Relationen gleichzeitig gelten und nicht gelten, so heisst das System 'kategorisch'. Konsistenten aber nicht-kategorischen Systemen entsprechen mindestens zwei wesentlich verschiedene d.h. nicht isomorphe Modelle, kategorische dagegen nur ein einziger 'Modell-Typus'.

136   G. H. MOORE

1 Vortrag Nr. 4: Endliche und unendliche Bereiche.

   Es kann der Fall eintreten, dass ein logisches System realisierbar ist durch
ein Modell, dessen Matrix vollständig und in extenso angegeben werden
kann, dessen Bereich dann also jedenfalls auch endlich und von begrenzter
Ausdehnung sein muss. Hier kann die Konsistenz des Systems und damit
auch die Widerspruchslosigkeit des zugrunde liegenden Axiomensystems
nachgewiesen werden. So wird das den allgemeinen Gruppen-Begriff
charakterisierende Postulaten-System schon realisiert durch ein aus einem
einzigen Elemente bestehendes Modell, desgleichen das System der für
wohlgeordnete Mengen gültigen Sätze durch eine solche [*sic*] von 2 Elementen, endlich auch das ganze System der Aussage- wie der Subsumptions-
Logik durch ein aus 1 bez. 2 Elementen gebildetes Modell. In den
angeführten Beispielen ist aber dieser direkte Nachweis der Konsistenz nur
dadurch möglich, dass die betrachteten Systeme nicht-kategorisch sind und
deswegen sowohl durch endliche wie durch unendliche Modelle realisiert
werden können. Für kategorische Systeme dagegen wie z.B. das der
Peano'schen Postulate oder überhaupt für alle Systeme, die ausschliesslich
durch unendlich Modelle realisiert werden können, bei denen also die
Unendlichkeit des Bereiches wesentlich ist, versagt dieses Verfahren der
sinnfälligen Repräsentation. Nun ist aber die ganze herkömmliche Arithmetik
auf die Annahme einer unbegrenzten Zahlenreihe, wie sie etwa durch die
Peano'schen Postulate definiert werden kann, gegründet, und selbst die
Theorien der endlichen Gruppen, der endlichen Mengen, der Kongruenzen
und der endlichen Körper von Primzahl-Charakteristik gewinnen ihre
eigentlich mathematische und nicht-triviale Bedeutung allein durch den
Umstand, dass sie auf endliche Bereiche von beliebiger Elementen-Zahl
angewendet werden können, also tatsächlich immer nur im Rahmen einer
umfassenden, selbst unendlichen Arithmetik entwickelt werden Eine reine
'finitistische' Mathematik, in der man eigentlich nichts mehr zu beweisen
bräuchte, weil doch alles schon am endlichen Modell verifiziert werden
könnte, wäre keine Mathematik mehr im wahren Sinne. Die wahre
Mathematik ist vielmehr ihrem Wesen nach infinitistisch und auf die
Annahme unendlicher Bereiche gegründet; sie kann geradezu als die 'Logik
des Unendlichen' bezeichnet werden.

MATHEMATICAL LOGIC AND AXIOMATIC SET THEORY   137

ACKNOWLEDGEMENTS

Earlier versions of this article were presented at the Erindale Logic Colloquium (University of Toronto, 19 April 1979) and at the Buffalo Logic Colloquium (SUNY Buffalo, 10 May 1979). For permission to publish material from Zermelo's *Nachlass*, I am grateful to the Handschriftenabteilung, Universitätsbibliothek, 78 Freiburg im Breisgau, Rempartstrasse 15, West Germany. I wish to thank Merrie Bergmann, Alejandro Garciadiego, and Calvin Jongsma for commenting on the penultimate draft. Most of all, I am indebted to John Corcoran, Ivor Grattan-Guinness and Michael Resnik for their many helpful suggestions and insights. Lastly, I thank Vello Sermat, whose house in the woods gave the peace that midwifed this article.

[2]

LESLIE H. THARP

WHICH LOGIC IS THE RIGHT LOGIC?

0. INTRODUCTION*

It has been generally accepted in the philosophy of mathematics that elementary logic (*EL*), also known as *the predicate calculus*, or *first order logic*, yields a stable and distinguished body of truths, those which are instances of its valid formulas. I am concerned with presenting and examining evidence relevant to such a claim. In sentential logic there is a simple proof that all truth functions, of any number of arguments, are definable from (say) 'not' and 'and'. Thus one has not overlooked any truth-functional connectives, even though one started with the few which naturally presented themselves. Operators such as 'for infinitely many x', or for an arbitrary cardinal $\aleph$, 'for at least $\aleph$ x', are in some ways analogous to the standard quantifier 'for at least one x'. If these operators are counted as quantifiers, there are many more such quantifiers than there are formulas of elementary logic; so on rather trivial grounds, there can be no theorem that all possible quantifiers are already definable in *EL*. This observation does not, however, rule out the possibility that there might be a narrower notion of quantifier for which such a theorem holds. If so, and to the extent that the narrower notion is significant, one will have evidence that *EL* is no arbitrary stopping point. I will argue, in Section 5, that natural and satisfying criteria are suggested by the standard quantifiers which characterize arbitrary formulas of elementary monadic logic. The full logic of relations, however, appears to be more problematical.

1. BACKGROUND

Whether or not it is a satisfactory picture of interpreted language, we shall retain the analysis which resolves interpreted language into a semantical part (models), a syntactical part (formulas), and the relation of a model satisfying (or being a model of) a formula. We do not contemplate altering the notion of model as it is used in elementary logic. A model still consists

Synthese **31** (1975) 1–21. *All Rights Reserved*
Copyright © 1975 *by D. Reidel Publishing Company, Dordrecht-Holland*

2   LESLIE H. THARP

of a universe and a finite sequence of relations over the universe (for simplicity we shall ignore constants and functions). What we do contemplate is enriching the set of formulas, and thereby of course extending the satisfaction relation. Let us suppose that the set of formulas of a possible logic is a countably infinite set, and that each formula contains only finitely many letters. Restrictions of effectiveness will be introduced when they are relevant. It hardly needs to be argued that these are reasonable conditions to impose on any potential competitor of elementary logic.

As a concrete example, let us take elementary logic and define a new logic $L(I)$ by adding the symbol (Ix) which is read 'for infinitely many x'. It is clear how to define the resulting set of formulas, and the corresponding satisfaction relation. This logic extends EL in a definite sense, since the class of models satisfying $\neg (Ix) (x=x)$ is just the class of all models with finite universes, which cannot be defined by a formula of EL. (We are taking EL and other logics to contain identity.) We can also characterize the natural numbers with a formula of $L(I)$, since we can say that each number has finitely many predecessors.

These examples suggest a general framework for comparing logics. Without specifying the inner workings of a logic L, we may take it to be a collection of L-classes; each L-class may be thought of as a class of models of the form $\{M : M \operatorname{Sat}_L \mathscr{A}\}$ where $\mathscr{A}$ is a (closed) formula of L and Sat_L is the satisfaction relation for L.[1] Then we may say that a logic L_1 is contained in L_2 if every L_1-class is an L_2-class. The relations of equivalence and (proper) extension are then defined in the obvious way from containment. In the previous example EL is contained in $L(I)$, but they are not equivalent, since a certain $L(I)$-class is not an EL-class.

Within this framework one can define, for a given logic L, a notion of logical implication between a set of formulas X and a formula $\mathscr{A}$. X logically implies $\mathscr{A}$ (X l.i. $\mathscr{A}$, for short) in case all models satisfying all members of X satisfy $\mathscr{A}$. This is the concept, going back to Bolzano, which is dealt with by Tarski in 'On the Concept of Logical Consequence',[2] and which we take to be a satisfactory formulation of the notion of a formula $\mathscr{A}$ following from X on 'purely logical grounds'. It should be remarked that our overall concern is basically the problem discussed by Tarski towards the end of his paper: Is there a sharp division of terms into logical and extra-logical?

Elementary logic is *axiomatizable*. That is, there is a proof procedure

⊢ such that X l.i. $\mathscr{A}$ if and only if $X \vdash \mathscr{A}$. The proof procedure is such that a proof can involve only finitely many formulas, so if $X \vdash \mathscr{A}$ then $\Gamma \vdash \mathscr{A}$ for some finite subset Γ of X. That ⊢ provides an axiomatization is equivalent to the conjunction of two conditions which are frequently singled out. The first is that ⊢ is *complete* in the sense that the formulas which are valid (i.e. true in all models) are exactly those which are provable without hypotheses. In other words, to say ⊢ is complete is to say ϕ l.i. $\mathscr{A}$ if and only if $\phi \vdash \mathscr{A}$. It follows easily from completeness that for finite sets of formulas Γ, Γ l.i. $\mathscr{A}$ if and only if $\Gamma \vdash \mathscr{A}$. The second condition is *compactness*, which says that if every finite subset of X has a model, then X has a model. Compactness is equivalent to saying that if X l.i. $\mathscr{A}$ then, for some finite subset Γ of X, Γ l.i. $\mathscr{A}$. Putting completeness and compactness together, one has axiomatizability. The reader should be warned that we have chosen terminology convenient for our purposes but which is by no means universally adopted. For example, 'axiomatizability' is frequently used to mean our completeness.

The notions above were defined for the logic *EL*, and we wish to formulate them for other logics. Compactness is a purely model-theoretic notion, and, as stated, it clearly makes sense for the most general logics. Completeness can also be defined in a fairly general setting. Identify the formulas with a recursive set of natural numbers and take completeness to mean that the valid formulas are effectively enumerable. Then define a proof procedure ⊢ for the logic: $\{\mathscr{B}_1 ..., \mathscr{B}_n\} \vdash \mathscr{A}$ in case $\mathscr{B}_1 \to (\mathscr{B}_2 \to ... ...(\mathscr{B}_n \to \mathscr{A})...)$ is in the enumeration of valid formulas; for an arbitrary set of formulas X, let $X \vdash \mathscr{A}$ in case $\Gamma \vdash \mathscr{A}$ for some finite Γ included in X. (We are assuming that the logic has effective constructions corresponding to the sentential connectives.) Again in the general case compactness and completeness give axiomatizability: X l.i $\mathscr{A}$ if and only if $X \vdash \mathscr{A}$.

A very elegant proof of axiomatizability for *EL*, due to Henkin, shows that every consistent set of formulas has a model. Since in fact the model produced is countable (i.e. finite or countably infinite), one has a further corollary: If X has a model, X has a countable model. This corollary is one version of a well known and somewhat controversial theorem, the Löwenheim-Skolem theorem. As a special case, if a single formula $\mathscr{A}$ has a model, then $\mathscr{A}$ has a countable model. We take this special case[3] and say that a logic L has the Löwenheim-Skolem property if every nonempty L-class has a countable member.

4   LESLIE H. THARP

This leads us to the first basic technical result, an intrinsic characterization of elementary logic due to Lindström:[4] Suppose L contains EL and is either complete or compact; then if L has the Löwenheim-Skolem property, L is equivalent to EL. As an example, the logic $L(I)$ mentioned before is easily shown to have the Löwenheim-Skolem property, using the submodel proof of Tarski and Vaught.[5] Thus since $L(I)$ is not equivalent to EL, $L(I)$ is neither complete nor compact. (Of course, to show that $L(I)$ was different from EL we in effect pointed out that it was not compact.) In Lindström's theorem one does have to make a few general assumptions about the logic L. For example the L-classes must be closed under intersection and complement. This of course means that L has the sentential connectives. It is not necessary, however, to assume that L has something like quantification.[6] A few more basic restrictions are needed, e.g., that isomorphic models lie in the same L-classes. Also, for completeness one needs to bring in the obvious stipulations of effectiveness.

2. WHAT MUST A LOGIC DO?

Lindström's result gives an exceedingly sharp characterization of elementary logic as a maximal solution to certain general conditions. Elementary logic has for many years been taken as the standard logic, with little explicit justification for this role. Since it appears not to go beyond what one would call logic, the problem evidently is whether it can be extended. Thus one is tempted to look to Lindström's theorem for a virtual proof that logic must be identified with elementary logic. The success of this enterprise obviously depends on the extent to which one can justify, as necessary characteristics of logic, the hypotheses needed in the theorem, principally completeness (or compactness) and the Löwenheim-Skolem property.

A complete logic has an effective enumeration of the valid formulas. The proof procedures proposed for elementary logic were clearly *sound* in the sense that they proved only valid formulas. Gödel established completeness by showing that all valid formulas were provable by a standard proof procedure. Second order logic, which is an extension of EL formed by allowing quantification over predicate letters, is a classical example of a logic with sound proof procedures, but which demonstrably admits no complete proof procedure. That is, there is no proof procedure complete with respect to the intended semantics. The standard proof procedures can be shown complete with respect to certain semantics – but such non-standard semantics have little independent interest.

Soundness would seem to be an essential requirement of a proof procedure, since there is little point in proving formulas which may turn out false under some interpretations. But it is trivial to provide sound proof procedures: take the null procedure, or take some finite set of valid formulas. Of course the proof procedures suggested for logics such as second order logic enumerate an infinite number of valid formulas, and perhaps appear to yield, in some further sense, a large and useful set of valid formulas; in particular one may incorporate the comprehension schema which says, for each formula $\mathscr{A}(x)$, that there exists the class of those individuals x such that $\mathscr{A}(x)$.

The question is, should one demand that a logic have a complete proof procedure? In order to answer this, it seems essential to be somewhat more precise about the role logic is expected to play. One can distinguish at least two quite different senses of logic.[7] The first is, as an instrument of demonstration, and the second can perhaps be described as an instrument for the characterization of structures. In the present context, a logic L will have this latter ability if, for example, there are L-classes consisting, up to isomorphism, of a single structure of mathematical interest. Second order logic is striking in this respect. Such central theories as number theory and the theory of the real numbers seem to involve in their concepts a quantifier over all subsets of the domain of elements, and in fact they can be characterized up to isomorphism in second order logic. Even rather large portions of set theory can be described categorically by this logic.

Interesting as such a notion of logic is, it seems perfectly reasonable to distinguish the other sense of logic as a theory of deduction.[8] Elementary logic cannot characterize the usual mathematical structures, but rather seems to be distinguished by its completeness. Thus one is inevitably led to ask whether it is a necessary stopping point, or whether it can be extended to a richer logic which is still a theory of deduction in the same sense. Completeness, after all, is not just another nice property of a system. When a deductive system of whatever sort is presented, one of the most immediate questions is whether it is (in the relevant sense) complete. If all valid (or true) formulas can be proven by the rules, then apart from

6   LESLIE H. THARP

practical limitations such as length and complexity, they can be *known* to
be valid (or true). This seems to be as interesting and significant a criterion
as one could propose. Until the modern development of logic, it was
generally assumed that mathematical systems had this property. One now
knows from Gödel's work that the condition is too stringent even for
number theory, and it is conceivable that it could have been too stringent
for logic. Thus one should not claim to see *a priori* that a logic must be
complete in order to be a theory of deduction. The point is rather that
when it is discovered that the best known candidate satisfies such a condition, that tends to establish a sense of logic and a standard to be applied
to competitors.

Two points concerning completeness should be mentioned. The mere
existence of an effective enumeration of the valid formulas does not,
by itself, provide knowledge. For example, one might be able to prove
that there is an effective enumeration, without being able to specify one.
Normally one will exhibit axioms known to be valid and rules known to
preserve truth; preferably these axioms and rules will be more or less
self-evident. Unfortunately these further conditions do not appear amen-
able to exact treatment. A second point, frequently noted, is that the
completeness proof for *EL* actually shows that a somewhat vague intuitive notion, 'valid *EL* formula', coincides with formal provability. It
appears from inspection of the axioms and rules that formally provable
formulas are intuitively valid; and intuitively valid formulas are certainly
true in (say) all arithmetical models. Since the completeness theorem
demonstrates that all formulas true in all arithmetical models are provable
in *EL*, one must conclude that all three notions coincide in extension.

It is not out of the question that even in a theory of deduction completeness might be sacrificed for other advantages, such as greater expressive power. The strongest example historically is perhaps second order
logic. One might claim that this is in some sense an acceptable theory of
deduction, since in particular it yields all of the inferences of *EL*. But in
fact it is not accepted as the basic logic. The expressive power of this logic,
which is too great to admit a proof procedure, is adequate to express
set-theoretical statements. Typical open questions, such as the continuum
hypothesis or the existence of big cardinals, are easily stated as questions
of the validity of second order formulas. Thus the principles of this logic
are part of an active and somewhat esoteric area of mathematics. There

seems to be a justifiable feeling that this theory should be considered mathematics, and that logic – one's theory of inference – is supposed to be more self-evident and less open.

Of course not all incomplete extensions of *EL* are as strong as second order logic. But the other known examples also tend not to look as natural, and they, like second order logic, invite further extension. In general, if completeness fails there is no algorithm to list the valid formulas; so one can expect many of the principles of the logic to be unknowable, or determinable only by means of *ad hoc* or inconclusive arguments. Clearly one will hesitate to substitute other desirable features for completeness in a theory of deduction. The negative evidence, together with the epistemological appeal of the completeness condition, make it seem reasonable to suppose that completeness is essential to an important sense of logic.

Strangely, compactness seems to be frequently ignored in discussions of the philosophy of logic. It is strange since the most important theories have infinitely many axioms. With only completeness it seems possible, *a priori*, that a logic might not prove all logical consequences of these theories. Compactness amounts to the condition that if X l.i $\mathscr{A}$ then Γ l.i. $\mathscr{A}$ for some finite subset Γ of X. Since completeness ensures that if Γ l.i. $\mathscr{A}$ then $\Gamma \vdash \mathscr{A}$, one may conclude that if the system is both compact and complete, all logical consequences of a set of hypotheses are provable. We claim that that is the philosophical point at issue: if something follows, it can be known to follow.

However, the primary question is whether compactness by itself can be given a better justification than was given for completeness, since we are concerned to justify Lindström's hypotheses, which require either completeness or compactness (plus the Löwenheim-Skolem condition). It seems to me that it is not all clear that compactness, *per se*, can be defended. The compactness condition in effect states that if $\mathscr{A}$ is implied by an infinite set of assumptions, $\mathscr{A}$ is already implied by a finite subset. The notion of implication here is of course the semantical one, not provability. The condition thus seems to state some weakness of the logic (as if it were futile to add infinitely many hypotheses), without yielding a compensating reward – such as knowledge that $\mathscr{A}$ is implied. To look at it another way, compactness also immediately entails that formalizations of (say) arithmetic will admit non-standard models.

8   LESLIE H. THARP

A second question, not strictly relevant to the present purpose, is whether, having accepted completeness, compactness can be seen to be an essential property of a theory of deduction. Perhaps one should first remark that to some extent completeness implies compactness,[9] so in many cases the question is dissolved, if not answered. We have noted that compactness is *sufficient* in conjunction with completeness to yield the desired consequence: if X l.i. $\mathscr{A}$ then $X \vdash \mathscr{A}$. It would appear that if one had a proof procedure of the usual sort the natural further condition to demand would be compactness, in order to get a reduction from an infinite set of hypotheses to a finite subset. The trouble is, it is not quite clear that one need divide the labor in exactly this way. For example, one might have a condition like: if X l.i. $\mathscr{A}$ then $\exists e(W_e \subseteq X$ and W_e l.i. $\mathscr{A})$, where W_e are the recursively enumerable sets (in a standard coding). If one also had an effective method for enumerating the pairs $\langle e, \mathscr{A} \rangle$ such that W_e l.i. $\mathscr{A}$, these two properties would do much the same work as completeness plus compactness. However, it is easy to show that they in fact imply compactness. This is no proof that compactness is necessary, but it seems highly likely that if completeness is required, compactness will be accepted. As for the main point, though, compactness by itself seems much less defensible than completeness.

This leaves the Löwenheim-Skolem property to be considered. On the face of it, this property seems to be undesirable, in that it states a limitation concerning the distinctions the logic is capable of making. Such a logic fails to make those distinctions intended in the usual theories, for example that there are uncountably many reals ('Skolem's paradox'). It is true that completeness also entails a limitation on the power of a logic to discern structures – a complete logic cannot, for example, determine the natural numbers by a single formula. But unlike completeness, the Löwenheim-Skolem condition does not express any clearly desirable property of a theory of deduction. It might follow from some other conditions which express desirable properties; in that case, it would be those conditions which one should attempt to defend. As it happens, it does follow from certain conditions which I shall attempt to justify in Section 5. (The existence of true but unprovable sentences in arithmetic is undesirable; it follows, however, from a defensible condition, namely that the axiom system be effective.)

There are positions, perhaps some kinds of finitism or countabilism, where the Löwenheim-Skolem property is, if not desirable, simply true. Obviously, under the general assumptions we have made, this neatly resolves the question of what logic is. However these positions are not merely minority positions, but positions which ignore, or drastically reinterpret, the overall body of science and mathematics. Further, I doubt whether the arguments that there are only countably many things are very cogent in their own right. I cannot pursue a discussion of these related positions here, but will simply accept mathematics as it exists, and conclude that there is no *a priori* reason to impose the Löwenheim-Skolem condition on a logic.

3. SOME COMPLETE COMPETITORS

Two examples of axiomatizable logics have been discussed in the literature. The first, call it $L(U)$, adds an additional quantifier (Ux), which is read 'for uncountably many x'. Some time after the logic was known to be axiomatizable, Keisler[10] proved that the following elegant set of axioms is adequate:

(0)   Axioms of *EL*.

(1) $(\forall x)(\forall y)\neg(Uz)(z=x \vee z=y)$, 'The uncountable is bigger than 2'.

(2) $(\forall x)(\mathscr{A} \to \mathscr{B}) \to [(Ux)\mathscr{A} \to (Ux)\mathscr{B}]$, 'If all $\mathscr{A}$s are $\mathscr{B}$s, and there are uncountably many $\mathscr{A}$s, then there are uncountably many $\mathscr{B}$s.'

(3) $(Ux)\mathscr{A}(x) \leftrightarrow (Uy)\mathscr{A}(y)$, 'Changing variables'.

(4) $(Ux)(\exists y)\mathscr{A} \to [(\exists y)(Ux)\mathscr{A} \vee (Uy)(\exists x)\mathscr{A}]$, 'If uncountably many things are put into boxes, either some box gets uncountably many members, or else uncountably many boxes are used'.

Modus ponens and generalization are the rules of inference. Notice that all of the axioms remain valid if the quantifier 'for infinitely many x' is substituted for (Ux). It is easy, however, to show that there are formulas valid under this interpretation which are not valid under the original interpretation. Also note that the axioms do not assert that there are uncountably many things.

By considering the contrapositive of Axiom (4), one sees that it ex-

10   LESLIE H. THARP

presses the principle that a countable union of countable sets is countable. This may be considered a weak form of the axiom of choice. Dropping choice, one can give models for Zermelo-Fraenkel set theory in which a countable union of countable sets is uncountable. On the other hand, if one assumes certain forms of the axiom of choice, such as the principle that a countable set of nonempty sets has a choice function, then (4) follows readily.

Since a version of the axiom of choice is expressed by the validity of a formula of $L(U)$, if one accepts $L(U)$ one accepts this form of the axiom of choice as a principle of logic. This is especially interesting in view of the rather controversial history of this axiom. The objections against it, however, now seem simply mistaken. Probably the main objection was that it was supposed that the choice function had to be given by some law or definition. But since sets are completely arbitrary collections, this requirement is irrelevant. Except for formalists, who regard it as meaningless, set theorists generally regard the axiom of choice as true, and indeed practically obvious. Whether or not it may be taken as a logical principle is another matter – its infinitistic nature might give pause.

The second example, $L(C)$, has the quantifier (Cx), the Chang quantifier:[11] $(Cx) \mathscr{A}(x)$ is satisfied in a model just in case $\{x : \mathscr{A}(x)$ is satisfied$\}$ has the same cardinality as the universe of the model. This logic, however, is complete and compact only if one rules out finite models. One does not have a simple and explicit axiom system for the Chang quantifier, but there is one for this logic with identity deleted.[12] Curiously, it too assumes a form of the axiom of choice. It would be interesting to know whether there is some deeper connection between axiomatizable extensions of EL and the axiom of choice. It is possible to give rather trivial extensions of EL and prove them complete and compact without using the axiom of choice. Whether this is true for any interesting extensions seems to be an open question.

Recently other examples of axiomatizable logics have been found which serve also to illustrate another generalization of quantifiers. Besides quantifiers of one argument, one might want quantifiers of two arguments. For example $(Wx, y) \mathscr{A}(x, y)$ might mean that $\mathscr{A}(x, y)$ defines a well ordering. This is a useful concept, but like (Ix), it will enable one to characterize arithmetic categorically and so cannot give a complete logic. Shelah[13] has shown that there is a whole category of quantifiers similar to

(Wx, y) which give axiomatizable logics. Unfortunately they involve technical notions of set theory: for any regular cardinal λ let ($S^{\lambda}x, y$) $\mathscr{A}(x, y)$ mean that $\mathscr{A}(x, y)$ defines a linear ordering of cofinality λ. The quantifiers S^{λ} give axiomatizable logics, and moreover one can generalize to certain sets of cardinals, and to the quantifiers 'cofinality less than λ', and still get axiomatizable logics.

One result to hope for is that there might be a maximum logic L, that is, an axiomatizable logic containing all others. This is immediately seen to be impossible since any logic L has only countably many L-classes, while the various logics of Shelah define an unlimited number (i.e. a proper class in number) of classes of models. More interesting is the observation that our first two examples are incompatible: $L(U)$ and $L(C)$ have no common extension.[14]

One might wonder how these extended logics could or would be used. Consider number theory. An obvious axiom to state using (Ux) is that there are only countably many numbers. Or, using the Chang quantifier, one can say that the set of predecessors of a number is of smaller cardinality than the universe. It seems possible that these axioms might yield new theorems in the original language of number theory, but they do not. The proof is easy for the first axiom, but is not at all trivial for the second.[15] In number theory there is no reason to suppose that all such proposals will yield conservative extensions. However, in set theory it is clear that one will never get new theorems, at least if the axioms of the logic are set theoretical principles. This is because any proof in the strong logic can be translated into an ordinary proof which uses principles of set theory. Although it is conceivable that some logical axiom independent of set theory could be seen to be true, it is probable that it would be clearly a set theoretical principle. Thus extended logics are probably best regarded as changes in the boundary which demarcates as logic a part of set theory.

4. CAN THE COMPETITORS BE REJECTED?

There is a serious objection against the quantifier 'for uncountably many x'. One would expect that if this were a legitimate quantifier, the quantifier 'for infinitely many x' would also be acceptable; but no complete logic can contain the latter quantifier. Specifically, the cardinal $\aleph_1$ plays a role in the model theory of any complete logic containing the quantifier (Ux)

12   LESLIE H. THARP

which cannot be played by the smaller cardinal $\aleph_0$. One can say 'The universe has cardinality less than $\aleph_1$', but one cannot say 'The universe has cardinality less than $\aleph_0$'. One cannot even have a formula, using various predicate letters, which is satisfiable exactly in finite universes. The fact that one can say so much more about $\aleph_1$ than $\aleph_0$ seems to me to be a state of affairs sufficiently unnatural to discredit (Ux) as a logical notion.

If the uncountable is no logical notion, a line of argument is suggested which invokes Lindström's theorem. This theorem entails that any complete extension of elementary logic must have something like an 'axiom of uncountability', that is, it must have a formula with uncountable models but no countable models. If the uncountable is no logical concept, one is tempted to regard this consequence as a proof that there are no complete logics extending elementary logic. However that would be a mistake. Elementary logic already has 'axioms of infinity' in a similar sense. That is, there are formulas with models of all infinite cardinalities but with no finite models, even though 'infinite' is not a logical concept in the sense of there being a quantifier expressing 'for infinitely many x'.

Being able to distinguish the uncountable by means of axioms of uncountability is evidently a much weaker property than having a quantifier 'for uncountably many x'. Further, if one allows even weaker criteria, elementary logic is already able to distinguish the uncountable from the countable. Call a theory with infinite models categorical in the cardinal $\aleph$ if all models of cardinality $\aleph$ are isomorphic. It is known that there also theories categorical in $\aleph_0$ but in no uncountable cardinal. There are also theories categorical in all uncountable cardinals but not in $\aleph_0$; theories categorical in all infinite cardinals; and theories categorical in no infinite cardinals. Morley's theorem demonstrates that for infinite cardinals, these are the only possible categories. In this sense, elementary logic cannot distinguish two uncountable cardinals, although it can distinguish the countably infinite from the uncountably infinite.

From this extensional viewpoint, Shelah's quantifier 'cofinality ω' fares considerably better than the quantifier (Ux). One has axioms of uncountability in precisely the same sense one has axioms of infinity, namely formulas with predicate letters satisfiable in, and only in, the uncountable universes. Other gross properties seem reasonable also: just as one cannot have formulas satisfiable exactly in finite universes, this logic has no formulas satisfiable exactly in countable universes. The obvious objection against the Shelah logic is that one would never have supposed that a technical notion like cofinality was a logical concept. Unless it could be shown equivalent to some more palatable notion, this must remain a serious objection.

5. CONTINUITY OF THE STANDARD QUANTIFIERS

If one considers, instead of the entire logic *EL*, the standard quantifiers, it would appear that there must be some sense in which $\forall$ and $\exists$ are very simple and primitive. It may be possible to state some natural condition which expresses this simplicity, and which, at the same time, rules out quantifiers one feels are no part of logic. To start with, $\forall$ and $\exists$ may be regarded as extrapolations of the truth functional connectives $\wedge$ and $\vee$ to infinite domains. To see how one can make such extrapolations, take the infinite list of sentential letters $P_0, P_1, P_2, \ldots$, and suppose one is given an arbitrary truth assignment t, that is, a function which assigns a truth value $t(P_i)$ to each letter P_i. By the truth table rules, t may be extended to assign a value $t(\mathscr{A})$ to each formula $\mathscr{A}$ of sentential logic.

Consider, for increasing n, the values $t(P_0 \wedge P_1 \wedge \ldots \wedge P_n)$ assigned to the finite conjunctions. One sees at once that no matter what assignment t is, the limit $\lim n \to \infty \, t(P_0 \wedge P_1 \wedge \ldots \wedge P_n)$ has a clear value: there is a finite point N such that the conjunction $(P_0 \wedge P_1 \wedge \ldots \wedge P_N)$ is assigned a value $t(P_0 \wedge P_1 \wedge \ldots \wedge P_N)$, and for all m greater than N, $t(P_0 \wedge P_1 \wedge \cdots \ldots \wedge P_m)$ is the same as $t(P_0 \wedge P_1 \wedge \ldots \wedge P_N)$. It is easily seen that exactly the same property is true of finite disjunctions.

Compare another familiar binary connective, the biconditional $\leftrightarrow$. It is also commutative and associative so that parentheses may be dropped, and $(P_0 \leftrightarrow P_1 \leftrightarrow \cdots \leftrightarrow P_n)$ considered to be a formula. However, in this case there is no evident limit of the values $t(P_0 \leftrightarrow P_1 \leftrightarrow \cdots \leftrightarrow P_n)$ as n increases. This is not merely due to the unfamiliarity of the construction. If, for example, $t(P_i)$ is $\bot$ (falsehood) for all i, then the value of $t(P_0 \leftrightarrow P_1 \leftrightarrow \leftrightarrow \cdots \leftrightarrow P_n)$ is $\top$ for n even, and $\bot$ for n odd. There is simply no well-defined limit value to be used for an infinite quantification.

These considerations can be restated in the language of quantifiers and predicates. Suppose a model M is given which interprets a one-place letter F. In each finite submodel J with universe $\{a_0, \ldots, a_n\}$, $\forall x F(x)$ has a truth value, the same value as $(F(a_0) \wedge F(a_1) \wedge ... \wedge F(a_n))$ has in J. $\forall$ is continuous in the sense that for each model M there is a finite submodel J, in which $\forall x F(x)$ takes a truth value and holds that truth value for all submodels K between J and M, including M. Thus if one thinks of the model M as being revealed step by step, there is a finite portion at which $\forall x F(x)$ assumes a truth value, and holds that truth value no matter how much the model is further revealed, and even if it is totally revealed.

This is the continuity condition we wish to isolate, and of course the quantifier $\exists$, as well as $\forall$, is continuous in this sense. However if (Qx) is any quantifier such that $(Qx) F(x)$ agrees with $(F(a_0) \leftrightarrow F(a_1) \leftrightarrow \cdots \leftrightarrow F(a_n))$ in models with universe $\{a_0, a_1, ..., a_n\}$, then (Qx) must exhibit a sort of discontinuity: For certain M, $(Qx) F(x)$ will oscillate back and forth in truth value as one considers larger and larger finite submodels, and the value of $(Qx) F(x)$ in infinite models will be no direct extrapolation of its value in finite submodels. It should be noted in passing that there is another possible kind of discontinuity which occurs, for example, with the quantifier 'for infinitely many x'. Since $(Ix)(x=x)$ comes out false in all finite models, it has a definite limiting value – the trouble is that its value in an infinite model is not equal to this limit.

We have exhibited a type of continuity for the basic quantifiers $\forall$ and $\exists$ in terms of the behavior of the formulas $\forall x F(x)$ and $\exists x F(x)$, where F is any one-place letter. Not only are the basic quantifiers continuous, but every formula of elementary monadic logic, which is the subsystem of *EL* using one-place relations, is continuous in this sense. Moreover one has an exact converse: Given any quantifier[16] of monadic type, if it is continuous in the sense sketched above, it is definable in elementary monadic logic.

It is clear that the standard quantifiers can be *tested* for truth in certain ways. Let us examine more closely the exact sense in which $\exists$ has operational meaning. Suppose one has an intuitively decidable one-place predicate F, interpreted (say) as 'is a frog'. One wants to determine the truth of $\exists x F(x)$ in the standard model – the actual world. One proceeds along, examining specimens, and at some finite point one reaches a portion of the model in which $\exists x F(x)$ comes out true; and at that point one knows that it is true in the whole model no matter what the rest of the model is like. (We are assuming, for purposes of the example, that the world is infinite.) It is easy to formulate this criterion precisely and show that any such quantifier is continuous.[17]

This test does not work for $\forall$; instead, if $\forall x F(x)$ is ultimately false, its falsehood can be known at a finite point. Another way to put it is that one has only half a test for $\exists$: if $\exists x F(x)$ is true, one can actually find it out in finitely many steps; but if $\exists x F(x)$ is false, one may never be sure, short of exhausting the entire model. One might consider the stronger demand that the correct truth value always be ascertainable at a finite point, even though $\forall$ and $\exists$ do not satisfy such a stringent condition. But it is easy to show that only trivial quantifiers result.

In summary, those quantifiers which are partially decidable in the same sense as $\exists$ are all continuous; their complements are the quantifiers partially decidable in the same sense as $\forall$, and are also continuous. The notion of continuous quantifier seems to be the most natural symmetrical condition containing both $\forall$ and $\exists$. We have remarked that any continuous *monadic* quantifier is definable in elementary monadic logic, hence in *EL*. For non-monadic continuous quantifiers, however, one must appeal to completeness of the resulting logic to conclude that they are definable in *EL*. In fact, one can give binary quantifiers which are partially decidable in the same sense as $\exists$, but which give incomplete logics. It would seem, in conclusion, that the major properties of the standard quantifiers of epistemic significance can be precisely captured, and any such quantifier shown to be definable in *EL*, invoking completeness only for non-monadic quantifiers.[18]

6. WHAT ABOUT COMPLEX FORMULAS?

One sees at once that no such simple criteria as continuity apply to an arbitrary formula of the full elementary logic of relations. For example, there is a formula $\mathcal{A}$, with a single binary letter G, which says G is a linear ordering without last element. $\mathcal{A}$ has only infinite models, so given a model M of $\mathcal{A}$, $\mathcal{A}$ must be false in each finite submodel of M. One can also exhibit other kinds of discontinuity: There is a formula $\mathcal{B}$ which says that G is a function mapping part of the universe one-one onto the remainder; $\mathcal{B}$ has infinite models, and for any such model N, $\mathcal{B}$ will be true in some finite submodels, and false in others.

Continuity does not apply to arbitrary complex formulas of *EL*, and the best one can say is that a complete logic based on continuous quanti-

16   LESLIE H. THARP

fiers does not extend *EL*. Whether this formulation is good enough is open to question. How does one, for example, know that a given logic can be defined by adjoining quantifiers to *EL*? Put this way, the question does not pose a problem. We have taken a quantifier to be an arbitrary class of models of a fixed type, closed under isomorphism of models, which amounts to saying that a totally arbitrary formula can be taken to be a quantifier (see Note 18). The assumption that the logic be based on quantifiers seems to involve no great restriction, for, given a logic *L*, one can take each formula of *L* as a quantifier and adjoin it to *EL*, getting *L**. Trivially, $L \subseteq L^*$, and it is reasonable to suppose $L = L^*$, because otherwise *L* would not be closed under the familiar constructions with connectives and the iteration of quantifiers.

The problem, then, is not that it is unreasonable to assume a given logic *L* is constructed by adjoining quantifiers to *EL*. Rather, the problem is to give some intrinsic justification for the particular method by which complex formulas are built up from the primitives. That is, suppose one is presented with a logic *L*, defined by adjoining quantifiers to *EL*, but which cannot be defined by the adjunction of continuous quantifiers to *EL*. How does one know that there might not be some other acceptable way of generating the logic from simple quantifiers? Second order logic is in a sense generated from the standard quantifiers – but one applies them to predicate letters as well as to individual variables.

Complex formulas with relations no longer have the operational simplicity of the basic quantifiers, or the finitistic nature of monadic formulas. Thus it is not easy to see a suitable criterion to apply directly to arbitrary formulas. And, in the absence of such a criterion, it is not easy to predict what primitives and operations might have a clear enough meaning to be used in constructing a logic.

These seem to me to be serious objections which pinpoint the weakness in this argument for the primacy of *EL*. I do not know how to overcome them, nor am I confident that there is a completely watertight argument for *EL*. Everything considered, I think it must be conceded that the considerations which gave a rather satisfactory characterization of elementary monadic logic do not provide a comparably definitive characterization of the full logic. But they do, I believe, give some insight into the reasons *EL* has been taken as standard, and the reasons it appears natural and primitive in comparison with the known extensions.

WHICH LOGIC IS THE RIGHT LOGIC? 17

7. LOGIC AND ONTOLOGY

The reasons for taking elementary logic as standard evidently have to do also with certain imprecise – but I think *vital* – criteria, such as the fact that it easily codifies many inferences of ordinary language and of informal mathematics, and the fact that stronger quantifiers can be fruitfully analyzed in set theory, a theory of *EL*. It is not surprising that in practice elementary logic has been taken as logic. Yet the criteria which justify this choice do not justify the use to which elementary logic is frequently put. It should be recalled that a tremendous amount of weight has been thrown on the alleged distinction between logic (i.e. elementary logic) and mathematics. Perhaps the most extreme example is Skolem,[19] who deduces from the Löwenheim-Skolem theorem that 'the absolutist conceptions of Cantor's theory' are 'illusory'. I think it clear that this conclusion would not follow even if elementary logic were in some sense the true logic, as Skolem tacitly assumed. From the absolutist standpoint, elementary logic is not able to preserve in the new structure all significant features of the initial structure. For example, the power set of ω is intended to contain *all* sets of integers. It should also be noted that one has a similar problem for ordinary number theory, and even for certain weak decidable subtheories of number theory: there are nonstandard models, countable as well as uncountable.

Elementary logic has also been invoked in connection with ontology by Quine, who has argued that one is to look to the range of the quantifiers to uncover one's ontological commitments. There are difficulties in interpreting this prescription which we shall not dwell on here, but at a very basic level one can question his doctrine by proposing different logics. One challenge he considers[20] is a logic due to Henkin, which has formulas with 'branching quantifiers' such as $\begin{smallmatrix} \forall u \exists v \\ \forall x \exists y \end{smallmatrix} F(u, v, x, y)$, which is to be interpreted, using Skolem functions, as $\exists f \exists g \forall u \forall x F(u, fu, x, gx)$. This looks very much like the Skolem form of a formula of *EL*, the difference being that v is a function of u alone, and y is a function of x alone. To express such a formula in *EL* one naturally quantifies over functions. Thus such a formula as $\begin{smallmatrix} \forall u \exists v \\ \forall x \exists y \end{smallmatrix} F(u, v, x, y)$ appears to blur the distinction between those objects

18   LESLIE H. THARP

which are quantified over, and those which are not. Quine rejects the
Henkin logic primarily[21] because it is not complete (it turns out, sur-
prisingly, that in it one can express the quantifier 'for infinitely many x').
But we have a number of 'deviant' logics which are axiomatizable and
allow us to do exactly this sort of thing: we can say that there are countably many helium atoms without quantifying over anything except physical objects. Quine does concede that some extensions of *EL* are complete,
and mentions the example of a logic which adds finitely many valid formulas. However, no indication is given how the complete extensions of
EL are to be ruled out.

To go to another extreme, one can formulate *EL* with modus ponens
as the sole rule of inference. By means of a simple translation one may
consider *EL*, and theories based on it, to be theories in *sentential* logic.
Sentential logic certainly has many attractive technical properties to recommend it. And perhaps one's ontological commitment would be to
the two truth values.

This is clearly an implausible suggestion, and it is not hard to see why.
Evidently our conceptual scheme is such that we think of the world in
terms of objects and relations. Sentential logic deals with whole sentences
and, unlike *EL*, suppresses this prior analysis and *prior commitment*. Of
course *EL* has quantifiers as well as individual variables and relation
letters. The particular choice of quantifiers must be explained, and we
have attempted to give illuminating reasons why the standard quantifiers
are singularly primitive. One can consider stronger quantifiers, but one
does not have as clear a grasp of their meaning, and they usually seem to
demand further explanation.

This is not to claim that it is impossible to operate with certain quantifiers, such as 'for infinitely many x', or equivalently, 'for finitely many x'.
It is true that in this case the resulting logic is not complete and invites
further extension. These are good reasons to conclude that the logic is not
as satisfactory as *EL*, but they do not seem to bear on the ontological
issue. Perhaps one could understand 'for finitely many x' in some intuitive sense as a primitive notion and work quite well within this logic.
Many of the rules would be sufficiently clear – for example if $\mathscr{A}(x)$ and
$\mathscr{B}(x)$ are each true of finitely many x, so are $(\mathscr{A}(x) \wedge \mathscr{B}(x))$ and
$(\mathscr{A}(x) \vee \mathscr{B}(x))$. The user of this logic might not have in mind any particular analysis of 'finite'. If so, it would seem incorrect to attribute to

him some such analysis in terms of an ontology of sets, or of numbers and functions. Which of several possible analyses should be attributed to him?

It may even be fair to say that mathematicians in effect used such quantifiers before the development of set theory. Since Cantor and Dedekind, one has a reasonably clear and quite general theory of 'finite', 'infinite', and related notions. In view of the existence of such a general theory, there is little point in taking notions such as 'finite' as primitive. The conceptual scheme of set theory deals with objects (sets), relations (membership and identity), and uses the standard quantifiers. Formulas of *EL* directly codify this scheme, and, so interpreted, they reflect ontological assumptions correctly. But to appeal to criteria such as completeness to justify the logic, and then mechanically use the logic to 'assess a theory's ontological demands', seems to stand the matter on its head. It also leads, via the Löwenheim-Skolem theorem, to pointless puzzles about ontological reduction. These, however, have been discussed at length elsewhere.[22]

The Rockefeller University

NOTES

* I am heavily indebted to Jonathan Lear, D. A. Martin and Hao Wang with whom I discussed many of these matters. I also wish to express thanks to a number of other colleagues and students who contributed valuable criticism and comments.
[1] An *L*-class has models of a fixed finite type, e.g., with one binary and four ternary relations.
[2] See A. Tarski, *Logic, Semantics, Metamathematics*, Clarendon Press, Oxford, 1956, pp. 409–420.
[3] For monadic logics it does matter whether one assumes the Löwenheim-Skolem property for single formulas or for infinite sets of formulas. See my paper, 'The Characterization of Monadic Logic', *The Journal of Symbolic Logic* 38 (1973), 481–488.
[4] See Per Lindström 'On Extensions of Elementary Logic', *Theoria* 35 (1969), 1–11. It should be noted that Harvey Friedman later rediscovered these theorems and pointed out their philosophical interest.
[5] See 'Arithmetical Extensions of Relational Systems', *Compositio Mathematica* 13 (1957), 81–102.
[6] Something like quantification is used with regard to the 'upward Löwenheim-Skolem property' (Theorem 3) and in Corollary 2 of Lindström's paper cited above.
[7] D. A. Martin has emphasized this distinction to me. Of course there are other senses of logic, for example, as "a science prior to all others, which contains the ideas and principles underlying all sciences". See Kurt Gödel, 'Russell's Mathematical Logic', in

20   LESLIE H. THARP

Philosophy of Mathematics, Benacerraf and Putnam (eds.), Prentice-Hall, Englewood Cliffs, N. J., 1964, p. 211.

[8] I concentrate only on certain gross features of a theory of deduction. This is not to deny that finer structure – the particular rules of inference and axioms – may be highly important.

[9] Lindström has shown in unpublished work that for a wide class of logics, if the logic is complete it is compact for recursively enumerable sets of formulas. The result holds for logics using finitely many 'generalized quantifiers'. Lindström's 'First Order Predicate Logic with Generalized Quantifiers' (*Theoria* 32 (1966), 186–195) discusses such quantifiers; see also Note 17 below.

[10] See H. J. Keisler, 'Logic with the Quantifier "There Exist Uncountably Many"', *Annals of Mathematical Logic* 1 (1970), 1–93. This logic is *countably* compact; this is the sense in which we are using 'compact', since sets of formulas are taken to be countable.

[11] Named after C. C. Chang. It is sometimes called the 'equicardinal quantifier', but this suggests the quantifier (Qx) which operates on a pair of formulas, $(Qx) [\mathscr{A}(x); \mathscr{B}(x)]$ being true in M just in case $\mathscr{A}(x)$ and $\mathscr{B}(x)$ are true of the same number of things in M. Although similar to (Cx), this quantifier is too strong to be axiomatizable. This follows because one can say, 'G is a linear ordering with no last element, and any two distinct points have a distinct number of predecessors'. This formula characterizes the ordering of the natural numbers.

[12] See Bell and Slomson, *Models and Ultraproducts*, North-Holland, Amsterdam, 1969, p. 283.

[13] Personal communication. These results are related to his 'On Models with Power-Like Orderings', *The Journal of Symbolic Logic* 37 (1972), 247–267.

[14] To prove $L(U)$ and $L(C)$ have no common extension, consider the formula which says 'G is a linear ordering with no last element and $(\forall x) \neg (Cy) G(y, x)$'. The only countable model of this formula is isomorphic to $\langle \omega, \langle \rangle$. Since one can say 'the universe is countable' in $L(U)$, any logic containing $L(C)$ and $L(U)$ characterizes $\langle \omega, \langle \rangle$. Martin pointed out that one can modify the quantifier C to a quantifier C' which is C in universes of cardinality $\leqslant \aleph_1$, and is $\forall$ in larger universes. The generalized continuum hypothesis was used to prove only special cases of the axiomatizability of C, and it turns out it is not needed for C'. Also C' is axiomatizable without restriction to infinite models. Our argument above still holds for $L(U)$ and $L(C')$.

[15] See Bell and Slomson, *op. cit.*, pp. 284–285.

[16] A *type* is a finite sequence of predicate letters; a *monadic* type has only monadic letters. A *quantifier* $\mathscr{Q}$ is a class of models of some fixed type, such that if $M \in \mathscr{Q}$ and N is isomorphic to M, then $N \in \mathscr{Q}$. For example, suppose $\mathscr{Q}$ is the class of models of type $\langle G \rangle$, where G is binary and is interpreted in the model as a well-ordering of the universe. One adjoins this $\mathscr{Q}$ to EL by introducing a symbol Q, and adding to the definition of formula the clause: $Qv_i, v_j \mathscr{A}(v_i, v_j)$ is a formula if $\mathscr{A}(v_i, v_j)$ is. $M \vDash Qv_i, v_j \mathscr{A}(v_i, v_j)$ is defined to mean $M^* \in \mathscr{Q}$ where the universe of M^* is the universe of M, and the binary relation of M^* is $\{ \langle x_i, x_j \rangle : M \vDash \mathscr{A}(x_i, x_j) \}$. Thus $M \vDash Qv_i, v_j \mathscr{A}(v_i, v_j)$ just in case $\mathscr{A}(v_i, v_j)$ defines a well ordering over the universe of M; this Q is just the quantifier W mentioned in passing in Section 3. For further discussion see my 'Continuity and Elementary Logic', *The Journal of Symbolic Logic* 39 (1974), 700–716.

[17] Let $\mathscr{P}$ be a class of finite models satisfying: if $M \in \mathscr{P}$ and N is isomorphic to M then $N \in \mathscr{P}$. A continuous quantifier $\mathscr{Q}$ is defined by: $M \in \mathscr{Q} \Leftrightarrow (\exists J) [J \subseteq M \wedge J \in \mathscr{P}]$. To bring in considerations of effectiveness, encode each finite model J with universe $\{0, 1, ..., n\}$

WHICH LOGIC IS THE RIGHT LOGIC?   21

as a number $\#(J)$. Given a recursive set R, define $\mathcal{P} = \{K:(\exists J)\,[K$ is isomorphic to J and $\#(J) \in R]\}$, and define $\mathcal{Q}$ as above. Then given a model M, suppose that one examines finite submodels K, checking whether there is an isomorphic J with $\#(J)$ in R. M is in $\mathcal{Q}$ if and only if this procedure eventually leads to a positive answer. Conversely, any quantifier which is partially decidable in the same sense as $\exists$ appears to fall under this definition, for suitably chosen recursive R.

[18] A logic based on continuous quantifiers satisfies the Löwenheim-Skolem condition, so if it is complete, it is *EL*. Note that if one demands a uniform finite bound (uniform continuity) one need not invoke completeness. See especially Theorems 2, 4, 5 and 7 of my paper cited in Note 17.

[19] See p. 47 of Thoralf Skolem, *Abstract Set Theory* (Notre Dame Mathematical Lectures No. 8, Notre Dame, 1962).

[20] See pp. 89–91 of W. V. Quine, *Philosophy of Logic*, Prentice-Hall, Englewood Cliffs, N. J., 1970.

[21] However, in an earlier paper Quine also notes that a certain typical construction which one would formulate with the Henkin quantifier "is not after all very ordinary language; its grammar is doubtful". See p. 112 of 'Existence and Quantification', in *Ontological Relativity and Other Essays*, Columbia University Press, New York, 1969.

[22] See my 'Ontological Reduction', *The Journal of Philosophy* **68** (1971), 151–164.

[3]

THE JOURNAL OF PHILOSOPHY

VOLUME LXXII, NO. 16, SEPTEMBER 18, 1975

ON SECOND-ORDER LOGIC *

I SHALL discuss some of the relations between second-order logic, first-order logic, and set theory. I am interested in two quasi-terminological questions, viz., the extent to which second-order logic is (or is to be counted as) logic, and the extent to which it is set theory. It is of little significance whether second-order logic may bear the (honorific) label 'logic' or must bear 'set theory'. What matter, of course, are the reasons that can be given on either side. It seems to be commonly supposed that the arguments of Quine and others for not regarding second- (and higher-) order logic as logic are decisive, and it is against this view that I want to argue here. I shall be concerned mainly with Quine's critique of second-order logic and with some of the reasons that can be offered in support of applying neither, one, or both of the terms 'logic' and 'set theory' to second-order logic.[1]

The first of Quine's animadversions upon second-order logic that I shall discuss is to be found in the section of his *Philosophy of Logic*[2] called "Set Theory in Sheep's Clothing." Much of this section is devoted to dispelling two confusions which we can easily agree with

* I am grateful to Richard Cartwright, Oswaldo Chateaubriand, Fred Katz, and James Thomson for helpful criticism.

[1] My motive in taking up this issue is that there is a way of associating a truth of second-order logic with each truth of arithmetic; this association can plausibly be regarded as a "reduction" of arithmetic to set theory. [It is described in Chapter 18 of *Computability and Logic* by Richard Jeffrey and myself (New York: Cambridge, 1974).] I am inclined to think that the existence of this association is the heart of the best case that can be made for logicism and that unless second-order logic has *some* claim to be regarded as logic, logicism must be considered to have failed totally. I see the reasons offered in this paper on behalf of this claim as part of a partial vindication of the logicist thesis. I don't believe we yet have an assessment that is as just as it could be of the extent to which Frege, Dedekind, and Russell succeeded in showing logic to be the ground of mathematical truth.

[2] Englewood Cliffs, N. J.: Prentice-Hall, 1970; parenthetical page references to Quine are to this book.

510   THE JOURNAL OF PHILOSOPHY

Quine in deploring: that of supposing that '($\exists F$)' and '(F)' say that some (all) predicates (i.e., predicate-expressions) are thus and so, and that of supposing that quantification over attributes has relevant ontological advantages over quantification over sets. What I wish to dispute is his assertion that the use of predicate letters as quantifiable variables is to be deplored, even when the values of those variables are sets, on the ground that predicates are not *names* of their extensions. Quine writes, "Predicates have attributes as their 'intensions' or meanings (or would if there were attributes) and they have sets as their extensions; but they are names of neither. Variables eligible for quantification therefore do not belong in predicate positions. They belong in name positions" (67).

Let us grant that predicates are not names. Why must we then suppose, as the "therefore" in Quine's sentence would indicate we must, that variables eligible for quantification do not belong in predicate positions? Quine earlier (66/7) gives this argument:

> Consider first some ordinary quantifications: '($\exists x$)(x walks)', '(x)(x walks)', '($\exists x$)(x is prime)'. The open sentence after the quantifier shows 'x' in a position where a name could stand; a name of a walker, for instance, or of a prime number. The quantifications do not mean that names walk or are prime; what are said to walk or to be prime are things that could be named *by* names in those positions. To put the predicate letter 'F' in a quantifier, then, is to treat predicate positions suddenly as name positions, and hence to treat predicates as names of entities of some sort. The quantifier '($\exists F$)' or '(F)' says not that some or all predicates are thus and so, but that some or all entities of the sort named by predicates are thus and so.

If Quine had argued:

> Consider some extraordinary quantifications: '($\exists F$)(Aristotle F)', '(F)(Aristotle F)', '($\exists F$)(17 F)'. The open sentence after the quantifier shows 'F' in a position where a predicate could stand; a predicate with an extension in which Aristotle, for instance, or 17 might be. The quantifications do not mean that Aristotle or 17 are in predicates; what Aristotle or 17 are said to be in are things that could be had *by* predicates in those positions. To put the variable 'x' in a quantifier, then, is to treat name positions suddenly as predicate positions, and hence to treat names as predicates with extensions of some sort. The quantifier '($\exists x$)' or '(x)' says not that some or all names are thus and so, but that some or all extensions of the sort had by names are thus and so.

we should have wanted to say that the last two statements were false and did not follow from what preceded them. It seems to me that the same ought to be said about the argument Quine actually gives.

To put '*F*' in a quantifier may be to treat '*F*' as having a *range*, but it need not be to treat predicate positions as name positions nor to treat predicates as names of entities of any sort. Quine seems to suppose that because a variable of the more ordinary sort, an individual variable, always occurs in positions where a name but not a predicate could occur, the same must hold for every sort of variable. We may grant that the ordinary quantifications mean what Quine says they mean. But we are not thereby committed to any paraphrase containing 'name' (or any of its cognates) that purports to give the meaning of our extraordinary quantifications. Perhaps someone might suppose that variables must always *name* the objects in their range, albeit only "indefinitely" or "temporarily." However, we have no reason not to think that there might be a sort of variable, a predicate variable, that ranges over the objects in its range (these will be extensions) but does not *name* them "indefinitely" or any other way; rather, predicate variables will *have* them "indefinitely," as (constant) predicates have their extensions "definitely." Such variables would not be names of any sort, not even "indefinite" ones, but would have a range containing those objects (extensions) which could be had by predicates in predicate positions.

It may be that a suggestion is lurking that an adequate referential account of the truth conditions of sentences cannot be given unless it is supposed that all variables act as names that (indefinitely) name the objects in their range. But this is not the case. Although variables must have a range containing suitable objects, it need not be that variables of every sort indefinitely name the objects in their ranges. '(∃*F*)' does not have to be taken as saying that some entities of the sort named by predicates are thus and so; it can be taken to say that some of the entities (extensions) had by predicates contain thus and such. So some variables eligible for quantification might well belong in predicate positions and not in name positions. And taking '*Fx*' to be true if and only if that which '*x*' names is in the extension of '*F*' in no way commits us to supposing that '*F*' names anything at all.

In the same section of *Philosophy of Logic* Quine has some advice for the logician who wants to admit sets as values of quantifiable variables and also wants distinctive variables for sets. The logician should not, Quine says, write '*Fx*' and thereupon quantify on '*F*', but should instead write '*x* ∈ *α*' and then, if he wishes, quantify on '*α*'. The advantage of the new notation is thought to be its greater

512   THE JOURNAL OF PHILOSOPHY

explicitness about the set-theoretic presuppositions of second-order logic. There is an important distinction between first- and second-order logic with regard to those presuppositions, which may be part of the reason Quine insists on regarding 'F', 'G', etc. in first-order formulas as schematic letters and not quantifiable variables. In order to give a theory of truth for a first-order language which is materially adequate (in Tarski's sense) and in which such laws of truth as "The existential quantification of a true sentence is true" can be proved, it is not necessary to assume that the predicates of the language have extensions, although it does appear to be necessary to make this assumption in order to give such a theory for a second-order language.

There are reasons for not taking Quine's advice, however. One is that the notation Quine recommends abandoning represents certain aspects of logical form in a most striking way.[3] Another, and more important, reason is that the usual conventions about the use of special variables like 'α' guarantee that rewriting second-order formulas in Quine's way can result in the loss of validity or implication. For example, '$\exists F \forall x\, Fx$' is valid, but '$\exists \alpha \forall x\, x \in \alpha$' is not; and '$x = z$' is implied by '$\forall Y(Yx \rightarrow Yz)$' but not by '$\forall \alpha(x \in \alpha \rightarrow z \in \alpha)$'.

Quine disparages second-order logic in two further ways: reading him, one gets the sense of a culpable involvement with Russell's paradox and of a lack of forthrightness about its existential commitments. "This hypothesis itself viz., '$(\exists y)(x)(x \in y \equiv Fx)$' falls dangerously out of sight in the so-called higher-order predicate calculus. It becomes '$(\exists G)(x)(Gx \equiv Fx)$', and thus evidently follows from the genuinely logical triviality '$(x)(Fx \equiv Fx)$' by an elementary logical inference. Set theory's staggering existential assumptions are cunningly hidden now in the tacit shift from schematic predicate letter to quantifiable set variable" (68). Quine, of course, does not assert that higher-order predicate calculi are inconsistent. But even if they are consistent,[4] the validity of '$\exists X \forall x(Xx \leftrightarrow\, \sim x \in x)$', which certainly looks contradictory, would at any rate seem to demonstrate

---

[3] For instance, writing out the definition of the ancestral aR_*b in this notation:

$$\forall F(\forall x(aRx \rightarrow Fx)\ \&\ \forall x\, \forall y(Fx\ \&\ xRy \rightarrow Fy) \rightarrow Fb)$$

shows it to be obtained from an ordinary first-order formula by prefixing a universal quantifier, and suggests an interesting question: Is there an *existential* quantification of a first-order formula that is a satisfactory definition of the ancestral? (The answer is no.)

[4] Gentzen showed that the problem of their consistency had a very easy positive solution. See "Die Widerspruchsfreiheit der Stufenlogik," *Mathematische Zeitschrift*, XLI, 3 (1936): 357–366. An English translation, "The Consistency of the Simple Theory of Types," is contained in M. E. Szabo, ed., *The Collected Papers of Gerhard Gentzen* (Amsterdam: North-Holland, 1969).

that their existence assumptions must be regarded as "vast." A problem now arises: although '$\exists X \, \exists x \, Xx$' and '$\exists X \, \forall x \, Xx$' are also valid, '$\exists X \, \exists x \, \exists y (Xx \,\&\, Xy \,\&\, x \neq y)$' is not valid; it would thus seem that, despite its affinities with set theory and its vast commitments, second-order logic is not committed to the existence of even a two-membered set. Both of these difficulties, it seems to me, can be resolved by examining the notion of validity in second-order logic. This examination seems to show a certain surprising weakness in second-order logic.

When is a sentence valid in second-order logic? When it is true under all its interpretations. When does it follow from others? When it is true under all its interpretations under which all the others are true. What, then, is an interpretation of a second-order sentence? If we are considering "standard" second-order logic in which second-order quantifiers are regarded as ranging over *all* subsets of, or relations on, the range of the first-order quantifiers,[5] we may answer: exactly the same sort of thing an interpretation of a first-order sentence is, viz., an ordered pair of a non-empty set D and an assignment of a function to each nonlogical constant in the sentence. The domain of the function is the set of all n-tuples of members of D if the constant is of degree n, and the range is a subset of D if the constant is a function constant and a subset of $\{T, F\}$ if it is a predicate constant. [Names (sentence letters) are function (predicate) constants of degree 0; functions from the set of all 0-tuples of members of D into an arbitrary set E are of course members of E.] We need not explicitly mention separate ranges for the second-order variables that may occur in the sentence. An existentially quantified sentence $\exists \alpha \, F(\alpha)$ is then true under an interpretation I just in case $F(\beta)$ is true under some interpretation J that differs from I (if at all) only in what it assigns to the constant β, which is presumed not to occur in $\exists \alpha \, F(\alpha)$ and presumed to be of the same logical type[6] as the variable α. The other clauses in the definition of *truth in an interpretation* are exactly as you would suppose them to be. Notice that in this account no mention is made of what sort (individual, sentential, function, or predicate) of variable α is; α may be any sort of variable at all. Notice also that, if only individual variables are allowed, the account is just a paraphrase of one standard definition of *truth in an interpretation*. The definition changes neither the conditions under which a first-order sentence is true in an interpretation nor the account of what an interpretation is, but

---

[5] Only "standard" or "full" second-order logic is considered in this paper.

[6] Two symbols are of the same logical type if they are of the same degree and are either both predicate symbols or both function symbols.

514   THE JOURNAL OF PHILOSOPHY

merely extends in the obvious way the account given in (say) Mates's *Elementary Logic*[7] or Jeffrey's *Formal Logic*[8] to cover the new sorts of quantified sentences that arise in second-order logic. Quine has stressed the discontinuities between first- and second-order logic so emphatically and for so long that the obvious and striking continuities may be forgotten. In Mates's book, for example, nineteen laws of validity are stated, of which all but one (the compactness theorem) hold for second-order logic. Thus there is a standard account of the concepts of validity and consequence for for first-order sentences, and there is an obvious, straightforward, non–ad hoc way of extending that account to second-order sentences.[9]

We can now see what is shown by the validity of

$$\exists X \, \forall x (Xx \leftrightarrow \sim x \, \epsilon \, x).$$

First of all, the sentence *is* valid : given any I, we can always find a suitable J in which '$\forall x (Bx \leftrightarrow \sim x \, \epsilon \, x)$' is true by assigning to 'B' the set of all objects in the domain of I that do not bear to themselves the relation that I assigns to 'ϵ'. Since the domain of I is a set, one of the axioms of set theory (an *Aussonderungsaxiom*) guarantees that there will always be such a subset of the domain. But without a guarantee that there is a set of all sets, we cannot conclude from the validity of '$\exists X \, \forall x (Xx \leftrightarrow \sim x \, \epsilon \, x)$' that there is a set of all non-self-membered sets. And we have guarantees galore that there is no set of all sets. We do, of course, land in trouble if we suppose that 'x' ranges over all sets, that 'X' ranges over all sets of objects over which 'x' ranges, and that 'ϵ' has its usual meaning; for then '$\exists X \, \forall x (Xx \rightarrow \sim x \, \epsilon \, x)$' would be false. But that it would then be false does not show it to be invalid; for there is no interpretation whose domain contains all sets.

Our difficulty is thus circumvented, but at some cost. We must insist that we mean what we say when we say that a second-order sentence is valid if true under all its interpretations, and that an interpretation is an ordered pair of a *set* and an assignment of functions to constants.

There is thus a limitation on the use of second-order logic to which first-order logic is not subject. Examples such as '$\exists X \, \forall x (Xx \leftrightarrow \sim x \, \epsilon \, x)$'

[7] 2d ed., New York: Oxford, 1972.
[8] New York: McGraw-Hill, 1967.
[9] In Part IV of *Methods of Logic*, 3d ed. (New York: Holt, Rinehart & Winston, 1972), Quine extends the notion of validity to first-order sentences with identity and discusses higher-order logic at length, but does not describe the extension of the notion of validity to second-order logic.

and '$\exists X \forall x\, Xx$', both valid, seem to show that it is impermissible to use the notation of second-order logic in the formalization of discourse about certain sorts of objects, such as sets or ordinals, in case there is no *set* to which all the objects of that sort belong. This restriction does not apply, as it appears, to first-order logic: ZF (Zermelo-Fraenkel set theory) is couched in the notation of first-order logic, and the quantifiers in the sentences expressing the theorems of the theory are presumed to range over all sets, even though (if ZF is right) there is no set to which all sets belong. In the case of '$\exists X \forall x\, Xx$', we cannot assume, for example, that the quantifier '$\forall x$' ranges over all ordinals, for then '$\exists X \forall x\, Xx$' would be true iff there were a set to which all ordinals belong, and there is no such set. Nor can we assume that it ranges over all the sets that there are, for it would then be true iff there were a set of all sets. Thus if we wish (as we do) to maintain that both sentences are true (because valid) and also wish to preserve the standard account of the conditions under which sentences are true, we cannot suppose that all sets belong to the range of '$\forall x$' in either, or that all ordinals belong to the range of '$\forall x$' in '$\exists X \forall x\, Xx$'. There is of course a step from supposing that the quantifier '$\forall x$' in '$\exists X \forall x\, Xx$' may not be assumed to range over all sets to supposing that all members of the range of first-order quantifiers in second-order sentences used to formalize a certain discourse must be contained in some one set (which depends upon the discourse), and there might be ways of not taking it. But all the difficulties do appear to have the same source, and seem to point to the impermissibility of second-order discourse about all sets, all ordinals, etc.

(We have been assuming all along that ZF is correct and that *sets* are the only "set-like" objects there are, the only objects to which membership is borne. If, however, as certain extensions of ZF assert, there are also certain *classes*, which are not sets, but which sets may be members of, then of course we are free to interpret '$\exists X \forall x\, Xx$' as saying that there is a class to which all sets belong and thus to suppose that '$\forall x$' ranges over all sets in '$\exists X \forall x\, Xx$'. But even if classes do exist, there is again a distinction between first- and second-order notation that is significantly like the distinction just described: we may use the former but not the latter to discuss *all members of the counterdomain* (the right field) *of 'ϵ'*. One of the lessons of Russell's paradox is that if we read 'Xx' as '(OBJECT) X bears R to (object) x', then the range of first-order quantifiers in second- but not first-order sentences may not contain all OBJECTS.)

516   THE JOURNAL OF PHILOSOPHY

There is a similar, but less significant, restriction on the use of the notation of first-order logic. One who uses it to formalize some discourse is committed (in the absence of special announcements to the contrary) to the non-emptiness of the ontology of the discourse and also to the presence in the ontology of references of any names that occur in the formalization. The use of names in formalization can be avoided, however, as Quine has pointed out, and various formulations of first-order logic exist in which the empty domain is permitted. But there is a striking difference between the commitment to non-emptiness of an ontology and the commitment to sethood: we believe that our own ontology is non-empty, but not that it forms a set! The contradictions appear, therefore, to teach us not that second-order logic may be inconsistent (as Quine perhaps intimates), but that it seems impossible that any "universal characteristic" should be couched in the notation of second-order logic.

What now of the existence assumptions of second- and higher-order logic, which Quine calls both "vast" and "staggering"? *Set theory* (ZF) certainly makes staggering existence *claims*, such as that there is an infinite cardinal number κ that is the κth infinite cardinal number (and hence that there is a set with that many members). Quine maintains that higher-order logic involves "outright assumption of sets the way [set theory] does."[10] Of course there are differences between set theory and higher-order logic: all set theories agree that there is a set containing at least two objects, but, as noted, '$\exists X \, \exists x \, \exists y (Xx \, \& \, Xy \, \& \, x \neq y)$' is not valid, for it is false in all one-membered interpretations. Let us try to see what the way*s* are in which second-order logic involves assuming the existence of sets.

First of all, "in second-order logic one quantifies over sets." There are certain (second-order) sentences of any given language that will be classified by second-order logic as logical truths (i.e., as valid), even though they assert, under any interpretation of the language whose domain forms a set, the existence of certain sorts of sub*sets* of the domain. (The sort depends upon the interpretation.) '$\exists X \, \forall x (Xx \leftrightarrow \sim x \, \epsilon \, x)$' and '$\exists X \, \forall x (Xx \leftrightarrow x = x)$' are two examples. Thus, unless there exist sets of the right sorts, these sentences will be false under certain interpretations.

Now one may be of the opinion that no sentence ought to be considered as a truth of *logic* if, no matter how it is interpreted, it asserts that there are *sets* of certain sorts. Similarly, one might hold that the truth of '$\exists f \, \forall x \, Rf(x)x$' ought not to *follow* from that of

[10] *Set Theory and Its Logic*, 2d ed. (Cambridge, Mass.: Harvard, 1969), p. 258.

'$\forall x\ \exists y\ Ryx$' (even if the axiom of choice is true), or one might think that it is not *as a matter of logic* that there is a set with certain closure properties if Smith is not an ancestor of Jones (i.e., not a parent, not a grandparent, etc.).

The view that logic is "topic-neutral" is often adduced in support of this opinion: the idea is that the special sciences, such as astronomy, field theory, or set theory, have their own special subject matters, such as heavenly bodies, fields, or sets, but that logic is not about any sort of thing in particular, and, therefore, that it is no more in the province of logic to make assertions to the effect that sets of such-and-such sorts exist than to make claims about the existence of various types of planet. The subject matter of a particular science, what the science is about, is supposed to be determined by the range of the quantifiers in statements that formulate the assertions of the science; logic, however, is not supposed to have any special subject matter: there is neither any sort of thing that may not be quantified over, nor any sort that must be quantified over.

I know of no perfectly effective reply to this view. But, in the first place, one should perhaps be suspicious of the identification of subject matter and range. (Is elementary arithmetic really not *about* addition, but only *about* numbers?) And then it might be said that logic is not so "topic-neutral" as it is often made out to be: it can easily be said to be about the notions of negation, conjunction, identity, and the notions expressed by 'all' and 'some', among others (even though these notions are almost never quantified over). In the second place, unlike *planet* or *field*, the notions of *set, class, property, concept*, and *relation*, etc. *have* often been considered to be distinctively logical notions, probably for some such very simple reason as that anything whatsoever may belong to a set, have a property, or bear a relation. That some set- or relation-existence assertions are counted as logical truths in second- or higher-order systems does not, it seems to me, suffice to disqualify them as systems of logic, as a system would be disqualified if it classified as a truth of logic the existence of a planet with at least two satellites. Part 3 of the *Begriffsschrift*, for example, where the definition of the ancestral was first given, is as much a part of a treatise on logic as are the first two parts; the first occurrence of a second-order quantifier in the *Begriffsschrift* no more disqualifies it from that point on as a work on logic than does the earlier use of the identity sign or the negation sign. Poincaré's wisecrack, "La logique n'est plus sterile. Elle engendre la contradiction," was cruel, perhaps, but not unfair. And many of us first learned about the ancestral and other matters from a work not unreasonably entitled *Mathematical Logic*.

Another way in which second- but not first-order logic involves existential and other sorts of set-theoretic assumptions is this: via Gödelization and because of the completeness theorem, elementary arithmetic ("Z") is a suitable background theory for the development of a significant theory of validity of first-order formulas. A notion of "validity," coextensive with the usual one (truth of the universal closure in all interpretations), can be defined in the language of Z via Gödelization, and the validity of each valid formula (and no others) can then be proved in the theory, as can many general laws of validity. Moreover, the invalidity of many invalid sentences can also be demonstrated. In contrast, not only is there no hope of proving the validity of each valid second-order sentence in elementary arithmetic, the notion of second-order validity cannot even be *defined* in the language of *second-order* arithmetic. We can effectively associate with each first-order sentence a statement of arithmetic of a particularly simple form that is true if and only if the first-order sentence is valid, but no such association is even remotely possible for second-order sentences.[11] Worse, for many highly problematical statements of set theory (such as the continuum hypothesis) there exist second-order sentences that are valid if and only if those statements are true. Thus the metatheory of second-order logic is hopelessly set-theoretic, and the notion of second-order validity possesses many if not all of the epistemic debilities of the notion of set-theoretic truth.

On the other hand, although it is not hard to have some sympathy for the view that no notion of validity should be so extravagantly distant from the notion of proof, we should not forget that validity of a first-order sentence is just truth in all its interpretations. (The equation of first-order validity with provability effected by the completeness theorem would be miraculous if it weren't so familiar.) And, as we shall see below, there are notions of (first-order) logical theory which, unlike *validity*, can be adequately treated of only in a background theory that is stronger than elementary arithmetic.

While comparing set theory and second-order logic, we ought to remark in passing that the definability in set theory of the notion of second-order validity at once guarantees both the nonexistence of a reduction of the notion of set-theoretical truth to that of second-order validity and the existence of a reduction in the opposite direction: no effective—indeed no set-theoretically definable—function that assigns formulas of second-order logic to sentences of set theory

---

[11] There is a precise sense in which the set of valid second-order sentences is *staggeringly* undecidable: it is not definable in *n*th-order arithmetic, for any *n*. Its "Löwenheim number" is also staggeringly high.

assigns second-order logical truths to all and only the truths of set theory (otherwise set-theoretical truth would be set-theoretically definable). However, the function that assigns to each formula of second-order logic the sentence of set theory that asserts that the formula *is* a second-order logical truth reduces second-order validity to set-theoretical truth. Thus each of the notions in the series ⟨first-order validity, first-order arithmetical truth, second-order arithmetical truth, second-order validity, set-theoretical truth⟩ can always be reduced via effective functions to later ones but never to earlier ones; the notions are thus in order of increasing strength of one certain sort.

Quine writes (66) that "the logic capable of encompassing [the reduction of mathematics to logic] was logic inclusive of set theory." If second-order logic is "inclusive of set theory," it would seem to have to count as valid some nontrivial theorems of set theory, and if, among those counted as valid, there were some to the effect that certain kinds of set existed, second-order logic might seem to involve excessive ontological commitments in yet another way. And it may easily seem that second-order logic involves such commitments. For '$\exists X \forall x \sim Xx$' and '$\exists X \forall y (Xy \leftrightarrow y \subseteq x)$' are both valid and might be thought to assert that the null and power sets exist, just as all set theories say.

It seems, however, that there is a serious difficulty in supposing that *any* second-order sentence asserts, for example, that there is a set with no members; it seems that no second-order sentence asserts the same thing as any theorem of set theory, and hence that not even the smallest fragment of set theory is, in this sense, included in second-order logic.

Consider the question "What does '$\forall x \; x = x$' assert?" One may answer, "Why, that everything is identical with itself." But if one answers thus, one must realize that one's answer has a determinate sense only if the reference (range) of 'everything' is fixed. A more cautious answer might be "Why, that everything in the domain (whatever the domain may be) is identical with itself." If the natural numbers are in question '$\forall x \; \exists y \; y < x$' is false; if the rationals, true. (It seems to me that the ordinary Peano-Russell notation is less than ideal in not representing in a sufficiently vivid way the partial dependence of truth-value upon domain. In some ways it would be nicer if each quantifier were required to wear a subscript that indicated its range. It seems that the design of standard notation is influenced by the archaic view that logic is about some one fixed domain of *objects* or *individuals*, and that a logical truth is a sentence that is true no matter what relations on that domain are assigned to the predicate letters in the sentence.)

Thus the correct answer to the question, "What does '$\exists X \forall x \sim Xx$' assert?" would seem to be something like "That depends upon what the domain is supposed to be (and also upon how that domain is 'given' or 'described'). But, whatever the domain may be, '$\exists X \forall x \sim Xx$' will assert that there is a subset of the domain to which none of its members belong."

It should now appear that no valid second-order sentence can assert the same thing as any theorem of set theory. For a second-order sentence, whether valid or not, asserts something only with respect to an interpretation, whose domain may not be taken to contain all sets. But if the sentence were to assert what any particular set-theoretic statement asserts, its domain, it would seem, would have to contain all sets. '$\exists X \forall x\, Xx$' is valid, but does not assert that there is a universal set, which, if ZF is correct, is false; rather, it asserts that there is a subset of the domain (whichever set that may be) to which everything in the domain belongs. The quantifiers in the first-order sentences that express the assertions of ZF range over objects that do not together constitute a set. We have argued that the ranges of the variables in second-order sentences must be sets. If so, it is hard to see how any second-order sentence could express or assert what any theorem of ZF does, or that second-order logic counts as valid some significant theorems of set theory.

There is a clear sense, however, in which second-order logic can at least be said to be committed to the assertion that an empty set exists. For since the empty set is a subset of the domain of every interpretation whatsoever and is the only set to which no members of any domain belong, '$\exists X \forall x \sim Xx$' may be taken to assert the existence of the empty set independently of any interpretation, and second-order logic may thus be regarded as committed to its existence too. Moreover, higher- and higher-order logics will be committed in the same way to more and more sets.[12] In the case of second-order logic, though, the commitment is exceedingly modest; the null set is the only set to whose existence second-order logic can be said to be committed.

One sense, already noted, in which the use of second- but not first-order logic commits one to the existence of sets in this: If L_1 is the first-order fragment of an interpreted second-order language L_2 whose domain D contains no sets, then there are many logical

---

[12] I owe this point to Oswaldo Chateaubriand.

truths of L_1 that claim the existence of objects in D with certain properties, but there are none that claim the existence of subsets of D; however, among the logical truths of L_2 there are many such: for each predicate of L_2 with one free individual variable, there is a logical truth of L_2 that asserts the existence of a subset of D that is the extension of the predicate.

We have already seen definitions of validity and consequence for second-order sentences which bring out the obvious continuity of second- with first-order logic: validity and consequences are, as always, truth in all appropriate interpretations; the definition of an interpretation remains unchanged, as does the account of the conditions under which a first-order sentence is true in an interpretation. The account needs only to be *supplemented* with new clauses for the new sorts of sentence that arise in second-order logic. The supplementation may be given in separate clauses for each new sort of quantifier, which will be perfectly analogous to those for individual quantifiers. It may also be given in a general account of the conditions under which a sentence beginning with a quantifier is true in an interpretation, which applies uniformly to all sorts of quantifier, and of which the clauses for sentences beginning with individual quantifiers are special cases. The existence of such a definition provides *a* strong reason for reckoning second-order logic as logic. We come now to a second virtue of second-order logic, the well-known superiority of its "expressive" capacity.

If we conjoin the first two "Peano postulates," replace constants by variables, and existentially close, we obtain

$$\exists z\, \exists S(\forall x\, z \neq S(x)\, \&\, \forall x\, \forall y(S(x) = S(y) \to x = y))$$

a sentence true in just those interpretations whose domains are (Dedekind) infinite. If we do the same for the induction postulate, we obtain

$$\exists z\, \exists S\, \forall x(Xz\, \&\, \forall x(Xx \to XS(x)) \to \forall x\, Xx)$$

which is true in just those interpretations with countable domains. Thus the notions of infinity and countability can be characterized (or "expressed") by second-order sentences, though not by first-order sentences (as the compactness and Skolem-Löwenheim theorems show). Although first-order logic's expressive capacity is occasionally quite surprising, there are many interesting notions such as *well-ordering*, *progression*, *ancestral*, and *identity* that cannot be characterized in first-order logic (first-order logic without '=' in the case of *identity*!), but that can be characterized in second-. And

the second-order characterizations of notions like these offer a way
of regarding as inconsistent certain apparently inconsistent (infinite)
sets of statements, each of whose finite subsets is consistent—a way
that is not available in (compact) first-order logic. Four examples of
such sets are {'Smith is an ancestor of Jones', 'Smith is not a parent
of Jones', 'Smith is not a grandparent of Jones', . . .}, {'It is not
the case that there are infinitely many stars', 'There are at least two
stars', 'There are at least three stars', . . .}, {'R is a well-ordering',
'a_1Ra_0', 'a_2Ra_1', 'a_3Ra_2', . . .}, and, of course, {'x is a natural
number', 'x is not zero', 'x is not the successor of zero', . . .}.[13]

Compare these four sets with {'Not: there are at least three stars',
'Not: there are no stars', 'Not: there is exactly one star', 'Not: there
are exactly two stars'} and {'R is a linear ordering', 'a_0Ra_1', 'a_1Ra_2',
'Not: a_0Ra_2'}. There is a translation into the notation of first-order
logic under which the latter two sets of statements are *formally* inconsistent. Moreover, the translation, together with an explanation
of the conditions under which the translations are true in interpretations, provides an important part of the explanation of the inconsistency of the two sets. One would have hoped that the same
sort of thing might be possible for the four former sets. It seems
impossible, on reflection, that all the statements in any one of these
four sets should be true; it also seems that the reasons for this
impossibility would have to be of the same character as those which
explain the inconsistency of the latter two sets, the kind of reason
it has always been the business of logic to give. That the logic taught
in standard courses demonstrably cannot represent the inconsistency
of our four sets of sentences shows not that they are consistent after
all, but that not all (logical) inconsistencies are representable by
means of that logic. One may suspect that the second-order account
of these inconsistencies is not the "correct" account and that
perhaps some sort of infinitary logic might more accurately reflect
the logical form of the sentences in question; in any event, second-
order logic does not muff these cases altogether. In addition, then,
to there being a "straightforward" extension of the definitions of
valid sentence and *consequence of* from first- to second-order logic,
another reason for regarding second-order logic as logic is that there
are notions of a palpably logical character (*ancestral, identity*), which
can be defined in second-order logic (but not first-) and which figure
critically in inferences whose validity second-order logic (but not
first-) can represent.

[13] Alfred Tarski, "On the Concept of Logical Consequence," in *Logic, Semantics,
Metamathematics* (New York: Oxford, 1960), p. 410.

ON SECOND-ORDER LOGIC 523

Let us turn now to the failure of the completeness theorem for second-order logic, which can hardly be regarded as one of second-order logic's happier features. The existence of a sound and complete axiomatic proof procedure and the effectiveness of the notion of proof guarantee that the set of valid sentences of first-order logic is effectively generable; Church's theorem shows that it is not effectively decidable. There are decidable fragments of first-order logic, e.g., monadic logic with identity, but decidability vanishes if even a single two-place letter is allowed in quantified sentences. However, in a 1919 paper called "Untersuchungen über die Axiome des Klassenkalkuls . . ."[14] Skolem showed that the class of monadic second-order sentences, in which only individual and one-place predicate variables and constants may occur, is also decidable.

Discussing the contrast between classical first-order quantification theory and an extension of it containing "branching" quantifiers, Quine writes,

> . . . there is reason, and better reason, to feel that our previous conception of quantification . . . is not capriciously narrow. On the contrary, it determines an integrated domain of logical theory with bold and significant boundaries, designate it as we may. One manifestation of these boundaries is the following. The logic of quantification in its unsupplemented form admits of complete proof procedures for validity (90).

The extension is then noted not to admit of complete proof procedures.

> A remarkable concurrence of diverse definitions of logical truth . . . suggested to us that the logic of quantification as classically bounded is a solid and significant unity. Our present reflections on branching quantification further confirm this impression. It is at the limits of the classical logic of quantifications, then, that I would continue to draw the line between logic and mathematics (91).

Completeness cannot by itself be a sufficient reason for regarding the line between first- and second-order logic as the line between logic and mathematics. We have seen, first, that monadic logic differs from full first-order logic on the score of *decidability*, every bit as significant a property as *completeness*; we have further seen that this difference persists into second-order logic; and we have discussed at length the fact that we can extend to second-order sentences the definition of *truth in an interpretation* without change

[14] Reprinted in Th. Skolem, *Selected Works in Logic* (Oslo: Universitetsforlaget, 1970), pp. 67–101, and especially pp. 93–101.

524   THE JOURNAL OF PHILOSOPHY

in the notation of an *interpretation*. How, then, can the *semi-*effectiveness of the set of first-order logical truths be thought to provide much of a reason for distinguishing logic from mathematics? Why *completeness* rather than *decidability* or *interpretation*? Of course there is a big difference between second- and first-order logic; there are many. There are also big differences among various fragments of first-order logic, between second- and third-order logic, and between second-order logic and set theory.

Quine does not state that the completeness theorem by itself provides sufficient reason for drawing the line, however. Another reason, or more of the reason, is given by what he calls the "remarkable concurrence of diverse definitions of logical truth." One of these diverse definitions is the usual one: a sentence (or "schema," in Quine's terminology) is a logical truth if it is satisfied by every model, i.e., if it is true under all its interpretations. The other is that a sentence of a reasonably rich language is a logical truth if truths alone come of it by substitution of (open) sentences for its simple component sentences. The languages in question are interpreted languages (otherwise the notion of truth of a sentence of a language, used in the definition, would be incomprehensible), and their grammar has been "standardized," i.e., put into the notation of the first-order predicate calculus, *without* function signs or identity. As usual, "reasonably rich" has to do with arithmetic. For Quine's purposes, a language may be taken to be reasonably rich if its ontology contains all natural numbers (or an isomorphic copy) and its ideology contains a one-place predicate letter true of the natural numbers (or their copies) and two three-place predicate letters representing the sum and product operations.

By appealing to a generalization of Löwenheim's theorem that is due to Hilbert and Bernays—any satisfiable schema is satisfied by a model whose domain is the set of natural numbers and whose predicates are assigned relations on natural numbers *that can be defined in arithmetic*—Quine proves a result he calls remarkable: a schema is provable (in some standard system) if and only if it is valid (true in all its interpretations), if and only if every substitution instance of it in any given reasonably rich object language is true. Dually, a schema is irrefutable if and only if it is satisfiable (true in at least one interpretation), if and only if some substitution instance of it in the object language is true. (The equivalence of validity and provability, and of satisfiability and irrefutability, is guaranteed by the completeness theorem.)

For the purposes of this theorem, Quine cannot count the identity sign as a logical symbol: '$\exists x \, \exists y \sim x = y$' is a schema and also a

ON SECOND-ORDER LOGIC 525

sentence whose only substitution instance is itself (if '=' counts as a logical symbol), which is true (since there exist at least two objects in the domain of the object language), but which is not a logical truth according to the usual definition, for it is false in all one-membered interpretations.

A second minor point about the definition is that it just does not work if the object language is not reasonably rich.[16] But the language of arithmetic, interpreted in the usual way, is certainly reasonably rich, or becomes so when '+' and '·' are supplanted by three-place predicate letters.

The theorem may be remarkable, but it is not, I think, remarkably remarkable. A distinction can be drawn between two kinds of completeness theorem that can be proved about systems of logic: between weak and strong completeness theorems. A weak completeness theorem shows that a sentence is provable whenever it is valid; a strong theorem, that a sentence is provable *from a set* of sentences whenever it is a logical consequence of the set. Most of the usual proofs of the weak completeness of systems of first-order logic can be expanded quite easily to proofs of the strong completeness of those systems. The strong completeness of first-order logic can be expressed: a set of sentences is satisfiable if it lacks a refutation. (A refutation of a set of sentences is a proof of the negation of a conjunction of members of the set.)

It seems to me that the concurrence of the two accounts of the concept of logical truth cannot be called remarkably remarkable if their extensions to the relation of logical consequence do not concur. If there is a reasonably rich language and a set of sentences in that language which is satisfiable according to the usual account but which cannot be turned into a set of truths by (simultaneous, uniform) substitution of open sentences of the language, then the interest of the alternative definition of logical truth is somewhat diminished, for it is a definition that cannot be extended to kindred logical relations in the correct manner. And, as it happens, there is a satisfiable set of sentences of a reasonably rich language with this property. Proof is given in the appendix.

The compactness theorem might be thought to provide a way out of the difficulty. Since a set is satisfiable if and only if all its finite subsets are satisfiable, we might propose to define satisfiability by saying that a set is satisfiable just in case every conjunction of its members has a true substitution instance. So there turn out to be

[16] See Peter G. Hinman, Jaegwon Kim, and Stephen P. Stich, "Logical Truth Revisited," this JOURNAL, LXV, 17 (Sept. 5, 1968): 495–500.

three accounts of satisfiability of sets of sentences, the account just mentioned, truth in some one model, and irrefutability.

But this concurrence is not in the least remarkable. The strong completeness theorem is remarkable; and the Löwenheim-Hilbert-Bernays theorem is remarkable. The concurrence of the two definitions of validity of single sentences—truth in all interpretations and truth of all instances—is remarkable too, *because both definitions have some antecedent plausibility as correct explications of a pretheoretical notion of logical validity* ("truth regardless of what the nonlogical words mean"). The definition of satisfiability of a set as "truth of some instance of each conjunction of schemata in the set" has no such plausibility as an account of satisfiability. It even sounds wrong.

One ought then to be wary of the claims that the concurrence of diverse definitions of logical truth is remarkable and that this concurrence suggests that classical quantificational logic is a "solid and significant unity." One of the definitions is a definition of logical truth only in virtue of a remarkable theorem about first-order logic; another cannot be generalized properly. Does classical quantificational logic then fail to be a significant and solid unity? Certainly not.

GEORGE S. BOOLOS

Massachusetts Institute of Technology

APPENDIX

We consider two first-order languages (without '='), L and M, whose predicate letters are F, Z, S, P, T, and G. The variables of both languages range over the natural numbers, and both specify that F is true of all natural numbers, that Z is true of zero alone, and that S, P, and T are predicate letters for successor, sum, and product, respectively. L specifies that G is true of all natural numbers. L is a reasonably rich language. Let A be the set of Gödel numbers of truths of L. A is not definable in L. Finally, M specifies that G is true of all and only the members of A.

Let B be the set of truths of M. B is satisfiable. But B cannot be turned into a set of truths of L by substitution of open sentences of L for the predicate letters F, Z, S, P, T, and G. For, if it could, A would be recursive in the extensions in L of the open sentences substituted for Z, S, and G, and hence A would be definable in L; for the extensions would certainly be definable in L, and *definable in L* is closed under *recursive in*.

Let '$E(\theta)$' abbreviate 'the extension in L of the open sentence substituted for θ'. The reason that A would be recursive in $E(Z), E(S), E(G)$ is that, for each natural number n,

$$\exists x_0 x_1 \cdots x_{n-1} x_n (Zx_0 \,\&\, Sx_0 x_1 \,\&\, \cdots \,\&\, Sx_{n-1}x_n) \text{ is in } B;$$

if $n \in A$, then

$$\forall x_0 x_1 \cdots x_{n-1} x_n (Zx_0 \,\&\, Sx_0 x_1 \,\&\, \cdots \,\&\, Sx_{n-1}x_n \rightarrow Gx_n) \text{ is in } B; \text{ and}$$

if $\sim n \, \epsilon \, A$, then

$\forall x_0 x_1 \cdots x_{n-1} x_n (Z x_0 \, \& \, S x_0 x_1 \, \& \, \cdots \, \& \, S x_{n-1} x_n \rightarrow \, \sim G x_n)$ is in B.

Then, to determine whether $n \, \epsilon \, A$, we may use "oracles" for $E(Z)$ and $E(S)$ to find an $(n+1)$-tuple $a_0, a_1, \cdots, a_{n-1}, a_n$ of natural numbers such that a_0 is in $E(Z)$ and the n pairs $a_0, a_1, \cdots,$ and a_{n-1}, a_n are in $E(S)$, and then use an oracle for $E(G)$ to determine whether a_n is in $E(G)$. a_n is in $E(G)$ iff $n \, \epsilon \, A$. The procedure is recursive in $E(Z)$, $E(S)$, $E(G)$.

We have thus shown that B is a satisfiable set of sentences of the reasonably rich language L which cannot be turned into a set of truths by (simultaneous, uniform) substitution of open sentences of L for the predicate letters of L which occur in the sentences in B.

NEW BOOKS IN PAPERBACK

ADAM, MICHEL: *Essai sur la bêtise.* Paris: Presses Universitaires de France, 1975. 195 p.

AHLUWALIA, JASBIR SINGH: *Marxism and Contemporary Reality.* New York: Asia Publishing House, 1973. 59 p. $5.25.

AIMONETTO, ITALO: *Le Antinomie logiche e matematiche.* Turin, Italy: Filosofia, 1975. 79 p. L 2,200.

AMBACHER, MICHEL: *Les Philosophies de la nature.* Paris: Presses Universitaires de France, 1974. 126 p.

ANSART-DOURLEN, MICHELE: *Dénaturation et violence dans la pensée de J. J. Rousseau.* Paris: Klincksieck, 1975. 302 p. F 75.

ARNHEIM, RUDOLF: *Entropy and Art: An Essay on Disorder and Order.* Berkeley: University of California Press, 1974. 64 p. (Cal 275.) $1.65.

BARION, JAKOB: *Was ist Ideologie?* Bonn: Bouvier, 1974. 164 p. DM 19,80.

BERLINGER, RUDOLPH: *Philosophie als Weltwissenschaft.* (Elementa. Band 2.) Amsterdam: Rodopi, 1975. 240 p. $20.00.

BERTMAN, MARTIN A.: *Research Guide in Philosophy.* Morristown, N.J.: General Learning Press, 1974. viii, 252 p.

BLANCHÉ, ROBERT: *L'Induction scientifique et les lois naturelles.* Paris: Presses Universitaires de France, 1975. 170 p.

BLANCHÉ, ROBERT: *Le Raisonnement.* Paris: Presses Universitaires de France, 1973. 263 p.

BLUMENTHAL, ALBERT: *Moral Responsibility: Mankind's Greatest Need. Principles and Practical Applications of Scientific Utilitarian Ethics.* Santa Ana, Calif.: Rayline Press, 1975. 500 p. $7.00.

BOOTH, WAYNE C.: *Modern Dogma and the Rhetoric of Assent.* Chicago: University Press, 1974. xvii, 235 p. $3.95.

BOSLEY, RICHARD: *Aspects of Aristotle's Logic.* New York: Humanities, 1975. 137 p. $10.00.

BRENNAN, JOSEPH GERARD: *Ethics and Morals.* New York: Harper & Row, 1973. ix, 398 p. $6.95.

[4]

THE JOURNAL OF SYMBOLIC LOGIC
Volume 50, Number 3, Sept. 1985

SECOND-ORDER LANGUAGES AND MATHEMATICAL PRACTICE

STEWART SHAPIRO

There are well-known theorems in mathematical logic that indicate rather profound differences between the logic of first-order languages and the logic of second-order languages. In the first-order case, for example, there is Gödel's *completeness* theorem: every consistent set of sentences (vis-à-vis a standard axiomatization) has a model. As a corollary, first-order logic is *compact*: if a set of formulas is not satisfiable, then it has a finite subset which also is not satisfiable. The *downward Löwenheim-Skolem theorem* is that every set of satisfiable first-order sentences has a model whose cardinality is at most countable (or the cardinality of the set of sentences, whichever is greater), and the *upward Löwenheim-Skolem theorem* is that if a set of first-order sentences has, for each natural number n, a model whose cardinality is at least n, then it has, for each infinite cardinal κ (greater than or equal to the cardinality of the set of sentences), a model of cardinality κ. It follows, of course, that no set of first-order sentences that has an infinite model can be categorical. Second-order logic, on the other hand, is inherently incomplete in the sense that *no* recursive, sound axiomatization of it is complete. It is not compact, and there are many well-known categorical sets of second-order sentences (with infinite models). Thus, there are no straightforward analogues to the Löwenheim-Skolem theorems for second-order languages and logic.[1]

There has been some controversy in recent years as to whether "second-order logic" should be considered a part of logic,[2] but this boundary issue does not concern me directly, at least not here. The present approach is to assess the adequacy of first-order *languages* in formalizing actual mathematical practice. This problem is one that occupied mathematicians and logicians earlier this century (see Moore [1980]), but seems to have received less attention recently. My main conclusion, in agreement with Bernays, Hilbert, and Zermelo (and in disagreement with Gödel and Skolem), is that no first-order language is sufficient for axiomatizing such branches

---

Received May 16, 1984; revised June 27, 1984.

[1] A detailed exposition of the technical differences between first-order logic and second-order logic is found in Boolos and Jeffrey [1980, Chapter 18].

[2] Quine [1970] suggests that the set-theoretic presuppositions of (the standard semantics of) second-order logic actually disqualify it as part of logic. On the other hand, Corcoran [1973] and, more extensively, Boolos [1975] argue against this, that second-order logic is logic. From a different perspective, Tharp [1975] suggests that since first-order logic has certain properties (such as completeness and compactness) one would *want* a logic to have, ceteris paribus it is preferable.

as arithmetic, real and complex analysis, and set theory—branches that (each) deal with a particular (infinite) structure or, in other words, branches whose languages have "intended interpretations". The argument presented rules out any language whose logic is either complete or compact. I go on to suggest that nothing short of a language with second-order variables will do.

The considerations brought against first-order languages are *semantical*. That is to say, I argue that the semantics of first-order languages is not adequate for the pre-formal semantics of mathematical practice. A few brief comments elaborating this perspective are in order.

First, the standpoint of this article might best be characterized as a "neutral realism". The "realism" indicates that mathematical discourse is taken at face value. Contra formalism, (most) mathematical assertions are regarded as meaningful assertions about mathematical entities. Mathematical *truth* is determined by the subject matter of mathematics and, thus, "truth" is synonymous with neither (real/ideal) "knowledge" nor "provability". Contra intuitionism and logicism, no attempt is made to criticize the bulk of mathematical practice. Rather, actual mathematical practice is taken to be the data for the considerations of this paper. The "neutral" in "neutral realism" indicates that at present, I have no view on the makeup or the ontological status of the subject matter of mathematics. I only hold that there is such a subject matter. In fact, I have no *a priori* objection to any interpretation of mathematics as long as the integrity of the bulk of mathematical discourse is preserved.[3]

It might be noted that much (but not all) of the literature on both sides of the first-order/second-order issue takes or presupposes a viewpoint of realism. For example, Gödel, an avowed platonist, was one of the strongest (and probably the most influential) proponents of first-order languages. Moreover, the recent arguments (see footnote 2) against higher-order languages are directed at the semantics of such languages and, consequently, seem to presuppose a realism of sorts. Indeed, if one is concerned *only* with codifying mathematical proof (for example, if one is a formalist of the Hilbert or Curry school), then there is little to object to concerning any effectively specified language and deductive system. In short, if one is not concerned with the interpretation of a language of mathematics *as such*, then, *a fortiori*, one will not worry about such things as excess ontological commitment and inconvenient semantic properties.

Higher-order languages are considered here with *standard semantics* in which, for a given interpretation, the second-order predicate variables range over *all* of the subsets of the domain, the second-order function variables range over all of the functions from the domain to the domain, etc. There is, of course, an alternate semantics for second-order languages, developed originally in Henkin [1950], in which, for a given interpretation, the predicate variables range over a *fixed subset* of the power-set of the domain, etc. For this alternate semantics (and the usual deductive system), second-order logic is sound, complete, and compact. Although it will not always be demonstrated directly, it is easily seen that most of the present

---

[3] Resnik [1980, Chapter 5] uses the term "methodological platonist" to refer to a similar, if not identical, position.

716   STEWART SHAPIRO

considerations against first-order languages apply to second-order languages with Henkin semantics.

Finally, the completeness theorem indicates that for first-order languages, the proof theory corresponds in a direct way with the semantics, or model theory. Consistency and satisfiability are coextensive, as are deductive consequence and semantic consequence. This, of course, is not the case with second-order languages. Thus, present considerations concerning the semantics of mathematics do not shed much light on the question of which deductive systems are appropriate for codifying mathematical *proof*. To put it differently, my thesis is that *reference* to the predicates or subsets of given domains is necessary to capture the semantics of mathematical practice, but I have little to say concerning the particular axioms or assumptions about such subsets necessary to codify normal proof techniques.[4]

As noted, first-order languages do not allow categorical characterizations of infinite structures. I take this as their main shortcoming. §1 deals with the importance of categoricity in understanding and communicating mathematics. This involves the relevance of the Löwenheim-Skolem theorems to the present issue and the epistemic presuppositions of second-order languages. §2 concerns further inadequacies of first-order versions of arithmetic, analysis, and set theory, concluding that such theories do not capture important, perhaps crucial, aspects of those fields. §3 is a discussion of the adequacy of several alternate languages. The first subsection concerns languages "intermediate" between first-order and second-order; the second subsection concerns the language of first-order set theory. The final §4 is a brief comparison of the semantics and proof-theoretic strength of standard first-order logic with that of standard second-order logic.

One of the purposes of logic is to codify correct inference. Thus, if my major conclusions are correct, the underlying logic of many branches of mathematics is (at least) second-order: one cannot codify the correct inferences of a second-order language with a first-order logic. It follows that the inconvenient technical properties and presuppositions of second-order logic must be accepted. The correct conclusion, I believe, is that there is no sharp distinction between logic and mathematics. The study of correct inference, like almost any other science, involves some mathematics and some mathematical presuppositions.

§1. Categoricity and the Löwenheim-Skolem theorems. In broad terms, one major purpose of axiomatizing a branch of mathematics is to codify the practice of that branch. Historically, this has two distinct, but related aspects, one involving the deductive system of the language of axiomatization, the other the semantics.

One purpose of axiomatization is to organize and systematically present the truths and correct inferences of the branch.[5] The goal of this aspect is *completeness*:

---

[4] The problem concerning the proof theory of mathematical practice is treated in some detail in Feferman [1977].

[5] As is well known, the organization of mathematical assertions was one of the chief aims of the Hilbert program:

> When we are investigating the foundations of a science, we must set up a system of axioms which contains an exact and complete description of the ... ideas of that science ... no statement within the realm of the science ... is held to be true unless it can be deduced from the axioms by means of a finite number of logical steps. (Hilbert [1900, Problem 2])

an axiomatization—language and deductive system—is *complete* iff it has as theorems all (and only) the truths of that branch. Of course, the incompleteness theorem (and subsequent work in the theory of computability) shows this goal to be unattainable even for arithmetic in *any* suitable language.

The other purpose of axiomatization is to *describe* a particular structure, an intended interpretation of a branch of mathematics.[6] At least one goal of this aspect is categoricity: an axiomatization—language and *semantics*—is *categorical* iff any two of its models are isomorphic. One of the first writers to discuss categoricity was Oswald Veblen in his axiomatization of geometry:

> In as much as *point* and *order* are undefined, one has a right, in thinking of the propositions, to apply the terms in connection with any class of objects of which the axioms are valid propositions. It is part of our purpose, however, to show that there is *essentially only one* class of which the ... axioms are valid ... a system of axioms is categorical if it is sufficient for the complete *determination* of a class of objects or elements. (Veblen [1904, 346–347])

In this passage, Veblen seems to take categoricity as the primary aim of his axiomatization. This is underscored with a remark that if one has a categorical axiomatization, then "any further axiom would have to be considered redundant", to which a footnote is added "... even were it not deducible from the [other] axioms by a finite number of syllogisms" (p. 346). A potential axiom (i.e., a true sentence) which is not deducible from the other axioms can be considered redundant only for the purpose of *describing* a structure. Such a sentence is certainly not redundant for the purpose of organizing and presenting the truths of a branch of mathematics.

The role of categoricity in the history of mathematics and logic is discussed in Corcoran [1980]. For present purposes, a simple, but important, point is that this aspect of the enterprise of axiomatization involves a distinction between a mathematical *structure* itself, and the *language* used to describe a structure. It is clear (at least with hind-sight) that if an axiomatization correctly describes a structure, then it also correctly describes any isomorphic structure. Thus, for the purpose of description, a categorical axiomatization is the best one can do. In Corcoran's words:

> The insight that truth in a formal language depends solely on the form of the interpretation (and is independent of content ...) is partly reflected in

---

[6] A similar distinction applies to the enterprise of logic (and metalogic) and roughly corresponds to the distinction between proof theory and model theory. Let *L* be a language. One purpose of the logic of *L* is to codify the correct inferences of *L*. Gödel's *completeness* theorem shows that this purpose can be achieved for first-order languages and, as noted, this purpose cannot be completely achieved for second-order languages (with standard set-theoretic semantics in both cases). A second purpose of the logic of *L* is to describe what correct inference in *L* amounts to. For the languages presently under consideration, this involves an account of the set-theoretic semantics. I suggest that the second purpose does not depend on the first—if anything, it is the other way around. That is, I maintain that (for second-order languages in particular) it is possible to characterize "correct inference" without being able to determine the correctness of every proposed inference. This stance concerning logic is part of my "neutral realism", but it seems to be shared by several nominalists. (See Field [1980] and Gottlieb [1980].)

the fact that isomorphic interpretations have the same same set of truths … Moreover, it has been clear since the turn of the century … that given any interpretation *i*, there are other interpretations isomorphic with *i* but having no content in common with *i*. The existence of such isomorphic "images" implies, of course, the impossibility of uniquely characterizing an interpretation by means of a set of sentences in a formal language. Accordingly, it is sometimes said that the best possible characterization of an interpretation would be a "characterization up to isomorphism" … (Corcoran [1980, 190])

Of course, the Löwenheim-Skolem theorems imply that no set of sentences in a first-order language can be a categorical description of an infinite structure. Thus, for example, any first-order axiomatization *A* of the natural numbers has "unintended" interpretations in the sense that there are models of *A* which are not isomorphic to the natural numbers. Thus, with a first-order language, a completely successful description of the natural numbers is not possible. Moreover, it is not difficult to see that a language and logic that *does* contain a categorical axiomatization of arithmetic is neither compact nor complete. Similar remarks apply to any infinite structure.

As is well known, much has been written on the philosophical significance of the Löwenheim-Skolem theorems. A general survey of this literature would go well beyond the scope of the present paper, but the remarks of two authors, Skolem [1923] himself and John Myhill [1951], are of particular interest.

Skolem's [1923] celebrated conclusion is that all set theoretic "notions" are "unavoidably relative". For example, one cannot claim that a given domain *D* is uncountable *simpliciter*, but only uncountable relative to a certain model (containing *D*) of set theory. For any such *D*, one cannot rule out the possibility that *D* may be countable relative to a (richer) model, one that contains a function from the natural numbers onto *D*. For a second example, Skolem holds that this relativity applies to the Dedekind notions of "finite" and "simply infinite sequence" (or model of the natural numbers). It seems that if one accepts the modern trend of regarding all mathematical notions as set-theoretic, then Skolem's conclusion entails the relativity of virtually all mathematical notions, including those of natural number and real number. Thus, it appears that Skolem rejects the above distinction between a mathematical structure and a formal axiomatization and, in particular, rejects the notion of an intended interpretation of an axiomatization.

The conclusions of Myhill [1951] are less extreme. He accepts the distinction between structures themselves and formal languages involved in description, but he argues that the Löwenheim-Skolem theorems indicate an inadequacy in the enterprise of formalization. His argument begins with the suggestion that mathematical structures are first apprehended by "intuition". Part of the purpose of formal axiomatization is to describe, communicate, and thus help study a particular structure. Myhill agrees that although deducing the properties of a structure only requires a sound axiomatization, the description and communication of a structure is possible only to the extent that an axiomatization uniquely characterizes it. He claims, however, that the Löwenheim-Skolem theorems show that for infinite structures, this is impossible. His first conclusion is that mathematicians cannot dispense entirely with the original intuitions which determined the intended interpretation of the axiomatization:

> [In a formalism] we operate with symbols which keep their shape rather than with ideas which fly away from us. All real mathematics is made with ideas, but formalism is always ready in case we grow afraid of the shifting [of our ideas]. The Skolem "paradox"... proclaims our need never to forget completely our intuitions. (Myhill [1951])

If this analysis is correct, then our ability to use a formalism to understand others is somewhat limited. Even if two mathematicians agree on an axiomatization of, say, arithmetic, real analysis, or set theory, they cannot be sure that they have in mind the same (or even isomorphic) interpretations of their agreed-on axiomatization. Concerning set theory, Myhill wrote:

> ... there seems to be no *formal* means of assuring that our concept of membership is the same as another person's ... For no ... number of formal assertions agreed on by us both could be evidence that his set-theory was not in my sense denumerable ... The second philosophical lesson of the Löwenheim-Skolem theorem is that the formal communication of mathematics presupposes an informal community of understanding.

Both Skolem and Myhill thus speak of the limits of formalization and formal languages. Since the Löwenheim-Skolem theorems only apply to first-order axiomatizations, their considerations seem to presuppose that all legitimate formalizations of mathematical practice employ first-order languages.[7] I propose here that their arguments actually represent a *reductio ad absurdum* against this presupposition.

There is virtually universal agreement among mathematicians that arithmetic, real analysis, and complex analysis each deals with a single, specific structure—the intended interpretation of the axiomatization. There is also some (but not universal) agreement that the same holds for set theory. Of course, the philosophical issues at hand are not to be settled by popularity, but this data is striking. Those working in arithmetic, for example, know that any first-order axiomatization has nonstandard interpretations, yet they believe both that there is a standard interpretation of arithmetic (or, at any rate, a class of standard interpretations, all of which are isomorphic) and that every other mathematician has in mind the same (or an isomorphic) interpretation. I take it as undisputed that every mathematician does have in mind the same (or an isomorphic) structure of natural numbers. Similar remarks apply to real analysis and complex analysis, if not set theory.

A question naturally arises as to how these structures are apprehended and communicated. Several philosophers (such as Myhill, as above, and Gödel) suggest that at least some mathematical structures are apprehended through a faculty of intuition. Postulating or suggesting such a faculty, however, does not completely solve the present problems. At best, a faculty of intuition can account for how a *single* mathematician apprehends, say, the natural number structure and then

---

[7] See Moore [1980] for an extended discussion of Skolem's views on this issue.

describes it in a formal or informal language. A question remains as to how a *second* mathematician could know *which* structure the first is describing. Presumably, she also has a faculty of intuition, but she can just as well intuit several structures including, perhaps, that of a nonstandard model of arithmetic. The problem is one of communication.

Myhill's point is that (first-order) formal languages are inadequate to insure communication. He suggests that the fact that mathematical structures *are* communicated presupposes an "informal community of understanding". This underscores the present problem but does not solve it.

One could, I suppose, postulate a faculty of mental telepathy between mathematicians to account for the communication of structures; but, without this, all communication is mediated by language. This is where categoricity is important.

To reiterate, different mathematicians do understand "the natural number structure", "the real number structure", and "the complex number structure" the same way (or, at least, in isomorphic ways). That is, mathematicians succeed in communicating these structures to each other. The *informal* language of mathematics is thus sufficient to insure this communication. Second, the purpose of formal axiomatization is to codify mathematical practice, one of whose purposes is the description and communication of structures. I conclude that a language *and semantics* of formalization should be sufficient to insure this communication. That is, the language of formalization should allow categorical characterizations. It follows that first-order axiomatizations are inadequate.

A related conclusion is found in Montague [1965]; he shows that the notion of a "standard model" of arithmetic, real analysis, or even set theory can be understood as a model of the respective *second-order* thoery. Thus, it is proposed that second-order languages are *sufficient* to insure the description and communication of these structures.[8]

This proposal, however, can be challenged and, in response, must be qualified. The assertion that a given structure is described (up to isomorphism) and communicated by a second-order language depends on a premise that the second-order language is itself unambiguously understood. That is, the categoricity of the second-order theories in question depends on there being a unique and clearly understood interpretation of such second-order quantifiers as "all subsets". Let T be a second-order theory and D a structure with domain d. The statement that D is a model of T has quantified variables ranging over the (entire) collection of subsets of d. It is thus possible (perhaps) for two mathematicians to disagree whether D is a model of T if they have in mind different "powersets" of d.[9] It is thus conceded that

---

[8] §3 below concerns the extent to which second-order languages are also necessary for the description and communication of structures.

[9] Similar remarks apply to the collections of relations on d and functions from d to d. This, in effect, is the insight behind the Henkin semantics for second-order languages. This objection was suggested to me by Mike Resnik and Nicolas Goodman. A similar consideration is found in Weston [1976].

Actually, the only role of the second-order terminology in standard claims and proofs of categoricity is to apply prenex universally quantified sentences to particular subsets (which may not be definable in the basic first-order theory alone). Thus, to grasp, understand, and follow the claims of categoricity, one need only grasp this kind of inference. More on this below.

the proposal concerning the expressive ability of second-order languages pre-supposes that the second-order quantifiers in the relevant definitions are understood unambiguously. A few remarks are in order.

First, understanding the second-order quantifiers of a given theory is not the same as grasping the set-theoretic hierarcy. In a given theory, the quantifier "all subsets" ranges over the collection of subsets of *a fixed domain*. In general there is no powerset operator to be iterated. Second-order arithmetic, for example, only has variables ranging over natural numbers and sets (and relations) of natural numbers. There are no variables ranging over such items as sets of sets of numbers and functions from numbers to sets of numbers. The set-theoretic hierarchy, on the other hand, is a proper class that contains the result of iterating the powerset operator into the transfinite.

I would suggest, then, a distinction between the "logical" conception of set and the "iterative" conception of set. The iterative conception refers to the set-theoretic hierarchy, itself an extremely large mathematical structure. The logical conception, on the other hand, occurs only in the context of a fixed domain. Thus, a second-order theory of a given domain has variables ranging over the collection of *logical* sets vis-a-vis that domain. With this terminology, the presupposition at hand is that for a fixed domain, the second-order quantifier "all logical subsets" is unambiguously understood.[10]

As above, let T be a theory and D a structure with domain d. I submit that what it means for a collection c to be a (logical) subset of d is clear and unambiguous: c is a subset of d if and only if every member of c is a member of d. What is at issue here is whether the totality (or range) of subsets of d is itself clear and unambiguous. Let $P1$ and $P2$ be two candidates for the range of the second-order predicate quantifiers of T (via-à-vis D). That is, let $P1$ and $P2$ be two candidates for the logical powerset of d. I suggest that if $P1 \neq P2$, then there is a clear sense in which (at least) one of them is not the powerset of d. Indeed, suppose that there were a collection c such that $c \in P1$ but $c \notin P2$. I take it that (for a classical mathematician) it is determinate whether

---

[10] There is a tradition, originating, perhaps, with Boole and including Peirce and Schroder, which takes the subsets of a domain to be under the purview of logic. Moreover, much of Gödel [1944] is a defense of the claim that mathematical logic deals with "classes, relations,... instead of numbers, functions, geometrical figures, etc." (reprint, p. 211). This tradition, of course, predates the development of the set-theoretic hierarchy (as a subject of mathematics itself) by Cantor, Zermelo, etc. (See Moore [1980, especially 97–101].) Cantor's distinction between sets and "inconsistent totalities" may also be closely related to the present distinction between iterative sets and logical sets (in the context of set theory). See Wang [1974, Chapter VI] for a detailed discussion of the historical distinction between the "limitation of size" doctrine of sets and the logical doctrine of sets. One may also compare a distinction made by Frege and (the early) Russell between sets-as-composed-of-their-elements and classes-as-extensions-of-concepts. (The differences between these authors notwithstanding. See, for example, Chapter VI and Appendix A of Russell's *Principles of Mathematics*.)

In the sequel, when the word "set" is employed, the context usually indicates whether "logical set" or "iterative set" is meant. An ambiguity can result only in the context of set theory because, in this case, the fixed domain in question (vis-à-vis the logical conception) is the set-theoretic hierarchy. In other words, in set theory, a "logical set" is a collection of iterative sets. Here I follow the custom of designating such collections *classes*. The upshot of Russell's paradox is that in the context of set theory, there are logical sets that are not iterative sets.

every element of *c* is an element of *d*. If every element of *c* is in *d*, then *P2* is not the powerset of *d*; otherwise, *P1* is not the powerset of *d*.

Finally, my presupposition/suggestion that the locution "all logical subsets" is unambiguously understood entails only that when a mathematician uses the phrase in connection with a domain, he and his listeners understand what he means the same way. I do not make the absurd claim that any or all of the properties of the powerset (such as its cardinality, or whether it contains a nonconstructible element) are known.

§2. First-order axiomatizations. This section deals with specific inadequacies of first-order axiomatizations of branches of mathematics. §2.1 concerns three concepts—finitude, minimal closure, and well-foundedness—which form an important part of general mathematical practice, but which cannot be formulated in first-order languages. §2.2 is a comparison of the standard first-order versions of arithmetic, real analysis, and set theory with their second-order counterparts.

2.1. Let *T* be a first-order theory and Φ a formula with one free variable. For each natural number *n*, there is, of course, a first-order formula that asserts that the extension of Φ (or the domain of discourse of *T*) has at most *n* members. There are, however, circumstances in which one wishes to assert that a given extension (or domain) is *finite* without specifying a fixed (numerical) bound on the cardinality of this extension. Consider, for example, the theory of finite groups in which it is stated that the domain of discourse is finite, or the thoery of computability in which each Turing machine is characterized by a finite number of instructions (but it being crucial that for each natural number *n*, there are Turing machines with at least *n* instructions).

I submit that *finitude* is a clear and unambiguous concept. If, for example, a mathematician asserts that a given extension is finite, then his listeners understand what he means. Therefore, a language used to formalize mathematical practice must be capable of expressing this property. No first-order language can do this. As is well known, if, for each natural number *n*, there is a model of *T* in which the extension of Φ (or the domain of discourse) has at least *n* members, then there is a model of *T* in which the extension of Φ (or the domain of discourse) is infinite.

Notice that the second-order formula

$$\forall f [(\forall x (\Phi(x) \rightarrow \Phi(fx)) \,\&\, \forall y \forall z (fy = fz \rightarrow y = z)) \rightarrow \forall y (\Phi(y) \rightarrow \exists x (fx = y))]$$

is satisfied by all those, and only those, models of *T* in which the extension of Φ is finite.

In a similar vein, Boolos [1981] shows that theories formulated in first-order languages cannot express such simple cardinality comparisons as "the extension of Φ is at least as large as the extension of Ψ".

Moving to the second example, it is common to describe a particular set (or structure) through what may be called a *minimal closure* property. The construction is usually carried out in a background theory *T* concerning a model *M* with domain *d*. To describe a set $A \in d$ by this technique, one first gives a *basis* subset $B \in d$ and a set *F* of functions or operations (or, perhaps, relations) on *d*. The set *A* is then characterized as the "smallest" set which contains *B* and is closed under the operations in F. The set A is sometimes described as *the* set obtained from B by closure under the operations in F, or one says that y is in A just in case y is the result of an iteration of the operations in F on members of B finitely many times.

There are numerous examples of the minimal closure construction. To mention a few, if R and S are two rings, $R \subseteq S$ and $a \in S$, then the ring $R[a]$ can be described as a minimal closure (within S) with basis $R \cup \{a\}$ and the functions of S-addition and S-multiplication. The *rational subfield* of a field is a minimal closure with basis $\{1\}$ under the field functions and their inverses. In analysis, the statement that the real numbers are *Archimedian* amounts to a claim that for every positive real number r, the minimal closure of $\{r\}$ under addition is unbounded. In proof theory, collections of terms, well-formed formulas, and theorems are defined as minimal closures on sets of strings, and, in model theory, *elementary submodels* are often constructed as minimal closures under a set of Skolem functions.

From the practice of mathematicians, I submit that the use of minimal closure is well-understood and that there is no ambiguity concerning the constructed set. It is, therefore, a requirement on languages of axiomatization that they be capable of expressing minimal closures. As with finitude, a straightforward compactness argument shows that no collection of first-order formulas can successfully define any nontrivial minimal closure.[11]

Notice that the following second-order formula $\Theta(x)$ does characterize the minimal closure of the extension of Φ under the function denoted by f:

$$\forall X \{\forall y [(\Phi(y) \to Xy) \& (Xy \to Xfy)] \to Xx\}.$$

That is, $\Theta(x)$ holds (in any model of the background theory) just in case x is in the minimal closure of the extension of Φ under the function denoted by f.

The final example is that of a *well-founded* relation. A binary relation E is well-founded iff there is no infinite sequence $\langle a_i \rangle$ such that $a_1 E a_0, a_2 E a_1, \ldots, a_{n+1} E a_n, \ldots$ all hold. Informally, it is sometimes stated that E is well-founded just in case there are no "infinitely descending E-chains". Once again, I submit that the notion of well-foundedness is both clear and unambiguous and, once again, the well-foundedness of a relation E cannot be characterized in a first-order language (provided only that for each natural number n, there is a model of the background theory in which E is well-founded and which contains $n + 1$ elements $a_0, \ldots, a_n$ such that $\langle a_1, a_0 \rangle, \ldots, \langle a_n, a_{n-1} \rangle$ all satisfy E).

A second-order formulation is straightforward:

$$\forall X [\exists x Xx \to \exists x (Xx \& \forall y (Xy \to \neg yEx))].$$

---

[11] Fix a first-order theory T and model M with domain d. Let Γ be a set of formulas, containing a new predicate Px, that purports to characterize the minimal closure of the extension of a formula Φ under a unary function p. Under the following assumptions, there are models of $T + \Gamma$ in which the extension of P is not the minimal closure of the extension of Φ under p: (1) the given model M can be extended to a model of $T + \Gamma$; (2) in each model of $T + \Gamma$, the extension of P both contains the extension of Φ and is closed under p; and (3) in the given model M, the minimal closure of the extension of Φ under p contains infinitely many members not in the extension of Φ. Actually, under these conditions, there are models of T that can be extended to models of $T + \Gamma$ in more than one way and, thus (following the main result of Corcoran [1971]) Γ fails to be a legitimate definition at all, let alone a definition of a minimal closure.

In arithmetic, proofs by induction and definitions by recursion presuppose that the predecessor relation and the "less-than" relation are both well-founded on the natural numbers. The major use of well-foundedness, of course, is in set theory where it is important that the membership relation be well-founded.

In [1980], Hilary Putnam argues against realism in set theory. In one section (pp. 468–469) he claims that the Löwenheim-Skolem theorems indicate that there is no "fact of the matter" concerning whether all sets are constructible or even whether a *given* countable set of real numbers is constructible. To support this, he introduces a theorem that for every countable set *s* of real numbers, there is an ω-model *M* of set theory which contains *s* and satisfies "every set is constructible".[12] Of course, a realist will maintain that the given set *s* may nevertheless be nonconstructible "in reality". Putnam replies:

> But what on earth can this mean? It must mean, at the very least, that ... the model [*M*] we have described [which contains *M* and satisfies "all sets are constructible"] would not be *the intended model*. But why not? (p. 469).

He then goes on to argue that there are no grounds to claim that the model *M* is "unintended" since *M* satisfies all of the "theoretical constraints" that have been placed on the notion of intended model. The only "theoretical constraints" considered, however, involve the structure of the finite ordinals and the satisfaction of the axioms of first-order set theory. But one can surely claim that the well-foundedness of the membership relation is a "theoretical constraint" on (intended) models of set theory: one would hardly consider a structure with a non-well-founded membership relation to be an "intended model". Yet it follows (from a result that Putnam indicates) that if a given set *s* is nonconstructible, then any model (containing *s*) that satisfies "*s* is constructible" is not well-founded. Thus, against Putnam, in the above theorem, if *s* is not constructible, then there is a clear sense in which *M* is an unintended model of first-order set theory.[13] I take these results as further evidence that a first-order language is not adequate to formalize set theory or, to borrow a phrase, to formulate the "theoretical constraints" on intended models of set theory.

2.2.   In most formulations of arithmetic, the only second-order axiom is the statement of mathematical *induction*:

(I) $\forall P([P0 \;\&\; \forall x(Px \rightarrow Psx)] \rightarrow \forall x Px).$

In common formulations of real analysis, the only second-order axiom is that of *completeness*, the statement that every bounded subset of the domain has a least

---

[12] A structure *M* is an ω-*model* of set theory iff the extension of "finite ordinal" under the relation of membership (in *M*) is isomorphic to the natural numbers under "less than".

[13] It might be noted, as an aside, that even if the present considerations concerning higher-order languages are correct, they do not affect the bulk of Putnam's conclusions concerning reference in [1980] and [1981]. I suggest that it is not the Löwenheim-Skolem theorems that are relevant, but the fact that isomorphic structures satisfy the same set of sentences or, in other words, that a language—of any order—cannot distinguish among isomorphic structures.

upper bound:

(C) $\qquad \forall P(\exists x \forall y(Py \to y \leq x)$

$\qquad\qquad \to \exists x[\forall y(Py \to y \leq x)\,\&\,\forall z(\forall y(Py \to y \leq z) \to x \leq z)]).$

In second-order Zermelo-Fraenkel set theory (ZFC) the only second-order axiom is that of *replacement*, which states that for every function f, the image of any (iterative) set under f is a set. Here, I give a formulation which contains a predicate (class) variable P (thought of as ranging over collections of ordered pairs), rather than a variable ranging over functions on the domain:

(R) $\qquad \forall P[\forall x \forall y \forall z(P\langle x, y\rangle\,\&\,P\langle x, z\rangle \to z = y)$

$\qquad\qquad \to \forall x \exists y \forall z(z \in y \leftrightarrow \exists w(w \in x\,\&\,P\langle w, z\rangle))].$

The usual first-order axiomatization of each of these theories is obtained by replacing the respective second-order axiom by a scheme. In each case, the second-order variable P is replaced by an "arbitrary" first-order formula $\Phi(x)$, with x free. The result is an infinite number of axioms, one for each suitable formula Φ of the respective first-order language:

(I-Φ) $\qquad\qquad \Phi(0)\,\&\,\forall x(\Phi(x) \to \Phi(sx)) \to \forall x \Phi(x),$

(C-Φ) $\qquad\qquad \exists x \forall y(\Phi(y) \to y \leq x) \to \exists x[\forall y(\Phi(y) \to y \leq x)$

$\qquad\qquad\qquad \&\,\forall z(\forall y(\Phi(y) \to y \leq z) \to x \leq z)],$

(R-Φ) $\qquad\qquad \forall x \forall y \forall z(\Phi\langle x, y\rangle\,\&\,\Phi\langle x, z\rangle \to z = y)$

$\qquad\qquad\qquad \to \forall x \exists y \forall z(z \in y \leftrightarrow \exists w(w \in x\,\&\,\Phi\langle w, z\rangle)).$

The difference between, say, second-order real analysis and first-order real analysis is that in the former it is asserted that the completeness property applies to every subset of the domain, whether it can be defined in the language of real analysis or not; whereas in the latter, it can only be shown that the completeness property applies to subsets of the domain that are definable in the given first-order language.

The purpose of this subsection is to argue that this restriction on the first-order theories is artificial—it does not conform to mathematical practice. This, of course, is not to deny the substantial utility of the metamathematical study of the first-order theories,[14] but it is to deny that the first-order theories adequately express the mathematical practice of the respective fields. I begin with three considerations suggested by Kreisel [1967] concerning arithmetic and real analysis. This is followed by a discussion of second-order set theory.

2.2.1. The first consideration is epistemic. As indicated in §1 above, a basic presupposition of the present paper is that arithmetic and real analysis each has an intended interpretation independent of the language used to describe it. The theories in question are not taken as hypothetical or logistic systems. One can therefore inquire as to why a given axiom (or other statement) is believed or

---

[14] The rich developments of nonstandard arithmetic and analysis are but two examples of the utility of the metamathematical study of first-order theories.

accepted, either by a particular mathematician or by the mathematical community as a whole. Kreisel wrote:

> A moment's reflection shows that the evidence of the first-order schema derives from the second-order [axiom]; the difference is that when one puts down the first-order schema, one is supposed to have convinced oneself that the specific formulae used ... are well-defined in any structure one considers. (Kreisel [1967, 148])

Kreisel's point, I take it, is that a given mathematician *believes* or *accepts* the instances of the first-order scheme only because she (already) believes or accepts the second-order axiom. Suppose, for example, that one is given an instance of the completeness scheme (C-Φ), perhaps one in which the indicated formula Φ is rather complicated, and suppose that the person is asked whether it is true of the real number structure and, if it is, why she believes it. On the basis of the first-order axiomatization, the answer would be something like "I accept this formula because it is an axiom—it has the form of the completeness scheme", or perhaps, "this formula is one of the *basic* or *defining characteristics* of the real number structure". The present suggestion is that a more accurate, or (at any rate) more natural response would be: "The subformula $\Phi(x)$ determines a set of real numbers (this follows from the comprehension axiom of second-order logic). The given formula (in the form (C-Φ)) asserts that if *this* set is bounded, then it has a least upper bound. I accept this because it is an application of the completeness axiom—an axiom that characterizes the real number structure".

As is well known, some philosophers hold that a given second-order language has undesirable or dubious commitments beyond those of its first-order counterpart. A like-minded mathematician might argue that by using first-order analysis, he is sacrificing epistemic clarity or simplicity for more acceptable ontological commitments. The onus on such a mathematician is to show why he accepts each instance of the scheme. A separate justification for each axiom is, of course, out of the question—there are infinitely many. Moreover, this mathematician cannot claim that he believes the instances of the scheme because they are the "safe" or "reasonable" consequences of the second-order axiom—the consequences which do not have the undesirable or dubious commitments. If one rejects the second-order axiom on ontological grounds, then one cannot use it as a premise to justify other sentences. A second possibility, perhaps, would be to formulate the completeness scheme informally as "every appropriately definable, bounded set of real numbers has a least upper bound". As it stands, however, this statement involves quantification over sets of real numbers and, thus, is prima facie second-order. Moreover, it presupposes a concept of definability which probably is to be characterized as a minimal closure. I do not claim here that a plausible justification of the first-order scheme that does not involve a second-order language is impossible. I do suggest that such a justification has yet to be given.

2.2.2. Since the second-order axiomatizations of arithmetic, real analysis, and set theory do not contain schemes, the second-order theories are somewhat independent of the nonlogical terminology available in the language. For example, the characterization of the natural numbers in a second-order language containing only the constant 0 and a name for the successor function is essentially the same as the characterization in a language containing names for other functions, such as addition and multiplication (the only difference being that the latter contains axioms to define those functions). This is not the case with the first-order versions. In arithmetic, the extent of the scheme (I-Φ) is determined by the available formulas Φ, and this, of course, depends on the nonlogical terminology of the formalizing language. Kreisel notes:

> The choice of the first-order schema is not uniquely determined by the second-order axiom! Thus, Peano's own axioms mention explicitly only the constant 0 and the successor function..., not addition nor multiplication. The first-order schema built up from 0 and [the successor function] is a weak ... subsystem of classical first-order arithmetic... and quite inadequate for formulating current informal arithmetic. (Kreisel [1967, 148])

Suppose, for example, that in the course of a treatise on the natural numbers, a mathematician decides to introduce a new function f. She proceeds by adding a function letter, giving a description (e.g., a recursive derivation) of the function, and proving that a unique function is thereby described. The attitude is that the mathematician has introduced a new function on the same domain and, thus, that she is working in the *same theory* as she was before the function was introduced. A theorem (which may not mention the function f) in the "extended" theory is taken to be true *of* the natural numbers, even if it could not be deduced from the previous axioms alone.

This attitude is reflected in the second-order axiomatization of arithmetic. In this case, the introduction of the function f does not alter the basic description of the natural numbers. Moreover, the indicated proof that a unique function has been introduced amounts to a demonstration of what may be called "unique extendibility"—a demonstration that each model of the original axiomatization can be extended to a model of the new theory in exactly one way. It follows from the main result of Corcoran [1971] that the characterization of f is semantically eliminable and noncreative or, in other words, that the requirements of an acceptable definition have been met.[15] In short, in the second-order theory, all is as it should be.

This is not the case with first-order axiomatizations. In this case, the introduction of a new function letter extends the language and, thus, extends the set of formulas in the form $\Phi(x)$. It follows that the induction scheme (I-Φ) is itself extended. That is, the introduction of a new function letter results in a change in the basic description or axiomatization of the original theory. It is as if one is working in a new theory. Moreover, the new theory may not even have the "same" models as the original. To elaborate Kreisel's example, first-order Peano arithmetic formulated with only 0 and the successor function is *not* uniquely extendible vis-à-vis addition. There are models of the original theory that cannot be extended to models of arithmetic-with-addition, and there are models of the original that can be so extended in more than one way. To reapply the result of Corcoran [1971], it follows that the introduction

---

[15] In the second-order case, if the new function f is defined by primitive recursion, then the eliminability and noncreativeness can be proven within the standard metatheory of the axiomatization.

of addition to first-order Peano arithmetic cannot be accomplished with a legitimate definition. (It follows, incidentally, that in the case at hand, the "informal" proof that a unique function is characterized cannot be formulated in a first-order language.)

2.2.3. With the use of schemes which depend on the language of the theory, first-order arithmetic and first-order analysis are presented as isolated theories which are independent of both each other and the rest of mathematics. This is not in accord with the practice of viewing such mathematical structures as interrelated. This practice is manifest in the common technique of "embedding" or "modeling" one structure in another. In Kreisel's words:

> ... very often the mathematical properties of a domain D become only graspable when one embeds D in a larger domain D'. Examples: (1) D integers, D' complex plane; use of analytic number theory. (2) D integers, D' p-adic numbers; use of p-adic analysis. (3) D surface of a sphere, D' 3-dimensional space; use of 3-dimensional geometry. Non-standard analysis [also applies] here ... (Kreisel [1967, 166])

To take a simple example, when one realizes that the set-theoretic hierarchy (or, for that matter, the complex plane) contains the natural numbers—or an "isomorphic copy" of the natural numbers—then one can use set theory (or complex analysis) to shed light on the natural numbers. That is, since isomorphic structures have the same set of truths, a theorem of set theory that refers only to the "natural-numbers-of-set-theory" is true of the natural numbers.

It is well known that this technique can produce results that are not obtainable in the original theories. In the indicated example, this happens because there are *subsets* of the natural numbers which are definable in set theory, but are not definable in arithmetic alone. Thus, one who accepts first-order arithmetic as adequate cannot make the straightforward claim that some theorems of set theory reflect truths of the natural numbers. Indeed, many of the indicated arithmetic statements are false in some models of first-order arithmetic.

The practice of embedding structures indicates a further area in which categoricity is important. The point is that in order to embed a structure D into a structure E, one must have a means of recognizing a substructure of E as isomorphic to D. Otherwise, one cannot be certain that D really is a substructure of E. The formal analogue of this requirement is a categorical characterization of D. Suppose, for example, that someone believes that a structure M "contains" the natural numbers and, thus, that the study of M may produce (new) theorems of arithmetic. In attempting to verify this, he defines a certain substructure N of M and shows that this structure satisfies the axioms of *first-order* arithmetic. Since the latter is not categorical, the mathematician cannot conlcude that N is isomorphic to the natural numbers, nor can he conclude that all theorems of M whose quantifiers are restricted to the domain of N are true of the natural numbers. For all he knows (so far), the substructure N may be a nonstandard model of arithmetic. The situation is perhaps analogous to an observation that a certain set-theoretic structure is a group. Since the group axioms are not categorical, it does not follow that the properties of this structure are "truths" of group theory. In the example at hand, of course, the situation would be different if the mathematician showed that N satisfies the axioms of second-order arithmetic. In this case, he *can* conclude that N is isomorphic to the natural numbers and, thus, that any theorem of M whose quantifiers are restricted to the domain of N is true of the natural numbers.

2.2.4. This section concludes with some remarks on second-order set theory (ZFC). Of course, the intended interpretation of ZFC is not itself an iterative set:[16] the set-theoretic hierarchy is not a member of itself. Thus, one who believes that all legitimate collections are (isomorphic to) sets may balk at the range of the second-order variables and, consequently, may have trouble envisioning second-order ZFC as a theory about the set-theoretic hierarchy[17] (see Boolos [1975]). Of course, even the totality of the (intended) range of the variables of first-order set theory is not a set, but at least every element thereof—every element referred to *by* the theory—is a set.

It might be noted that it is common for set-theorists to speak of the set-theoretic hierarchy itself, at least in informal language. For example, for a given formula $\Phi(x)$ and structure M with domain d, it is stated that Φ is *absolute* in M just in case for each $x \in d$, $M \models \Phi(x)$ iff $\Phi(x)$ holds in the set-theoretic hierarchy.

Moreover, there is at least no formal antinomy involved in using second-order ZFC to describe the set-theoretic hierarchy. Indeed, second-order ZFC is deductively equivalent to the so-called Morse-Kelley set theory (MK), a first-order theory with two variable sorts, one ranging over sets, the other over "classes". It follows from a well-known theorem concerning MK that second-order ZFC is consistent if the theory consisting of first-order ZFC and an axiom asserting the existence of one inaccessible cardinal is consistent. Following the theme of this article, however, I suggest that what makes the second-order version attractive is not its deductive strength,[18] but rather its semantics.

The considerations of the previous §§ 2.2.1–2.2.3 apply to set theory, but only to a limited extent. The best case can be made for the epistemic point of §2.2.1: one accepts the instances of the first-order scheme of replacement (R-Φ) only because one accepts the second-order version (R). The possibility of introducing new terminology, which would extend the scheme (R-Φ) is moot (at present)—I do not

---

[16] To follow footnote 10, in this subsection (only) the word "set" is taken as "iterative set" or, in other words, as "member of the set-theoretic hierarchy". Also, the terms "model" and "interpretation" refer to structures whose domains are (isomorphic to) iterative sets. As above, in the context of set theory, logical sets are designated "classes".

[17] It might be noted, as an aside, that when one envisions a (set-theoretic) *model* M of set theory, these "commitments" are unexceptionable. The domain d of such a model is itself a *set*. The statement that M satisfies the second-order replacement axiom (R) contains variables ranging over subdomains of d. Such items are themselves sets. Thus, the statement that "M is a model of second-order ZFC" only requires first-order variables.

[18] There are, perhaps, a few items of interest concerning the relative deductive strength of second-order ZFC (i.e. MK) over first-order ZFC. Let T be the collection of "set" theorems of second-order ZFC: $\Phi \in T$ iff Φ is a first-order sentence provable in second-order ZFC. It is easily seen that T contains sentences not provable in first-order ZFC. Moreover, the "natural" first-order extensions of first-order ZFC whose theorems contain T are much stronger (vis-à-vis relative consistency) than second-order ZFC. Second, one can prove in second-order ZFC that there is a countable set d which is first-order elementarily equivalent to the set-theoretic hierarchy. This theorem represents a rather natural (informal) application of the downward Löwenheim-Skolem theorem. Yet the result in question cannot be stated, much less proved, in first-order ZFC.

know of any relevant cases. Concerning the embedding of theories, one can perhaps view the construction of Boolean-valued models as an embedding of the set-theoretic hierarchy in a richer structure, but, of course, Boolean-valued models can be reinterpreted in the set-theoretic hierarchy.

As noted above, it is essential to the concept of set that the membership relation be well-founded. It is useful at this point to look at the axiom that "asserts" this, the axiom of Foundation. It is a first-order sentence:

(F) $\forall x(\exists y(y \in x) \rightarrow \exists y(y \in x \,\&\, y \cap x = \varnothing))$.

The axiom (F) asserts that every nonempty set has a member that is disjoint from it. In first-order set theory, this axiom does prevent *definable*, infinitely descending ω-chains (such as finite, closed ω-sequences), but, as noted above, neither (F) nor any first-order improvement precludes non-well-founded models. The important point here is that in *second-order* ZFC the same axiom (F) does insure the well-foundedness of membership. That is, the axiom of foundation, a first-order sentence, only "works" in second-order set theory.

More can be said. There is a well-known theorem that a structure $M = \langle d, E \rangle$ is a model of second-order ZFC iff there is an inaccessible cardinal κ such that M is isomorphic to $V\kappa$.[19] Thus, up to isomorphism, the models of second-order ZFC are certain "initial segments" of the set-theoretic hierarchy. It may be concluded that the second-order theory characterizes the set-theoretic universe in every respect except the "size" of the class of ordinals. Because of this, Weston [1976] calls second-order set theory "almost categorical" (although he denies that this is significant).

Of course, any model of second-order ZFC is also a model of first-order ZFC. I suggest that if $M = \langle d, E \rangle$ is any *other* model of first-order ZFC—that is, if M is not isomorphic to an inaccessible rank—then there is a clear sense in which M is *nonstandard*. There are three possibilities. First the relation E may not be well-founded. This, by itself, would rule out M as standard. If, on the other hand, E is well-founded, then M is isomorphic to a transitive set. Thus, E may be taken as the membership relation on transitive d. In this case, a second possibility is that M is not closed under the subsets of its elements. That is, there are sets $b \in d$ and $a \subseteq b$ such that $a \notin d$. I take it that this also disqualifies M—standard models should contain all of the subsets of each element, whether they are definable or not. The third possibility is that M fails to satisfy the replacement axiom, even on its ordinals: There is an ordinal γ in d and a sequence of (γ-many) ordinals $\langle \alpha_\beta \rangle_{\beta \in \gamma}$ such that $\sup \alpha$ is not in d. I submit that this also disqualifies M as a standard model.

§3. Other languages. Previous sections have been devoted to the claim that axiomatizing the various branches of mathematics with separate first-order theories does not reflect important aspects of mathematical practice. One may be convinced

---

[19] The usual definition of "κ is inaccessible" is "κ is regular and for every $\alpha \in \kappa$, $2^\alpha \in \kappa$". Although in practice this is a rather useful definition, an equivalent one that perhaps illustrates the "inaccessibleness" would be "$V\kappa$ is a model of second-order ZFC". Informally, the former definition indicates that κ cannot be "obtained" from smaller ordinals by the operations of powerset and α-fold union (for any $\alpha \in \kappa$); the latter definition, that κ cannot be "obtained" from any sets of smaller rank by any operation or function implied by the axioms of second-order ZFC.

of this, of course, without believing that it is necessary to provide second-order theories. This section examines several alternatives. §3.1 concerns three "intermediate" languages—infinitary languages, ω-languages, and free-variable versions of second-order languages. The conclusion is that, among these, only the latter substantially overcomes the deficiencies of first-order languages. §3.2 concerns the program of using the language of first-order set theory to axiomatize the various branches of mathematics.

3.1. It is easily seen that most of the above considerations against first-order languages apply to any language whose semantics is compact. Thus, this subsection is limited to languages with noncompact semantics. Of course, in such cases, any "sound" deductive system is not complete: there are consistent sets of sentences that are not satisfiable.

3.1.1. Infinitary languages are easily dismissed. There is little doubt that the study of such languages has proven to be a fruitful and insightful branch of mathematical logic, but it need hardly be mentioned that infinitary languages are not serious candidates for the underlying language of mathematics. One of the chief purposes of language is to facilitate communication. Minimally, to be successful for communication, a given sentence must be capable of being spoken or written in a finite amount of time, using a finite amount of materials, etc.

Of course, in the language of informal mathematics, it is possible to *describe* the formation rules and some of the formulas of an infinitary language (such as infinite conjunctions and disjunctions). Thus, one may claim that the language of mathematical practice is best formulated as a (finitary) *metalanguage* for an infinitary *object* language whose subject matter is one of the various structures.[20] Notice, however, that since the intended interpretation of the metalanguage in question is itself an infinite structure—an infinitary language—many of the above considerations apply. In short, the conceived metalanguage cannot be first-order.

3.1.2. More serious candidates, perhaps, are languages that allow quantification over the (standard) natural numbers: an *ω-language* is a language that contains two variable sorts, one of which ranges over the intended domain, the other over the natural numbers (see Barwise [1977, 42–44]). It is required that in every interpretation, the range of the "natural number variables" be isomorphic to the natural numbers. In such languages, functions and relations involving the natural numbers and the intended domain can be introduced by primitive recursion. In what follows, let $m, n, \ldots$ be variables ranging over the natural numbers and $x, y, \ldots$ be variables ranging over the intended domain.

To begin with, first-order ω-languages do not have all of the shortcomings of first-order languages discussed in §2.1 above. In particular, such languages can characterize individual minimal closures. For example, if $\Phi(x)$ is a formula and p denotes a unary function on the domain, one can introduce a relation $R(m, x)$ between natural numbers and the domain by primitive recursion as follows:

$$R(0, x) \quad \text{iff} \quad \Phi(x); \qquad R(sn, x) \quad \text{iff} \quad \exists y(R(n, y) \ \& \ x = py).$$

---

[20] Such an account was developed by Zermelo [1931] (see also Moore [1980, 124–127]). For Zermelo, the "metalanguage" for this infinitary object language is, in effect, a second-order set theory.

The minimal closure of the extension of Φ under the function denoted **by** p is then characterized by the formula $\exists n R(n, x)$.

Well-foundedness, however, cannot be characterized by a first-order ω-language. This is indicated by the fact that there are ω-models of first-order ZFC **that** are not well-founded. It is easily seen that such structures are also models of virtually any version of ZFC formulated in a first-order ω-language.

Finally, for any *given* formula $\Phi(x)$ with only x free, it can be **stated** that the extension of Φ is finite. If a new function letter f (from the natural **numbers** to the domain) is introduced, then the sentence:

$$\forall n[\exists x(\Phi(x) \; \& \; \forall m < n(x \neq fm)) \to \Phi(fn) \; \& \; \forall m < n(fm \neq fn)]$$
$$\& \; \exists n(\neg \Phi(fn) \lor \exists m < n(fn = fm))$$

is satisfied by all those, and only those, interpretations of the rest of **the** theory in which the extension of Φ is finite.

Similarly, if g and h are new function letters, then the sentences

$$\forall n \forall m(\Phi(gn) \; \& \; (m \neq n \to gn \neq gm)) \quad \text{and} \quad \forall x(\Phi(x) \to \exists n(hn = x))$$

are satisfied by all those, and only those, interpretations of the rest of **the** theory in which the extension of Φ is, respectively, infinite and countable.

This, however, is the limit of the ability of first-order ω-languages to characterize the cardinality of an extension. Indeed, the semantic version of the *downward* Löwenheim-Skolem theorem applies: For a (countable) ω-language, any structure whose domain is infinite has a countable substructure that is elementarily equivalent to it.

On the other hand, the *upward* version of the Löwenheim-Skolem theorem does not apply to first-order ω-languages. It is possible, in particular, to provide categorical axiomatizations of some countable structures. Of course, one can easily characterize the natural numbers (up to isomorphism) with a first-order ω-language, but if anything has the advantages of theft over toil, this does. Less trivially, the rational numbers can be characterized as an infinite field whose universe is the minimal closure of $\{1\}$ under the field functions and ther inverses.

Moreover, even the theory of real analysis formulated in a first-order ω-language is an improvement over the usual first-order version. In the former, one can apply the completeness scheme to enough sets defined as minimal closures to insure that all the models are Archimedian. It follows that each such model is isomorphic to a subset of the real numbers.[21]

---

[21] Another group of languages that may be considered are ω-languages that have variables ranging over functions from the natural numbers to the intended domain. For infinite domains, such languages are equivalent to those with variables ranging over *countable* subsets of the domain. In this context, the above "partial" characterization of finitude can be extended to a full characterization, and well-foundedness can be straightforwardly characterized. Moreover, in real analysis, the completeness scheme would imply that every bounded, countable subset of the domain has a least upper bound. This, of course, entails that the characterization is categorical. Finally, even the characterization of set theory is a significant improvement in the sense that every model thereof is well-founded and closed under its countable subsets.

From the present point of view, the major shortcoming of ω-languages is that they assume or presuppose the natural numbers. Therefore, such a language cannot be used to show, illustrate, or characterize how the natural number structure is itself understood, grasped, or communicated. The attractiveness of such languages is perhaps that, by studying their model theory, one can learn which structures can be characterized *in terms of* the natural numbers. Thus, such a language might be useful to someone who accepts a categorical axiomatization of arithmetic, but is skeptical of the characterizations of other, richer domains. (Analogous remarks apply to the so-called M-languages, where M is any fixed structure.)[22]

3.1.3. The final "intermediate" language considered is one that contains free, but not bound, second-order variables. Such a language L can be obtained from a first-order language by adding a list $X, Y, \ldots$ of (free) predicate variables and a list $f, g, \ldots$ of (free) unary function variables (and modifying the formation rules accordingly). A formula of the extended language L is called a *sentence* if it has no free first-order variables. If $\Phi(x_1, \ldots, x_n, f_1, \ldots, f_m)$ is such a sentence all of whose second-order variables are indicated, and M an interpretation with domain d, then $M \vDash \Phi$ just in case M satisfies Φ under every assignment of subsets of d to the variables X_i and functions on d to the variables f_j. Thus, Φ is treated as a universally quantified formula (equivalent to $\forall x_1 \cdots \forall x_n \forall f_1 \cdots \forall f_m \Phi$). We call such languages *diminished second-order* languages.

The diminished second-order languages differ from full second-order languages in several ways: (1) The only second-order quantification permitted in the former is, in effect, *prenex* universal quantification. That is, instead of arbitrarily many quantifiers arbitrarily embedded, the diminished versions allow only universal quantifiers whose scope is the entire formula.[23] (2) The diminished second-order languages have neither n-ary function variables nor n-ary relation variables for any $n > 1$. (3) Diminished second-order languages have no higher-order predicate, relation, or function constants.

Similar languages are explicitly formulated and studied in Corcoran [1980, 192ff.] as "slightly augmented first-order languages" (the difference being that Corcoran's languages have only a single free predicate variable and no function variables). Church [1956, §48] implicitly classifies diminished second-order languages as "applied" first-order languages.

It is easily seen that virtually none of the above shortcomings of first-order axiomatizations are shared by diminished second-order axiomatizations. The above

---

[22] It might be noted that there is a sense in which the expressive power of first-order ω-languages is "equivalent" to that of first-order languages augmented with an "ancestral operator": If Rxy is a binary relation (such as parenthood) we say that b is an R-*ancestor* of c just in case there is a finite sequence $a_0, \ldots, a_n$ such that $a_0 = c$, $a_n = b$ and $Ra_{i+1}a_i$ for each i, $0 \leq i \leq n$. An *ancestral operator* is a variable-binding operator Γ such that for each formula $\Phi(x, y)$ with x and y free, $(xy\Gamma\Phi)zw$ is a formula equivalent to "w is an ancestor of z under the extension of Φ". By a construction similar to that for minimal closures, an ancestral operator can be formulated in an ω-language. Conversely, the natural numbers can be characterized up to isomorphism in a first-order language augmented with an ancestral operator.

[23] A sentence $\Phi(X)$ of L is equivalent to the second-order "every subset of the domain satisfies Φ". In general, there is no sentence of L that is the *contradictory* of this. The "contrary" $\neg \Phi(X)$ is equivalent to "*no* subset of the domain satisfies Φ".

characterizations of minimal closure, finitude, and well-foundedness, as well as the standard axiomatizations of arithmetic, real analysis, and set theory involve only prenex universal quantification. In these cases, at least, no other higher-order quantification is needed.[24]

In the present article, no stand is taken on the issue of whether diminished (or even full) second-order languages are *sufficient* to axiomatize branches of mathematics.[25] The only claim made is that some second-order variables are necessary. In short, the present thesis is that diminished second-order languages serve as a "lower bound" on languages to formulate mathematical theories, and not necessarily a "greatest lower bound".

3.2. Perhaps another alternative to second-order languages would be to formulate mathematical theories in the language of first-order set theory. In general, with any theory T formulated in a second-order language by a finite number of axioms, there corresponds (in a straightforward manner) a formula $T(x)$, of first-order set theory, which amounts to "x is a model of T". Every sentence Φ of the language of T can then be "translated" into a sentence $\Phi' = \forall x(T(x) \to \Phi_x)$ in the language of set theory, where Φ_x is obtained from Φ by replacing all first-order variables by (set) variables ranging over (i.e., relativized to) the "domain" of x, replacing all predicate variables by variables ranging over the subsets of the domain of x, etc.

Thus, if Φ is a sentence in the language of T, then the set-theoretic Φ' amounts to "Φ is true in all (set-theoretic) models of T". Since the set-theoretic hierarchy is usually taken to provide the semantics of second-order languages, Φ' amounts to "Φ is a semantic consequence of T". Also, under normal conditions concerning the relative strengths of deductive systems, a proof of Φ in T can be routinely translated into a proof of Φ' in set theory. That is, concerning object language *proofs*, first-order set theory can do anything a second-order language can do, usually more.[26]

It might even be added that the concepts and properties discussed in §2.1 above have straightforward characterizations in (first-order) set theory. For example, if b is a set and c is a set of functions, then "x is in the minimal closure of b under the members of c" is characterized by the following formula:

$\text{MC}(b, c, x)$: $\quad \forall y([b \subseteq y \ \& \ \forall z \forall s \forall w(z \in y \ \& \ s \in c \ \& \ \langle z, w \rangle \in s \to w \in y)] \to x \in y)$.

---

[24] Actually, for the theories whose domains are infinite, the function variables may be eliminated by introducing constants for pairing and unpairing functions. In such cases, predicate variables can "play the role" of function variables.

[25] Kreisel [1967] seems to suggest that some third-order axiomatizations may be required. Some writers seem to envision αth order languages, where α is any ordinal. There is also an issue as to whether the deductive strength of diminished second-order languages is adequate for mathematical practice.

[26] It might be added that many metalanguage statements can be formulated in first-order set theory. For example, the sentence $\exists x T(x)$ amounts to "T is satisfiable". The categoricity of T is asserted by a set-theoretic formula $C(T)$ of the form $\forall x \forall y(T(x) \ \& \ T(y) \to (x \text{ and } y \text{ are isomorphic}))$; if T has been proven categorical, then—up to the relative strength of set theory over the metatheory of T—the sentence $C(T)$ is a theorem of set theory.

Notice that $\forall b \forall c \exists! y \forall x (x \in y \leftrightarrow MC(b, c, x))$ is a theorem of set theory. That is, it is provable that for every b and c there is a unique minimal closure of b under c.

Thus, it might seem that first-order languages have been revived. I submit, however, that any formulation of the various branches of mathematics in first-order set theory does not reflect mathematical practice. It is not sufficient for an axiomatization to get the appropriate theorems; the semantics must also be correct.

Different reasons are given for this negative judgement, depending on whether the background language of first-order set theory is taken as an interpreted language or an uninterpreted language. In short, if the language is considered to be *uninterpreted*, then the above advantages are, in a certain sense, merely formal and illusory: they do not apply to the overall semantics. If the language of set theory is considered to be *interpreted*, then the advantages are those of theft over toil. Indeed, if M is the intended interpretation, then the question remains as to how M is itself grasped, understood, and communicated. Moreover, there is a clear sense in which the presuppositions of an interpreted set theory are greater than those of the semantics of second-order languages.

Consider first the case in which the background language of set theory is taken as uninterpreted. Since *this* language is first-order, it has nonisomorphic models, some of which are clearly nonstandard. There are models, for example, in which the membership relation is not well-founded. This observation, in effect, undermines the above "advantages" to the present program.

To focus on an example, consider the formula $N(x)$ of set theory that asserts that "x is a model of the natural numbers". The fact that second-order arithmetic is known to be categorical corresponds to a set-theoretic *theorem* of the form

$$\forall x \forall y (N(x) \,\&\, N(y) \to (x \text{ and } y \text{ are isomorphic})).$$

It follows that for each model M of the background set theory, if a and b are in the domain of M and $M \vDash N(a)$ and $M \vDash N(b)$, then $M \vDash (a \text{ and } b \text{ are isomorphic})$. In other words, the above theorem entails that *within the same model* of set theory, any two sets satisfying $N(x)$ are isomorphic (in that model). This is not enough. The preformal understanding of the categoricity of arithmetic is that *any* two models of arithmetic are isomorphic, not just any two within the same model of set theory. By a straightforward compactness argument, it is easy to see that if the set theory is consistent, then it has a model M' with an element b (of the domain thereof), such that $M' \vDash N(b)$, but the collection of M' elements of b is not isomorphic to the natural numbers. This, of course, is a variant of the Skolem "paradox". Similar considerations apply to any theory that has an infinite model.

Virtually the same considerations apply to the (first-order) set-theoretic versions of the items of §2.1. Consider, for example, the formula $MC(b, c, x)$ corresponding to "x is in the minimal closure of b under c". There is a model M' of set theory containing elements b, c, d such that $M' \vDash \forall x (x \in d \leftrightarrow MC(b, c, x))$, but the collection of M'-elements of d is not the required minimal closure.

Thus, the practice of formulating mathematical theories in an uninterpreted first-order set theory does not preclude nonstandard or unintended interpretations. Such interpretations occur in nonstandard or unintended interpretations of the background set theory. Once again, I agree with Skolem that the result is an unavoidable relativity of all mathematical notions. However, I submit that such a relativity does not reflect mathematical practice.[27]

I turn to the case in which the (first-order) background language of set theory is taken as interpreted. The background theory need not be ZFC, of course; it may be a formalized version of informal set theory. Let M be the intended interpretation. Presumably, M is standard in the sense that it is well-founded, extensional, and closed under the subsets of its elements. For simplicity of treatment, assume that M does not contain urelements. It follows that M is isomorphic to a limit rank $V\lambda$. If M is to be adequate for arithmetic, geometry, analysis, functional analysis, etc., the ordinal λ must be at least 2ω—otherwise, there are no models of the above theories in M. If M is to be adequate for ZFC (and perhaps category theory), then λ must be larger than an inaccessible cardinal.

Notice, first, that in the present situation, the considerations against uninterpreted languages do not apply. For example, the formula $N(x)$ is interpreted as "x is a set *in* M that is a model of the natural numbers". Since M is standard, all such sets x are isomorphic to the natural number structure.

Recall that from the present perspective, the major shortcoming of the ω-languages is that they *presuppose* an understanding of the natural numbers. A similar, and perhaps more serious, problem is found in the present use of interpreted set theory. To say that a structure P is characterized up to isomorphism by the language of set theory as interpreted is only to say that P can be characterized in terms of M or "up to M". The problem as to how M is itself grasped, understood, or communicated is left open. Moreover, I suggest that this latter problem is more difficult than the original problem of accounting for how the natural number structure, the real number structure, etc. are grasped, understood, or communicated. That is, the present program is a case of reducing one problem to a more difficult one. Without an independent characterization of M, it is not clear how the language of set theory overcomes the problems of characterizing structures in first-order languages.

At this point, perhaps, it might be suggested that one need not actually *characterize* the structure M. It is sufficient to let M be any fixed model of the background set theory. It needs to be pointed out, however, that not just any model will do. To correctly characterize the requisite structures and concepts, M must be a *standard* model of set theory. At a minimum, M should be well-founded and closed under the subsets of its elements. As above, these are not first-order concepts.

It is instructive to compare the presuppositions of a given theory as formulated in a second-order language with those of the same theory as formulated in an interpreted language of set theory. The usual examples of arithmetic, real analysis, and set theory are considered.

---

[27] Notice that none of these considerations apply if the uninterpreted background language of set theory is second-order (considered, as usual, with standard semantics). For example, if M is a model of second-order set theory and $M \vDash N(b)$, then the collection of M-elements of b is in fact isomorphic to the natural numbers under the indicated successor relation. This holds even though $N(x)$ is a first-order formula.

It seems safe to say that a theory formulated in a second-order language presupposes at least one model. Formalism and logicism aside, if there is no model of a theory (that is, if the theory is not satisfiable), then it has no *possible* subject matter. It is hard to imagine someone devoting time to studying a theory if he did not believe that it has an interpretation.[28] Thus, arithmetic presupposes a denumerable structure with a successor function, real analysis presupposes a complete ordered field of cardinality $2^{\aleph_0}$, and set theory presupposes a hierarchy of (at least) inaccessible cardinality.

I follow the Quinean view that the ontological commitments of a theory are the values of its variables. Thus, a second-order version of a theory T has presuppositions beyond those of its first-order counterpart. Indeed, the former has more variables. If Q is an intended model of T, and q the domain of Q, then the second-order theory presupposes the existence of each element of q, each subset of q, each function from q to q, and each relation on q. Since the domain of Q, together with its subsets, functions, and relations, exhausts the variable-ranges of the second-order formulation of T, it follows that this list exhausts the indicated ontological presuppositions.

The important point here is that the second-order formulation of T only presupposes the subsets of, and functions and relations on, a domain that is already presupposed.[29] Indeed, the elements of q are presupposed by the acceptance of any version of T: these elements are the values of the first-order variables. In particular, note that a second-order theory short of set theory does not presuppose a set-theoretic hierarchy.

The presuppositions of the formulation of T in an interpreted language of set theory are thus much greater than the presuppositions of the formulation of T in a second-order language (or, for that matter, the formulation of T in an nth order language, for virtually any n). As above, let M be the given interpretation of the background set theory and let d be the domain of M. Presumably, each element of d is presupposed by the acceptance of the background language as interpreted. If the standard models of T are infinite, then to be adequate for T, the domain d must contain an infinite set c. Presumably, the background theory has a powerset axiom (or theorem). Thus, the overall commitments of the set-theoretic program include the powerset of c, the powerset of the powerset of c, etc. In many set theories, this

---

[28] It might be noted that the same presupposition applies to a theory formulated in a first-order language, and, in this case, it could be added that a *standard* model is presupposed. Suppose, for example, that it should happen that a dramatic event causes that mathematical community to believe that although first-order analysis is consistent, it does not have a standard model. (Suppose, for example, that the community comes to believe that every model of the theory has an undefinable, bounded subset that does not have a least upper bound.) Of course, the consistency of the first-order theory implies that it is satisfiable. In fact, the theory could be reinterpreted as a complicated structure of natural numbers. I suggest, however, that this fact would give little comfort to one who has devoted her life to studying a nonexistent real number structure.

[29] The principle of accepting collections (as opposed to functions and relations) of previously accepted entities is held by such nominalists as Nelson Goodman [1972] and Hartry Field [1980]. For these philosophers, the objection to mathematical theories lies with the acceptance of the original first-order versions.

process can be iterated into the transfinite. In any case, these presuppositions are much more than one needs for arithmetic, real analysis, or just about any theory short of set theory.

Of course, it is to be conceded that despite the greater presuppositions, there are advantages to using a background language of set theory to formulate mathematical theories. Probably the most important of these is that the set-theoretic program provides a single, uniform "foundation" for all (or most) of mathematics. Indeed, as stated above, it is common to take mathematical theories and structures to be interrelated. This presupposes a common semantics for different branches and, hence, may indicate the propriety of a common language—a language of set theory. Consider, for example, the technique of embedding one structure, such as the natural numbers, into another, the complex plane. In the present framework, the embedding has a straightforward account in the set-theoretic object-language common to the two theories. It amounts to a theorem that for any model of the natural numbers and any model of the complex plane, there is a one-to-one homomorphism from the former into the latter.

With this uniform foundation, however, comes a uniform group of presuppositions—the intended model M of the background set theory. This, I submit, is unnatural. Informally at least, the presuppositions of arithmetic are less than those of real analysis. The former has the relatively modest commitment to a denumerable set (plus, perhaps, its subsets, etc.) while the latter is committed to an uncountable continuum (plus, perhaps, *its* subsets, etc.). One who is doing arithmetic *alone* does not appear to be committed to, say, the real number line (or the set-theoretic hierarchy). From this perspective, a *decision* to use real analysis to study the natural numbers involves as expanded commitment.[30] Kreisel [1967] observes that the desire to keep track of presuppositions is implicit in the practice of at least some mathematicians and, moreover, that this desire is reflected in the program of providing separate (second-order) theories of each branch:

> ... Bourbaki [for example] is extremely careful to isolate the assumptions of a mathematical theorem, but never the axioms of set theory implicit in a particular deduction ... This practice is quite consistent with the assumption that what one has in mind when following Bourbaki's proofs is the second-order axioms ... (Kreisel [1967, 151])

It is agreed, of course, that there is a common semantics for the various second-order languages. It does not follow, however, that there are substantial presuppositions and commitments for this semantics that apply uniformly to any theory

---

[30] As indicated in footnote 26, a related "advantage" of the set-theoretic program is illustrated by the fact that for a given theory T, the "object language" statements of T are interpreted as statements about the models of T or, in effect, as statements about the *semantics* of T. Thus, the set-theoretic program provides a clear and straightforward connection between a theory and its semantics. It follows from Tarski's theorem, however, that, in some sense, the metatheory or semantics of a branch of mathematics is stronger (or, at any rate, substantially different) than the object language theory of that branch. Once again, for the purpose of keeping track of presuppositions, it is worthwhile to keep theory and metatheory separate.

formulated in a second-order language. Notice that in any case, the common metatheory, or semantics, of second-order languages in weaker than standard set theory, say ZFC. As observed by Boolos [1975], there is a straightforward "interpretation" of second-order logic in (first-order) set theory. That is, for each (effectively presented) second-order language L, there is a formula $\Theta(x)$ in the language of first-order set theory, such that for each natural number n, $\Theta(\bar{n})$ is true (in the set-theoretic hierarchy) just in case n is the Gödel number of a valid sentence of L. From Tarski's theorem, it follows that there is no set-theoretically definable translation T from the language of set theory to L such that for each sentence Φ of set theory, Φ is true iff $T(\Phi)$ is valid.

§4. First-order logic and second-order logic.

The main conclusion of this article is that an adequate formalization of such branches of mathematics as arithmetic, real analysis, and set theory must involve (at least) a second-order language. I suggest, then, that the natural underlying *logic* of these branches is (at least) second-order.

As indicated above (see footnote 2), there has been some work in recent years aimed at showing that second-order logic is not logic at all. Such arguments usually focus on either the ontological presuppositions of second-order languages or the "inconvenient" semantic properties—incompleteness and noncompactness—of second-order logic. I suggest that the considerations of the present article preempt such reasoning. It can be agreed that, all things equal, it would be desirable to have a recursively axiomatized, compact logic with fewer presuppositions. I take it, however, that the purpose of logic is to study and codify correct inference. Since one cannot codify the correct inferences of a second-order language with a first-order logic, it follows that the logic of mathematics cannot be first-order.

Thus, I suggest that the presuppositions (and inconveniences) of second-order logic must be accepted. The purpose of this section is to briefly assess the presuppositions of second-order logic *vis-à-vis* first-order logic. The interested reader is referred to Boolos [1975] for more detailed considerations.

As noted by Boolos [1975], the present evaluation is simplified by the fact that in one important respect, the semantics of a given first-order language $L1$ is the same as that of a corresponding second-order language $L2$: a model or interpretation of *either* language consists of a nonempty domain together with a function giving appropriate assignments to the nonlogical constants. Since, in the cases at hand, the nonlogical terminologies are identical, the respective languages have exactly the same classes of interpretations. (Of course, this does not mean that a given theory in $L2$ will have the same *models* as a counterpart in $L1$.) The *logics* of $L1$ and $L2$ can therefore be evaluated in terms of the presuppositions of each concerning the common semantics.

4.1. Ontology.
The main difference between the languages is that the second-order $L2$ has variables ranging over the subsets of the domain, functions from the domain to the domain, etc. Thus, for a given interpretation, the first-order $L1$ presupposes only the elements of the domain, while $L2$ also presupposes the subsets, functions, and relations on the same domain.

I consider first the presuppositions of "pure" logic or, in other words, the presuppositions of uninterpreted languages. The logical truth $\exists x(x = x)$ of $L1$

corresponds to a commitment to at least one element (of each domain). Similarly, the pair of logical truths $\exists X(\forall x(Xx))$ and $\exists X\forall x(\neg Xx)$ of $L2$ corresponds to a commitment to at least two subsets (of each domain), an empty set and a "universal" set. However, since there are interpretations whose domains have only a single element, the following is not a logical truth of $L2$:

$$\exists X\exists x\exists y(Xx\ \&\ Xy\ \&\ x\neq y).$$

That is, second-order logic uninterpreted does not presuppose a two-element set. Thus, the existential assumptions of uninterpreted second-order logic are rather weak and, moreover, are not much greater than those of uninterpreted first-order logic.

Of course, the concern with the ontological commitments of second-order languages is focused on the commitments of *interpreted* theories. For example, the concern with second-order arithmetic is directed at the commitment to sets of numbers. In general, let d be the domain of an interpretation of both $L1$ and $L2$. Assume that d has infinite cardinality κ. Concerning this interpretation, $L1$ is committed to the elements of d or, in other words, to κ-many items. The further commitments of the second-order $L2$ are the subsets, etc., of d, which total 2^κ items. Thus, the second-order language as interpreted does have greater presuppositions, but the difference is (in effect) limited to a single "powerset" operation.[31]

4.2. Completeness and satisfication. I turn now to the relative presuppositions of various metatheorems for first-order logic and second-order logic. For $L1$, the completeness theorem is that every consistent set of sentences is satisfiable. This theorem and its proof depend on the axiom of infinity (in the metalanguage). Indeed, if the semantics of $L1$ contained only finite domains, then the respective logic would *not* be complete. The reason for this is that there are consistent first-order sentences which are satisfiable only in infinite domains. An example of such a sentence is:

(In) $\forall x\exists yRxy\ \&\ \forall x\forall y\forall z(Rxy\ \&\ Ryz\rightarrow Rxz)\ \&\ \forall x(\neg Rxx).$

It is important to be clear as to what is, and what is not, presupposed here. First-order languages, by themselves, do not presuppose infinite domains. As above, a language itself hardly presupposes anything. However, first-order languages make it *possible* to presuppose infinite domains. That is, when one adopts or asserts a sentence like (In) (or accepts the metatheory needed to prove the completeness theorem), then, and only then, one is committed to an infinite domain.

The situation concerning second-order languages is similar, but less modest. It is easy to see that there are consistent second-order sentences which are satisfiable only in very large domains. For example, if Z is the conjunction of the axioms of second-order ZFC, then Z is satisfiable only in domains of inaccessible cardinality.

---

[31] Although present concern is not with deductive systems, it might be noted that the soundness theorem for standard second-order logic does not require extensive set-theoretic presuppositions. Most of the axioms (and rules of inference) are either those of first-order logic or second-order versions of the quantifier axioms (which involve virtually no set theory). The only exceptions are the comprehension scheme and, perhaps, the axiom of choice. The soundness of these, of course, involves the axiom of separation and the axiom of choice, respectively, in the metalanguage.

I suggest that this is not a defect of second-order logic, but rather a result of the expressive ability of second-order languages. In other words, the present observations are the result of what I take to be the main *advantage* of second-order languages. As with the first-order case, the second-order $L2$ does not, of itself, presuppose large domains, but does make it possible to presuppose large domains. The presuppositions come when one adopts or asserts a sentence like Z. In such a situation, and only then, one is committed to an inaccessible domain.

4.3. Logical truth. I close with an admission that there are statements of second-order logic that amount to rather substantial set-theoretic propositions. An example follows.

It is easily seen that there is a second-order formula $C(X)$ which (in any interpretation) is equivalent to "X is either finite or denumerably infinite", and that there is a second-order formula $E(X, Y)$ equivalent to "X has the same cardinality as Y". It follows that the formula

$$(O(X)) \qquad\qquad C(X) \& \forall Y(Y \subseteq X \rightarrow C(Y) \vee E(X, Y))$$

is equivalent to "X has cardinality $\aleph_1$". There is also a second-order formula $P(X)$ equivalent to "$\exists x \exists y \exists f \exists g(\langle X, x, y, f, g \rangle$ is a model of real analysis)". Of course, $P(X)$ amounts to "X has the cardinality of the continuum".

Consider the sentences

$$(\text{CH}) \qquad\qquad \forall X(O(X) \equiv P(X)),$$

$$(\text{NCH}) \qquad\qquad \forall X(P(X) \rightarrow \neg O(X)).$$

Notice that (CH) is a logical truth if and only if the continuum hypothesis is true, and that (NCH) is a logical truth if and only if the continuum hypothesis is false.

Of course, one does not normally think of the continuum hypothesis or its negation as a logical truth. The fact that one of them is a second-order logical truth is, again, a result of the expressive power of second-order languages: substantial statements about the semantics of $L2$ can be made *by* sentences of $L2$. Once again, this expressive power is here taken to be the main strength of second-order languages. The proper conclusion, I suggest, is not to reject second-order languages and second-order logic, but rather to reject the notion of a sharp distinction between mathematics and the logic of mathematics.

Acknowledgments. I would like to thank the following people for many helpful suggestions on earlier versions of this paper: John Corcoran, Haim Gaifman, Nicolas Goodman, Charles Kielkopf, George Kreisel, Timothy McCarthy, Jon Pearce, Michael Resnik, Michael Scanlon, George Schumm, Craig Smoryński, and an anonymous referee. I would also like to thank as a group the Ohio State philosophy of mathematics seminar for devoting several sessions to this project.

REFERENCES

J. BARWISE [1977], *An introduction to first-order logic*, **Handbook of mathematical logic** (J. Barwise, editor), North-Holland, Amsterdam, pp. 5–46.

G. BOOLOS [1975], *On second-order logic*, **Journal of Philosophy**, vol. 72, pp. 509–527.

―――― [1981], *For every A there is a B*, **Linguistic Inquiry**, vol. 12, pp. 465–467.

742   STEWART SHAPIRO

G. BOOLOS and R. JEFFREY [1980], *Computability and logic*, 2nd ed., Cambridge University Press, Cambridge.

A. CHURCH [1956], *Introduction to mathematical logic*, Princeton University Press, Princeton, New Jersey.

J. CORCORAN [1971], *A semantic definition of definition*, this JOURNAL, vol. 36, pp. 366–367.

―――― [1973], *Gaps between logical theory and mathematical practice*, **The methodological unity of science** (M. Bunge, editor), Reidel, Dordrecht, pp. 23–50.

―――― [1980], *Categoricity*, **History and Philosophy of Logic**, vol. 1, pp. 187–207.

S. FEFERMAN [1977], *Theories of finite type related to mathematical practice*, **Handbook of mathematical logic** (J. Barwise, editor), North-Holland, Amsterdam, pp. 913–971.

H. FIELD [1980], *Science without numbers*, Princeton University Press, Princeton, New Jersey.

K. GÖDEL [1944], *Russell's mathematical logic*, **The philosophy of Bertrand Russell** (P. A. Schilpp, editor), Northwestern University, Evanston and Chicago, Illinois, pp. 123–153, reprinted in **Philosophy of mathematics** (P. Benacerraf and H. Putnam, editors), Prentice-Hall, Englewood Cliffs, New Jersey, 1964, pp. 211–232.

NELSON GOODMAN [1972], *Problems and projects*, Bobbs-Merill, Indianapolis, Indiana.

D. GOTTLIEB [1980], *Ontological economy: substitutional quantification and mathematics*, Oxford University Press, Oxford.

L. HENKIN [1950], *Completeness in the theory of types*, this JOURNAL, vol. 15, pp. 81–91.

D. HILBERT [1900], *Mathematische Problems*, English translation, **Bulletin of the American Mathematical Society**, vol. 8 (1902), pp. 437–479; reprinted in **Mathematical developments arising from Hilbert problems**, Proceedings of Symposia in Pure Mathematics, vol. 28, American Mathematical Society, Providence, Rhode Island, 1976, pp. 1–34.

G. KREISEL [1967], *Informal rigour and completeness proofs*, **Problems in the philosophy of mathematics** (I. Lakatos, editor), North-Holland, Amsterdam, pp. 138–186.

R. MONTAGUE [1965], *Set theory and higher-order logic*, **Formal systems and recursive functions** (J. Crossley and M. Dummett, editors), North-Holland, Amsterdam, pp. 131–148.

G. MOORE [1980], *Beyond first-order logic: the historical interplay between logic and set theory*, **History and Philosophy of Logic**, vol. 1, pp. 95–137.

J. MYHILL [1951], *On the ontological significance of the Löwenheim-Skolem theorem*, **Academic freedom, logic and religion** (M. White, editor), American Philosophical Society, Philadelphia, Pennsylvania, pp. 57–70.

H. PUTNAM [1980], *Models and reality*, this JOURNAL, vol. 45, pp. 464–482.

―――― [1981], *Reason, truth and history*, Cambridge University Press, Cambridge.

W. V. O. QUINE [1970], *Philosophy of Logic*, Prentice-Hall, Englewood Cliffs, New Jersey.

M. RESNIK [1980], *Frege and the philosophy of mathematics*, Cornell University Press, Ithaca, New York.

T. SKOLEM 1923, *Einige Bemerkungen zur axiomatischen Begründung der Mengenlehre*, **Wissenschaftliche Vorträge gehalten auf dem Fünften Kongress der Skandinavischen Mathematiker in Helsingfors vom 4. bis 7. Juli 1922**, Akademiska Bokhandeln, Helsinki, 1923, pp. 217–232.

L. THARP [1975], *Which logic is the right logic?*, **Synthese**, vol. 31, pp. 1–31.

O. VEBLEN [1904], *A system of axioms for geometry*, **Transactions of the American Mathematical Society**, vol. 5, pp. 343–384.

H. WANG [1974], *From mathematics to philosophy*, Routledge and Kegan Paul, London.

T. WESTON [1976], *Kreisel, the continuum hypothesis and second-order set theory*, **Journal of Philosophical Logic**, vol. 5, pp. 281–298.

E. ZERMELO [1931], *Über Stufen der Quantifikation und die Logik des Unendlichen*, **Jahresbericht der Deutschen Mathematiker-Vereinigung**, vol. 41, pp. 85–88.

DEPARTMENT OF PHILOSOPHY
OHIO STATE UNIVERSITY AT NEWARK
NEWARK, OHIO 43055

[5]

HISTORY AND PHILOSOPHY OF LOGIC, 7 (1986), 143–154

What are Logical Notions?

ALFRED TARSKI

Edited by
JOHN CORCORAN
Department of Philosophy, State University of New York at Buffalo,
Buffalo, New York 14260, U.S.A.

Received 28 August 1986

In this manuscript, published here for the first time, Tarski explores the concept of logical notion. He draws on Klein's Erlanger Programm to locate the logical notions of ordinary geometry as those invariant under all transformations of space. Generalizing, he explicates the concept of logical notion of an arbitrary discipline.

1. Editor's introduction

In this article Tarski proposes an explication of the concept of logical notion. His earlier well-known explication of the concept of logical consequence presupposes the distinction between logical and extra-logical constants (which he regarded as problematic at the time). Thus, the article may be regarded as a continuation of previous work.

In Section 1 Tarski states the problem and indicates that his proposed explication shares features both with nominal (or normative) definitions and with real (or descriptive) definitions. Nevertheless, he emphasizes that his explication is not arbitrary and that it is not intended to 'catch the platonic idea'. In Section 2, in order to introduce the essential background ideas, Klein's Erlanger Programm for classifying geometrical notions is sketched using three basic examples: (1) the notions of metric geometry are those invariant under the similarity transformations; (2) the notions of descriptive geometry are those invariant under the affine transformations; and (3) the notions of topological geometry (topology) are those invariant under the continuous transformations. This illustrates the fact that as the family of transformations expands not only does the corresponding family of invariant notions contract but also, in a sense, the invariant notions become more 'general'. In Section 3 Tarski considers the limiting case of the notions invariant under all transformations of the space and he proposes that such notions be called 'logical'. Then, generalizing beyond geometry, a notion (individual, set, function, etc) based on a fundamental universe of discourse is said to be *logical* if and only if it is carried onto itself by each one-one function whose domain and range both coincide with the entire universe of discourse.

Tarski then proceeds to test his explication by deducing various historical, mathematical and philosophical consequences. All notions definable in *Principia mathematica* are logical in the above sense, as are the four basic relations introduced by Peirce and Schröder in the logic of relations. No individual is logical: all numerical properties of classes are logical, etc. In Section 4 Tarski considers the philosophical question of whether all mathematical notions are logical. He considers two construals of mathematics—the type-theoretic construal due to Whitehead and Russell, and the set-theoretic construal due to Zermelo, von Neumann and others. His conclusion is that mathematical notions are all logical relative to the type-theoretic construal but not relative to the set-theoretic construal. Thus, no answer to the philosophical question of the reducibility of mathematical notions to logic is implied by his explication of the concept of logical notion.

2.   Editorial treatment

The wording of this article reveals its origin as a lecture. On 16 May 1966 Tarski delivered a lecture of this title at Bedford College, University of London. A tape-recording was made and a typescript was developed by Tarski from a transcript of the tape-recording. On 20 April 1973 he delivered a lecture from the typescript as the keynote address to the Conference on the Nature of Logic sponsored by various units of the State University of New York at Buffalo. I made careful notes of this lecture and from them wrote an extended account which was published in the University newspaper (*The reporter*, 26 April 1973). Copies of the newspaper article were sent to Tarski and others. It was Tarski's intention to polish the typescript and to publish it as a companion piece to his 'Truth and proof' (*1969*). Over the next few years I had several opportunities to speak with Tarski and to reiterate my interest in having the lecture appear in print. In 1978 I began work on editing the second edition of Tarski's *Logic, semantics, metamathematics*, which finally appeared in December 1983 shortly after Tarski's death. During the course of my work with Tarski for that project, he said on several occasions that he wanted me to edit 'What are logical notions?', but it was not until 1982 that he gave me the typescript with the injunction that it needed polishing.

For the most part my editing consisted in the usual editorial activities of correcting punctuation, sentence structure and grammar. In some locations the typescript was evidently a transcript written by a non-logician. Occasionally there was a minor lapse (e.g. in uniformity of terminology). The bibliography and footnotes were added by me. The only explicit reference in the typescript is in Section 3 where the 1936 article by Lindenbaum and Tarski is mentioned. Of course, the greatest care was taken to guarantee that Tarski's ideas were fully preserved.

For further discussion and applications of the main idea of this paper see the book by Tarski and Steven Givant (*1987*), especially section 3.5 in chapter 3.

WHAT ARE LOGICAL NOTIONS?

Alfred Tarski

1.   The title of my lecture is a question; a question of a type which is rather fashionable nowadays. There is another type of question you often hear: what is psychology, what is physics, what is history? Questions of this type are sometimes answered by specialists working in the given science, sometimes by philosophers of

science; the opinion of a logician is also asked from time to time as an alleged authority in such matters. Well, let me say that specialists working in a given science are usually the people least qualified to give a good definition of the science. It is a domain where you would normally expect an intelligent discussion from a philosopher of science. And a logician is certainly not an authority—he is not specially qualified to answer questions of this type. His role and influence are rather of a negative character—he offers criticism, he points out how vague a certain formulation is, how indefinite an account of a certain science is. In view of his negative approach to discussing definitions of other sciences, a logician must certainly be especially cautious when he discusses his own science and tries to say what logic is.

Answers to the question 'What is logic?' or 'What is such and such science?' may be of very different kinds. In some cases we may give an account of the prevailing usage of the name of the science. Thus in saying what is psychology, you may try to give an account of what most people who use this term normally mean by 'psychology'. In other cases we may be interested in the prevailing usage, not of all people who use a given term, but only of people who are qualified to use it—who are expert in the domain. Here we would be interested in what psychologists understand by the term 'psychology'. In still other cases our answer has a normative character: we make a suggestion that the term be used in a certain way, independent of the way in which it is actually used. Some further answers seem to aim at something very different, but it is very difficult for me to say what it is; people speak of catching the proper, true meaning of a notion, something independent of actual usage, and independent of any normative proposals, something like the platonic idea behind the notion. This last approach is so foreign and strange to me that I shall simply ignore it, for I cannot say anything intelligent on such matters.

Let me tell you in advance that in answering the question 'What are logical notions?' what I shall do is make a suggestion or proposal about a possible use of the term 'logical notion'. This suggestion seems to me to be in agreement, if not with all prevailing usage of the term 'logical notion', at least with one usage which actually is encountered in practice. I think the term is used in several different senses and that my suggestion gives an account of one of them.[1] Moreover, I shall not discuss the general question 'What is logic?' I take logic to be a science, a system of true sentences, and the sentences contain terms denoting certain notions, logical notions. I shall be concerned here with only one aspect of the problem, the problem of logical notions, but not for instance with the problem of logical truths.

2.   The idea which will underlie my suggestion goes back to a famous German mathematician, Felix Klein. In the second half of the nineteenth century, Felix Klein did very serious work in the foundations of geometry which exerted a great influence on later investigations in this domain.[2] One problem which interested him was that of distinguishing the notions discussed in various systems of geometry, in various geometrical theories, e.g. ordinary Euclidean geometry, affine geometry, and topo-

1   It would be instructive to compare these remarks with those that Tarski makes in connection with his explications of truth in his *1935a* and of logical consequence in his *1936*, especially p. 420. See also Corcoran *1983*, especially pp. xx-xxii.
2   See, e.g., Klein *1872*.

logy. I shall try to extend his method beyond geometry and apply it also to logic. I am inclined to believe that the same idea could also be extended to other sciences. Nobody so far as I know has yet attempted to do it, but perhaps one can formulate using Klein's idea some reasonable suggestions to distinguish among biological, physical, and chemical notions.

Now let me try to explain to you very briefly Klein's idea. It is based upon a technical term 'transformation', which is a particular case of another term well known to everyone from high school mathematics—the term 'function'. A function or a functional relation is, as we all know, a binary relation r which has the property that whatever object x we consider there exists at most one object y to which x is in the relation r. Those x's for which such a y actually exists are called 'argument values'. The corresponding y's are called 'function values'. We also write $y = r(x)$; this is the normal function notation. The set of all argument values is called the 'domain of the function', the set of function values is called in *Principia Mathematica* the 'counter-domain', more often 'the range', of the function. So every function has its domain and its range. We often deal in mathematics with functions whose domain and range consist of numbers. However, there are also functions of other types. For instance we may consider functions whose domain and range consist of points. In particular, in geometry we deal with functions whose domain and range both coincide with the whole geometrical space. Such a function is referred to as a 'transformation' of the space onto itself. Moreover we often deal with functions which are one-one functions, with functions which have the property that to any two different argument values the corresponding function values are always different. We say that such a function establishes a one-one correspondence between its domain and its range. So a function whose domain and range both coincide with the whole space and which is one-one is called a one-one transformation of the space onto itself (more briefly, 'a transformation'). I shall now discuss transformations of ordinary geometrical space.

Now let us consider normal Euclidean geometry which again we all know from high school. This geometry was originally an empirical science—its purpose was to study the world around us. This world is populated with various physical objects, in particular with rigid bodies, and a characteristic property of rigid bodies is that they do not change shape when they move. Now every motion of such a rigid body corresponds to a certain transformation because a rigid body occupies one position when it starts moving and as a result of this motion occupies another position. Each point occupied by the rigid body at the beginning of the motion corresponds to a point occupied by the same body at the end of the motion. We have a functional relation. It is true that this is not a functional relation whose domain includes all points of the space, but it is known from geometry that it can always be extended to the whole space. Now what is characteristic about this transformation is that the distance between two points does not change. If x and y are at a certain distance and if $f(x)$ and $f(y)$ are the final points corresponding to x and y, then the distance between $f(x)$ and $f(y)$ is the same as that between x and y. We say that distance is *invariant* under this transformation. This is a characteristic property of motions of rigid bodies—if it did not hold, we would not call the body a rigid body.

As you see, we are naturally led in geometry to consider a special kind of transformation of this space, transformations which do not change the distance between points. Mathematicians have a bad habit of taking a term from other domains—from physics, from anthropology—and ascribing to it a related but different meaning. They have done this with the term 'motion'. They use the term 'motion' in a mathematical sense, in which it means simply a transformation in which distance does not change. So the motion of a particular physical object, a rigid body, results in a certain transformation; but to a mathematician motions are simply transformations which do not change distance. Such transformations are more properly called 'isometric transformations'.

Now, Klein points out that all the notions which we discuss in Euclidean geometry are invariant under all motions, that is, under all isometric transformations. Let me say again what we mean when we say that a notion is invariant under certain transformations. I use the term 'notion' in a rather loose and general sense, to mean, roughly speaking, objects of all possible types in some hierarchy of types like that in *Principia mathematica*. Thus notions include individuals (points in the present context), classes of individuals, relations of individuals, classes of classes of individuals, and so on. What does it mean, for instance, to say that a class of individuals is invariant under a transformation f? This means that x belongs to this class if and only if $f(x)$ also belongs to this class, in other words, that this class is carried onto itself by the transformation. What does it mean to say that a relation is invariant under a transformation f? This means that x and y stand in the relation if and only if $f(x)$ and $f(y)$ stand in the relation. We can easily extend the notion of invariance in a familiar way to classes of classes, relations between classes, and so on.

Now a close analysis of Euclidean geometry shows that all notions which we discuss there are invariant not only under motions, under isometric transformations, but under a wider class of transformations, namely under those transformations which geometers call 'similarity transformations'. These are transformations which do not all preserve distance, but which so to speak increase or decrease the size of a geometrical figure uniformly in all directions. More precisely, some similarity transformations do not preserve distance, but all preserve the ratio of two distances. If you have, for instance, three points, x, y, z, and if the distance from y to z is larger by 25% than the distance from x to y, then the result of a similarity transformation is again three points, $f(x)$, $f(y)$, $f(z)$, where the distance between $f(y)$ and $f(z)$ is 25% larger than the distance between $f(x)$ and $f(y)$. In other words, a triangle is transformed into a triangle which is similar to it, with the same angles and whose sides are proportionally larger or smaller. And it turns out that all properties which one discusses in Euclidean geometry are invariant under all possible similarity transformations. This means, incidentally, that we cannot discuss in Euclidean geometry the notion of a unit of measure. We should not ask such a geometer whether from the point of view of his discipline the metric system or a non-metric system is preferable. In Euclidean terms we cannot distinguish a metre from a yard; we cannot even distinguish a centimetre from a yard. Any two segments are "the same", since you can always transform them one into another by means of a similarity transformation.

Every Euclidean property that belongs to one segment belongs to every other segment as well.

Now Klein says that invariance under all similarity transformations is the characteristic property of the notions studied in metric geometry,[3] which is another term for ordinary Euclidean geometry. We can express this as a definition: a *metric* notion, or a notion of metric geometry, is simply a notion which is invariant under all possible similarity transformations. We could certainly imagine a discipline in which we would be interested in a narrower class of transformations, for instance only in isometric transformations, or only in transformations which preserve the distinction between being to the right and being to the left (a distinction which we are unable to make in our normal geometry), or between a motion which is clockwise from a motion which is counter-clockwise (again a distinction we cannot make in normal Euclidean geometry). But by narrowing down the class of permissible transformations we can make more distinctions, i.e. we widen the class of notions invariant under permissible transformations. The extreme case in this direction in geometry would be to single out four points, give them names, and to consider only those transformations which would leave these four points invariant. This would mean introducing a co-ordinate system, and we would be at a limit of the domain of geometry, i.e. at what is called analysis. Actually in this case there would be no permissible transformations except one "trivial" identity transformation.

On the other hand one can go in the opposite direction; instead of narrowing down the class of permissible transformations, and in this way widening the class of invariant notions, we can do the opposite, and widen the class of transformations. We can for instance include also transformations in which distance may change, but what is unchanged is mutual linear position of points. More precisely, if three points are on one line, then their images, after the transformation, are also on one line. If one point is between two other points, then its image is between the images of the two other points. One calls such transformations 'affine transformations'. Collinearity and betweenness are just two of the notions which are invariant under all transformations of this kind. The part of geometry where such notions are used is called affine geometry.[4] In this geometry we cannot distinguish, for example, one segment from another, indeed we cannot make any distinctions among triangles. Any two triangles are so to speak equal, that is, indistinguishable from the point of view of affine geometry. This means that we cannot point out any property in affine geometry which

---

3   Terminology in this field is not uniform, and Tarski's usage may not be familiar to some readers. The present terminology derives from Tarski *1935b*, where the term 'descriptive geometry' is used to indicate the part of ordinary Euclidean geometry based only on 'point' and 'between' (which Tarski refers to as 'the descriptive primitive'). The term 'metric geometry' is used to indicate all of ordinary Euclidean geometry (which, as Tarski notes, can be taken to be based only on 'point' and 'congruence'—a notion that Tarski calls 'the metric primitive'). In the same article Tarski indicates that descriptive geometry is a proper part of metric geometry in the sense that 'between' is definable from 'point' and 'congruence' while 'congruence' is not definable from 'point' and 'between'.

4   'Affine geometry' is in current use in exactly this sense. What Tarski calls 'affine geometry' here, he called 'descriptive geometry' in *1935b*. An affine transformation that is not a similarity can be obtained in plane geometry by a parallel projection of the plane onto a non-perpendicular, intersecting "copy" of itself. Concretely, the image of a suitably placed isosceles right triangle is scalene, but all images of triangles are triangles.

is possessed by one of the triangles but not by all others. In metric geometry we know many such properties, for example, the property of being equilateral, or of being right-angled. In affine geometry we cannot make any such distinctions. What we can distinguish is a triangle from a quadrangle, because no affine transformation could start with a triangle and lead to a quadrangle. So here we have an example of a wider class of transformations, and as a result of this, a narrower class of notions which are invariant under this wider class of transformations; the notions are fewer, and of a more "general" character.

We can go a step further. We can include, for instance, transformations in which even the betweenness relation is not preserved, and even transformations where points which lie on the same straight line are transformed into points lying on different lines. The characteristic thing which is preserved here is, roughly speaking, connectedness or closedness. A connected figure remains connected. A closed curve remains closed. Sometimes it is said, putting things "negatively" so to speak, that these transformations are those which do not "break up" or "tear apart". This is a very imprecise way of formulating it, but some of you probably have guessed what I have in mind; I have in mind the so-called continuous transformations, and the part of geometry, the geometrical discipline which deals with notions invariant under such transformations, is topology. In metric geometry we can distinguish one triangle from another; in affine geometry we cannot do so, but we can still distinguish between a triangle and, let us say, a quadrangle. But in topology we cannot distinguish between two polygons, or even between a polygon and a circle, because given a polygon, if we imagine it to be made of wire, we can always bend it in such a way as to obtain a circle, or any other polygon. Such a transformation will be continuous: we do not separate anything which was connected. What we can distinguish in topology is, for example, one triangle from two triangles. For a triangular wire can be bent into two triangles only if we break it into two parts and form a triangle from each part—and this would not be a continuous transformation.

3.   Now suppose we continue this idea, and consider still wider classes of transformations. In the extreme case, we would consider the class of *all* one-one transformations of the space, or universe of discourse, or 'world', onto itself. What will be the science which deals with the notions invariant under this widest class of transformations? Here we will have very few notions, all of a very general character. I suggest that they are the logical notions, that we call a notion 'logical' if it is invariant under all possible one-one transformations of the world onto itself.[5] Such a suggestion perhaps sounds strange—the only way of seeing whether it is a reasonable suggestion is to

5   Apart from Mautner *1946*, which Tarski seems not to have known, this is, I believe, the first attempted application in English of Klein's Erlanger Programm to logic. However, in Silva *1945*, which is written in Italian, we find applications which anticipate essential elements of later model theory. Keyser (*1922*, 219) and Weyl (*1949*, 73) indicate in more or less vague terms the possibility of connections between logic and the Erlanger Programm. Tarski's papers from 1923 to 1938 (collected in Tarski *1983*) do not mention Felix Klein. The history of the influence of the Erlanger Programm on the development of logic remains to be written. Also needing investigation is the role of the Erlanger Programm in physics, especially relativity.

discuss some of its consequences, to see what it leads to, what we have to believe if we agree to use the term 'logical' in this sense.

A natural question is this: consider the notions which are denoted by terms which can be defined within any of the existing systems of logic, for instance *Principia mathematica*. Are the notions defined in *Principia mathematica* logical notions in the sense which I suggest? The answer is yes; this is a rather simple meta-logical result, formulated a long time ago (*1936*) in a short paper by Lindenbaum and myself. Though this result is simple, I think that it should be included in most logic textbooks, because it shows a characteristic property of what can be expressed by logical means. I am not going to formulate the result in a very exact way, but the essence of it is just what I have said. Every notion defined in *Principia mathematica*, and for that matter in any other familiar system of logic, is invariant under every one-one transformation of the 'world' or 'universe of discourse' onto itself.[6]

Next we look for examples of logical notions in a systematic way, starting with the simplest semantical categories[7] or types, and going on to more and more complicated ones. For instance, we can start with individuals, with objects of the lowest type, and ask: What are examples of logical notions among individuals? This means: What are examples of individuals which would be logical in the above sense? And the answer is simple: There are no such examples. There are no logical notions of this type, simply because we can always find a transformation of the world onto itself where one individual is transformed into a different individual. The simple fact that we can always define such a function means that on this level there are no logical notions.

If we proceed to the next level, to classes of individuals, we ask: What classes of individuals are logical in this sense? It turns out, again as a result of a simple argument, that there are exactly two classes of individuals which are logical, the universal class and the empty class. Only these two classes are invariant under every transformation of the universe onto itself.

If we go still further, and consider binary relations, a simple argument shows that there are only four binary relations which are logical in this sense: the universal relation which always holds between any two objects, the empty relation which never holds, the identity relation which holds only between "two" objects when they are identical, and its opposite, the diversity relation. So the universal relation, the empty relation, identity, and diversity—these are the only logical binary relations between individuals. This is interesting because just these four relations were introduced and discussed in the theory of relations by Peirce, Schröder, and other logicians of the nineteenth century. If you consider ternary relations, quaternary relations, and so on,

6   In his Buffalo lecture Tarski indicated that the present remarks apply to 'notions' taken in the narrow sense of sets, classes of sets, etc. but that the truth-functions, quantifiers, relation-operators, etc. of *Principia mathematica* can be construed as notions in the narrow sense and, so construed, the present remarks apply equally to them. For example, construing the truth-values T and F as the universe of discourse and the null set leads immediately to construing truth-functions as (higher-order) notions. Construals of this sort are familiar and natural to mathematicians, but they involve philosophical questions of the sort investigated by contemporary philosophers of logic.

7   In Tarski *1935a*, 'The Wahrheitsbegriff', there is an extended discussion of semantical categories (which properly include the 'types' treated by Whitehead and Russell). On p.215 Tarski attributes the concept of semantical categories to Husserl.

the situation is similar: for each of these you will have a small finite number of logical relations.

The situation becomes a little more interesting if you go to the next level, and consider classes of classes. Instead of saying 'classes of classes' we can say 'properties of classes', and ask: What are the properties of classes which are logical? The answer is again simple, even though it is quite difficult to formulate in a precise way. It turns out that the only properties of classes (of individuals) which are logical are properties concerning the number of elements in these classes. That a class consists of three elements, or four elements ... that it is finite, or infinite—these are logical notions, and are essentially the only logical notions on this level.

This result seems to me rather interesting because in the nineteenth century there were discussions about whether our logic is the logic of extensions or the logic of intensions. It was said many times, especially by mathematical logicians, that our logic is really a logic of extensions.[8] This means that two notions cannot be logically distinguished if they have the same extension, even if their intensions are different. As it is usually put, we cannot logically distinguish properties from classes. Now in the light of our suggestion it turns out that our logic is even less than a logic of extension, it is a logic of number, of numerical relations. We cannot logically distinguish two classes from each other if each of them has exactly two individuals, because if you have two classes, each of which consists of two individuals, you can always find a transformation of the universe under which one of these classes is transformed into the other. Every logical property which belongs to one class of two individuals belongs to every class containing exactly two individuals.

If you turn to more complicated notions, for instance to relations between classes, then the variety of logical notions increases. Here for the first time you come across many important and interesting logical relations, well known to those who have studied the elements of logic. I mean such things as inclusion between classes, disjointness of two classes, overlapping of two classes, and many others; all these are examples of logical relations in the normal sense, and they are also logical in the sense of my suggestion. This gives you some idea of what logical notions are. I have restricted myself to four of the simplest types, and discussed examples of logical notions only within these types. To conclude this discussion, I would like to turn to a question which has probably already occurred to some of you as you listened to my remarks.

4.   The question is often asked whether mathematics is a part of logic. Here we are interested in only one aspect of this problem, whether mathematical notions are logical notions, and not, for example, in whether mathematical truths are logical truths, which is outside our domain of discussion. Since it is now well known that the whole of mathematics can be constructed within set theory,[9] or the theory of

---

8 See Whitehead and Russell *1910*, III (2).

9 Tarski is using the term 'set theory' here in a vague and general sense in which several distinct concrete theories may all qualify as set theory. In particular, the Whitehead–Russell theory of types and the (first-order) Zermelo–Fraenkel theory both qualify as set theory. It is to the point here to note that Tarski regarded the current variety of 'set theories' as only a small sample of what can usefully be developed in this field. In the Editor's introduction, 'set theory' is used in a narrower sense that contrasts with type theory.

classes, the problem reduces to the following one: Are set-theoretical notions logical notions or not? Again, since it is known that all usual set-theoretical notions can be defined in terms of one,[10] the notion of belonging, or the membership relation, the final form of our question is whether the membership relation is a logical one in the sense of my suggestion. The answer will seem disappointing. For we can develop set theory, the theory of the membership relation, in such a way that the answer to this question is affirmative, or we can proceed in such a way that the answer is negative.

So the answer is: 'As you wish'! You all know that as a result of the antinomies, basically Russell's Antimony, which appeared in set theory at the turn of the century, it was necessary to submit the foundations of set theory to a thorough investigation. One result of this investigation, which is by no means complete at this moment, is that two methods have been developed of constructing what can be saved from set theory after the crushing blow which it had suffered. One method is essentially the method of *Principia mathematica*, the method of Whitehead and Russell—the method of types. The second method is the method of people such as Zermelo, von Neumann, and Bernays—the first-order method. Now let us look to our question from the point of view of these two methods.[11]

Using the method of *Principia mathematica*, set theory is simply a part of logic. The method can be roughly described in the following way: we have a fundamental universe of discourse, the universe of individuals, and then we construct out of this universe of individuals certain notions, classes, relations, classes of classes, classes of relations, and so on. However, only the basic universe, the universe of individuals, is fundamental. A transformation is defined on the universe of individuals, and this transformation induces transformations on classes of individuals, relations between individuals, and so on. More precisely, we consider the universal class of the lowest type, and a transformation has this universal class as its domain and range. Then this transformation induces also a transformation whose domain and range is the universal class of the second type, the class of classes of individuals. When we speak of transformations of the 'world' onto itself we mean only transformations of the basic universe of discourse, of the universe of individuals (which we may interpret as the universe of physical objects, although there is nothing in *Principia mathematica* which compels us to accept such an interpretation). Using this method, it is clear that the membership relation is certainly a logical notion. It occurs in several types, for

10 This remark presupposes the convention that a given notion can be said to be definable in terms of one fixed notion if there is a definition (of the given notion) which uses no notions other than the following: (i) the one fixed notion, (ii) the universe of discourse, (iii) other notions already accepted as being logical. It is obvious, e.g., that there is no way to define the null set using the membership relation and absolutely nothing else. It should also be noted that Tarski says 'all *usual* set-theoretic notions' and not 'all set-theoretic relations'; there are uncountably many of the latter but only countably many definitions.

11 Tarski takes the first method to involve a higher-order underlying logic and the second method to involve a first-order underlying logic. It is possible of course to reconstrue type theory in a many-sorted first-order underlying logic, but this would be incompatible with the spirit and letter of this lecture. Likewise, it is possible to develop Zermelo's set theory in a higher-order logic. This too is incompatible with the spirit of this lecture—despite the historical fact that Zermelo may have done so himself. Incidentally, the historic papers establishing the two methods were published in the same year, 1908.

individuals are elements of classes of individuals, classes of individuals are elements of classes of classes of individuals, and so on. And by the very definition of an induced transformation it is invariant under every transformation of the world onto itself.

On the other hand, consider the second method of constructing set theory, where we have no hierarchy of types, but only one universe of discourse and the membership relation between its individuals is an undefined relation, a primitive notion. Now it is clear that this membership relation is not a logical notion, because as I mentioned before, there are only four logical relations between individuals, the universal relation, the empty relation, and the identity and diversity relations. The membership relation, if individuals and sets are considered as belonging to the same universe of discourse, is none of these relations; therefore, under this second conception, mathematical notions are not logical notions.

This conclusion is interesting, it seems to me, because the two possible answers correspond to two different types of mind. A monistic conception of logic, set theory, and mathematics, where the whole of mathematics would be a part of logic, appeals, I think, to a fundamental tendency of modern philosophers. Mathematicians, on the other hand, would be disappointed to hear that mathematics, which they consider the highest discipline in the world, is a part of something so trivial as logic; and they therefore prefer a development of set theory in which set-theoretical notions are not logical notions. The suggestion which I have made does not, by itself, imply any answer to the question of whether mathematical notions are logical.

Editor's acknowledgements

I am grateful to Mr. Jonathan Piel, Editor of *Scientific American*, and to the Tarski family, especially Dr. Jan and Mrs. Maria Tarski, respectively Alfred Tarski's son and widow, for concurring with Alfred Tarski's permission to publish an edited form of this article. Professors Stephen Schanuel and Scott Williams of the SUNY/Buffalo Department of Mathematics gave advice on mathematical matters. Professor Ivor Grattan-Guinness deserves thanks for historical and editorial advice, as does Professor Michael Scanlan of Oregon State University, Ms. Rosemary Yeagle and Mr. Patrick Murphy, both of SUNY/Buffalo. Important refinements resulted from suggestions by Dr. Jan Tarski and by members of the Buffalo Logic Colloquium which devoted its September 1986 meeting to discussion of this paper. My largest debt is to Professor George Weaver, Chairman of the Philosophy Department of Bryn Mawr College, for substantial help in every phase of this work.

Editor's bibliography

Corcoran, J. *1983* 'Editor's introduction to the revised edition', in Tarski *1983*, xvi–xxvii.
Keyser, C. J. *1922 Mathematical philosophy*, New York.
Klein, F. *1872* 'A comparative review of recent researches in geometry', English trans. by Haskell, M. W., *Bulletin of the New York Mathematical Society*, 2 (1892–93), 215–249.
Lindenbaum, A. and Tarski, A. *1936* 'On the limitations of the means of expression of deductive theories', in Tarski *1983*, 384–392.
Mautner, F. I. *1946* 'An extension of Klein's Erlanger Programm: logic as an invariant theory', *Amer. j. maths.*, **68** (1946), 345–384.

154 *What are Logical Notions?*

Silva, J. S. *1945* 'On automorphisms of arbitrary mathematical systems', English trans. by de Oliveira, A. J. F. *History and philosophy of logic*, **6** (1985), 91–116.

Tarski, A. *1935a* 'The concept of truth in formalized languages', in Tarski *1983*, 152–278.

—— *1935b* 'Some methodological investigations on the definability of concepts', in Tarski *1983*, 296–319.

—— *1936* 'On the concept of logical consequence', in Tarski *1983*, 409–420.

—— *1969* 'Truth and proof', *Scientific American* **220**, no. 6, 63–77.

—— *1983* Logic, semantics, metamathematics, 2nd ed. (ed. Corcoran, J.), Indianapolis. [1st ed. (ed. and trans. Woodger, J. H.), Oxford, 1956.]

Tarski, A. and Givant, S. *1987 A formalization of set theory without variables*. Providence (Rhode Island)

Weyl, H. *1949 Philosophy of mathematics and natural science*, Princeton.

Whitehead, A. and Russell, B. *1910 Principia mathematica*, vol. 1, Cambridge.

[6]

GEORGE BOOLOS

A CURIOUS INFERENCE*

The inference is:

$$I$$

(1) $\forall n \; fn1 \;=\; s1$

(2) $\forall x \; f1sx \;=\; ssf1x$

(3) $\forall n \forall x \; fsnsx \;=\; fnfsnx$

(4) $D1$

(5) $\forall x \; (Dx \rightarrow Dsx)$

∴

(6) $Dfsssss1ssss1$

I is an inference in the first-order predicate calculus with identity and function signs. (s is a 1-place, f a 2-place function sign.) I is small: it contains 60 symbols or so, fairly evenly distributed among its five premisses and conclusion. And I is logically valid; the Frege–Russell definition of natural number enables us to see that there is a derivation of (6) from (1)–(5) in any standard axiomatic formulation of second-order logic, e.g. the one given in Chapter 5 of Church's *Introduction to Mathematical Logic.*[1] A sketch of a second-order derivation of (6) from (1)–(5) is given in the appendix, and it should be evident from the sketch that there is a derivation of (6) from (1)–(5) in any standard axiomatic system of second-order logic *whose every symbol can easily be written down.*

But it is well beyond the bounds of physical possibility that any actual or conceivable creature or device should ever write down all the symbols of a complete derivation in a standard system of *first-order* logic of (6) from (1)–(5): there are far too many symbols in any such derivation for this to be possible. Of course in every standard

Journal of Philosophical Logic **16** (1987) 1–12.

2   GEORGE BOOLOS

system of first-order logic there is (in the sense in which "there is" is used in and out of mathematics) a derivation of (6) from $(1)-(5)$, for every standard system of first-order logic is complete. But as we shall see, no such derivation could possibly be written down in full detail, in *this* universe.

For definiteness, we shall concentrate our attention on the system M of Mates' book *Elementary Logic*.[2] It is because Mates' book is a standard text and its system M is a perfectly standard system of natural deduction that we have chosen to focus on it. (The rules of M are: premiss introduction, conditionalization, truth-functional consequence, universal instantiation, universal generalization, the usual identity rules, and existential quantification $(-\forall\alpha - \phi/\exists\alpha\phi)$.) A result similar to the one we shall obtain for M can be gotten for any other standard formulation of first-order logic, e.g. the axiomatic system of first-order logic contained in Monk's or Shoenfield's *Mathematical Logic*,[3] or any of the systems found in Quine's *Methods of Logic*.[4]

What we shall show is that the number of symbols in any derivation of (6) from $(1)-(5)$ in M is at least the value of an exponential stack

$$
2^{2^{\cdot^{\cdot^{\cdot}}{2^{(2^{(2^{\cdot^{\cdot^{\cdot}}{(2^{2)}\cdots))}}}}}}}}
$$

i.e.

$$
2^{2^{\cdot^{\cdot^{\cdot}}{2^{(2^{(2^{\cdot^{\cdot^{\cdot}}{(.}}}}}}}
$$

containing 64 K, or 65 536, "2"s in all. Do not confuse this number, which we shall call $f(4, 4)$, with the number $2^{64\,\mathrm{K}}$. The latter number is minuscule in comparison, not even containing as many as 20 000 (decimal) digits. (It is the value of a stack containing only 5 "2"s.) The so-called Skewes' number, which is of interest in prime number theory and has been described as "the largest number found in science", is

$$
10^{10^{10^{34}}}
$$

A CURIOUS INFERENCE   3

Skewes' number is readily seen to be less than the value of a stack of 7 "2"s. It is not hard to show that if "#" denotes Skewes' number, then for some $N < 10$, $f(4, 4) >$ the value of a stack of $64 K - N$ "#"s. A proof that any derivation in M of (6) from (1)–(5) contains at least $f(4, 4)$ symbols is also given in the appendix.

In the intended interpretation of I, the variables range over the positive integers, 1 denotes one and s denotes the successor function. There is no particular interpretation intended for D. f denotes an Ackermann-style function $n, x \mapsto f(n, x)$ defined on the positive integers: $f(1, x) = 2x$; $f(n, 1) = 2$; and $f(n + 1, x + 1) = f(n, f(n + 1, x))$. Here are some of the early values of f:

x:	1	2	3	4	5	6	
n:							
1	2	4	6	8	10	12	$2x$
2	2	4	8	16	32	64	2^x

$$
\begin{array}{llllllll}
 & & & & & & 2 & \\
 & & & & 2 & 2 & 2 & \\
 & & & 2 & 2 & 2 & 2 & \\
 & & 2 & 2 & 2 & 2 & 2 & \text{the value} \\
 & 2 & 2 & 2 & 2 & 2 & 2 & \text{of a stack} \\
3 & 2 & 4 = 2 & 16 = 2 & 64\,K = 2 & 2 & 2 & \text{of } x \text{ "2"s} \\
\end{array}
$$

$$
\begin{array}{lllll}
 & & 2 & & \\
 & & 2 & & 2 \\
 & & 2 & & \cdot\quad 64\,K \\
4 & 2 \quad 4 & 64\,K = 2 & & 2 \quad \text{"2"s} \\
 & & 2 & & 2 \quad \text{in all} \\
\end{array}
$$

$$
\begin{array}{lllll}
 & & & \cdot\quad 64\,K & \\
 & & 2 & \text{"2"s} & \\
5 & 2 \quad 4 & 2 & \text{in all} \quad \cdot & f(5, 5) \quad \cdot \\
\end{array}
$$

So $f(n, 2) = 4$ (all n); $f(2, x) = 2^x$; $f(3, x) =$ the value of an exponential stack containing x 2s; $f(4, 3) = f(3, 4) = 64 K$; $f(4, 4) =$ the value of a stack containing $64 K$ 2s.

4   GEORGE BOOLOS

Thus by pursuing the obvious strategy of appending a definition of
a well-known sort of fast-growing function to a formalization of the
premisses of the paradox of heap and employing the function to construct a short conclusion to the paradox mentioning a very large
number, we obtain an inference which we can see to be valid by
means of a simple argument that cannot be replicated in any standard
system of first-order logic. I assume that (it is evident that) no formal
derivation containing at least $f(4, 4)$ symbols can count as *replicating*
this argument, or indeed any argument that *we* can comprehend.
Indeed, a shorter and even more extravagant conclusion than (6)
follows from $(1)-(5)$: $Dffs1ss1ss1; f(8, 3) \gg f(5, 5)$. One might
wonder whether there is any valid inference interestingly simpler than
I whose shortest derivation in some standard system of first-order
logic is significantly greater.[5]

In brief, I is a simple and natural example of a valid first-order
inference the conclusion of which cannot feasibly be derived from
the premisses in any standard system of first-order logic; but there
is a short and simple argument that demonstrates the validity of I,
which can be formalized in any standard system of second-order
logic.

Of course, it has been known since Gödel's "On the length of
proofs"[6] that the use of higher types can drastically reduce the minimum length of derivations in formal systems. In that paper, it will be
recalled, Gödel states (without proof) that for any recursive function
ϕ and any i, there are infinitely many arithmetical theorems F of both
i^{th} and $(i + 1)^{st}$ order logic[7] such that if k and l are the lengths of the
shortest proofs of F in i^{th} and $(i + 1)^{st}$ order logic, respectively, then
$k > \phi(l)$. (The length of a proof is the number of formulae of which
it consists.) And it was shown by Statman[8] that there is no function ϕ
provably recursive in second-order arithmetic such that whenever a
first-order formula F is derivable in a certain standard system of
second-order logic with length $\leqslant l$, then F is derivable in a certain
standard system of first-order logic with length $\leqslant \phi(l)$. Noteworthy
investigations of speedup have recently been carried out by Harvey
Friedman. One of Friedman's theorems is that a certain "finitization"
of a combinatorial theorem due to J. Kruskal concerning embeddings
of trees can be proved in ZFC in a few pages, but not in the system

A CURIOUS INFERENCE   5

of second-order arithmetic called ATR (for Arithmetical Transfinite Recursion) in under $f(3,1000)$ pages.[9]

But our aim is neither to prove a general speedup theorem nor to demonstrate the "practical incompleteness" of first-order logic; rather we are interested in showing that this incompleteness can be demonstrated by means of an inference like I that is remarkably elementary.[10] Indeed, I arises in quite a natural way: the first three premises of I can be taken as defining a kind of function very well known to logicians and computer scientists; the last two premises and conclusion can be used to formalize an ancient and completely familiar logical paradox involving large numbers. Without exaggeration, it may be said that I or a close relative might well be the first inference one would think of if one were *trying* to show first-order logic practically incomplete.

Since Skolem's discovery of non-standard models of arithmetic, it has been well known that there are simple and fundamental logical concepts, e.g., the *ancestral*, that cannot be expressed in the notation of first-order logic. It is also well known that there are notions of a logical character expressible in natural language that cannot be expressed in first-order notation. And it is increasingly well understood that it is neither necessary nor always possible to interpret second-order formalisms as applied first-order set theories in disguise. Thus although the existence of a simple first-order inference whose validity can be feasibly demonstrated in second- but not first-order logic cannot by itself be regarded as an overwhelming consideration for the view that first-order logic ought never to have been accorded canonical status as *Logic*, it is certainly one further consideration of some strength for this view.

On the other hand, the fact that we so readily recognize the validity of I would seem to provide as strong a proof as could be asked for that no standard first-order logical system can be taken to be a satisfactory idealization of the psychological mechanisms or processes, whatever they might be, whereby we recognize (first-order!) logical consequences. "Cognitive scientists" ought to be suspicious of the view that logic as it appears in logic texts adequately represents the whole of the science of valid inference.

It may be remarked in passing that the second-order derivation of

6   GEORGE BOOLOS

(6) from (1)−(5) given in the appendix has a certain foundational interest. If we interpret the variables in I as ranging over a set (species) containing the positive integers and possibly other objects, 1 as denoting one, s as denoting a one-place function whose restriction to the positive integers is the usual successor function, f as denoting an (unspecified) 2-place function, and D as denoting the set N of positive integers, then (4) and (5) are true. Consider the following argument, which shows that (6) is true (relative to the choice of the domain and the denotations of s and f) if (1)−(3) are.

We first show by induction on n that for every n in N, for every x in N fnx is in N. By (1), $f11 = s1 \in N$. Suppose that $x \in N$ and that $f1x \in N$. Then by (2), $f1sx = ssf1x \in N$. By induction on x, for every x in N $f1x \in N$. Now suppose that $n \in N$ and that for every x in N $fnx \in N$. By (1), $fsn1 = s1 \in N$. Suppose that $x \in N$ and $fsnx \in N$. By the i.h., $fnfsnx \in N$. By (3), $fsnsx = fnfsnx \in N$. By induction on x for every x in N, $fsnx \in N$. By induction on n, for every n in N, for every x in N $fnx \in N$. Since $5 \in N$, $f55 \in N$, and (6) is true.

This argument, a simple modification of the derivation of (6) from (1)−(5) given below, is evidently intuitionistically acceptable. But because of the presence of the unbounded universal quantifier "for every x in N" in the induction hypothesis, it cannot be regarded as finitistically acceptable.[11] Thus the notions of intuitionist and finitist acceptability may readily be seen to diverge.

The details of the proof that any derivation of (6) from (1)−(5) in M must contain at least $f(4, 4)$ symbols are tedious, but an outline of the reasoning is easily given: We translate derivations in M into derivations in a modification of S of the system of Schwichtenberg's article in the *Handbook of Mathematical Logic*,[12] a system for which the proof of a cut-elimination theorem is readily available.[13] Any *cut-free* derivation of (6) from (1)−(5) must contain roughly $f(5, 5)$ symbols, for in a cut-free system one has to take an instance of premiss (5) for every integer between 1 and $f(5, 5)$ to derive the conclusion. Because of the presence of the unanalyzed rule T (tautological inference) in M, translating a derivation in M into one in S may result in an exponential increase in the length of the derivation;

and eliminating cuts from a derivation in S may increase its length super-exponentially, of the order of the value of a stack of "2"s; but such increases are as nothing when compared with the difference between $f(4, 4)$ and $f(5, 5)$. More of the details of the proof are contained in the following appendix.

APPENDIX

We first present a sketch of a second-order derivation of (6) from $(1)-(5)$ of which a complete formalization in any standard axiomatic formulation of second-order logic can easily be written out:

By the comprehension principle of second-order logic,
$\exists N \forall z(Nz \leftrightarrow \forall X[X1 \;\&\; \forall y(Xy \rightarrow Xsy) \rightarrow Xz])$, and then for some N,
$\exists E \forall z(Ez \leftrightarrow Nz \;\&\; Dz)$.

LEMMA 1: $N1$; $\forall y(Ny \rightarrow Nsy)$; $Nssss1$; $E1$; $\forall y(Ey \rightarrow Esy)$; $Es1$.

LEMMA 2: $\forall n(Nn \rightarrow \forall x(Nx \rightarrow Efnx))$.

Proof. By comprehension, $\exists M \forall n(Mn \leftrightarrow \forall x(Nx \rightarrow Efnx))$. We want $\forall n(Nn \rightarrow Mn)$. Enough to show $M1$ and $\forall n(Mn \rightarrow Msn)$, for then if Nn, Mn.

$M1$: Want $\forall x(Nx \rightarrow Ef1x)$. By comprehension, $\exists Q \forall x(Qx \leftrightarrow Ef1x)$. Want $\forall x(Nx \rightarrow Qx)$. Enough to show $Q1$ and $\forall x(Qx \rightarrow Qsx)$.

$Q1$: Want $Ef11$. But $f11 = s1$ by (1) and $Es1$ by Lemma 1.

$\forall x(Qx \rightarrow Qsx)$: Suppose Qx, i.e. $Ef1x$. By (2) $f1sx = ssf1x$; by Lemma 1 twice, $Ef1sx$. Thus Qsx and $M1$.

$\forall n(Mn \rightarrow Msn)$: Suppose Mn, i.e. $\forall x(Nx \rightarrow Efnx)$. Want Msn, i.e. $\forall x(Nx \rightarrow Efsnx)$. By comprehension, $\exists P \forall x(Px \leftrightarrow Efsnx)$ Want $\forall x(Nx \rightarrow Px)$. Enough to show $P1$ and $\forall x(Px \rightarrow Psx)$.

$P1$: Want $Efsn1$. But $fsn1 = s1$ by (1) and $Es1$ by Lemma 1.

$\forall x(Px \rightarrow Psx)$: Suppose Px, i.e. $Efsnx$; thus $Nfsnx$. Want $Efsnsx$. Since $Nfsnx$ and Mn, $Efnfsnx$. But by (3) $fnfsnx = fsnsx$; thus $Efsnsx$. By Lemma 1, $Nssss1$. By Lemma 2, $Efssss1ssss1$. Thus $Dfssss1ssss1$, as desired.

We now show that any derivation in M of (6) from $(1)-(5)$ contains at least $f(4, 4)$ symbols. We assume that the reader is familiar

8   GEORGE BOOLOS

with Section 2 of Schwichtenberg's *Handbook* article.[14] Following
Shoenfield, call a formula a quasi-tautology if it is a tautological
consequence of instances of the identity and equality axioms. Call a
finite set Δ of atomic formulae and negations thereof an *axiom set* if
the disjunction of its members is a quasi-tautology. Revise Schwich-
tenberg's rule A to read: Γ, Δ if Δ is an axiom set, and take the p.f.
in A to be the members of Δ. It is routine to verify that the cut-
elimination theorem and the other results of Section 2 of Schwichtenberg's article hold for the system thus revised. (For Case 2.1 of the
reduction lemma, note that if Δ_0, ϕ and Δ_1, $-\phi$ are axiom sets, so is
Δ_0, Δ_1.)

Define a translation T from formulae of M to formulae of
S(chwichtenberg's system). We suppose $\rightarrow$, $\leftrightarrow$, etc. to be defined from
$-$, &, and $\vee$ in any one of the usual ways.

If ψ is atomic, $T\psi = \psi$ and $T - \psi = -\psi$;

if ψ is (ψ_0 & ψ_1), $T\psi = (T\psi_0$ & $T\psi_1)$ and
  $T - \psi = (T - \psi_0 \vee T - \psi_1)$;

if ψ is ($\psi_0 \vee \psi_1$), $T\psi = (T\psi_0 \vee T\psi_1)$ and
  $T - \psi = (T - \psi_0$ & $T - \psi_1)$;

if ψ is $\forall\alpha\psi'$, $T\psi = \forall\alpha T\psi'$ and $T - \psi = \exists\alpha T - \psi'$, and

if ψ is $\exists\alpha\psi'$, $T\psi = \exists\alpha T\psi'$ and $T - \psi = \forall\alpha T - \psi'$.

As in Schwichtenberg, $|\psi|$ is the maximum number of nested logical
symbols in ψ other than $-$ or $=$. Note that $|T\psi| = |\psi|$.

The width of a derivation d in S is max $\{|\phi|: \phi$ is in some set occur-
ring in $d\}$, the width of a derivation d in M is max $\{|\psi|: \psi$ is on some
line of $d\}$, and the length of a derivation in M is the number of lines
it contains.

We want to show that if there is a derivation in M of ψ from
$\psi_1, \ldots, \psi_s$, of length $\leqslant h$ and width $\leqslant w$, then there is a derivation
in S of $\{-T\psi_1, \ldots, -T\psi_s, T\psi\}$ of width $\leqslant w$ and length $\leqslant 3wh + 2^w h(h + 1)/2 + h(h - 1)/2$. Write ϕ for $T\psi$, ϕ_0 for $T\psi_0$, etc.

Now let d be a derivation in M (possibly the null derivation, which
has 0 lines) of width w. Let B_i be the set of formulae occurring on the
lines mentioned in the set of line numbers of line i, let $\Gamma_i = \{-\phi:$

A CURIOUS INFERENCE 9

$\psi \in B_i$} and let ψ_i be the formula occurring on line i. Suppose that for each $i < n$ there is a derivation in S of Γ_i, ϕ_i of length $\leqslant h$ and width $\leqslant w$.

By means of a laborious case-by-case analysis of the seven rules of M, it can be shown that there is a derivation in S of Γ_n, ϕ_n of length $\leqslant h + (3w + n2^w + (n - 1))$ and width $\leqslant w$. It follows that if there is a derivation in M of ψ from $\psi_1, \ldots, \psi_s$, of length $\leqslant h$ and width $\leqslant w$, then there is a derivation in S of $\{- T\psi_1, \ldots, - T\psi_s, T\psi\}$ of width $\leqslant w$ and length $\leqslant$

$$\sum_{n=1}^{h} (3w + n2^w + (n - 1)) = 3wh + 2^w h(h + 1)/2$$

$$+ h(h - 1)/2.$$

We now look at certain cut-free derivations in S.

Let

(1′) be $\exists n \, fn1 \neq s1$,

(2′) be $\exists x \, f1sx \neq ssf1x$,

(3′) be $\exists n \exists x \, fsnsx \neq fnfsnx$,

(4′) be $- D1$, and

(5′) be $\exists x(D1x \, \& \, - D1sx)$.

Note that $(1)′ = - T(1), \ldots, (5)′ = - T(5)$, and $(6) = T(6)$.

Let $\Gamma = \{(1′), \ldots, (5′), (6)\}$. We show that any cut-free derivation of Γ contains at least $f(5, 5) - 1$ sentences of the form $(Dt \, \& \, - Dst)$. Suppose d is a counterexample. We shall obtain a contradiction.

Assign to 1, s, and f their standard denotations. Assign to any other function symbol in d a function of the appropriate number of places. Write 'den(t)' for the denotation of the term t under this assignment. Since d contains fewer than $f(5, 5) - 1$ sentences of the form $(Dt \, \& \, - Dst)$, for some positive integer $i < f(5, 5)$ there is no term t such that den(t) $= i$ and $(Dt \, \& \, - Dst)$ occurs in d. (If den(t) $= k \neq j = $ den(u), then $t \neq u$ and $(Dt \, \& \, - Dst) \neq (Du \, \& \, - Dsu)$.)

Define a model M: The domain of M is the set of positive integers. M assigns to 1, s, and f their standard denotations. Thus the value of the term t under M is den(t). M specifies that D is true of x iff $x \leqslant i$. It is evident that (1′), (2′), (3′), (4′), and (6) are all false in M.

10   GEORGE BOOLOS

And although (5′) is true in M, every instance $(Dt \& -Dst)$ of (5′) *that occurs in d* is false in M. For suppose $(Dt \& -Dst)$ is true in M. Let $j = \text{den}(t)$. Then Dt is true, $j \leqslant i$, $\text{den}(st) = j + 1$, $-Dst$ is true, and $j + 1 > i$; thus $j = i$, and $(Dt \& -Dst)$ does not occur in d.

Define a path p through d: The bottom set (of formulae) on p is Γ. Suppose that a set B is on p. If (an occurrence of) B is inferred by the rule $\exists$, then the set of premisses of B is also on p. Note that the m.f. in any application of $\exists$ is false in M. If B is inferred by the rule $\&$, then the p.f. of the inference will be a conjunction $(\phi \& \psi)$. If $(\phi \& \psi)$ is false in M, put the premiss set on p whose m.f. is ϕ if ϕ is false in M; otherwise put the other premiss set on p. Thus all sentences other than (5′) in any set on p are false in M. Since the set at the top of a path always includes a set of quantifier-free formulae whose disjunction is a quasi-tautology, we have the desired contradiction.

No sentence of the form $(Dt \& -Dst)$ belongs to Γ. Each such sentence occurring in any cut-free derivation of Γ is thus the (unique) m.f. of an application of rule whose p.f. is (5′). Thus any cut-free derivation of Γ contains at least $f(5, 5)$ sets of formulae.

Since the rules of inference of S are binary, any cut-free derivation of Γ has length $\geqslant \log_2(f(5, 5))$.

We can now show that any derivation in M of (6) from $(1)-(5)$ must be large. Define $r, h \mapsto 2(r, h)$ by: $2(0, h) = h$; $2(r + 1, h) = 2^{2(r,h)}$. According to Corollary 2.7.1 of Schwichtenberg, if there is a derivation in S of cut-rank r and length h, then there is a cut-free derivation with the same conclusion of length $2(r, h)$. If $h > 0$, then $2(r, h) \leqslant f(3, r + h)$.

Let d be a derivation in M of (6) from $(1)-(5)$, and let $h = \text{length}(d)$ and $w = \text{width}(d)$. $h > 0$. We want to see that $h \geqslant f(4, 4)$ or $w \geqslant f(4, 4)$.

We have seen that there is a derivation in S of Γ of width $\leqslant w$ and length $\leqslant 3wh + 2^w h(h + 1)/2 + h(h - 1)/2$. Let $p = \max(h, w)$. Since the cut-rank of a derivation in S is $\leqslant$ its width, there is a cut-free derivation of Γ of length $\leqslant f(3, (p + 3p^2 + 2^p p(p + 1)/2 + (p - 1)/2))$. The value of this expression is $\leqslant f(3, 2^{4p}) = f(3, f(2, f(1, f(1, p))))$.

Assume (for *reductio*) that $p < f(4, 4)$. Since any cut-free derivation in S of Γ has length $\geqslant \log_2(f(5, 5))$,

A CURIOUS INFERENCE   11

$$f(3, f(2, f(1, f(1, f(4, 4)))))) \geq \log_2(f(5, 5)), \text{ whence}$$

$$f(2, f(3, f(2, f(1, f(1, f(4, 4))))))) \geq f(5, 5), \text{ and}$$
therefore

$$f(3, f(3, f(3, f(3, f(3, f(4, 4))))))) \geq f(5, 5).$$

But this is absurd. For $f(5, 5) = f(4, f(5, 4)) = f(3, f(4, f(5, 4) - 1)) = f(3, f(3, f(4, f(5, 4) - 2))) = \cdots = f(3, f(3, f(3, f(3, f(3, f(4, f(5, 4) - 5)))))),$ whence $4 \geq f(5, 4) - 5$, which is patently absurd.

Thus $p \geq f(4, 4)$, and therefore any derivation of (6) from (1)–(5) in Mates' system must have length or width $\geq f(4, 4)$, and therefore contain at least $f(4, 4)$ symbols.

NOTES

* I am grateful to Rohit Parikh, Scott Weinstein, and a referee for the JPL for helpful comments.
[1] Alonzo Church, *Introduction to Mathematical Logic, Volume I*, Princeton University Press, Princeton, N.J., 1958.
[2] Benson Mates, *Elementary Logic*, second edition, Oxford University Press, New York, 1972.
[3] J. Donald Monk, *Mathematical Logic*, Springer-Verlag, New York, 1976; Joseph R. Shoenfield, *Mathematical Logic*, Addison-Wesley Publishing Company, Reading, Massachusetts, 1967.
[4] W. V. Quine, *Methods of Logic*, fourth edn., Harvard University Press, Cambridge, Massachusetts, 1982.
[5] Of course, (1)–(5)/*Dffs1ss1ss1* is not simpler in an *interesting* way.
[6] Kurt Gödel, 'On the length of proofs', trans. Martin Davis, in Martin Davis (ed.), *The Undecidable*, Raven Press, Hewlett, New York, 1965.
[7] Comparison with Gödel's earlier papers on incompleteness makes it reasonable to suppose that the systems S_i considered in 'On the length of proofs' contain the Peano axioms for successor; "i^{th} order arithmetic" might thus be a more apt term for S_i than "i^{th} order logic".
[8] R. Statman, 'Bounds for proof-search and speed-up in the predicate calculus', *Annals of Mathematical Logic* 15, 1978, 225–287.
[9] Anil Nerode and Leo A. Harrington, 'The work of Harvey Friedman', *Notices of the American Mathematical Society* 31, 1984, 563–566.
[10] An analogy with miniature Universal Turing Machines was suggested to me by Rohit Parikh: miniature UTMs are of interest not in showing the halting problem unsolvable but in showing that unsolvability arises in *such* simple structures.
[11] W. W. Tait, 'Constructive reasoning', in B. van Rootselaar and J. F. Staal (eds.), *Logic, Methodology, and Philosophy of Science III*, North-Holland Publishing Company, Amsterdam, 1968 and 'Finitism', *The Journal of Philosophy* 78, 1981, 524–546.

12   GEORGE BOOLOS

Helmut Schwichtenberg, 'Proof theory: Some applications of cut-elimination',
in *Handbook of Mathematical Logic*, Jon Barwise (ed.), North-Holland Publishing
Company, Amsterdam, 1977, 867–895.
[13] The cut-elimination theorem for this system is due to W. W. Tait, 'Normal derivability in classical logic', in Jon Barwise (ed.), *The Syntax and Semantics of Infinitary
Languages. Lecture Notes in Mathematics* **72**, Springer-Verlag, Berlin, 1972, 204–236.
[14] *Op. cit.*

Department of Linguistics and Philosophy,
Massachusetts Institute of Technology,
77 Massachusetts Avenue,
Cambridge, MA 02139,
U.S.A.

[7]

Notre Dame Journal of Formal Logic
Volume 28, Number 1, January 1987

The Rationalist Conception of Logic

STEVEN J. WAGNER*

The failure of Frege's foundation for mathematics in [9] led to an endur-
ing tension in the philosophy of logic. If Frege had succeeded, almost everyone
would have granted his system the title of logic in a favored, primary sense.[1]
First-order logic (FOL) would, like sentential logic, have been considered an
interesting special case. Stronger systems might have been called logics in view
of their similarities to Frege's. But anything beyond what was needed for the
general formalization of mathematics would have borne the name logic by
courtesy — particularly if its principles were less evident than Frege's axioms I–V.
Unfortunately, however, Russell discovered that there were no stronger systems;
a generation later, Gödel showed that the truths of any RE logic could only
make up a tiny part of classical mathematics. Logicians after Frege have therefore had to consider a proliferation of systems sharing to various extents the
attractions of his paradigm. Since one may differ over which features, if any,
can serve to pick out a logic from among the many alternatives, the "scope of
logic" ([23], ch. 5) has remained in dispute. Broadly speaking, the disputants fall
into two camps, one emphasizing strength as a criterion for the title of logic,
the other conceptual simplicity. Stronger systems are more nearly adequate for
the job of founding mathematics, yet increasing strength yields less elementary,
transparent notions of logical validity and proof. Frege's system, of course, was
both elementary and strong, but that was too good to be true.

This paper deals with the problem of characterizing logic that we have
inherited from Frege. I will also consider a problem of Quine's. Verbally, it is
the same — 'what is logic?' — but Quine's motives and philosophical framework
are so different that the relation of his question to Frege's is not obvious. For
Quine, the determination of a speaker's ontology, which depends on a choice

*I thank Tim McCarthy for very valuable conversations. I also thank Michael Detlef-
sen for his advice and, particularly, for his encouragement.

Received November 20, 1985; revised December 13, 1985

4   STEVEN J. WAGNER

of logic, replaces the logicist project as a reason for caring about the scope of
logic, and his empiricism is deeply opposed to Frege's outlook. The connections
between Frege's and Quine's concerns must therefore be investigated. I will argue
that the two problems are in fact closely linked, and that Frege would endorse
Quine's identification of logic with FOL. But the argument itself will not be
Quine's. Indeed, it would appear suspect to him as well as many other philosophers. My aim is to defend, with qualifications, a version of the traditional view
of logic as an instrument for reasoning. FOL is best for that, although other
logics have other virtues. That view, however, is at home in a Fregean rationalism, not in Quine's empiricism and naturalism.[2] We should not be surprised
to find that taking a Fregean view of logic means siding with him against the
viewpoints that he most strenuously resisted; this may encourage us to reconsider these now prevailing tendencies.

1   Let me clarify some issues and review the present situation. Assume that we
are given a semantically interpreted language L. To select a logic for L is to fix
a relation R, holding among sets of sentences of L, that is to be understood as
logical consequence. (Of course our real interests are in consequence relations
in classes of languages. I follow the textbook convention of leaving this generality implicit.) What conditions must R meet to justify this understanding? Presumably, a relation associating even-numbered sentences with odd-numbered
ones under an arbitrary coding will not do. Useful restrictions, however, are hard
to find without making major decisions about the nature of logic.

We may begin with the abstract conditions of Gentzen's [10]:

(1)   reflexivity: $R(\Sigma, \Sigma)$, where Σ is any set of sentences in L.
(2)   dilution: if $R(\Sigma, \Sigma')$, then $R(\Sigma \cup \Sigma'', \Sigma')$ and $R(\Sigma, \Sigma' \cup \Sigma'')^3$
(3)   transitivity: for any sentence S, if $R(\Sigma, S)$ and $R(S, \Sigma')$, then
      $R(\Sigma, \Sigma')$.

Although these are trivial for FOL and many other logics, only the first holds
for everything that might reasonably be called logic: without question, everything follows logically from itself. (2) is already stronger: dilution rules out
inductive logics, because adding premises can weaken inductive arguments. Since
we are not assuming at the start that any logic must be deductive, dilution is not
immediately acceptable. Transitivity, too, can be challenged. One plausible condition on logical inference might be preserving justification: roughly, if
$R(\Sigma, \Sigma')$, and if belief in Σ is justified to degree d, then belief in Σ' is justified
to the same degree. (This cannot be right for inductive logics, but transitivity
would have to be modified for the inductive case anyway.) The problem here
is that our fallibility may prevent even perfect, deductive justification from trans-
mitting down chains of inferences. Suppose there is a series of justification-
preserving steps from S_1 to S_2, S_2 to S_3, S_3 to...to S_n. Assuming that R
preserves justification, it is questionable whether $R(S_1, S_n)$ holds if the chain is
too long and complex, the relation of S_1 to S_n too subtle, for anyone to grasp.
One might, for example, think that any justified step from one statement to a
second must be an instance of a generally reliable pattern of reasoning; and we
might be as likely to go wrong as right in cases of this type. It would follow that

RATIONALIST CONCEPTION OF LOGIC   5

S_1 could be justly believed in the absence of good reasons to believe S_n, thus that the latter is no logical consequence of the former. Of course this can be disputed in various ways, but the point is that if our notion of logic is suitably general, transitivity cannot simply be taken for granted.

Because we do not want to prejudge the issue against inductive logic, we can also not immediately require R to preserve truth. Induction can lead from truths to falsehoods. On the other hand, the close connection between logical consequence and justification, which underlies the idea of an inductive logic, is disputable even if one overlooks the problem about transitivity. Second-order consequence is often considered to be logical, but correct second-order steps may be unjustified. The continuum hypothesis (CH) provides a standard illustration. There is a formula CH^* true in any second-order model M iff M contains no set larger than the natural numbers and smaller than the real numbers ([25] indicates the construction). If CH is true, CH^* is a second-order validity; analogously for $\sim CH^*$ if CH is false. $R(S, C)$ therefore holds in second-order logic (SOL) for any S and the valid member C of the pair $(CH^*, \sim CH^*)$. But S may have no epistemic relevance to C. In fact, there may be no way at all to come to hold justified belief in C—if this seems implausible for C, even more mysterious propositions of set theory will generate similar examples. So second-order consequence is quite divorced from epistemic relations.

Another approach would start with the claims that every logical validity—every logical consequence of any sentence whatever—is true, in fact necessary, and that logical consequences are necessary consequences. This by itself is consistent with taking R to be the null relation, but R becomes nontrivial if we can designate a significant body of truths as being logical. Logicism offers such a result by making the class of validities include all mathematical truths. Unfortunately, there are no obvious independent reasons to accept logicism. The most natural approach would be to introduce a notion of form, then to claim that logical truth and consequence can be defined in terms of form alone. But no *intuitive* notion of form seems to guarantee the truth of every sentence with the same form as a given mathematical truth. (Consider, e.g., CH or CH^*, or their negations.) So one can neither appeal to logicism to fix the extent of logic, nor readily derive logicism from any natural characterization of logic.

We see that very little concerning logic can be taken for granted. The result has, in recent years, been agnosticism and pluralism. Various writers have considered a variety of dimensions, formal as well as epistemic and pragmatic, along which logics may be distinguished. No consensus selecting a particular logic has emerged, and philosophers concurring on a choice of logic (most often on FOL) frequently disagree on the reasons. (For a representative sample of views, see [2], [5], [14], [15], [20], [25], [27].) Only relative to a particular context or set of interests is the title of logic seen as more than a conventional label. In a way, this assessment is impossible to dispute. The range of more or less acceptable, plausible accounts of logic proves that the concept is flexible, so that sheer intuition about what counts as logic cannot rule out a diversity of candidates. Yet I want to promote a single conception as far as this indeterminacy will allow.

Frege took over from Aristotle and Leibniz a view of logic as the basis of idealized cognition. Logic may not describe actual reasoning, but in setting out a logical language and its rules of inference, we are defining a system of thought

6   STEVEN J. WAGNER

for us to approximate as best we can. Ideally, thought (or its linguistic expression) is formulated in logical notation, and steps of reasoning follow the rules of logic. This traditional view was sometimes ambiguous between reasoning as an instrument of discovery and as an instrument of justification, hence unclear about the role of logic. Frege, however, is well-known for having distinguished sharply between discovery and justification, indeed between everyday justification and justification showing the "deepest grounds" for holding something true ([8], pp. 2-3). This was an historically important clarification, permitting a more definite classification of logic. For Frege, logic is part of the theory of ideal justification. It has no direct connection with the description of mental processes, nor with the problem of finding the truth. It is also not concerned with what should count as sufficient justification in practice. Instead logic provides the language and rules for the best justifications possible in abstraction from the limits of our intelligence. The project of [8] and [9] is to find the propositions to which we can reduce arithmetic and to give fully rigorous, explicit derivations. It is no objection if this cannot increase the certainty of arithmetic, or if humans can hardly maintain the chosen level of rigor. And it is just for this project that [7] provides the logic, that is, the standard of rigorous argument.

Thus, Frege's conception makes logic part of epistemology — of the theory of justification. What logic is depends on what counts as ideal justification. Notice that the move to *idealized* reasoning takes care of our problem about transitivity. If failures of memory and attention are not a factor, then transitivity is preserved, and we can regard logical proof as a way of obtaining conclusions that are exactly as credible as their premises. This, of course, clarifies the importance epistemologists have historically attached to logic.[4]

Many contemporary authors seem to share Frege's epistemic orientation, for epistemic considerations are often brought to bear on questions of choice of logic. Infinitary "logics," for example, are mathematically interesting, but it is widely claimed that they cannot serve as *logics* for finite beings, who cannot in general write infinite formulas or reason with infinite arguments. Moreover, this inability remains under all reasonable idealizations. But the widespread interest in SOL and other systems with nonelementary consequence relations limits our agreement with Frege, as does the specific character of his epistemic concerns. Since Frege does not seem to be after justifications in an ordinary sense, we cannot assume that a logic appropriate to his epistemic goals would suit ours. This will be discussed further below. But we can provisionally connect Frege's project with rationalist epistemology. Characteristic of rationalism are its interests in perfect rigor and in setting out our knowledge in an order reflecting not the demands of everyday exposition and argument, but rather some kind of ideal systemization. Descartes, Spinoza, and other classical rationalists pursued these aims in abstraction from practical considerations; the same seems to hold for Frege. (In Frege's case, the project is limited to systemizing mathematical knowledge, but I think he could have accepted its extension to natural science.) So I suggest that we regard Frege's conception of logic as being determined by the needs and goals of rationalistic system-building; I will call it a rationalist conception of logic (RCL). Of course, this raises a number of questions about the nature of Frege's project and about the precise connection between one's epistemology and one's logic. Although I cannot fully investigate these issues here,

RATIONALIST CONCEPTION OF LOGIC   7

I hope that the rest of this paper may somewhat clarify them. Perhaps the preceding characterization of Frege will suffice for the moment, so that I may sketch the course of my discussion.

I will first explain why Frege's RCL encourages an identification of logic with FOL. Basically, the reasons consist in the familiar epistemic arguments against nonelementary logics, but we must address several difficulties before the identification can be confirmed. Here I should note that FOL will be defended only against a restricted class of alternatives. My main concern is with its relation to stronger competitors, e.g., SOL, not with whether (classical) FOL is already too much. Arguments to that effect, e.g., Dummett's case for intuitionistic logic, involve epistemic and semantical assumptions that cannot be dealt with here; we will instead take the usual view that FOL is transparently in order. I will also not deal with modal systems, such as alethic modal logics and tense logics. Although I think my viewpoint discourages their elevation to the status of genuine logics, I will leave this implicit. In any case, the question whether any modal logic is really logic may well contain enough of a conventional element to reduce its importance. Finally, I consider only deductive logics.[5] This is a more serious omission, for views of logic as an instrument for justification do not merely admit a generalization to the inductive case, they positively invite it. If logic defines the language and rules of argument, one would naturally suppose that there should also be nondeductive logics. But their possibility lies outside the scope of this discussion. The proposed identification of logic with FOL should therefore be viewed as holding for a special case. Ideally, we should want our conception of logic to cover appropriate inductive logics as well.

After developing a Fregean view of logic, I want to defend it in two ways. First, I will try to eliminate some competitors: attempts to identify logic with FOL for non-Fregean reasons (Quine) or to give the title to SOL or something similar (various authors). I think each of these can be shown to be untenable on internal grounds, without begging questions about the nature of logic. This appears significant, for although I have conceded that the nature of logic may be open to interpretation, the failure of rival interpretations suggests that our concept is fairly determinate after all. Second, I will examine the presuppositions of the RCL more carefully. Trying to characterize logic is pointless unless one has a viewpoint from which it matters what logic is. Of course, Frege's logicism provides one, but if we reject it, it not obvious that he could offer any deeper reasons for caring about the nature of logic. Here is where the viability of a form of rationalism becomes critical. I will defend Frege on this point and thus maintain that we have good reason to regard FOL as logic on broadly Fregean grounds—even if other viewpoints remain possible. This will amount to a substantial vindication of Frege.

2   I think that for Frege, logic must turn out to be FOL. One should not dismiss this proposal because FOL is too weak to validate logicism; or brand it as trivial because the inconsistency of Frege's views about logic (as well as in it) might lead to an identification of logic with anything one pleases. For clearly, the truth of logicism was not built into Frege's notion of logic, but was rather to be confirmed by his constructions. That is why Russell's paradox led him to

8   STEVEN J. WAGNER

abandon logicism and to seek, late in his life, geometrical foundations of mathematics. A prior conception of logic had been shown to be inadequate to the foundational task.

The main features of this conception are clear. Frege holds that logical knowledge is a priori and analytic, where the latter means at least that it is not based on any kind of spatial intuition. Further, as van Heijenoort has stressed [16], he views the *Begriffsschrift* as providing at once a *calculus ratiocinator* and a *lingua characterica*: a deductive calculus and a universal language for the expression of thought. These features are evidently tied to the rationalist programme of a comprehensive, rigorous development of our system of beliefs. We want a logic that can express all elements of this system and clarify the logical connections among them.[6]

The identification of logic with FOL then suggests itself naturally. Such fragments as propositional and monadic logic are ruled out by their expressive, representational weakness. Language exists for the expression of thought, and Fregean thoughts have no less structure than formulas of FOL. Any use of one of these fragments as a language would therefore be parasitic on the use of a full first-order language. If we can express an arbitrarily complex thought, say, that high interest rates help large corporations drive out smaller ones, using just the sentence symbol '*p*', that is only because of our ability to translate from a richer background language into the sentential notation. The latter could not stand on its own. And in any case, the sentential logic prevents us from formalizing inferences universally agreed to be logical. So logic is at least FOL. Now a salient property of FOL, the existence of a complete set of rules, is implicit in the idea of a deductive calculus and the apriority condition. Although a rigorous grasp of mechanical calculation was made possible only by Turing and Post, the notion was tolerably clear to Leibniz, whom Frege follows in his concept of formal inference. Any deductive consequence C of a set Σ of statements can, on Frege's conception of logic, be mechanically calculated: a finite series of steps leads from premises in Σ to C, with each step governed by a rule the applicability of which can be recursively determined. Underlying this condition is the traditional assumption of the apodictic character of a priori knowledge. If it is a priori whether something is a proof, then that question can be settled with certainty, not merely established with some degree of probability; thus calculability is properly understood as nothing less than recursiveness. But if it is mechanically determinable whether a given argument is in fact a proof of logic, there must also be a recursive characterization of the entire set of logical rules. This suffices for the recursive enumerability of the consequence relation, that is, for completeness. Thus it is natural to identify logic with FOL, since completeness is typically lost when we go beyond it.

These observations are commonplace (cf. [3], ch 0). It is also a commonplace that they do not force the identification of logic with FOL. One large problem is that the link between logic and proof, which underlies the completeness condition, can be denied in favor of a model-theoretic notion of consequence (see section 5). An immediate difficulty, however, is that numerous complete logics essentially extend FOL, for example, Keisler's logic with the quantifier (Ux) ("there exist uncountably many x"), or the logics with so-called equicardinality or cofinality quantifiers. It seems unintuitive to call these genuine logics—

one wants to say that they belong to set theory. Yet it would beg the question against various positions, including forms of logicism, to hold that set-theoretical principles can *ipso facto* not belong to logic, so it is unclear how to rule out these extensions. Tharp faces and, I think, does not solve these difficulties in [27]. The concept of cofinality, for example, is rejected simply for being "technical" and "unpalatable" (p. 13). But Tharp's reaction to the uncountability logic hints at a better reply:

> There is a serious objection against the quantifier "for uncountably many *x*."
> One would expect that if this were a legitimate quantifier, the quantifier "for
> infinitely many *x*" would also be acceptable; but no complete logic can con-
> tain the latter quantifier... [this] seems to me to be a state of affairs suffi-
> ciently unnatural to discredit (*Ux*) as a logical notion. (pp. 11–12)

This is unclear, but we can make sense of it along Fregean lines. To the previously mentioned features of logic we should add that it is *fundamental*: roughly, its principles and concepts cannot depend on nonlogical ones. Frege clearly intends such a condition. It may well be necessary to acquire empirical beliefs before grasping any logical notions clearly at all, or before coming to believe anything more than trivial truths of logic. Extralogical (e.g., geometrical) mathematical thought may also be needed. But in principle, logic is independent. The point concerns not the order of belief acquisition but the order of definition and justification. Logical beliefs need rest on no others, and logical concepts are primitive or defined from logical ones alone. This leaves it open how far up logic, the foundation, goes. An identification of logic with all of set theory would be consistent with the fundamentality criterion. But fundamentality rules out gaps that would need to be filled by nonlogical cognition, as in Tharp's examples. If we admit (*Ux*) as a quantifier, then it is "unnatural" to omit the analogous (*CTx*) — "for countably many *x*" — because one could not, it seems, formulate or understand the Keisler logic without using the concept of countability. Yet adding the capacity to express countability — or quantifiers (*Fx*) and (*Ix*) expressing finiteness and infinity — to the logic for (*Ux*) destroys completeness (see [27] for discussion.) Similar arguments apply to the complete logics for other generalized quantifiers. Consider (*Cx*), the Chang quantifier: (*Cx*)*Ax* is satisfied in a model just in case the set of things that satisfy *A*(*x*) has the same cardinality as the universe of the model. Evidently, understanding (*Cx*) presupposes the ability to grasp "there are as many *A*s as *B*s". Yet completeness again vanishes, as Tharp observes, when our logic is augmented to express this notion. The problem is even more obvious for the cofinality quantifiers, with their dependence on the theory of ordinals.

We are assuming that logic includes the laws governing logical concepts. If having a logic for (*Ux*) makes countability a logical concept (because of fundamentality), we will count as logical the principles governing it, which yields incompleteness. This is not quite trivial, but it falls naturally out of Frege's viewpoint. We must claim that grasping a logical concept (or any other) is inseparable from believing certain truths involving it. To understand 'and', for example, one must know the usual rules or axioms for conjunction. Similarly for (*Ux*) or (*CTx*). This seem reasonable in any case and certainly accords with Frege's views. Frege sees our grasp of concepts as being exhibited in our acceptance of principles; this is clearest in [8], where gaining a concept of number is a matter of finding laws that define one.[7] One must, then, understand fundamentality as entailing the right kind of closure. The axioms governing (*Ux*) do not rest on those for (*CTx*) in the sense of directly requiring them for their justification. But knowing the former involves knowing the latter. If logical thought as a whole is independent of nonlogical thought, then the axioms for (*CTx*) would also belong to logic. Hence, the completeness of logic rules out (*Ux*).

The fundamentality thesis needs further clarification. Since it posits definability relations among concepts, as well as necessary connections between having a concept and holding certain beliefs, it is involved in the troubles of analyticity. Still, these troubles are not fatal. It is hard to deny that anyone who understands, say, the notion of uncountability understands countability too and shares some of our basic beliefs about cardinality. The critique of analyticity should allow such facts, and with them the fundamentality condition. Nor is the condition *ad hoc*. The traditional conception of logic obviously includes something like an idea of independence and conceptual closure that the complete extensions of FOL violate.

We thus have a basis for identifying logic with FOL. By combining the ideas of a universal language and a deductive calculus with the fundamentality condition, we can argue that logic must be at least FOL while ruling out the suggested complete extensions. This identification, however, is still both conditional and provisional. It is conditional because it presupposes Frege's general view: that logic is an instrument for formulating and checking proofs. Otherwise our epistemic constraints are unmotivated. There would be no clear reason to suppose that any proof should be recognizable as such, hence none to restrict ourselves to complete logics. Nor is fundamentality plausible apart from some view separating logic from other areas of inquiry and giving it a kind of priority. But this conditional aspect is acceptable. We will presently consider alternatives to the RCL. Now we want merely to argue that on Frege's conception, at least, FOL turns out to be logic. The provisional character of the argument is more immediately troubling. We have hardly shown that nothing essentially stronger or more expressive than FOL can meet Frege's criteria. Let us consider some specific difficulties.

Lindstrom's theorem suggests the possibility of a single argument against a range of alternatives to FOL. Lindstrom showed that any logic (in a general sense) that is either complete or compact and satisfies the Löwenheim-Skolem (L-S) theorem must be FOL. Since we have already restricted ourselves to complete logics, we would only need to find a connection between the RCL and the L-S property to clinch the matter. The standard objection, however, is that the L-S property simply lacks epistemic motivation.[8] Why should it matter, intuitively, whether any (sets of) formulas in our logic have only uncountable models? This feature is clearly linked to something of epistemic interest: it means, roughly, that the resources of logic alone do not enable us to express differences between infinite cardinalities. The distinction between countability and uncountability falls outside of logic (unlike the finite–infinite distinction, in the sense that we can write formulas with only infinite models). But why consider this desirable?

I see no full answer, but perhaps the proof of the L-S theorem provides a clue. Its essence is that countably many formulas can only require the existence of countably many objects. Of course this depends on the specific character of FOL, where formulas are built up from atomic formulas using universal and existential quantifiers. The former do not posit objects, while the latter only introduce them singly: in the proof, one expands an initial domain by adding an object chosen to satisfy each existential quantification. Since this can only increase the domain by countably many things, we never get beyond a countable model. This suggests that what matters is the countability of the language together with the elementary nature of the quantifiers. Not that the proof works just for the quantifiers of FOL. Adding, e.g., (Ix) to the language permits the argument to go through. But a variety of ways in which one might extend FOL, and which might preserve completeness (unlike (Ix)), are sufficiently infinitistic to block the proof of the L-S theorem. (Ux) is one example. This suggests a kind of correspondence between the L-S property and the fundamentality condition.

We noted that although completeness allows a variety of nonelementary logics, conjoining it with fundamentality generally blocks them. The relation of completeness to the L-S condition is apparently similar: the devices complete, nonelementary logics introduce add enough expressive power to make the proof of the L-S theorem impossible. Now the L-S condition and the fundamentality condition are entirely different. One limits the relations among infinite models of a theory, the other requires a kind of definitional closure. One has clear epistemic motivation but is unformalizable, the other is formal but unmotivated. Two distinct conditions can, however, be equivalent in the presence of a third. This may be the case here. A complete extension L' of FOL will contain an operator O interfering with the model construction in the standard proof of the L-S theorem. In order to turn L' into a logic satisfying the fundamentality condition, we must (I conjecture) add further operators that destroy completeness. The presence of O violates fundamentality in one way and the L-S condition in another. This does not mean that the latter simply inherits the motivation of the former. The L-S condition in itself remains unmotivated. But suppose we have conjectured that fundamentality blocks any complete, nonelementary logic; and that we want a formal condition that has the same effect. In the presence of completeness, the appropriate condition should rule out the same devices that fundamentality does. If the L-S condition works, and if we can (by inspecting the standard proof) understand how, then its motivation is sufficient. The epistemic motivation for completeness and our interest in fundamentality together explain why adding the L-S condition is reasonable.

These considerations, however, are vague and call for backing by a general classification of the devices that can extend FOL; one could then try to go through these and show that, for complete extensions anyway, violation of the L-S criterion is systematically correlated with violation of fundamentality. Of course, that might not happen. It would then be dogmatic to insist on the L-S criterion: if some complete extensions of FOL violate the L-S condition but not fundamentality, one might well count them as logic. In a clash between the two conditions, the one with direct epistemic motivation would have priority. But this issue is hard to discuss until some concrete case is described. And since it is hard to see how such a case could involve more than a weak extension of FOL, we may conjecture that using the L-S condition and using the fundamentality condition to supplement completeness lead to roughly the same results.

We have not touched on the logical status of operators that may appear intuitively elementary, e.g., identity or nonclassical quantifiers such as 'many', 'most', and 'few'. Some of these decisions seem to be trivial. Identity, for example, is a useful, epistemically and formally harmless addition to FOL. Further, the absolutely general applicability of identity makes natural its inclusion in a universal language for conducting proofs. The other cases cannot be discussed here, but they are surely peripheral. It we can make a reasonable case for some form of FOL, the exact boundary of logic will not be a pressing issue.[9] Hence I will turn to some points of motivation glossed over so far.

I claimed that the idea of a deductive calculus involves an effectiveness requirement on proofs — but mightn't something weaker suffice? One natural suggestion is that we do not need effective procedures for determining whether something is a proof, since procedures that settle the question with high probability would do as well in practice. But this is mistaken, since it does not define a genuine alternative from its own practical viewpoint. In reality, so-called effective procedures only yield their results with high probabilities anyway, given the possibilities of error in their application. To speak of effective procedures is already to abstract from limitations on finite memory and attention and from the possibility of mistakes in calculation and the like. The RCL offers a view of logic which, under such an abstraction, yields an effective notion of proof. Our question must be whether this is appropriate, or whether we should instead accept a logic that places weaker conditions even on ideal argumentation.

Perhaps we can grant that we should at least have the positive half of a decision procedure for the class of proofs: a way to establish mechanically that something is a proof, if it is. This is not indispensable, but it has an attractive consequence that lies at the heart of rationalism. I will call it the Cartesian condition: once a proof is given, it cannot be overthrown on logical grounds. Its premises might be false, but nothing can force us to retract any of the inferential steps, because we have already determined their correctness mechanically. In terms of the construction of an ideal order of knowledge, this means that as long as we are working from true beliefs (and logic is not responsible for that), we can continue to build without having to take anything back. If this is possible without any great sacrifice, it is clearly desirable. But what about the other half of effectiveness: why not allow the status of some alleged proofs to remain open? This leads to no obvious practical difficulty. It may be annoying not to be able to tell whether something is a proof of P, but we are imagining a situation in which the correctness of certain methods of proof — rules of inference, say — is already given. Our available half of a decision procedure has taken care of that. We are therefore free to look for a proof of P using the methods already known to be acceptable, plus any others we might discover as we go along. As far as the establishment of new results is concerned, this is not clearly worse than the situation in FOL, our usual logic.

I think this challenge can be answered, but the answer suggests a subjective element in the choice of FOL. Such an element is already present, actually, in the Cartesian condition, since there is no knock-down reply to someone who doesn't mind reversing judgments of logical correctness. Although such reversals may seem clearly unfortunate, this is to some extent a matter of epistemic taste. And taste plays a similar role in the following argument for a full effectiveness requirement.

Church's theorem rules out a decision procedure for logical implication in FOL. But we can have something weaker that is still of interest and has traditionally been taken for granted: an effective way to tell whether Q follows from P *by an immediate step*. Views connecting logic to inference and argument give the consequence relation more structure than we have so far allowed. They recognize a decidable collection of immediate inferences (e.g., universal instantiation, modus ponens, conjunction introduction) such that whenever $R(Q, P)$, there is a chain $Q, Q_1, Q_2, \ldots, Q_n, P$ in which each link follows immediately from its predecessor. This condition, together with the decidability of immediate consequence, restricts us to complete logics; but its motivation is not at once evident. What underlies the belief in smallest logical steps, and why should it be decidable whether two sentences can be linked by such a step? The best answer depends on a further assumption from the logical tradition: that immediate steps — all instances of any immediate step-type — have the epistemic property of being self-evidently, transparently correct, as directly obvious as anything can be. A logic with a consequence relation based on immediate steps then offers an epistemic gain. Once we have a proof of Q from P in such a logic, we know that the argument for Q need not and cannot be made more persuasive or clear. (Given the truth of P, of course. We might want a proof of Q from something less controversial than P, but again this kind of consideration is not an issue for logic. Also, we might still want a shorter or simpler proof from the same premise. But we are here abstracting from pragmatic problems about length, surveyability, and the like, since we are considering the situation of an ideal prover.) Again taking a Cartesian viewpoint, this is an advantage, for it shows another way in which proofs will not be superseded. And clearly, this advantage has not just been conjectured *ad hoc* to back up the RCL, but is entirely consonant with traditional views. A fully explicit logical proof is the paradigm of an unimpeachable, unimprovable argument from its premises to its conclusion.[10]

But if this clarifies the advantage of reducing consequence to chains of immediate steps, it does not yet show why immediate inferability should be decidable. Again, we might hope to manage with just the positive half of decidability; with the ability to chain together immediate steps identified with certainty as such. Although we would lack a procedure for showing that a step is not immediate, we could simply eschew the use of steps whose status was still open. (Given the elementary character of immediate steps, this is hard to imagine, so that we might well dismiss this possibility out of hand. But it will be instructive to pursue the matter.) Note, however, that this idea would reduce the significance of the distinction between a decidable immediate consequence relation and a merely enumerable one. At any time, the collection of immediate steps *we could use* would be decidable, since it would consist of the steps enumerated up to that point. So we have no argument against always working with a decidable notion of proof based on immediate steps; the difference would lie in the possibility of extending our notion over time. Now nothing in the RCL as such counts against this. It is of course opposed to the classical view of logic as being

14   STEVEN J. WAGNER

fixed, and given all at once. But that view does not follow from the connection
between logic and reasoning, or even from rationalism. We could thus regard
the use of FOL, our complete logic, as a stage, perhaps to be superseded by a
later complete logic incorporating as yet unrecognized immediate steps. Our
arguments above for identifying logic with FOL would continue to hold under
a relativization to the present time.

The defender of the RCL and of our claims for FOL should be ready to
concede this possibility. Yet it is hard to grasp, since no plausible evolution out
of FOL suggests itself. Also, our considerations about fundamentality and Lindstrom's theorem still stand. We have no clue to the construction of a complete
logic stronger than FOL that does not sacrifice desirable features. Part of the
problem is that our concept of decidability seems to be absolute, even if logical consequence should turn out not to be. We can try to imagine a system S
that, while being logic, extends FOL, but the change in logic would not change
the class of effective procedures. Thus, if S is incomplete from our present perspective, it would remain so from the later one. Similarly, the new logic would
not affect our notions of infinity and countability, which are independent of
logic. To whatever extent our reasoning about extensions of FOL was plausible before, it should continue to have force. (Since we are considering only
extensions that leave earlier inferences in place, we need not worry about the
overthrow of this reasoning itself on logical grounds.) If it now appears that
complete extensions of FOL must involve notions definitionally dependent on
others that lead to an incomplete logic, then the change in logic will not alter
this. So the burden of proof is heavily on those who would claim that FOL could
be superseded by other logics with its epistemic properties.

This may indicate the epistemic significance of FOL. We will shortly turn
to alternative approaches to characterizing logic. First, however, some remarks
on a challenge to the association of the RCL with Frege.

Dummett has in effect asked whether the idea of logic developed here is
so much Frege's as Gentzen's. Gentzen made logical inference the focus of his
investigation, while Frege was in the first instance concerned with the extent of
logical truth. According to Dummett, this points up major differences between
Frege's and Gentzen's conceptions of logic, with the latter's being far superior
([4], pp. 432 ff.). Dummett might then add that my emphasis, in presenting the
RCL, on the relation of logical consequence makes my approach close to Gentzen's (and that of pre-Fregean writers). Although this is strictly irrelevant to the
evaluation of the RCL, it merits a few remarks on Frege's behalf.[11]

> (i)   Dummett contrasts the emphasis on logical truth with the emphasis
>   on logical consequence. But this difference may reflect nothing more
>   than Frege's and Gentzen's divergent foundational aims. A logicist,
>   whose thesis is that mathematical truths are logical, will naturally give
>   the generation of logical truths center stage. The central question
>   within Hilbert's programme, on the other hand, was consistency:
>   given a system, is a contradiction derivable in it? It was, then, nat-
>   ural for Gentzen to frame his project as a study of derivability. But
>   neither approach has the consequence Dummett alleges. Frege's is not
>   committed to the idea that logic is more concerned with truth than

with consequence, and nothing in Gentzen entails the opposite conclusion.

(ii)   Axiomatic and Gentzen-type formalizations correspond in obvious
ways. The latter yield a class of logical truths, the sentences derivable from null hypotheses, while axioms can straightforwardly be converted into rules. In spite of this, Dummett claims that Frege's choice
of formalization was pernicious. His reason seems to be that it led
to a preoccupation with logical truth as a special kind of truth, and
to related preoccupations with analytic truth, necessary truth, and
their opposites. But—setting aside the correctness of Dummett's
historical story and critique of the history—recognizing axioms of
logic carries no commitment at all to any particular view of logical
truth. Of course, unless 'logic' is simply an arbitrary designation, we
must say something about what makes a given truth logical. But
essentially the same problem will arise in Gentzen's framework, as a
question about the status of the rules and of the logical theorems they
generate. In any case, axiomatic formulations say nothing about
whether logical truths are analytic or necessary. Frege thought they
were both, and the same view harmonizes as well with Gentzen's logical theory as with Frege's. (Consider, on the question of analyticity,
the common idea that Gentzen-style rules give the meanings of the
connectives.) But this is simply another matter. Even less is there any
implication, in Frege's approach, that logical truth differs from other
kinds of truth in any more fundamental way—as one might hold if
one believed that truth was correspondence for nonlogical truths, but
essentially conventional or linguistic in logic. Many philosophers after
Frege did believe that, but he is blameless in this, and his use of
axioms particularly so.

(iii) In any case, the author of [7], [8], and [9] was plainly interested in
proof above all. This concern is inseparable from an interest in logical
consequence. Thus, even if Dummett is somehow right in stressing
the primacy of consequence over truth in logic, no criticism of Frege
on this score is appropriate.

Quite possibly, the RCL is more readily formulated in the light of post-
Fregean logical developments—a body of work in which Gentzen's writings are
prominent. But its main features can still be found in Frege.

To argue that an epistemic conception selects FOL as logic is not, of
course, to make an absolute case for FOL. I turn to other viewpoints.

3   Two routes to a nontraditional characterization of logic have been developed. The first is to look for technical criteria that pick out a certain logic or
class of logics in some natural way. For example, Quine has proclaimed FOL
a "solid and significant unity" for satisfying a remarkable group of
metatheorems ([20]; [23], ch. 6). But while Quine is clearly right in this, not
everyone reads the technical data the same way. Tharp's discussion is inconclusive (FOL seems to win, but not decisively) and Hacking argues that natural criteria, framed in terms of the sequent calculus, pick out an extension that includes

ramified type logic. Boolos finds a respectable case for drawing the line no lower than SOL. We will look at some of these issues below. For the moment, we note that the technical strategy at least tends to yield an identification of logic with FOL, so that there is a rough extensional agreement with the traditional view. Not so for the second main strategy. This alternative, best presented in [25], makes our choice of logic turn on adequacy of expressive power. (Some of [2] also invites this reading, and the idea is well-known among logicians.) The basic fact is that first-order axioms fail to characterize most mathematical notions, while second-order ones suffice. This is usually illustrated by the contrast between first- and second-order arithmetic: only the latter determinately expresses our concept of the number sequence. An important further example, on which Shapiro rightly dwells, is *set*. Although it is not quite true, at least without further restrictions, that second-order set theory is categorical, the second-order axioms improve by ruling out nonstandard models, leaving open only the height of the universe. Shapiro therefore concludes that mathematicians need second-order principles to express their theories. The first-order formulations are mere surrogates, the interest of which is parasitic on the second-order cases. Thus, Shapiro argues that the logic of mathematical practice should be (at least) second order.

Simply looking for technical differences among various logics is a strictly mathematical project. Moreover, technical features alone could not confer a distinguished status, for familiar, trivial reasons. Any logic will have numerous distinguishing features. A good many logics will have enough of them to count as "solid and significant unities". Unless one is content simply to classify a range of systems, one needs some prior selection of important criteria. Yet no such selection could be made on purely technical grounds, since the importance or interest of technical features is not itself a purely technical question. And in fact, the criteria at issue usually seem to be interesting in virtue of a connection to cognition. To review some examples, this is clear for completeness, with its relevance to determining validity, and compactness, which helps to guarantee the existence of finite proofs. Tharp's continuity condition in [27] suggests that the first-order quantifiers are conceptually elementary, minimal extensions of non-quantificational operators. Hacking [15] tries to ground the demarcation of logic in a view of how we come to understand certain operators. And although the epistemic significance of the L-S property is problematic, it does seem to be connected to the relatively finitary character of our logic, which is presumably important because our minds are finite. All this is obvious and may lead to the suspicion that the technical strategy is a straw man. Yet a number of remarks in the literature seem to suggest it.

Suppose, however, that a purely technical approach to characterizing logic is not really being pursued; that the real issue is whether any logic has conspicuously many features of epistemic interest. This could be motivated in two ways.

One might, first, deny that logic is epistemically significant, yet hold that traditional philosophers of logic did succeed in picking out *something*. One would then argue that what we call logic has interesting features that stand as technical analogues to epistemic properties, in the way that recursiveness is close to psychological notions of computation or calculation. Thus, one would see earlier philosophers as being on to something but misdescribing it — as mistaking

RATIONALIST CONCEPTION OF LOGIC   17

a technically important body of doctrine for an epistemically important one. The analogies between the mathematical concepts and their epistemic–psychological counterparts would explain this confusion. The characterization of logic would, then, still be a purely mathematical project, but one with interesting extramathematical roots.

One can find this viewpoint in Quine. Quine denies that logic is a priori or necessary, that it provides a foundation of knowledge, that it is in any sense a theory of inference, and so forth. Very little remains of the assumptions behind the RCL and other traditional views. But still, FOL is given a special place for its completeness and compactness, for the coincidence of various intuitive definitions of logical truth in its case, and for confining itself to relatively elementary concepts — all features related to presumably outdated reasons for selecting FOL.

As far as our *present* goals are concerned, this is no different from the purely technical approach. Defining logic serves no philosophical purpose. If one does not take this line, then one seems to be forced back to an epistemic characterization of logic, which then needs to be spelled out and justified. One must explain the cognitive significance of the properties of logic and connect them to a general view of the place of logic. To do less is to maintain an untenable intermediate position: to hold that one's interest is more than purely technical, yet decline to make sense of it. I think that in the end Quine avoids this trap. For in spite of his official relation to the logical tradition, important elements in his philosophy depend on a more philosophically committal view of logic. I will explain this with reference to his views on existence and quantification in [20] and [23].

4   Quine holds that to discern what a person *P* believes in — her ontology — we must first represent her beliefs in FOL. By way of illustration, he considers someone advancing a theory in Henkin's branched-quantifier logic (HL).[12] If this formulation uses only first-order quantifiers, the theorist may appear to believe only in the objects in their intended domain (whatever it might be), but for Quine this is an illusion. Since the equivalent formulation of her theory using linear quantifiers will quantify over functions on that domain, he holds that such functions likewise belong to her ontology. Quine bases this view on the remarkable metatheorems for FOL: the solid and significant unity of FOL recommends its choice as a measure of ontology. Quine adds, however, that should there be more or less reasonable alternatives, "it seems clearest and simplest to say that deviant concepts of existence exist along with them" ([20], p. 113).

For familiar reasons, Quine should not be able to speak of *P*'s having definite theoretical beliefs at all. His writings on translation imply, roughly, that any two representations of *P*'s theory alike in their observational consequences will be equally correct [22]; and it seems clear that wide variation in what would intuitively count as content is compatible with this kind of empirical equivalence. Thus, there should be no sense in asking what *P* believes in, except relative to an arbitrary (in principle) translation scheme. This is also a direct consequence of the thesis of ontological relativity. But just because these doctrines do away with the whole idea of a persons's ontology, it is pointless to consider them here. We are concerned with a thesis in the philosophy of logic that presupposes the

18   STEVEN J. WAGNER

sensicality and value of ontological inquires. Although elements in Quine's philosophy contradict this presupposition, the thesis has independent interest and
deserves to be evaluated on its own terms. Further, such an evaluation seems
to be required in order to do justice to the parts of Quine's thought that assume
the significance of ontological questions, notably his insistence on parsimony
and his attempts to simplify our ontology via reductions. So we may consider
these parts in partial isolation from their Quinian context.

These views certainly raise difficulties. It is, for example, hardly clear why
users of HL would have a deviant concept of existence if HL were a "reasonable alternative" to FOL. A deviant logic, perhaps, but why a deviant concept
of existence? And granting that for a moment, it is also puzzling why the *rea-
sonableness* of HL should matter. If the use of HL involves a deviant concept
of existence, that should be so even if preferring HL is somehow unreasonable.
In order to sort these problems out, it is best to begin with Quine's underlying
motivation.

I assume that Quine's interest in ontological improvement is basic here: in
getting empirically adequate theories with the smallest, clearest ontologies. A
dependence of ontology on logic would threaten this Occamite project. If one
could avoid belief in a class of objects not by refining one's theory, but simply
by changing logics, then moving to another logic would seem to offer the advantages of theft over honest toil. Consider, for example, a theorist who has posited
a certain domain of objects plus functions on that domain. If she suspects that
the functions may be unnecessary, she ought to look for a clever way to explain
the same data without them. If she can instead simply switch to HL and claim
to have reduced her commitments in that way, then the chance for genuine
insight has been lost and ontological reform threatens to become irrelevant.[13]
To be precise, we may distinguish two dangers:

(1)   the possibility of avoiding a commitment by changing logics, as just
      described.
(2)   a relativization of ontology to logic. We would say that our ontology
      is indeterminate absolutely; that it may contain *F*s relative to logic *L*
      but no *F*s (although perhaps *G*s instead) relative to *L'*. (Of course, this
      is close to what Quine does embrace in [19], but we are setting that
      aside.) On this view, one could still try to be parsimonious within an
      arbitrarily chosen logic, but the interest of that would hardly be clear.

I see two possible escapes from these problems. One would posit an association between logics and criteria of ontology that makes ontology invariant.
Thus, if the criterion (QC) identifying the ontology with the intended domain
of quantification were correct for theories formalized in FOL, while the criterion (HC) that adds functions on that domain were to apply to theories in HL,
then changing from one logic to the other would leave ontology the same. This
would block (1) and (2) and help to ensure that ontological progress could come
only though meaningful improvements in theory. But this idea faces difficulties.
Once we raise the question of how criteria for ontology are paired with logics,
it is mysterious why the answer should come out in Quine's favor. Of course it
seems intuitively right to rule out ontological reforms accomplished by changing one's logic, but this appeal to intuition begs the question against (1) and (2).

RATIONALIST CONCEPTION OF LOGIC   19

And even assuming invariance, why should QC go with FOL instead of, say, HL? Quine dwells on the attractive formal properties of FOL, but they look completely irrelevant here. FOL can be as solid and significant as you please, but that does not show why ontology should be read off of the quantifiers in this logic rather than in some other one. Also, the idea of a multiplicity of criteria leaves out of account Quine's insistence on the triviality of QC. He rightly regards the (objectual) existential quantifier as no more than the formal device for asserting existence, the logician's translation of 'there are...'. QC would then seem to be the only reasonable criterion. Trivially, what we believe there is is what we believe there is, so no matter what our logic, the intended range of our quantifiers should reveal our ontology. But then the suggested escape from (1) and (2) is blocked.

If this last point is taken seriously, then Quine needs to show that our logic is in fact FOL and that we cannot change it.[14] Although this is a strong thesis, it is not the absurd denial of transcriptions from one logic to another. One can, for example, obviously go back and forth between statements of a theory with linear and with branching quantifiers. But the thesis is that these transcriptions, while constituting different representations of various bodies of doctrine, do not change a speaker's underlying logic: that they occur within a fixed background logic. Suppose we try to axiomatize all of our theories in HL. That is, we present each theory T as a set A of HL-axioms plus the statement that its truths are the consequences of A. The Quinian should claim that in spite of this apparent commitment to HL, our logic is still FOL because our use of A is really embedded in a broader first-order theory that includes a discussion of A. The illusion of nonelementary discourse will vanish once this background is made explicit.

Quine's remarks on solidity and significance may be intended to explain our hypothesized adherence to FOL by showing why we would never give up this logic. His comment on reasonable alternatives at the end of [20] would then concede that differences over logic might make certain ontological disputes rationally intractable. For if our disagreement with P can be traced to her use of HL instead of FOL, and if there is no strong argument for preferring one logic over the other, then we must leave our dispute unresolved — which may be what the (obscure) attribution of a "deviant concept of existence" is supposed to mean. The trouble with this perspective is that the appeal to solidity and significance is far too weak, since many logics will be solid and significant unities in one way or another. Further, viewing our logic as a matter of choice opens the door to the relativity we want to avoid; it is hard to see how one could prove that nothing besides FOL could ever reasonably be chosen. Quine concedes this result with equanimity (maybe because he wants to affirm ontological relativity on other grounds anyway), but we can hardly follow him. So Quine seems in the end to have no satisfactory account of the status of FOL and its relation to ontology.

Quine would fare better with a Fregean defense. If we regard logic as the basis of idealized argument, then what counts as logic will depend on our ideal practices. The relevant idealizations are not arbitrary. Although they abstract from our usual memory limitations and the existence of competing demands on our time, they leave intact such deeper features as the finiteness of our minds. They also assume no changes in what might be called our fundamental capacity for logical insight. Creatures who can simply see whether a formula of FOL

20   STEVEN J. WAGNER

is valid, or to whom the truth value of CH is immediately clear, would presumably have a different logic; but since these powers are not simply a matter of having more time or memory, we do not attribute them to our idealized selves. So the idealization remains tied to the most general facts about our mental powers. Because we cannot even imagine how to alter these, a Fregean viewpoint makes changes of logic impossible in a strong sense. And if we can argue that, given these facts, our logic must be FOL, then we have just the result that Quine wants. The inescapability of FOL, plus the trivial correctness of QC, ensures a uniform standard of ontology.

By way of illustration, consider again the attempt to convert all our theories to HL. Let T be given via some axiomatization A. We may imagine a theorist P who, trying to avoid FOL, applies some (incomplete) set of HL-rules to A in order to find further elements of T. Our claim, however, is that everything she gets in this way could equally be obtained by first-order arguments from premises she necessarily accepts. Instead of applying HL-rules to A, she can reason from the statement that the elements of T are the sentences true in every HL-model of A. Since this is a first-order statement, and since the entire model theory of T can be given in a first-order language, there is no need for P to go beyond FOL in order to develop T. She can simply use first-order model theory and set theory to discover what besides A holds in every model of T. Moreover, the incompleteness of HL forces her to view T model-theoretically; incompleteness just means the absence of any substitute for the model-theoretic notion of consequence. So the introduction of rules in HL gains nothing. And since P must have a model theory for T, no possibility of ontological gain can be connected with the use of A. The model theory will reintroduce the functions that seemed to vanish with the change to HL.[15]

But why be certain that the first-order approach will yield everything that P could get by working in HL? The alternative would be to posit a set of rules R for HL and a sentence S such that

   (i) P can prove S from A, using R.
   (ii) P cannot prove that S is true in all models of A within her (first-order) model theory.

Note that (ii) plus the semantic counterpart to (i), the claim that A semantically entails S, is reasonable. Many semantic consequences of A can presumably be derived neither within our present model theory, nor from any we are ever likely to be justified in believing. But if (i) and (ii) hold, then P finds evident principles about HL that do not follow from her statable principles about models of HL. This would seem to amount to a kind of direct insight into consequence in HL. Now direct logical insight is not objectionable in principle. We must claim it ourselves for elementary logic. But we seem to lack it for HL, and that cannot be remedied by more computing time or any imaginable sharpening of our wits. If P is like us, then (i) and (ii) must be jointly impossible for her, too. She can therefore not — on the view I an taking — claim HL as her logic. Whatever fragments of HL she might use are reducible to, and derivable from, first-order statements and rules.

Quine might dislike this defense, with its involvement in psychology and traditional epistemology. But since his alternative seems to fail, we may conjec-

RATIONALIST CONCEPTION OF LOGIC   21

ture that FOL is special just insofar as we connect the nature of logic to questions about human reasoning and ideal systems of knowledge. For Quine as for Frege, this is what the issue must come down to, in spite of the differences in their assumptions and interests. Of course, much of Quine's motivation stems from a distrust of traditional views of logic, so that we can hardly dismiss his approach without a closer look at what a Fregean might be committed to. That is the business of section 6. First, let us take up the case for SOL sketched in section 3.

5   Boolos has convincingly rebutted various common objections to SOL [2]. I also think that the attempt to dismiss SOL (as logic) because of its existence assumptions (e.g., [6]) fails. Although logic can clearly not demonstrate the existence of any spatiotemporal individual, certain abstract objects might well be said to exist on logical grounds. If we recognize abstract things at all, at least some of them will appear to exist necessarily. Among the necessary existents, some will arguably exist as a matter of logic alone. Perhaps this is so for all properties, if their existence does not require their instantiation, or perhaps only for certain ones (e.g., "logically necessary" properties like thinking-or-not-thinking). I am not asserting anything of this sort (see instead [1]). But since such claims cannot be dismissed on intuitive grounds, the ontology of SOL does not seem to be a serious drawback.[16]

Since the positive reasons for preferring SOL are widely known and have been excellently detailed in [25], we may be brief. They are also epistemic, but instead of reflecting a concern with reasoning and argument, they focus on the problem of characterizing mathematical concepts. Consider the two versions of the induction axiom:

(1) $F(0)$ & $(x)(F(x) \to F(x + 1)) \to (x)F(x)$
(2) $(AF)(F(0)$ & $(x)(F(x) \to F(x + 1)) \to (x)F(x))$.

(1) is a schema. To obtain its instances, we replace 'F' by appropriate open sentences of first-order number theory. In (2), the variable 'F' ranges over all subsets of the domain of natural numbers. (Instead of (1) we could use an orthographic copy of (2) in which the main quantifier is restricted to first-order definable subsets of this domain.) The widespread use of (1) in logic is due to the technical fact that the well-developed, relatively tractable theory of first-order models applies to number theories employing (1). It seems, however, that any discussion of number must depend on a grasp of the notion defined by (2). (2) describes the numbers up to isomorphism, whereas (1) allows models with "nonstandard" numbers, i.e., elements that aren't numbers at all. Further, as Shapiro emphasizes, (1) suffers from language-relativity. Since it is schematic, any change in our language will yield a new axiom. Thus there is really a multiplicity of first-order induction schemas corresponding to the variously expressive number-theoretic languages we might use. Since we seem to have a single number concept, and since it seems arbitrary to select any one of these languages as being primary or basic, is mysterious how our understanding of number could reside in any form of (1). What links the different schemas is rather that they are all restrictions of (2). So (2), besides being categorical, is in an important way absolute.

22   STEVEN J. WAGNER

   Similar remarks apply to the axiomatizations of real number and set. One
might of course hold that the concepts they define are not genuinely absolute,
since they still depend on a notion of subset (or set) that may itself be relative.
Some mathematicians have thought that a model of a theory given one construal
of 'subset', hence of the second-order quantifier, may fail on another, equally
legitimate, interpretation. But most foundational workers find the notion of subset determinate. And the second-order formulations retain an advantage in any
case. If the notion of set is indeterminate, we may simply have to live with a fundamental ambiguity in our thinking, one infecting our concepts in all mathematical domains. The relativity that comes with the first-order axiom schemas,
however, is additional, since it derives from the need to make an arbitrary choice
of language even after we have settled on our notion of set.
   These observations certainly suggest that second-order formulations of
(many) mathematical theories are basic. From this we obtain an argument for
SOL. A second-order language and consequence relation must apparently be
used to set out principal mathematical concepts and their associated theories.
The truths of arithmetic, for example, must be described as the set of all consequences of a certain second-order sentence *PA*; similarly for analysis and set
theory. So mathematical discourse seems to be essentially second order. First-
order theories of numbers or sets are interesting objects of study, but they only
partially formalize our mathematical beliefs. This view forms a counterpart to
the one proposed on Quine's behalf at the end of section 4: it is a way of saying that we think in SOL, whatever notations we may use.
   This defense of SOL does not deal with arguments, thus differing from the
defense of FOL given in section 2. This omission is not obviously a weakness.
Although Shapiro says nothing about how to discover second-order consequences, nothing in his view prevents such discoveries. Since we know many
truths of arithmetic, we can indeed establish many consequences of *PA*. We lack
a mechanical procedure for generating all of them, but the defender of FOL cannot offer such a procedure either, so the two positions seem to be on a par. The
difference lies in the relation between derivation and logical consequence. For
Shapiro, of course, there is no method for deriving all the logical consequences
of any given sentence. Thus, there is no guarantee that any logical consequence
of *S* can be obtained from *S* by finitely many obvious, elementary steps. But
Shapiro can deny that logical implication must have this feature. A reply to him
cannot simply beg the question in favor of the RCL. As things stand, Shapiro
will not even allow that there are two reasonable notions of logic, one defined
by his criteria, the other inferential. He has, as just noted, an argument to the
effect that our discourse is fundamentally second order. From this he infers that
the logical consequences of our sentences are their second-order consequences,
so that there is no place for the RCL to get a foothold. Thus:

> I suggest that the considerations of the present article preempt such reason-
> ing [i.e., the arguments for a complete or compact logic] . . . all things equal,
> it would be desirable to have a recursively axiomatized, compact logic with
> fewer presuppositions . . . however, the purpose of logic is to study and codify
> correct inference. Since one cannot codify the correct inferences of a second-
> order language with a first-order logic, it follows that the logic of mathemat-
> ics cannot be first-order.[17]

RATIONALIST CONCEPTION OF LOGIC   23

Shapiro's objections to the usual first-order axiomatizations are convincing. The argument for SOL, however, has a large hole that I believe he cannot close. (2) need not be read as a second-order statement. One can equally well take it to be part of a first-order language allowing quantification over sets. Looking at the English that (2) formalizes — 'every set of numbers that contains 0 and is closed under successor contains all numbers' — makes it clear that the question whether (2) is first or second order is unreal. Inspection of this statement in isolation from its linguistic background tells us nothing. (Compare the situation in section 4, where the appearance of working in HL was removed when we considered the background theory.) For all that has been said so far, it appears possible to view Shapiro's representations not as supporting SOL, but as showing that mathematics must be presented in first-order theories with set quantifiers. And proponents of the RCL can then add that since the gains Shapiro wants can be obtained with first-order theories, the supposedly preempted arguments for FOL come back into play.

A Quinian scenario may clarify the situation. Suppose that we land on an island and hear a native N uttering (2) by way of characterizing number. What is her logic? If (2) characterizes number, we know that '(F)' must be a set quantifier, but that does not settle the question. Yet Shapiro's argument boils down to the fact that such a quantifier is needed to block nonstandard models. If that leaves the question of logic unanswered, then Shapiro's reasoning cannot bear on it, and we may take into account other facts about N, such as her inferential practices. If it seems that FOL adequately codifies her inferences, that is evidence that N is presenting her mathematics in a first-order framework. And as usual, considerations about the foreign native reflect on us. Shapiro claims that SOL is needed to formalize our mathematical discourse; but a first-order representation seems to be equally good from his viewpoint and better overall.

Shapiro replies that if we advocate the first-order reading, we cannot take the concept of set (or subset) to given by any formalization. First-order set theory admits ill-founded models, which adequate conceptions of set presumably rule out. So we must hold that when we present mathematics in FOL, set is a concept fixed and understood (in a "standard" way) in advance, to one to be elucidated. But this is no reply in itself, because Shapiro presupposes the same thing. There is, for him, no way to explain or elucidate the meaning of the second-order quantifiers. If second-order presentations of mathematical theories are basic, then these devices must simply be clear from the start — and they involve the concept of set.[18] Hence Shapiro tries to argue that despite these evident parallels, the presuppositions of the second-order approach are smaller and more elementary. He observes that a second-order axiomatization presupposes only subsets of (or functions and relations on) a given domain (the domain of numbers, for example, in the case of second-order arithmetic). In contrast, Shapiro claims, the first-order approach with standard set variables presupposes a grasp of a set theoretic hierarchy. When we use such an axiom system, we must be working within a background language L that includes a theory of sets. L will therefore contain a power set operator and will hence carry commitment to the power set of any set it discusses, the power set of that set, and so on. This is far more than one needs for any branch of analysis (assuming one starts with at least the domain of numbers). If L has the resources for transfinite iterations of the power set operation, as it very well may, so much the worse. Shapiro concludes that instead of a single layer of sets on top of a presumably well understood domain, the first-order approach introduces a large iterative universe.

This reply is unfair. The difference between the first- and second-order readings of (2) concerns the logic of the background language. (2) itself has exactly the same conceptual and ontological presuppositions against either background. In each case, we must grasp and believe in the ("absolute") power set of the natural numbers. If understanding set requires belief in power sets of sets one believes in, then the second-order theorist is committed to as many sets as her first-order rival. If set can be understood prior to any strong theory of sets, then the first-order theorist can equally well get by with minimal commitments. In short, both theorists presuppose a standard notion of set, whatever that may involve. There is absolutely no difference.

Having rejected Shapiro's position, we may ask what underlies it: what considerations, not necessarily operative for any particular recent defender of SOL, might make it seem natural? I will suggest an answer that leads us back to Frege.

For a Hilbert-style formalist, axiom systems are self-contained presentations of mathematical concepts. Understanding only logic and syntactical notions in advance, one should on this view be able to grasp set or number via an appropriate axiomatization; logic and axioms suffice to fix any mathematical subject matter. Incompleteness shows this to be impossible if the logic is first order. But if one holds to this ideal anyway, uncritically and perhaps unconsciously, one is led to identify SOL with logic.[19] Another formalist idea is also relevant here: that the process of understanding mathematical concepts should follow the order of the strength or complexity of the corresponding theories (e.g., that *counting number* (arithmetic) should be prior to *real number* (analysis), and this in turn prior to *set* (set theory)). From this it follows that one shouldn't need a standard notion of set in order to understand number. My first-order approach violates that. Of course, the second-order treatment of number theory likewise violates it, by presupposing a standard grasp of the set quantifiers. But if one has decided that the second-order quantifiers are logical, and also that logical concepts are admissible as foundations for our grasp of number, then one may find the second-order way superior. This would be confused, since one preserves one's view of the order of mathematical concepts only by building the principal concepts at issue into logic. Yet a confusion that is obvious enough when exposed can shape our views as long as it remains implicit.

The truth behind this confusion is that arithmetic is more elementary than set theory, conceptually as well as proof-theoretically. The concepts of arithmetic are simpler in any reasonable sense, and various basic principles about sets have turned out be far more debatable that any axiom of arithmetic. We may, however, still need to know something about sets to have any concept of number. Probably nothing of the principles that give set theory most of its power: replacement, choice, and strong power set and abstraction axioms. But arithmetic may be inseparable from thinking about relations between collections. Then it is perfectly natural to suppose that number might need a set-theoretic definition.

The connection between arithmetical and set-theoretic thinking was of course a main logicist insight. The "bottom up" tendency, on the other hand, has roots in an empiricist tradition opposed to the rationalistic and Kantian elements in logicism. Empiricists have been concerned to minimize assumptions and to show how simpler concepts or systems of thought can be built up without presupposing anything from more advanced levels.[20] We can now see that this difference also bears on the philosophy of logic. Where Frege's own views lead, as I have argued, to an identification of logic with FOL, the main current arguments for choosing SOL seem to rest on formalist and empiricist ideas. The fact that these arguments turn out to have no force would not have surprised Frege and can be seen as supporting his own conception of logic.

My response to Shapiro exploits a shared assumption about logic and reasoning. Without his agreement that "the purpose of logic is to study and codify correct inference," I could not have supported the first-order reading of (2) against him. Indeed, the quickest way to make a case for SOL is to assert the irrelevance of proof to logical consequence. If one starts from a semantic viewpoint, defining validity as truth in all models and admitting no reason why implications should be humanly computable, then FOL will look arbitrary. The flat rejection of an epistemic view of logic preempts counterarguments based on the epistemic advantages of FOL. Should Shapiro have taken this line? To my mind, his actual procedure indicates the strong conceptual tie between logic and inference. It would be easy to promote SOL by jettisoning epistemic conditions. Instead, Shapiro tries to show that his conception of logic satisfies them as well as one could hope. They appear hard to ignore. Perhaps the moral is that if one respects them at all, one cannot avoid choosing FOL; any concession to an inferential view blocks the move to a stronger logic.

Resisting all concessions, I think, simply yields a different conception of logic. I have nothing against that, but wonder whether the designation 'logic' still fits. The pull of the inferential view is universal: virtually every significant position in the philosophy of logic admits some connection between logic and proof. The purely model-theoretic alternative therefore has no serious claim to being part of *our* conception of logic.

We have outlined a Fregean view of logic and argued that it singles out FOL. Further, we have rejected Quine's way of singling out FOL, and the support for SOL has dissolved. But we need to reexamine our rationalist assumptions. The relation of FOL to inference is, as we shall see, open to attacks of a sort quite different from those considered so far.

6   Frege's view of logic clashes with the naturalism that now dominates epistemology. (Actually, the dominant perspective combines elements from naturalism, empiricism, and pragmatism. But I will emphasize the naturalism here.) One form of naturalism, found both in [21] and among Frege's contemporaries, rejects normative epistemology in favor of purely descriptive studies of cognition. Frege rightly saw this as a threat to the foundation of his enterprise: logic cannot be part of the theory of ideal justification if we are not supposed to talk about justification at all. But this challenge, important as it is, must fail. Normative thinking seems to be ineliminable, and [21] (the best source) offers no real argument against it in epistemology (cf. [18]). Less extreme naturalistic ideas, however, may lead to more plausible challenges. I will consider one due largely to Goldman [12],[13] and variously known as the "externalist" or "relia-

bilist" view of justification. Although I cannot fully describe this position here, some main points can fairly readily be brought to bear on the RCL.

The externalist takes justified beliefs to be those formed by reliable processes — processes likely to yield true beliefs. Notoriously, this statement alone is unclear and cannot distinguish externalism from the traditional views it is intended to oppose. There are problems about the identification and individuation of processes; and Descartes, too, thought that good reasoning should generally lead to the truth. In spite of such difficulties, a distinctive feature emerges when we ask how likelihood — the likelihood that a process will lead to the truth — is to be assessed. A Cartesian canon of method would ideally consist of ways to find the truth under all circumstances. Descartes recognized this to be impossible and needed a benevolent God to restrict the possible relations between thought and reality. Nonetheless, he aimed for methods of scientific discovery that would work in all circumstances short of the global forms of deception represented by madness, the Evil Demon, and the like. In contrast, externalism strongly ties reliability to the context of a particular inquirer (or maybe a community or species of inquirers). Someone's methods are taken to be reliable, hence justification-conferring, if they lead her and persons rather like her to the truth in her circumstances and rather similar ones (cf. Goldman's famous example of the barns in [11]). Of course this is still vague, but it definitely departs from the Cartesian viewpoint. The idea of a universal canon of method is replaced by the search for guidelines sensitive to the character of their user and context of use. This brings out the naturalism in Goldman's approach. If he is right, we need knowledge of our epistemic situation to find the correct theory of justification for us. Epistemology must therefore go hand in hand with cognitive psychology, the study of error in measurement and sampling, the sociology of science, perhaps even politics and economics. All of these contribute to the description of ourselves as knowers and the assessment of various epistemic policies. If this does not make epistemology itself another empirical science (e.g., because it remains normative), it certainly suggests that any a priori epistemology would have to be a sterile, limited discipline. In any case, it radically revises traditional views of the nature and point of epistemology.

These remarks focus on discovery and the processes by which beliefs are formed. The problems about justification commonly addressed by Goldman and others lie in that area. But the same ideas apply to justification as it concerned Frege: the providing of convincing, illuminating arguments. From an externalist viewpoint, a good method of argument would presumably tend to produce belief in truths, assuming that true premises were used. Arguments conducted by this method should indeed have true conclusions when they convince, while ones leading from truths to falsehoods should readily be seen to be fallacious. No doubt the externalist would add further conditions pertaining to the role of arguments in producing insight. For example, good arguments should be surveyable and should bring out relevant relations among the propositions appearing in them. An efficiency requirement will also be important: fallible methods tending to produce insight quickly on important questions might be better than more accurate but cumbersome alternatives. This is in line with the idea that our methods should be the best for us, relative to our epistemic abilities and goals, in worlds like our own.

RATIONALIST CONCEPTION OF LOGIC   27

It is this relativity, in particular, that casts doubt both on the RCL and on the status of FOL. Arguments in FOL seem largely irrelevant to our ordinary practices of justification. The notation of FOL is almost useless even for the conduct of mathematics, and the standards of rigor set in [7] are impossibly burdensome. Here it is not enough to reply that FOL is supposed to be a source of *ideal* patterns of justification, patterns optimal only insofar as we abstract from limits on memory, attention, and available time. To be sure, naturalism will permit us our interest in ideals, both in descriptive science and in normative theory. But it may change our ideals. The ideal of arguments conducted in FOL does seem to be associated with Cartesian inquiry. Descartes self-consciously set aside the pragmatic reasons for limiting the time and care spent on any piece of reasoning, and he was willing to pay a high price in efficiency. The recognition of many truths could be postponed a long time to avoid any false belief. Now the externalist will not only regard this as an unattainable goal, but will dispute it as an ideal. Ideals function as standards: we remain under practical constraints of course, but the closer we can get, the better. To an externalist, concerned to recognize our contingent limits and needs, the Cartesian ideal will appear irrelevant. It is perhaps appropriate for ideal cognizers, who have unlimited memory and attention and can take their time when necessary. They, too, will not always be able to carry out proofs according to the standards of FOL, since they also have practical concerns, but they can at least attain the ideal when circumstances warrant it. We, on the other hand, are quickly exhausted by the demands of rigorous proof and can seldom escape the pressures to increase efficiency, hence fallibility. As compensation for this, we can tolerate considerable error, particularly in theoretical areas where proof matters. These contrasts between ideal cognizers and ourselves suggest that only the former should be concerned with the *Begriffsschrift's* standard of proof.

Naturally, we are free to have multiple ideals, and to be guided by different ones at different times. Arguments in FOL, or close approximations to them, clearly do have a place. But the externalist can still ask why FOL should provide more than one (possibly not very important) ideal among many.

These externalist objections may appear overly concerned with the cumbersome notation of FOL, thus still allowing first-order consequence a special status. But we singled out first-order consequence just by defining it in terms of chains of elementary inferential steps. If their interest is called into question, then so is the relevance of FOL. It is not obvious what an externalist view of logic and logical consequence would be, but, offhand, I do not see that externalists must be particularly interested in first-order implication. If empirical research were to demonstrate a significant advantage to, say, intuitionistic consequence— which is readily conceivable—then intuitionistic logic might be preferred.[21] "Logic" might even turn out to be something quite unlike any of the systems logicians now study. We have seen something of the interplay between one's conception of logic and one's epistemology and philosophical projects. If externalism alters these significantly, it is unlikely to let our traditional choice of logic stand.

I suggest that these (and similar) doubts are implicit in our resistance to a Fregean conception of logic. That would be historically appropriate. I have argued that opposition to Frege from one direction is rooted in formalism and

28   STEVEN J. WAGNER

a view of mathematical concept formation that has formalist overtones. It now appears that naturalism, and the associated attempts to give epistemology a more empirical and pragmatic footing, underlie a second principal brand of opposition. Non-Fregean views of logic must be understood with reference to the broader issues separating him from other philosophers of the recent past. Conversely, Frege's philosophy of logic fits his overall perspective.

That is one mark of a great philosopher: his views turn out to be systematic in unobvious ways. But it raises a difficulty. I have tried to defend the RCL, but isn't our naturalism superior to Frege's viewpoint? How then can the RCL be maintained?

In fact I am sympathetic to Frege's philosophical tendencies. Such large questions as the status of naturalism or the dispute between empiricism and rationalism cannot, however, be taken up here.[22] Instead I want to suggest that Frege's view of logic can be defended with the help of one idea that, although foreign to him, can be grafted onto his philosophy without harm: a social conception of knowledge. By this I mean not the platitude that acquiring significant amounts of knowledge depends on social cooperation. Even Descartes believed that. It is rather the idea that knowledge is a property of social groups or cultures. This does not rule out parallel, individualistic conceptions of knowledge. But one may hold that some items of knowledge, particularly in well-developed disciplines such as the sciences, history, and philosophy, belong primarily to groups and only derivatively to their members. In the first instance, the group is the knower. I think the epistemic role Frege gives logic can be understood from such a viewpoint.

Any social account of knowledge will need considerable clarification and defense. Here, however, I will elaborate very briefly, assuming that such accounts are not unfamiliar to the reader and hoping for some degree of sympathy. I am interested in the knowledge of certain kinds of groups, of which modern scientific communities, for example, those of professional mathematicians or cognitive psychologists, are paradigmatic. Among their distinctive features are their historical character (they evolve and accumulate results over a period exceeding the lifetime of any individual), the existence of institutions for exchanging and pooling information (journals, books, and libraries), and the existence of recognized community boundaries and qualifications for membership (roughly, academic affiliation). We may identify the beliefs of such a community with the results its members recognize as being established. This leaves room for questions both about how widely a group belief needs to be accepted and about what counts as establishing a proposition, but the effect of this identification should be reasonably clear. The community of set theorists believes the usual theorems of set theory. It also believes in the truth, or correctness, of the axiom of choice, even though this is not a (nontrivial) theorem and even though it is sometimes questioned. It does not, I would say, believe in the falsity of CH, for even though CH is widely doubted, no argument against it meets normal mathematical standards for plausibility arguments, let alone for proof. Nor does the community believe unreported results of individual workers, or propositions about which there is significant controversy. Taking these remarks to suffice by way of illustration, we may add that some community beliefs will also be justified or known. The latter will be true, and both will satisfy suitable

RATIONALIST CONCEPTION OF LOGIC 29

analogs of the conditions for individual knowledge or justified belief. I assume that if psychological concepts can be extended to communities, the extension of these conditions will not bring major additional problems. Thus, an epistemology for scientific communities should be possible. And it appears that these communities, if not their members, are agents for which a Fregean view of logic and justification is appropriate, because they are sufficiently free from the relevant cognitive limits. Let us reconsider the externalist's doubts.

There was, first, the problem of memory and attention. FOL seemed not even to present an ideal of argumentation, because no human can follow more than a few moderately long proofs in FOL. But communities can exploit the combined resources of many individuals, and they can take their time. Of course they, like individuals, will recognize practical constraints. The mathematical community writes out very few of its proofs in FOL. But this ideal of proof is relevant for the community in a way that it is not for individuals. If there were a point in doing so, most of the business of proof could be conducted in rigorous first-order formalizations. Everything would take longer, and individual mathematicians would find it harder to survey their field, but it would be perfectly possible for the community to assemble the body of proofs that constitute established mathematics in this way. (Russell and Whitehead did something like this for much of the mathematics of eighty years ago.) So the community can approach the ideal of a system of proofs set out in FOL as nearly as it pleases. Similarly, factors limiting the rigor with which individuals can do mathematics tend to matter less to the community. Insofar as the community at any time is interested in the long-term growth of a field, not in immediate results, it can proceed with all due care and rigor. Barring historical upheavals of a rare order, the field should develop indefinitely, unimpeded by the distractions and limits of individual life. The community can therefore slow this development as much as seems appropriate, secure in the knowledge that the job will eventually get done. The length and complexity of FOL proofs, then, is no barrier to their use in justifying community beliefs.

A second main problem was efficiency — the fact that rigorous avoidance of error comes at too high a price in the acquisition of truths. It seems that individuals and communities will assess such trade-offs differently, for scientific communities adopt new beliefs slowly when any possibility of doubt attaches to them. Recall that community beliefs as I characterized them are accepted by almost all relevant scientists. Yet it is well known that controversial ideas in science may undergo decades of discussion before reaching this stage. Communities are conservative in their adoption of beliefs, much more so than individuals, and for good reason: this attitude makes possible the construction of a nearly cummulative body of scientific doctrine over time. Although the community, too, can err, and although occasionally the discovery of an error requires extensive backtracking, the once standard view of scientific progress still seems *roughly* correct to me: scientists build on the results of earlier generations. And only very conservative procedures for acceptance make this possible. In any case, since the conservativism of communal epistemic policy is a plain fact, the conservativism implicit in accepting FOL as an ideal of justification is no objection.

These remarks suggest a reply to a further challenge. The externalist might

30   STEVEN J. WAGNER

well accept normative concepts in principle, yet ask why justification should be very important to us. Its value in everyday life is unclear: individuals are notoriously poor at justifying their beliefs, and as long as one's beliefs are true, one need generally not worry about their justification anyway. Such considerations might raise doubts about the importance of logic in our overall scheme, even if they do not threaten our conception of logic. But justification matters much more to communities, due to their epistemic conservativism. If we will not adopt a belief unless we are fairly sure it is true, and if making sure means subjecting it to stringent criticism, then the community needs access to the grounds on which the belief might be held. Hence justifications must be publicly presented along with our intended conclusions. That, of course, is why evidence and methods matter so much more in science than in everyday life. Conservativism aside, there is another reason for communities to be more concerned with justification. An individual, having enough important beliefs to remember, could never manage to retain very many of their justifications too. But communities have the resources to keep track of reasons along with beliefs. Justification is more relevant for communities because it is cognitively far more affordable.

Human communities do not extend an individual's cognitive powers *too* far. They can handle infinite formulas and arguments no better and have no superior insight into nonelementary logical consequence. Thus, going social yields no case for drawing the limits of logic higher than FOL.[23] But if we are finite groups of finite beings, FOL might already seem to be too much. The communities possible in our universe can grasp only the beginning of the infinitude of proofs in FOL. Why, then, allow proofs of any finite length? Suppose, for example, that Σ is unsatisfiable, but that the shortest derivation of a contradiction from Σ is 10^{150} steps long. Since no community could give that derivation, let alone recognize it as a proof, why should Σ count as being *logically* inconsistent? There is no clear naturalist justification for such an empirically unconstrained notion of logical proof. Setting a bound on the lengths of derivations, however, would entail incompleteness, thus seeming to invite a renewed challenge from proponents of such incomplete logics as SOL.

But on inspection, the case for staying within FOL stands up. The argument from our finiteness does not undermine the reasons for doing mathematics in a first-order language, or for taking proofs to be chains of elementary, first-order steps. Nor has the connection of logic to inference been weakened. Sections IV–V above are therefore still in order. The real issue here is between FOL and some fragment L' admitting only finitely many proofs, perhaps only those that a physically possible community in our universe could grasp. (Let us call the proofs in question "short"; the precise bound and its precise justification do not matter.) In other words: is P a logical consequence of Σ if it is provable from Σ in FOL? Or must the proof be short? This question may not be critical. The answer makes no difference to the main arguments of this paper, as just noted. The status of any concrete proof will also be unaffected, since proofs we can actually give are acceptable either way. Thus, after having advocated FOL at considerable length, I would not seriously resist a naturalistically motivated restriction to L'. But the choice of FOL yields a more appealing classification. If we call Σ in the example above logically inconsistent, we see Σ as being similar to sets of sentences that allow short proofs of contradictions. If we decide

RATIONALIST CONCEPTION OF LOGIC   31

on any particular L', we are basing a significant division on a partly arbitrary cutoff. The former choice seems far more natural. One might wonder why — after all, ω is in principle as arbitrary a limit as any finite ordinal. Yet the finite proofs seem analogous to a natural kind. We find them essentially akin to each other and unlike their infinite counterparts. From our viewpoint, ω marks a major boundary. I think this reflects a sense of relative necessity. We are inclined to view it as accidental whether communities can grasp proofs of length 10^{10}, or 10^{100}, or 10^{1000}. Infinitary reasoning, on the other hand, seems more radically impossible, ruled out by fundamental aspects of reality. And in deciding what to call logic, we are guided by this sense, not by the natural limits to our capacities. So Quine's remark applies here: a perception of solidity and significant unity underlies our choice of logic. Whether this perception is available to a strict naturalist is another question.

I have conceded the naturalist little. The cognizers rationalist epistemology postulates must be possible; and we appeal to facts about scientific communities to establish that. This does not make epistemology, let alone logic, empirical in any way. The empirical fact is part of the explanation of why logic matters to us as Frege thought it should. And in making this reply, we do not offer what the naturalist will demand: a full naturalistic justification of our conception of logic and our interest in FOL. I have tried to indicate how the naturalist can make sense of standards of rigor and systematicity that are inappropriate for individuals. This does not explain our interest in developing a *calculus ratiocinator* and a *lingua characterica*, in a relatively Cartesian notion of proof, or in identifying a body of doctrine meeting the fundamentality condition. Yet it was just these ideas that provided Frege's starting point and led to our selection of FOL. Further work along the lines of this paper would therefore have to clarify the assumptions and motives underlying Frege's interests, and to evaluate their relation to naturalism. If they turn out to have no place in any naturalist programme, that might be more the naturalist's problem than Frege's. For although much remains to be said about the RCL, I think the power and interest of Frege's vision of logic is clear. These issues must be left for another occasion. In the meantime, we should not assume that a social conception of knowledge can fully reconcile rationalism with naturalism. A hard choice may lie ahead.[24]

I have admitted conventional elements in the designation of logic. But FOL stands up well in spite of that. FOL has interesting, important epistemic properties. Its choice follows from a natural elaboration of traditional ideas on logic and argument. Further, the best existing alternatives to this choice do not withstand scrutiny; we have no present reason to take the possibility of challenges to FOL seriously. I conclude that although Frege was wrong about the foundations of mathematics, he was right about logic. Yet he did not have the last word. Defending him against his naturalistic opponents may force us to draw on the tradition, due above all to Hegel and Marx, that subordinates individual to social consciousness — a viewpoint deeply alien to the one Frege inherited from Leibniz and Kant. Although this goes beyond Frege, it would be a happy result, a synthesis of two great contributions of the nineteenth century. The prospect of this marriage may make up for the brevity with which I have proposed it here.

32   STEVEN J. WAGNER

NOTES

1. I will sometimes mean logic in such a sense by 'logic', as the context should make clear.

2. My view of Frege's historical place generally follows [26]. I interpret his theory of number in this light in [29].

3. Following Gentzen's convention, this last formula means "at least one member of $\Sigma' \cup \Sigma''$ follows from Σ." Also, I write 'S' for '$\{S\}$' in (3).

4. From this epistemic viewpoint, the requirement of truth preservation is not rooted in our conception of logic as such. It rather has to do with the favored application of logic: to a discipline, mathematics, in which probabilistic argument is generally not accepted.

5. I am speaking intuitively, since we cannot take the notion of "deductive" logic for granted here. There are various plausible attempts to explain the distinction at issue without assuming too much. E.g., deductive logics are characterized by dilution and the necessitation of conclusions by premises, inductive logics by the use of probabilistic notions.

6. Frege also maintains that logic studies the laws of truth. This important idea is a side issue for us; it played a role in Frege's battle against psychologism (the view that logic describes thought) but cannot settle the dispute between FOL and its rivals.

7. See also the discussion of the priority of judgments over concepts in [26].

8. Here I am particularly indebted to Tharp.

9. Other approaches may be useful, e.g., the attempt to identify logic via "topic neutrality" and the like. See esp. [17]. I am neglecting them because they do not seem to help with the choice between FOL and its various standard extensions, such as SOL. But they may complement my approach and show things that it cannot.

10. Michael Detlefsen remarks in a letter that "Frege wanted rigor... not so much for reasons of 'certainty,' but rather to provide a strict control on the type of information used in a proof [as a prerequisite for establishing logicism]." That is basically right: Frege emphasizes the role of logical formalization in identifying assumptions (or gaps) that might otherwise be overlooked. Still, our Cartesian conditions are central to the standard view of proof—a view that hardly changed from Aristotle's time to 1900, and that Frege certainly shared (e.g., a Fregean proof is clearly intended not to be provisional in any way). These conditions are also relevant to Frege's explicit concerns. Suppose there were a question about the clarity or force of a step in an arithmetical proof. If geometrical intuition, and only geometrical intuition, could help there, then the claim of logicism would have to be qualified.

11. What follows is not a full critique of Dummett's discussion, which seems to me to misrepresent Frege in more than one way.

12. [20] gives a brief introduction to HL.

13. Cf. Quine's description of SOL as "set theory in sheep's clothing" and as "hiding staggering existential assumptions" ([23], pp. 66–68).

RATIONALIST CONCEPTION OF LOGIC   33

14. Note that "change of logic" is ambiguous between a change in the logical principles we accept, as when we decide to reject excluded middle or add an infinitary rule, and a change in what counts as *logic* for us. The latter is the main issue here.

15. Of course working in FOL will not allow an enumeration of the truths of *T*, in spite of the completeness of FOL. We will miss intended elements of *T* that fail only in nonstandard models of A when we try to generate *T* from the first-order model theory. This could be gotten round only if the notion of a standard model of an HL theory were fully formalizable, which it is not. (But we can do better and better by refining our first-order theories.) Similarly for the remarks below on doing arithmetic within a first-order set theory.

16. Unless one is a nominalist. But then one will reject SOL anyway, independently of any views about the connection between logic and existence.

17. These remarks are misleading in that Shapiro otherwise says very little about inference and how it is to be studied and codified. Also, it may be question-begging to define the correct *inferences* in a language in terms of second-order logical consequence.

18. Which must of course be taken to be standard, lest we admit the Henkin models and lose the categoricity Shapiro values.

19. Various passages in [2] and [25] echo the desire to characterize key mathematical notions within logic.

20. Compare [24], where Quine's project is to build up set theory step by step, using the smallest possible assumptions at each stage.

21. Here I am indebted to Richard Grandy.

22. I discuss these questions in a book in progress (working title: *Truth, Pragmatism, and Ultimate Theory*), on which this essay draws.

23. Reflection shows that infinitary reasoning would pose difficulties even for an infinite collection of finite beings. In any case, such collections are impossible in a strong sense, hence irrelevant to the limits of our logic.

24. [29] pursues questions about rationalism and communal knowledge.

REFERENCES

[1] Bealer, George, *Quality and Concept*, Oxford University Press, Oxford, 1982.

[2] Boolos, George, "On second-order logic," *Journal of Philosophy*, vol. 72 (1975), pp. 509–527.

[3] Church, Alonzo, *Introduction to Mathematical Logic*, Princeton University Press, Princeton, New Jersey, 1956.

[4] Dummett, Michael, *Frege: Philosophy of Language*, Harper & Row, New York, 1973.

[5] Field, Hartry, *Science without Numbers*, Princeton University Press, Princeton, 1981.

34   STEVEN J. WAGNER

[6] Field, Hartry, "Is mathematical knowledge just logical knowledge?", *Philosophical Review*, vol. 93 (1984), pp. 509–552.

[7] Frege, Gottlob, *Begriffsschrift*, Halle, 1879, translated in Heijenoort, Jan van (ed.), *From Frege to Gödel*, Harvard University Press, Cambridge, 1967.

[8] Frege, Gottlob, *The Foundations of Arithmetic* (*Die Grundlagen der Arithmetik*, with facing English text), translated by J. L. Austin, Northwestern University Press, Evanston, Illinois, 1974.

[9] Frege, Gottlob, *Die Grundgesetze der Arithmetik*, Olms, Darmstadt and Hildesheim, 1962.

[10] Gentzen, Gerhard, "Untersuchungen über das logische Schließen," *Mathematische Zeitschrift*, vol. 39 (1935), pp. 176–210, 405–431; translated in *The Collected Papers of Gerhard Gentzen*, ed. by M. E. Szabo, North Holland, Amsterdam, 1969.

[11] Goldman, Alvin, "Discrimination and perceptual knowledge," *Journal of Philosophy*, vol. 73 (1976), pp. 771–791.

[12] Goldman, Alvin, "What is justified belief?," in Pappas (ed.), *Justification and Knowledge*, Reidel, Dordrecht, 1979.

[13] Goldman, Alvin, "The internalist conception of justification," *Midwest Studies in Philosophy* V (1980), pp. 27–51.

[14] Grandy, Richard, "Philosophy of logic," ms., 1983.

[15] Hacking, Ian, "What is logic?", *Journal of Philosophy*, vol. 76 (1979), pp. 285–319.

[16] Heijenoort, Jan van, "Logic as calculus and logic as language," *Synthese*, vol. 17 (1967), pp. 324–330.

[17] McCarthy, Timothy, "The idea of a logical constant," *Journal of Philosophy*, vol. 78 (1981), pp. 499–523.

[18] Putnam, Hilary, "Why reason can't be naturalized," *Synthese*, vol. 52 (1982), pp. 3–23.

[19] Quine, W. V., *Ontological Relativity and Other Essays*, Columbia University Press, New York, 1969.

[20] Quine, W. V., "Existence and quantification," in [19].

[21] Quine, W. V., "Epistemology naturalized," in [19].

[22] Quine, W. V., "On the reasons for indeterminacy of translation," *Journal of Philosophy*, vol. 67 (1970), pp. 178–183.

[23] Quine, W. V., *Philosophy of Logic*, Prentice-Hall, Englewood Cliffs, New Jersey, 1970.

[24] Quine, W. V., *Set Theory and Its Logic*, Harvard University Press, Cambridge, 1970.

[25] Shapiro, Stewart, "Second-order languages and mathematical practice," forthcoming in *The Journal of Symbolic Logic*, vol. 50 (1985), pp. 714–742.

RATIONALIST CONCEPTION OF LOGIC 35

[26] Sluga, Hans, *Gottlob Frege*, Routledge and Kegan Paul, London, Henly, and Boston, 1980.

[27] Tharp, Leslie, "Which logic is the right logic?", *Synthese*, vol. 31 (1975), pp. 1–21.

[28] Wagner, Steven, "Cartesian epistomology," ms., 1983.

[29] Wagner, Steven, "Frege's definition of number," *Notre Dame Journal of Formal Logic*, vol. 24 (1983), pp. 1–21.

Department of Philosophy
University of Illinois, Urbana-Champaign
Urbana IL 61801

[8]

HISTORY AND PHILOSOPHY OF LOGIC, 14 (1993), 67–86

A Critical Appraisal of Second-Order Logic

IGNACIO JANÉ

Departamento de Lógica, Historia y Filosofía de la Ciencia, Universidad de Barcelona, 08028 Barcelona,
Spain

Received 23 June 1992

Because of its capacity to characterize mathematical concepts and structures—a capacity which
first-order languages clearly lack—second-order languages recommend themselves as a convenient
framework for much of mathematics, including set theory. This paper is about the credentials of
second-order logic: the reasons for it to be considered logic, its relations with set theory, and especially
the efficacy with which it performs its role of the underlying logic of set theory.

1. Introduction

It is often maintained that there is no proper distinction between logic and
mathematics—not because mathematics is reducible to logic, as the old logicists
claimed, but because any part of logic broad enough to meet the needs of
mathematical practice is inevitably intertwined with mathematics. There is a sense
of the word 'logic' in which this is doubtless true, namely the sense in which *every*
language has its logic, determined by its consequence relation, regardless of how
much content is carried by it. In this sense, second-order logic is certainly logic
(being the logic of second-order languages); but then so are appropriate
reformulations of, say set theory and real analysis. Indeed, any mathematical theory
can be embedded in the logic of some language just by treating the terms peculiar to
the theory as logical particles of the language, that is, roughly speaking, by not
allowing them to be reinterpreted. We can refer to this and similar situations by
saying that the logic of such a language holds the content of the theory.

There is, however, a narrower sense of 'logic'. In this sense, logic is still the logic
of a language and thus semantically determined, but now only those languages are
considered whose logic holds no substantial content. Logic, in the strict sense, is
contentless. The main attraction I see in a conception of logic along these lines is
that, when dealing with an axiomatized theory, logic does not add any mathematical
presuppositions through the back door, forcing us to be totally explicit about what
we accept regarding the objects of the theory. The point is not that these
presuppositions may be false, but that they are substantial. (Should we need them,
we could postulate them beside the proper axioms of the theory.)

First-order is clearly logic in this strict sense. That no substantial content is coded
in it is made clear by noticing that to reason from premises to conclusion within its
limits it suffices to follow well-understood rules concerning the use of the logical
constants.[1] Second-order logic, on the other hand, does not meet anything close to
these requirements, as no set of rules can be devised that yield all its correct
inferences. However, in sharp contrast to first-order logic, it is capable of sustaining

---

1 Thus, all that is involved in the proper use of the universal quantifier while reasoning about an
implicitly fixed domain of individuals is essentially that (1) whenever $\forall x\phi(x)$ is true, then $\phi(a)$ is also
true for any object a in the domain, and (2) whenever $\phi(a)$ is true for arbitrary a in the domain, then
$\forall x\phi(x)$ is also true.

0144-5340/93 $10.00 © 1993 Taylor & Francis Ltd

a good deal of mathematics. Not that in second-order logic we can show the existence of, say, natural or real numbers; but with second-order means we can characterize numbers, natural and real, by single sentences which, consequently, entail all truths about them. From second-order logic no significant existence claims follow (see Boolos *1975*), but given that there exist natural and real numbers, second-order logic enables us to synthesize all true facts about them in just one sentence (one for each). This affords no reduction of mathematics to logic, but it comes close to it, since it allows restating many propositions of classical mathematics as logical truths of *pure* second-order logic—i.e. as logically true sentences containing no non-logical constants (as explained in the last paragraph of §3).

The strength of second-order logic, together with the feeling that it is really *logic* (because the values of second-order variables—especially if they are thought of as classes, or properties, or extensions of concepts—have a distinctive logical flavor) makes of second-order logic a likely candidate for the underlying logic of large parts of mathematics, in particular of set theory. For not only the natural form of the axioms of separation (*Aussonderung*) and of replacement is second-order, but second-order set theory allows us to characterize large fragments of the cumulative hierarchy, whereas the usual first-order axiomatization falls very short of this goal. In this paper I assess this role of second-order logic and the grounds on which its strength rests.

I do not address at all the issue of the ontological commitments of second-order logic. Indeed, many of the points I raise against standard[2] second-order logic (especially those relating to its being used as a framework for set theory) apply with little modifications to the so-called plural quantification interpretation of a monadic second-order language, which presumably is not committed to more abstract entities than the corresponding first-order language is.[3] Where in standard second-order logic we presuppose the unambiguous determination of the power set of the individual domain of quantification *d*, in its plural quantification version we presuppose that the totality of ranges of the plural quantifier *there are some elements of d such that* is fully determined as well.[4] To any such range there corresponds a subset of *d*, and *vice versa*; a subset of *d* being the set of individuals in a range.

2. Languages and logic

In this paper we will not take a formal language to be a mere formalism, consisting just a meaningless sequences of symbols obtained according to well-defined rules of a combinatorial kind; we assume that, in addition to this syntactical component, all languages have a fixed semantic constituent as well. This decision is not just terminological, but it has some methodological interest. For if we want that some formulas of our languages count as suitable counterparts of certain informal mathematical sentences, then these formulas must have at least a rudiment

---

2   In this paper I only consider *standard* second-order logic. This is the logic associated (as explained in §1) to second-order languages so interpreted that the (monadic) predicate variables range over the *full* power set of the individual domain *d* under consideration. This contrasts with the *Henkin interpretation*, in which predicate variables may range over proper subsets of the power set of *d*. With Henkin semantics, second-order languages behave as first-order ones (see Shapiro *1991*, 73–76).

3   See Boolos *1984*. For the plural quantification version of the axioms of separation (*Aussonderung*) and replacement see Lewis *1991*, 101–102.

4   As, in plural quantification parlance, a set *x* belongs to the power set of *d* if and only if there are some elements of *d* which are precisely the members of *x*.

A Critical Appraisal of Second-Order Logic   69

of meaning: its logical terminology must be meaningful, and the semantic categories of their non-logical constants must be delimited. And, more important for our purposes, the semantics of a language determines its logic, which is thus something inherent to the language and not more or less arbitrarily superimposed on it.

Each language of the kind we are interested in has a fixed *similarity type*, which may be taken as the set of the non-logical constants of the language (predicate, function, and individual constants). To each similarity type there corresponds the class of *all* structures of its type, a structure consisting of (1) a non empty set d, the domain of the structure, and (2): for each n-ary predicate constant R, each m-ary function constant f, and each individual constant c, and n-ary relation R on, an m-ary operation f on, and an element c of the domain d, respectively.

Every language is thus associated to the class of all structures of its similarity type. Any such structure counts as a *possible interpretation* of the language. To round up the general description of the semantics, one must give the conditions for a sentence of the language to be true in any such possible interpretation. Some of these conditions are to be seen as articulating the meaning of the logical symbols of the language.

It will be convenient to distinguish between *fully* interpreted and *partly* interpreted languages. In a fully interpreted language, a particular structure is singled out as *the* (intended) interpretation of the language (so that the language is about that structure). In a partly interpreted language, no special structure is chosen. But notice that a partly interpreted language does *not* lack semantics, for, as indicated in the preceding paragraph, its logical constants have a meaning and the semantical categories of its non-logical components are fixed through the similarity type, which determines the class of all possible interpretations. It would seem natural to restrict the term 'language' to fully interpreted languages, but doing so would involve us in some awkward circumlocutions, specially in algebra (and, of course, in model theory). Thus, while in doing number theory a fully interpreted language seems to be called for, in group theory a partly interpreted language is more appropriate. Clearly, each sentence of a fully interpreted language is either true or false; while sentences of a partly interpreted languages are only true or false in a given structure, *i.e.*, in a given possible interpretation.

Every language (whether fully or partly interpreted) has its notions of logical truth and of logical consequence. A sentence of a language L is *logically true* if it is true in all possible interpretations (i.e. in all structures of the similarity type) of the language; a sentence σ is a *logical consequence* of a set of sentences Γ of L if σ is true in all *models* of Γ (i.e. in all structures in which all sentences of Γ are true). Thus, if L and L' are fully interpreted languages which differ only on the particular structure chosen as their interpretation (so that we can think of them as obtained one from another by a re-interpretation of their non-logical constituents), then L and L' share the same logical truths and have the same consequence relation. This is expected to express that logical truth and logical consequence are independent of the particular meaning of the non-logical constants of the language. But we should keep in mind that if this (Tarskian) rendering of these notions is to be convincing, we must have some assurance that *all possible* (re-)interpretations of the language have been taken into account.

Except in some examples, we shall limit ourselves to first- and second-order languages. Besides the non-logical constants (making up the similarity type) these languages have *logical constants*: the propositional connectives($\neg$, $\wedge$, $\vee$, $\rightarrow$, $\leftrightarrow$), the quantifiers (∀, ∃), and possibly the identity sign (≈); and *variables*. First-order languages have only individual variables, whereas second-order languages have predicate and function variables as well: for each positive integer n, there are infinitely many n-ary predicate and function variables, *i.e.* predicate and function variables of n arguments.

The recursive clauses fixing the truth value of every sentence in any given structure with domain d will not be repeated here. But it should be stressed that our second-order languages are endowed with the so-called *full* or *standard* semantics. The point to insist upon is the range of the second-order variables. Individual variables take arbitrary elements of d as values, *i.e.* they range over the domain of the structure. Monadic predicate variables take arbitrary subsets of d as values, *i.e.* they range over the full power set of d. Binary predicate variables take arbitrary binary relations (in extension) on d as values, and so on. Analogously, n-ary functional variables take arbitrary n-ary operations on d as values. All variables can be bound by quantifiers. This means than in a second-order language we quantify over *arbitrary* subsets of, and *arbitrary* relations and operations on, the domain of the structure the language is about.

3. Second-order logic

Second-order languages enjoy a considerable characterization strength. With second-order sentences we can both define mathematical concepts and characterize mathematical structures which are out of reach with first-order means.[5] This is not the place to demonstrate the strength of second-order languages (see Shapiro *1985*; or *1991*, 97–109). However, for the recond, some basic cases should be mentioned. Thus, the structure of *natural numbers*, or, properly speaking, the structure of the natural numbers with the successor operation is second-order characterizable. This is so mainly because we can express the principle of induction with a second order sentence: *Every set* of natural numbers containing zero (the unique number without a predecessor) and containing the successor of each of its members contains all natural numbers. (As we know, this cannot be done with first-order means, as any set of first-order sentences with a countably infinite model has uncountable models as well.)

Again, *the ordered field of the real numbers* is characterizable with a second-order sentence. We can express with a first-order sentence what it is to be an ordered field, but we need to quantify over subsets to say that such a field is complete: completeness for an ordered field means that *every subset* bounded above has a least upper bound. A first-order sentence or set of sentences will not work, because the set of reals is uncountable, and, by the Löwenheim-Skolem theorem, every set of first-order sentences with infinite models has a countable model as well.

We can also characterize with second-order sentences *the field of complex numbers* and the *Euclidean spaces* of each dimension. Indeed, all structures considered in classical mathematics (and many more besides) are singled out up to isomorphism with second-order means. As above, this can be shown piecemeal, but

---

5 We say that a sentence characterizes a class of structures if the structures in the class are exactly those in which the sentence is true, and we say that a sentence characterizes a structure if the structure is, up to isomorphism, the only model of the sentence.

we may see it in a uniform way by noticing that any such structure is definable in some $R(\omega + n)$—the set of all pure sets of rank less than $\omega + n$—and that each $R(\omega + n)$ is characterizable by a sentence in a second-order language.

This power of characterization confers a remarkable strength to the logic of second-order languages. Since second-order languages do not distinguish among isomorphic structures, every second-order sentence true in a second-order characterizable structure will be a logical consequence of the sentences characterizing it. In particular, every second-order sentence true in the structure of natural numbers will be a logical consequence of the so-called Peano axioms (consisting of the principle of induction plus a few first-order sentences stating some obvious properties of the successor operation). Since the arithmetical operations of addition, multiplication, exponentiation and so forth are definable by second-order formulas from the successor operation, among the logical consequences of these axioms there must be the solutions to all open problems of elementary number theory.

The situation with respect to the real numbers is analogous. Many important propositions of analysis can be expressed in the second-order language of the real field. In this language, whose primitive non-logical concepts are just the two basic arithmetical operations (plus and times) and the order relation, we can say what the limit of a sequence of real numbers is, we can define continuity, we can define differentiation, we can state many facts and conjectures about these and other concepts. Each such fact, each true conjecture must be a consequence of the axioms characterizing the field of real numbers.

There are certainly some important concepts of analysis which cannot be defined in the second-order language of the real field. These are concepts whose definition requires quantification over sets of sets of real numbers, or over sets of sets of sets of real numbers, etc. But all these can also be taken care of in second-order logic, as we can characterize with second-order sentences not only the reals, but the set of sets of reals, the set of sets of sets of reals, ... each with enough structure to embody any concept of analysis. Thus, the true solution to any problem of analysis is a logical consequence of the second-order sentence characterizing the right one of these sets.

This is not all. Second-order logic has the solution to all these problems in a more striking way: through logical truth. Suppose σ is a second-order sentence in the language of a structure A characterizable in second-order logic, and let Φ be a second-order sentence characterizing A. Since being true (in A) is the same as being a logical consequence of Φ, σ is true iff σ is a consequence of Φ iff the conditional $\Phi \rightarrow \sigma$ is a logical truth; and σ is false iff $\neg \sigma$ is a consequence of Φ iff the conditional $\Phi \rightarrow \neg \sigma$ is a logical truth. So, truth and falsity of any sentence of this language amounts to the logical truth of one of two sentences of the same language (one for truth, one for falsity) easily constructed from the original sentence.

We can go further still. Suppose $k_1, \ldots, k_n$ are all non-logical (predicate, function or individual) constants occurring in Φ (thus all non-logical constants occurring in σ will be among them) and let $v_1, \ldots, v_n$ be new variables of the same kind (predicate, function or individual) as $k_1, \ldots, k_n$. For any sentence α of this language, let α^* be the formula obtained from α by substituting $v_1, \ldots, v_n$ for $k_1, \ldots, k_n$. It is then clear that α^* is a formula with no non-logical symbols and that α is logically true iff $\forall v_1 \ldots \forall v_n \, \alpha^*$, the universal closure of α^*, is. Thus, for any second-order sentence σ in the language of a second-order characterizable structure, there are pure second-order sentences, *i.e.* sentences with no non-logical constants, σ^T and σ^F—namely,

$\forall v_1...\forall v_n \ (\Phi^* \rightarrow \sigma^*)$ and $\forall v_1...\forall v_n \ (\Phi^* \rightarrow \neg \sigma^*)$—such that σ is true iff σ^T is a logical truth, and σ is false iff σ^F is a logical truth.

4. The issue of underlying logic

A theory (or, if we wish, a given version of a theory) is formulated in a language, which we shall call 'the language of the theory'. Being a language, it is endowed with a semantics which, in turn, determines a consequence relation between sets of sentences and sentences; in short, such a language has a logic, to which we refer as 'the *underlying logic* of the theory' (see Corcoran *1973*, 27; and Church *1956*, 317, including note 520).

The role of the underlying logic is most conspicuous in axiomatic theories. In these a set of distinguished sentences (the axioms) is chosen from which the whole theory follows. The theorems of the theory are the consequences of this set of axioms.

We can distinguish two uses of axiomatic theories. In axiomatic number theory, for example, we are interested in the structure of the natural numbers with the usual arithmetical operations. The axioms and theorems are about them and the theory is supposed to provide us with an organized body of truths about the structure. This contrasts with the situation in algebra, where, *e.g.* an axiomatic theory of groups is not about any particular group: instead, its theorems are the sentences of the language of the theory which hold true in all groups, a group just being a model of the theory.

The difference between these two uses of axiomatic theories can be put this way: in one case (as in number theory) the language of the theory is fully interpreted and the axioms (and all theorems) are supposed to be true; in the other case (algebraic theories) the language is only partly interpreted and the axioms are used to characterize a class of structures.

For the time being, we restrict our attention to axiomatic theories of the first kind. Suppose we have an axiomatic theory in a full interpreted language whose non-logical constants are just the predicate constants $R_1,...R_n$. The language is, thus, about a structure A with a certain domain d and with certain distinguished relations $R_1,...R_n$ (of the appropriate number of places) on d.

The axioms can be seem as expressing certain connections holding among these relations and the domain of the structure. The relations are referred to by the non-logical constants, and the domain of the structure is the range of the individual variables. Being the logical consequences of the axioms, the theorems of the theory are true in every reinterpretation of the language in which the axioms hold. In some of these reinterpretations the range of the variables will be different from d and the non-logical constants will refer to other relations on the corresponding domain. Hence, what follows from the axioms (according to the logic of the language) is independent of the exact meaning of the non-logical constants; it depends only upon the connections among their referents as stated in the axioms.

Thus, although the meaning of the non-logical terminology is essential for the truth of the axioms (and for our taking them as axioms), we do not want this meaning to play any part in the further development of the theory. We choose the axioms because (among other things) their constants mean what they do, but we want to express in the axioms some basic truths with the aim of appealing only to what is explicitly stated in them in the development of the theory. This is a good policy, since it restricts the appeal to intuition to the finding of the axioms, so that two people who agree on the truth of the axioms may accept the theory even though they do not fully agree on the exact meaning of the non-logical terminology involved.

This account of axiomatization is, I guess, commonplace. But it can be seriously misleading if we do not pay closer attention to the nature of the language in which we formulate the theory. We said that only what is explicitly stated in the axioms is supposed to be accepted regarding the structure the theory is about, but it may so happen that some facts concerning this structure be implicitly coded in the underlying logic of the theory, through the logical terminology of the language.

Indeed, some languages are loaded with assumptions; that is, the logic of some languages is. The loading is affected through the choice of the logical terminology, insofar as treating a notion as logical amounts to embodying in the language many facts about it

It is easy to contrive languages of this kind. A whole class of them can be gotten by adding a new quantifier, Q, to first-order languages, and giving the truth-conditions for sentences involving it in such a way that a sentence $Qx\alpha(x)$ is to be true in a structure whenever the set of objects in its domain satisfying α has a certain fixed property. Such a property may be: being countable, being uncountable, having cardinality κ (for any given cardinal κ), being well-ordered (with respect to the interpretation of some fixed two-place predicate constant), being ordered in a given order type, or more generally, being isomorphic to some fixed structure (with respect to the interpretations of some chosen list of non-logical constants of the language). A particular example applying only to languages containing (possibly among others) the symbols $+$ and $\times$ has the following truth-condition clause:

> $Qx\alpha(x)$ is true in a structure iff $<A, +, \times>$ is isomorphic to the field of real numbers (where A is the set of elements of the domain of the structure satisfying α, and $+$ and $\times$ are the interpretations of $+$ and $\times$, respectively).

Languages of this kind can be very useful and natural, and there may be good reasons to recommend their use in some specific situations. For example, when developing the theory of vector spaces over the reals, we assume the real field as known. Thus if we want to axiomatize this theory, we may choose to do it in a language embodying this knowledge, a language whose logic includes what we may call $\mathbb{R}$-logic—for example through the quantifier last considered.

This may be a good way to model the informal theory of real vector spaces, because in it we are interested in the vectors, not in the scalars (which we take for granted). Accordingly, in the formal treatment the required knowledge of the field of scalars is not articulated in the axioms, but is implicitly assumed in the notation. To see how much knowledge is assumed, we just need to notice that every true proposition about the real field expressible in the language is a logical consequence of $Qx(x\approx x)$ (this sentence meaning, roughly speaking, that everything is a real number).

The single sentence $Qx(x\approx x)$ is a complete and categorical axiomatization of the real field. Thus, in this rendering of $\mathbb{R}$-logic, the commutativity of addition, say, follows from $Qx(x\approx x)$, while nothing at all is *stated* (as opposed to *assumed*) about addition in the axiom. Whatever intuitions we may have about what a proper axiomatization is are no doubt violated in this example. And it is in the $\mathbb{R}$-logic, the underlying logic of the axiomatized theory, where the fault must be found.

$\mathbb{R}$-logic may be a fine logic for the development of real vector spaces, but it is not a good logic for the study of the real field. In some sense, it is only *logic* in a generalized way: it is properly a piece of mathematics, only that not presented as a body of explicit propositions, but implicitly coded in a logomorph way. Logic, in a proper or stricter sense, should carry no substantive information, which ℝ-logic certainly does. I admit that it is not always a clear matter what to count as substantial information, and I know that it has often been denied that there is any real difference between logic and mathematics and that, therefore, it is pointless to try to differentiate them. That there is no *exact* border between them I do not deny, but we should not be hasty to conclude from this that there is no ground for their distinction. Indeed, there is no exact point in the visible spectrum where a color ends and another starts—but this does not allow us to conclude that there is no ground for distinguishing green from red.

I would like to make a suggestion concerning why and where it is important to distinguish between logic in a strict sense and logic in a generalized sense (the latter being the logic of any *arbitrary* language; the former of just some class of simple, straightforward languages, among which we find first-order ones). I agree that for the ordinary development of mathematics, such a distinction is unimportant; indeed, in developing one mathematical theory we freely presuppose and make use of concepts and results of some other theories. This procedure may be described as saying that some parts of the latter are taken as underlying logic of the former. Anyway, in the ordinary mathematical *practice* the role of logic is, to say the least, inconspicuous.

One place where the distinction between strict and generalized logic is important is in foundations, especially in set theory (in its role of a foundation of mathematics). Following Shapiro (*1991*, 26), we may consider two kinds of foundation of a mathematical theory: we may *axiomatize* it or we may *reduce* it to some other mathematical theory. As we have seen, when we axiomatize a theory T the consequences of the axioms depend on the language used, specifically on its logical terminology. If to understand this logical terminology we have to know something which is dealt with in some other theory T', then we would be possibly more accurate if we said that we have given an axiomatization of T only *modulo* T'. This should remind us that we cannot claim to understand T unless we understand (at least some relevant part of) T' as well. In some clear cases this is totally obvious: e.g. we cannot claim to understand the theory, whatever it be, of metric spaces while denying any knowledge of the real number field. The notion of a metric space is dependent on that of the reals, but the axioms for metric spaces do not include axioms about the behavior of the reals, they presuppose it).

If an axiomatization may be relative to a theory, a reduction always is, as a reduction is a reduction *to* some theory. As we know, much of mathematics is reducible to set theory, and we may here suppose that all mathematical theories we are interested in have in fact been reduced to it. This being supposed, we must now ask what is to count as a proper foundation for set theory. Certainly not a reduction to some other mathematical theory, since we do not want to be guilty of circularity (not because we look for a *secure* foundation for mathematics, but simply because we want to be *explicit* about what we assume.) Thus we are led to consider what an axiomatization of set theory (not as one more mathematical theory, but as the reductive foundation of the rest of mathematics) should be like.

Suppose AX is a set of axioms for set theory. The axiomatized theory (which need not capture its whole informal or semi-formal counterpart) is thus determined by AX and by the underlying logic, that is, by the logic of the language in which the axioms are couched. It is through the underlying logic that a *prima facie* acceptable axiomatization may be wanting, which will be the case if the logical notions of the language can only be made clear by appealing to (intuitive) set theory itself, or to a fragment thereof. For if the underlying logic involves set-theoretical notions, then the set-theoretical content they carry will be diffused through the theory without being articulated in the axioms. If this is so, then our axiomatization of set theory will be only modulo some fragment of informal set theory. Call this fragment F. Now, either (1) F can be isolated and fully articulated, in which case its content could be added to the theory in the form of new axioms or modifications of the old ones, thereby getting an unobjectionable axiomatization; or (2) F cannot be so sistematized, in which case there is a clear sense in which our axiomatization fails to capture the set-theoretical content buried in F, a content which we possibly fail to understand well (witness our inability to articulate it, as we are in case (2)), but which we nevertheless pretend to inject to the axiomatized theory through the workings of the underlying logic.

These remarks are meant as an argument for the view that the underlying logic of an acceptable (for foundational purposes) axiomatization of set theory should presuppose no set theory; indeed, that it should presuppose no mathematics either. By this I do *not* mean that the syntax and the semantics of the language of the axiomatization should be developed without appeal to mathematics—this I do not deem possible, and I do not find it interesting, either. What I mean is rather that the *use* of the language presuppose no mathematical knowledge.

5. Second-order logic versus set theory

In the last two sections of this paper (§§8,9) I will try to show in some detail how second-order logic fails to be a good candidate for the underlying logic of set theory. But now I would like to tackle in a general way the issue of the dependence of second-order logic on set theory. For it has been maintained that although in second order languages we quantify over subsets of the domain of the structure the language is about, the notion of set here involved should not be identified with that which constitutes the subject matter of set theory. This line of defence of the logical character of second-order logic is developed in Shapiro (*1985*, 721; *1991*, 18). He suggests that when we identify second-order logic with (a fragment of) set theory we overlook an important difference between two distinct conceptions of set, which he calls the *logical* conception of set and the *iterative* conception of set. Sets in the logical sense are the subsets of a given domain, whereas sets in the iterative sense are the members of the cumulative hierarchy. But I fail to see that we are faced with *two conceptions* of set, and I will presently argue that the iterative conception of set which unfolds in set theory does not differ from the conception of set needed to develop the ordinary semantics of second-order languages. To develop this semantics we need to have at our disposal the power set operation, which assigns to every domain the set of all its subsets. Yet, the set-theoretic hierarchy is but the result of the iteration along the ordinals of this same operation starting from a (possibly empty) set of individuals. And it should be insisted upon that our shortcomings in the understanding of this hierarchy do not concern only the process of transfinite iteration; they are also due to our poor knowledge of the contents of the power set of each infinite level of the hierarchy, indeed to our poor knowledge of the contents of the power set of an arbitrary infinite set.[6]

In the second-order language for a structure with domain d there are many sentences whose truth is sensitive to fine details about the contents of the power set of d. It often happens that for us to gain knowledge of the details of the contents of the power set of a set of relatively low rank we must turn to sets of much higher rank, *i.e.* to sets which appear in the cumulative hierarchy only after a large number of transfinite iterations. Certainly, if the power set of a set d is a well-determined totality (which I am now assuming it is), whatever is taking place at higher levels will not affect its contents. But one thing is for the contents of this power set to be determined and quite a different thing is for us to have the means to find out what these contents are. The truth of some second-order sentence we care about may depend on the existence of subsets of d which can only be defined either in terms of other sets appearing later in the hierarchy or through quantification over them. In such a case, we will only know about the former when we know the latter. More generally, to assess the truth of second-order sentences about structures with low-rank infinite domains we will have to have recourse to higher set theory. Thus, to prove that in every infinite game with perfect information over an arbitrary Borel set of reals one of the two players has a winning strategy we need to rely on uncountably many iterations of the power set operation (see Friedman *1971*), and to gain knowledge about definable sets of reals (say, whether they are Lebesgue measurable, or whether they have the property of Baire, of whether the uncountable ones include a perfect set) we work from measurable and other large cardinals (see Solovay *1969*). Indeed, the light they might throw upon the power set of sets of low rank (as the set of real numbers or the set of all sets of real numbers) lies behind our interest in strong axioms of infinity. Now, given that all these problems whose solution requires such amount of set theory can be naturally formulated in the second-order language of familiar structures, can we reasonably maintain that the conception of set relevant to second-order logic is a different conception from that of iterative set theory? I think not.

For the semantics of second-order languages we rely on sets; in set theory we make them the subject of our investigations. But in both cases, sets are taken to be arbitrary possible collections of given objects. In set theory, dealing with sets, we treat as actual these possible collections. We also treat them as objects, so that they can be members of new possible collections, which we again regard as objects. This procedure is to be iterated any possible number of times (the theory gives a partial working explanation of these terms through the concept of ordinal) to the effect that any possible collection have an analogous copy in the set-theoretical hierarchy. Since in set theory we can simulate relations by sets, in this hierarchy we will also find structures isomorphic to any possible structure, so that every mathematical structure has a representative in the set-theoretical universe.[7]

6. Conceptual *vs.* combinatorial sets

The conception of set presupposed in the semantics of second-order languages does not differ, I argued, from the conception of sets as members of the cumulative

---

6 Witness to that the many independence results from set theory relating to sets of reals.

7 Thus the set-theoretical universe is not a possible structure. This agrees with Cantor's conception of this universe as absolutely infinite, and thereby incapable of mathematical determination (see Hallett *1984*, ch. 1).

hierarchy. The opposition between *logical set* (a subset of a given domain) and *iterative set* (a member of the cumulative hierarchy) is not a substantive one, but is loosely analogous to the opposition between *parent* and *ancestor* or to the distinction between *immediate successor* and *less than* in the domain of natural numbers. On the other hand, I suspect that to regard as 'logical' this (partial) conception of set is not even historically sound, although '[t]here is a tradition, originating, perhaps, with Boole and including Peirce and Schröder, which takes the subsets of a domain to be under the purview of logic' (Shapiro *1985*, 721, note 10) For there are two well-differentiated acceptations of the term 'set' and it is precisely the one underlying both the iterative conception and the usual semantics of second-order languages to which Frege denied any logical relevance. According to one of these acceptations, a set is the extension of a concept; according to the other, a set is an arbitrary collection of objects of a previously determined domain. We may refer to sets in the former sense as *conceptual* and to sets in the latter sense as *combinatorial*. Given a domain *d*, *e.g.* the set of natural numbers, any conceptual subset of *d* must be specifiable in some way or other, not necessarily in a fixed language, whereas a combinatorial set must be seen 'as the result of infinitely many independent acts deciding for each number whether it should be included or excluded' (Bernays *1934*, 260). Leaving metaphors aside, a set in the combinatorial sense 'is conceived as something which exists in itself no matter whether we can define it in a finite number of words (so that random sets are not excluded)' (Gödel *1990*, 259 note 14).

In *Philosophy of logic*, Quine tells us that '[p]ioneers in modern logic viewed set theory as logic; thus Frege, Peano, and various of their followers, notably Whitehead and Russell' (Quine *1990*, 65). I am afraid that Quine would not accept the soundness of the distinction between conceptual and combinatorial sets, although he sometimes speaks as if he would favor the conceptual acceptation, especially when he insists that the only thing which distinguishes the realm of sets from the specious one of properties is the principle of extensionality: 'A class in the useful sense of the word is simply a property in the everyday sense of the word, minus any discrimination between coextensive ones' (Quine *1987*, 23; for a qualification see *1963*, 1–2). But normally at present, when we speak of sets without qualification we refer to the iterative conception, partially axiomatized in the system of Zermelo-Fraenkel, which is a theory of combinatorial sets. Frege is very explicit against this way to conceive sets: 'I do, in fact, maintain that the concept is logically prior to its extension; and I regard as futile the attempt to take the extension of a concept as a class, and make it rest, not on the concept, but on single things' (*1984*, 228), and he insists that combinatorial sets have nothing to do with logic (p. 226):

> This led [Schröder] to the domain-calculus, to the view that classes consist of single things, are collections of individuals; for in fact what else is there to consititute a class, if we ignore the concepts, the common properties! [...] All this is intuitively very clear, and indubitable; only unfortunately it is barren, and it is not logic. Only because classes are determined by the properties that individuals in them are to have [...] it becomes possible to express thoughts in general by stating relations between classes; only so do we get logic.

It is precisely this intimate connexion between Fregean concepts and sets what led to the view that set theory is a part of logic. Then, after the collapse of Frege's theory and its gradual shift towards the combinatorial conception of sets, we have kept referring to them, at least in certain contexts, as 'logical'. A clear manifestation of the uncomfortable coexistence of the two conceptions of set can be found in Russell's *Introduction to mathematical philosophy*. For Russell, sets are (provided they are something) conceptual. 'Every class [...] is defined by some propositional function which is true of the members of the class and false of other things' (*1919*, 183). However, in order to be able to carry on the logicist program of reducing mathematics to logic, he is forced to call (at least as hypotheses) on principles whose truth for his conception of set is, to say the least, suspect. These are the axiom of infinity, the multiplicative axiom and the axiom of reducibility. Russell explicitly acknowledges their lack of justification from a logical standpoint, and nevertheless, in the last chapter of the book, entitled 'Mathematics and logic', he writes: 'If there are still those who do not admit the identity of logic and mathematics, we may challenge them to indicate at what point, in the successive definitions and deductions of *Principia mathematica*, they consider that logic ends and mathematics begins' (pp. 194–195). But it is quite easy to take up the challenge, since Russell himself has shown us three breaking-points: the above mentioned axioms. In particular, the multiplicative axiom, *i.e.*, the axiom of choice.

The axiom of choice is the mark of the combinatorial conception of set. '[N]othing can express better the meaning of the term "class" than the axiom of classes [*Aussonderungsaxiom*] and the axiom of choice' (Gödel *1990*, 139). 'The axiom of choice is an immediate application of the quasi-combinatorial concepts in question' (Bernays *1934*, 260). 'Finally, the axiom of choice is just as evident as the other set-theoretical axioms for the "pure' concept of set' (Gödel *1990*, 255, note 2). For, how could there fail to be a set which has exactly one element in common with every non-empty set of a disjointed family? There must be one such set, if for 'set' we mean 'set in the combinatorial sense'; but there may be none if we just consider conceptual sets. Says Russell: 'Unless we can find a *rule* for selecting [...] we do not know that a selection is even theoretically possible' (*1919*, 126), and he adds: 'It is conceivable that the multiplicative axiom in its general form may be shown to be false' (p. 130). Insofar as he accepts the axiom of choice, he shifts from his avowed conception of set to the combinatorial one, and in so doing, his claim to have reduced mathematics to logic becomes less persuasive.

7. Second-order set theory

For all purposes of ordinary mathematics we can do without second-order logic by working in set theory. But this resort is not available for set theory itself.[8] My objective now is to appraise the adequacy of second-order logic as underlying logic for set theory.

Needless to say, when speaking of set theory I mean the theory of combinatorial sets. The notion of set in the conceptual sense is, I fear, ill-determined. The suspicion of indeterminateness can be brought to light by trying to answer the following question: faced with a reasonable description of a set of elements of a domain, we cannot fail to acknowledge that it is a set in the conceptual sense; but if we suspect that a given combinatorial set is not a conceptual one, how must we proceed in principle to ground our misgivings? I brought up the conceptual conception of sets because I surmise that the reasons that almost a century after

---

8 Not, at least, when we regard set theory as the theory to which the rest of mathematics is reduced. When dealing with set theory as one more mathematical theory we can certainly make use of any resources at our disposal.

Frege and Russell we still ponder whether second-order logic, and perhaps even set theory, may be regarded as part of logic in the strict sense must be found in the ambivalence of the notion of set. The Fregean view of sets was certainly related to logic—as concepts can be argued to be logical entities—and had it been consistently axiomatized we would probably count at present among the logical theories the now non-existent theory of concepts—or of sets in the conceptual sense.

This, however, does not mean that the usual theory of (combinatorial) sets is irrelevant to logic; on the contrary. As fundamental a notion as that of universal validity can be explicated with the help of sets in the combinatorial sense—but not of conceptual sets. As was pointed in §5, set theory can be understood as the theory of all possible collections; more to the point, as the theory of all possible structures. Thus, if a sentence holds in all structures of the cumulative hierarchy (not just if it is provably true in those structures) it must hold as well in every possible structure; it is universally valid. But there is no reason whatsoever to suppose that every sentence true in every structure whose domain and relations are conceptual sets be universally valid. Universal validity is truth in every possible model—not just in every model, which can be specified in any of a variety of ways. The fact that in some cases (e.g. in the field of first-order languages) it makes no difference what kind of structures we consider (as every first-order sentence true in some structure is also true in a structure on the set of natural numbers whose distinguished relations are definable in arithmetic) does not diminish the importance of the distinction. For one thing is to extensionally delimitate the set of universally valid sentences in a given class of languages, and quite another thing is to offer an analysis of the notion of logical truth as universal validity.

Let us now address the issue of how to develop set theory. The objective of this theory is to describe the pure cumulative hierarchy, the transfinite succession of rank-levels gotten by iterating the power set operation starting from the empty set. We must keep in mind that the universe we want to describe is to be absolutely maximal. There are two general kinds of problems we want to solve by lying down the axioms of set theory: (1) What is there in every rank-level? and (2) How many rank-levels are there? The first question is actually about the contents of the power set of an arbitrary set, whereas (2) inquires how many ordinals we must acknowledge. The natural answer to both questions, dictated by the intended maximality of the set-theoretical universe, is wholly useless; for it is *all*. What sets are there in the power set of set *a*? *All* possible subsets of *a*. What ordinals must we acknowledge? *All* possible ordinals. We must find axioms to give substance to both *all*.

Basic set theory is ZFC, the theory of Zermelo-Fraenkel with the axiom of Choice. It is normally presented as a first-order theory, although it can also be formalized in second-order logic, as Zermelo himself did in *1930*. The difference between both versions can be limited to two axioms: the axiom of separation (*Aussonderungsaxiom*) and the axiom of replacement. Informally, they would be rendered as:

Separation: If *a* is any set and *P* is any property, then there is a set whose members are the elements of *a* having the property *P*.

Replacement: If *a* is any set and *R* assigns a unique object to every element of *a*, then there is a set whose members are the objects assigned by *R* to elements of *a*.

In a first-order language Separation and Replacement must be both formulated

80   *Ignacio Jané*

as schemata, thus becoming in fact infinitely many axioms. Consider Separation. First-order languages allow us no means to refer globally to properties, so we have to deal independently with every single property expressible by a first-order formula. To have the property in question amounts to satisfying the formula. Thus the first-order separation-schema is:

$$\text{Sep}_1\colon \forall a \exists b \forall x (x \epsilon b \leftrightarrow (x \epsilon a \wedge \phi(x))),$$

where ϕ is a formula in which the variable 'b' does not occur. Every individual formula put in place of ϕ gives rise to a separation axiom. In second order, however, the separation axiom can be formulated as a single sentence:

$$\text{Sep}_2\colon \forall X \forall a \exists b \forall x (x \epsilon b \leftrightarrow (x \epsilon a \wedge Xx)).$$

Something analogous can be said about the replacement axiom.

Before discussing the propriety of the first- and second-order formulations of these axioms, I should mention a strong argument for the second-order version of Zermelo-Fraenkel set theory (to which we henceforth refer as 'ZF2', in opposition to ZF1).

Being a theory whose underlying logic is first-order, ZF1 has countable models. Thus, it is utterly unable to characterize the set-theoretical universe. ZF2 does not characterize the universe either.[9] But it can single out up to isomorphism fragments large enough to allow for a substantial part of ordinary mathematical practice. As Zermelo showed (see Zermelo *1930*, or Moore *1980*), ZF2 is quasi-categorical:[10] it does not have a unique model, but a series of models each of which is determined up to isomorphism by an inaccessible cardinal, *viz.* the order-type of their ordinal sequence. If M is a model of ZF2, then there is an inaccessible cardinal, κ, such that M is isomorphic to $R(\kappa)$ (the set of all sets of rank less than κ). Conversely, if κ is an inaccessible cardinal, then $R(\kappa)$, together with the membership relation, is a model of ZF2. It follows that if κ_0 is the least inaccessible cardinal, then every model of ZF2 includes an isomorphic copy of $R(\kappa_0)$. Since virtually all structures needed for ordinary mathematics (set theory excluded) can be found in $R(\kappa_0)$, we can obtain a useful extension of ZF2 by adding to its axioms a sentence stating that there are no inaccessible cardinals. This extension is fully categorical, as all its models are isomorphic to $R(\kappa_0)$. Thus, all true propositions (whether known or unknown) of virtually all ordinary mathematics are consequences of the few axioms of ZF2 plus an additional 'accessibility axiom'.

8. More on ZF2

Strictly speaking, however, ZF2 hardly yields more solutions to ordinary mathematical problems than ZF1 does. In a certain, perhaps metaphorical sense, it has all solutions to the problems of ordinary mathematics, but it keeps them to itself,

9   Strictly speaking, the set-theoretical universe is uncharacterizable for technical reasons as well-as only structures (having a *set* as domain) are subject to characterization. Besides, the use of a second-order language to formalize set theory has the additional drawback that some sentences one would count as universally valid (being true in all *set* structures) as '$\exists x \forall x (xx \leftrightarrow \neg x \epsilon x)$' or simply '$\exists x \forall x (Xx \leftrightarrow x \approx x)$', are false in the set-theoretical universe. (This drawback is not shared by the plural quantification version of second-order logic mentioned in §1.) I have chosen not to discuss these points because I think that the arguments given in §8 and §9 against the use of second-order logic as a framework for set theory are more substantial than these seemingly minor objections.

10   I partially borrow this term from Weston *1976.*.

there being in general not enough means to extract them. Nevertheless, there may be reasons to prefer ZF2 to ZF1. Two of them have already been mentioned but not discussed. They are (1) the naturalness of the second-order axiomatization; and (2) its quasi-categoricity.

With respect to (1), I will restrict myself to the separation axiom. The most direct argument in favour of the second-order formulation of this axiom is that it expresses exactly what we want to express, namely that given a property and a set, the elements of this set having that property make up a set as well. This is precisely what the second-order sentence

$$\forall X \forall a \exists b \forall x (x \epsilon b \leftrightarrow (x \epsilon a \wedge Xx))$$

is supposed to say.

It does not quite say this, however, although we usually read it this way. The catch is in the word 'property'. The variable 'X' ranges over sets, and while every (reasonable) property determines a set—its extension—there is no reason to believe that every set (in the combinatorial conception, the one underlying second-order semantics) is the extension of a property. But I do not want to dwell on this distinction, since I am wary of properties and I think I have a stronger argument against the propriety of the second-order rendering of the separation axiom.

We will best grasp the significance of the informal axiom of separation, the one we want to formulate in a first- or second-order language, by inquiring about its purpose. As already mentioned, the two fundamental operations which bring about the cumulative hierarchy are the power set operation and its transfinite iteration. For the present discussion of the separation axiom only the former is relevant. One of the axioms of ZF explicitly says that every set has a power set, or, with a little detail, that for every set a there is a set $P(a)$, whose elements are precisely the subsets of a.

$P(a)$ will, thus, contain all subsets of a. But we should be careful here. We must not forget that we are concerned about axiomatizing a theory -not just fixing the reference of the terms. To say that $P(a)$ contains all subsets of a can be virtually vacuous if there are no further axioms specifying what subsets does a have.

One of these additional axioms is the axiom of separation. From this standpoint, to specify part of the contents of $P(a)$, it says that among the subsets of a one will find the extensions of properties of members of a. And how do we express this with the second-order means at our disposal? 'Property' translates to 'set in the intuitive, preformal sense'. Thus what the second-order axiom of separation really says is that every (intuitive) set of elements of a is a subset of a.

This is practically vacuous. But we should not be surprised. When using a second-order language endowed with the standard semantics, we regard the concept of *arbitrary subset of* as perfectly understood. This being so, it is then only natural that when we come to axiomatize set theory we literally say that every arbitrary subset (in the preformal sense) of a is a subset (in the formal sense) of a. By so doing, we assure that the set $P(a)$ provided for by the theory will coincide with the true power set of a.

Thus, by axiomatizing set theory in a second-order language (with the usual semantics) we close the door to a thorough study of the power set of infinite sets, one of the main objectives of set theory.

Things are very different in first-order logic. The infinitely many axioms of separation of ZF1, one for each individual formula, explicitly say that the elements of *a* which have the property expressed by that formula make up a subset of *a*. The schema of separation is doubtless weak. It says less than the informal version means to say, since it is restricted to properties which can be expressed by first-order formulas in the specific language of the theory. But it is not at all vacuous.

9.   On the strength of quasi-categoricity

Let us now turn to the second reason to prefer ZF2 to ZF1, its quasi-categoricity. To prove that ZF2 is quasi-categorical we first show that if M is a model of ZF2, then M's ordinals (*i.e.* the elements of M which satisfy the formula expressing the property of being an ordinal) are well-ordered by M's membership relation and its order type is inaccessible. Then we show by transfinite induction that if M and N are two models of the theory, M's rank-levels and N's rank-levels coincide up to isomorphism until one of the models is exhausted. We need not enter into the details of the proof; we do not even have to be more precise about its main lines. But there is a point I do want to stress. For the proof to go forth it is essential that, roughly speaking, if M is a model of ZF2 and *a* is a set-in-M (*i.e.* *a* is a member of M), then all (true) subsets of *a* are sets-in-M—in other words, it is essential that M contain all true subsets of its members.[11] This critical property is conferred on M by the separation axiom.

Let us see how this goes in its essentials. Suppose that M is a model of ZF2 and *a* is a set-in-M. We must show that to M belong all subsets of *a*, *i.e.* that all true subsets of *a* are sets-in-M. The argument is this: let X be an arbitrary subset of *a*. Since the axiom of separation holds in M, the set $\{x\epsilon a : Xx\}$ belongs to M. But this set is X. Thus, X is a set-in-M.[12] In short, X belongs to M because X is a subset of *a* and, as such, it is a value of the monadic second-order variables.

So cheap a proof of seemingly so strong a fact looks suspicious. We are thus naturally led to assess the strength of quasi-categoricity.

I would like to point to some facts that suggest how misleading quasi-categoricity can be. The first concerns the axiom of choice (AC). AC is among the axioms of ZF1, but we do not need to assume it in ZF2. Indeed, we carry through the proof of quasi-categoricity without ever having to take a stand on whether AC holds in the models of the theory. At first sight this might seem surprising, since AC has a clear existential intent, but on a closer look we come to realize that the avoidability of AC is perfectly natural and should be expected. That we can do without the axiom of choice in ZF2 is clarified by the remarks we made about the axiom of separation as contributing to the description of the contents of the power set of an arbitrary set. In this respect the axiom of choice belongs together with the axiom of separation, as it guarantees the existence of certain subsets of any set: Given a disjointed family of non-empty subsets of a set *a*, there is a subset of *a* having exactly one member in common with every one of those subsets. This axiom is necessary in the first-order formulation of set theory because the sets whose existence it postulates may not be at all definable and, hence, may not be obtainable by means of the separation schema. But these subsets, whether definable or undefinable, will be among the

---

11   This is inaccurate because M's membership relation need not be the true membership relation. But it is easily mended. The accurate version is this: If E is M's membership relation, *a* is a set-in-M and $A = \{x\epsilon M : xEa\}$, then for any subset B of A there is a set-in-M, *b*, such that for all $c \epsilon M$, $c \epsilon B$ iff cEb. This follows immediately from the separation axiom by letting the variable 'X' take the value B.

12   This proof has some technical shortcomings, but it is basically correct. See the preceding footnote.

values of the variable '*X*' of the second-order separation axiom, so that this axiom already takes care of their existence. The axiom of choice is therefore superfluous in ZF2.

But is it? Suppose we have some misgivings about the axiom of choice; suppose, to fix matters, that we are uneasy about the existence of a well-ordering of the reals—an elementary consequence of AC. These doubts are not at all preposterous, as we know that no well-ordering of the reals can be exhibited (see Solovay *1970*). What should we do to reach a decision on this matter? The obvious answer is; turn to set theory.

In ZF1, the axiom of choice is independent of the remaining axioms, and these axioms do not entail either the existence or the inexistence of a well-ordering of the reals. Thus ZF1 is of no avail to assuage our doubts. But knowing of its independence, we are directed to a closer scrutiny of the axiom—be it through the hypothetical-deductive method, drawing conclusions from it with the help of the remaining axioms and then appraising these conclusions as a touchstone for its truth, or be it through a deeper analysis of the concept of set.

ZF2, in contrast, gives us a more reassuring answer. Because of its quasi-categoricity, all models of ZF2 (among whose axioms we do not find AC) can be said to contain the same set of reals and of whatever is needed to determine the existence or inexistence of a well-ordering of this set. Thus either all these models will contain a well-ordering of the reals or none will, and therefore one consequence of ZF2 is the existence of a well-ordering of the reals, if the former is the case, or its inexistence, if the latter is. To know which we just have to look among all its consequences.

But how? Let us suppose that AC is a consequence of ZF2. For us to find out that this is so we must use the same axiom of choice in the underlying logic! Only thus will we be able to conclude that the choice set we look for is among the values of the variable of the separation axiom. But then all misgivings we had about the axiom of choice become misgivings about logic.

This overturns the role of logic. We engage in set theory to discover what sets there are (*i.e.* what kinds of collections are possible), how sets relate to other sets, etc. The theory's underlying logic should take nothing about sets for granted, since we want the theory to tell us explicitly whatever we can know about them. To assume too much (implicitly, in the logic) may entice us into believing that the theory encompasses more than it explicitly says.

And this is precisely what happens when we formulate set theory in a second-order language endowed with the usual semantics. Its enormous fertility is but a mirage.

Because of quasi-categoricity, ZF2 has the answer to many mathematical questions. One of them, perhaps the most widely mentioned in this connection, is the continuum problem: How many real numbers are there? According to Cantor's continuum hypothesis (CH), its number is the least uncountable cardinal. This is equivalent to saying that every infinite set of reals can be mapped one-to-one either onto the set of natural numbers or onto the set of all real numbers. We know that CH is independent of ZF1, so we should say that ZF2 is stronger than ZF1, since ZF2 decides CH. Moreover, it decides it in the right way, because the set-theoretical region which is relevant to the determination of the truth value of CH is the same, up to isomorphism, in every model of ZF2, lying in the fragment of the cumulative hierarchy going up to the least inaccessible cardinal. But, what is CH's truth value?

Is CH a consequence of ZF2? Or rather its negation is? The only answer we get is oracular: If CH is true, then it is a consequence of ZF2; if it is false, then its negation is. Knowing this does not improve our knowledge. But such is the kind of answer we consistently get when faced with an ordinary mathematical problem whose solution ZF1 ignores and ZF2 assures to possess.

One more fact that may be mentioned in this connection is that any disagreement concerning the contents of the power set of an infinite set is bound to weaken the effectiveness of quasi-categoricity. If two people dissent (whether knowingly or not) about what a set of, say, natural numbers is, then, even though both of them will be able to prove (by means of the same proof!) that ZF2 is quasi-categorical, the models they will get may not be isomorphic, since their disagreement may affect what counts as the power set of ω in every model.

Shapiro seems to have an argument against this possibility of misunderstanding. His argument can be taken to show that if two mathematicians agree on which set a set d is they will also agree on what subsets d has.[13] He says (*1985*, 721–722):

> Let *P1* and *P2* be two candidates for the logical powerset of d. I suggest that if $P1 \neq P2$, then there is a clear sense in which (at least) one of them is not the powerset of d. Indeed, suppose that there were a collection c such that $c \in P1$ but $c \notin P2$. I take it that (for a classical mathematician) it is determinate whether every element of c is an element of d. If every element of c is in d, then $P2$ is not the powerset of d; otherwise, $P1$ is not the powerset of d.

But what does this argument prove? At best it shows that the power set of d is unique for each mathematician—not that any two mathematicians share the same power set of d. The difficulty with this argument I find in the phrase: 'suppose there were a collection c'. Naturally, *given* a collection c (and given d) it is determinate whether c is a subset of d. The trouble lies in *giving*, or in *considering*, the collection c.

For imagine that M1 and M2 are two mathematicians agreeing on what set the set d is, and also *verbally* agreeing on what the power set of d is: the collection of *all* (*combinatorial*) subsets of d. Suppose however, that M1 allows for a subset c of d which M2 fails to recognize.[14] How could M2 suspect that his power set of d is wanting? Only by leaving his conceptual universe. For if M2's conception of sets is elaborate enough, then it will satisfy all known statable conditions about them, so that there can be no room for c in it; for if there were, M2's accepted closure requirements would force c into his conceptual universe and, therefore, into the power set of d. In particular, c cannot be describable in M2's terms. Accordingly, either c is nameless or the names that M1 has for c name in M2's universe some subsets of d other than c.[15]

---

13   It *can be* so taken, even though Shapiro only takes it to show that 'the totality (or range) of subsets of d is [...] clear and unambiguous' (*1985*, 721). What follows, then, is not really a reply to Shapiro, but a reply to a use of his argument for a wider purpose.

14   To fix matters and to make the supposition more realistic, we can think that M1's universe contains a measurable cardinal whereas M2's is Gödel's constructible universe. In this setting, d could be ω, the set of natural numbers, and c could be $0^\#$ (see *e.g.* Drake *1974*, 257).

15   Or some name for c in M1's universe has no reference for M2, being an improper description. But we can assume that improper descriptions name the empty set.

In the former case, M1 cannot succeed in showing M2 that his alleged power set of d falls short of being the full power set. He just can utter something as 'I know there is a subset of d you fail to recognize!', but he will not be able to specify what subset this is. In the latter case, M2 can rightly reply that the set M1 is naming belongs already to his power set of d (it may even by the empty set!).[16] No agreement will be reached—not because M1 and M2 have a different conception of what it is to be a subset of d (about this they fully agree), but because of the (conceptual) impossibility of M2 recognizing c.

These remarks are not meant to be an argument against the determination of the power set of an infinite set; they are just a way to put forward that such determination cannot be proven—just presupposed. And I find that this is too strong a presupposition for logic to rest on.[17]

In set theory, however, this presupposition is harmless, for there we really make no use of it. In (first-order) set theory we certainly do assume the power set of every set to be determined, but we pull out nothing from this determinateness that we previously did not put in through the axioms.[18]

First-order logic, I conclude, is the right underlying logic for set theory. It offers us the means to extract all consequences from the axioms and does presuppose nothing about sets. Its obvious shortcomings regarding categoricity are irrelevant here, as we do not axiomatize set theory with a view to characterize the (uncharacterizable) universe of sets, but to investigate it and know it better. To insist on the fact that our theory has many spurious models, that it even has countable models, is wholly beside the point. We are not interested in models of set theory; we do not implicitly define sets through the axioms of the theory—just the opposite, we start from some relatively clear insights about sets and then proceed to construct a theory to deepen our understanding of them.

Acknowledgments
I would like to thank my colleagues Manuel García-Carpintero and Ramon Jansana and my former student Mario Gómez-Torrente for many helpful discussions on the contents of this paper. I must also thank Stewart Shapiro who, as a referee for this journal, made many comments and criticisms to an earlier version of this paper.

16 This would be the case in the example of the next to last footnote ($d = \omega, c = 0^{\#}$). Suppose M1 says: 'c is the set of Gödel numbers of all formulas satisfied in L by the first ω uncountable cardinals'. M2 will answer that he certainly recognizes the existence of such a set. But they do not mean the same set. This lack of communication cannot be dispelled by their learning each other language; it is deeper than that, as it depends not really on what it means to be a subset of ω, but rather on the conception of the whole set-theoretical universe. For the real disagreement is not about what to count as a subset of ω, but about the possibility of there being a class of ordinals indiscernible for L. (What the power set of ω is depends on whether this is a possibility).

17 This presupposition rests on the lack of ambiguity of the combinatorial conception of sets. But is it really inconceivable that some day we find that this conception is not one but really many—that is, that it can be developed along two or more differing and mutually contradictory lines?

18 With respect to the preceding footnote, if we ever discovered that the combinatorial conception is ambiguous, then such a discovery would probably result in a major step forward in set theory, being the source of new fundamental axioms and providing perhaps an explanation of the difficulty we have in solving some basic problems like CH.

86   *A Critical Appraisal of Second-Order Logic*

Bibliography

Bernays, P. *1934* 'On Platonism in mathematics', in P. Benacerraf and H. Putnam (eds.), *Philosophy of mathematics. Selected readings*, second ed., London (Cambridge University Press), 258–271.

Boolos, G. *1975* 'On second-order logic', *The journal of philosophy*, **72**, 509–527.

—— *1984* 'To be is to be the value of a variable (or to be some values of some variables)', *The journal of philosophy*, **81**, 430–449.

Boolos, G. and Jeffrey, R. *1980 Computability and logic*, second ed., London (Cambridge University Press).

Church, A. *1956 Introduction to mathematical logic*, Princeton (Princeton University Press).

Corcoran, J. *1973* 'Gaps between logical theory and mathematical practice', in M. Bunge (ed.), *The methodological unity of science*, Dordrecht (Reidel), 23–50.

Drake, F. *1974 Set theory. An introduction to large cardinals*, Amsterdam (North-Holland).

Frege, G. *1984 Collected papers*, Oxford (Basil Blackwell).

Friedman, H. *1971* 'Higher set theory and mathematical practice', *Annals of mathematical logic*, **2**, 326–357.

Hallett, M. *1984 Cantorian set theory and limitation of size*, Oxford (Clarendon Press).

Gödel, K. *1990 Collected works. Volume II*, Oxford (Oxford University Press).

Lewis, D. *1991 Parts of classes*. Oxford (Basil Blackwell).

Moore, G. *1980*, 'Beyond first-order logic: the historical interplay between logic and set theory', *History and philosophy of logic*, **1**, 95–137.

Quine, W. *1963 Set theory and its logic*, London (Cambridge University Press).

—— *1970 Philosophy of logic*. Englewood Cliffs, N. J. (Prentice-Hall).

—— *1987 Quiddities. An intermittently philosophical dictionary*, London (Harvard University Press).

Russell, B. *1919 Introduction to mathematical philosophy*, London (Allen & Unwin).

Shapiro, S. *1985* 'Second-order languages and mathematical practice', *The journal of symbolic logic*, **50**, 714–742.

—— *1990* 'Second-order logic, foundations and rules', *The journal of philosophy*, **87**, 234–261.

—— *1991 Foundations without foundationalism*, Oxford (Clarendon Press).

Solovay, S. *1969* 'On the cardinality of Σ_2^1 sets of reals', in J. Bulloff (ed.), *Foundations of mathematics*, Berlin (Springer), 58–73.

—— *1970* 'A model of set theory in which every set of reals is Lebesgue measurable', *Annals of mathematics*, **92**, 1–56.

Weston, T. *1976*, 'Kreisel, the continuum hypothesis and second-order set theory', *Journal of philosophical logic*, **5**, 281–298.

Zermelo, E. *1930*, 'Über Grenzzahlen und Mengenbereiche', *Fundamenta mathematica*, **16**, 29–47.

[9]

WHO'S AFRAID OF HIGHER-ORDER LOGIC?

Peter SIMONS
Universität Salzburg

1. *Introduction*

One of Henri Lauener's philosophical heroes is Quine, and one of Quine's best-known views is that for regimentation it is meet and right to use first-order predicate logic and set theory, but not higher-order logic. By higher-order logic I mean any logic using higher-order quantification, and by higher-order quantification I mean any which binds a variable other than a nominal or a propositional variable. Some may want to see propositional quantification also as higher-order, but since sentences, like names, are a basic syntactic category (as distinct from a functor category) I think this terminology is inappropriate. In any case, I shall not discuss propositional quantification at all in this paper. What should philosophers think about higher-order quantification?

This paper is frankly programmatic: I do not claim to solve the problems I raise, especially in such a short compass. But I think it is good now and again to set out as clearly as one can the reasons for being stubbornly attached to what appear to be hopeless positions, in the hope that one can see the way out oneself more clearly, or that someone else will see the way out for one, or finally that the forlornness of one's position becomes fully apparent. So be it.

As far as its pedigree goes, higher-order predicate logic is as old as predicate logic itself: Frege's *Begriffsschrift* is explicitly higher-order, since function variables are bound. The Russell-Whitehead ramified type theory and the simple type theories of Ramsey, Chwistek, Church and others are all higher-order. The isolation of the first-order part of what was, until then, just (modern) symbolic logic began with Löwenheim in 1915, and continues with the work of Skolem, Hilbert/Bernays, Gödel, Church, and Gentzen. The do-

254

minance of first-order predicate logic (I shall from now on abbreviate this to 'PL1') since the 1930s is based on three facts:

(1) its relative simplicity, the familiarity of its model theory, and its nice logical behaviour: it is sound, complete, and compact.
(2) the success of first-order set theory as a *lingua franca* and foundation in mathematics.
(3) the energetic lobbying of great logicians for PL1, and their authoritative example: I am thinking in particular of Skolem, Gödel, and Quine.

But who's afraid of higher-order logic? The question is not just rhetorical. Philosophers who tend to certain philosophical points of view will find higher-order logics both necessary and problematic. I shall list the philosophical views which put such people in this uncomfortable position and show how they lead to cognitive tension. It will emerge from this who need not be afraid of higher-order logic. I shall offer reasons for holding some, but not all of these theses. Since I find myself in the uncomfortable position of those who are both attracted and repelled by higher-order logic, I shall try to set out and justify the considerations which have put my mind at least partly at rest.

2. *Structure of the Argument*

The theses which serve as point of departure are listed here. Some of them are vaguely formulated. This is intentional, since in most cases there are several variants. Each thesis will be given an proper name, here printed in bold letters.

INADequacy: PL1 is inadequate for several philosophical tasks.
LOGIcism: Logicism is an attractive and plausible thesis.
NOSET: Set theory as an adjunct to logic is unattractive and unnatural.
FINitude: Humanly usable languages are lexically as well as syntactically finite.
TARSKI: A Tarskian semantics for mathematical discourse is worth striving for.

255

ANTIPLATonism: There are no abstract objects in the sense of Platonism.

How do these assumptions land us in difficulty? Since I do not think the best argument is deductively tight, despite its suasive power, here is an informal version, bringing as much out into the open as I think we can while retaining this power.

If PL1 is inadequate (**INAD**) then something stronger is needed. What is on offer? We could add set theory to PL1, or move to logics with infinitely long expressions, such as the logic $L_{\omega1\omega}$ allowing countably infinite conjunctions and disjunctions,[1] or to higher-order logic. If we accept **LOGI** then PL1 + Set Theory is ruled out, because set theory is generally held to be not a logical theory. This conclusion is reinforced by the premiss **NOSET**. So we are left with logics with infinitely long expressions and higher-order logic. But **FIN** enjoins us to remain with languages having only finite expressions. So only higher-order logic remains as an attractive and plausible extension of PL1. There one quantifies higher-order variables, and the question is how this quantification is to be interpreted. We seem to have to choose between referential and substitutional interpretations. But as **ANTIPLAT**onists we cannot accept a substitutional interpretation, since either we accept infinitely many substitutends (offending against **FIN**) or we accept abstract expression types as substituends. So it seems that only a referential interpretation remains. If we accept this semantics at face value, à la **TARSKI**, then we are led to ontological commitments corresponding to higher-order quantification, which is Platonistic, in contradiction to **ANTIPLAT**.

From this argument, one can see who will have little time for the attractions of higher-order logic, and so will need to have no *Angst* in the face of these charms:

(1) Someone convinced of the adequacy of PL1.
(2) Someone who finds set theory natural and appropriate and who

---

1. Cf. Van Benthem & Doets 1983, p. 279. This essay has been a valuable source of information about the metalogical properties of higher-order logic, supplemented by Barwise 1977, pp. 41-45, and more recently Shapiro 1991.

256

rejects logicism.
(3) Someone who is prepared to accept languages with expressions of infinite complexity.
(4) Someone who accepts the substitutional interpretation of quantifiers (whatever the price).
(5) Someone who rejects a face-value (Tarskian) semantics for forms of discourse such as mathematics.
(6) A Platonist.

If you belong to one or more of these categories, you can afford to breathe out for the moment. In fact I think it might be possible to ruffle that confidence, but that would require a lot more argument than I have space for here.

I want first to go back and cover some of the less well-known issues a little more, in particular the thesis **INAD** and the interpretation of the quantifiers. But first, it may be noticed that I have not mentioned two issues which are very frequently raised in conjunction with higher-order logic, namely the incompleteness of higher-order logic and the problem of the identity conditions for intensional objects like properties and relations. In fact I think both apparent defects of higher-order logic are not real defects. Completeness is very nice if you can get it, but if the price is expressive inadequacy, it seems to me we should be prepared to accept incompleteness. After all, the primary reasons for wanting higher-order logic are semantic ones, and the standard semantics for higher-order logic engenders completeness not because of its own inherent problems but because we cannot recursively enumerate the sentences valid with respect to this semantics. As far as intensional objects are concerned, I think a discussion of their identity conditions, though important in itself for the ontology of such (purported) entities, is beside the point when we are considering a purely extensional logic. The interpretation of second-order letters as ranging over properties and relations is itself a disputable one, and in an extensional second-order logic we are entitled to say that we are only interested in such items corresponding to these letters as have extensional identity conditions – whatever they are.

257

3. *Why Higher-Order Logic?*

What forces us to leave the familiar territory of PL1 or PL1 + Set Theory? Here are some reasons for believing the thesis **INAD**.

(a) PL1 is too weak for the adequate treatment of diverse mathematical concepts and structures.
(b) PL1 is too weak to show of many logical constants *that* they are logical constants.
(c) PL1 is too weak for the analysis of natural languages.
(d) PL1 is too weak to support commonsense realism and scientific realism.

In addition there are reasons for not being satisfied with a (first order) set theory S (i.e. for supporting **NOSET**):

(e) PL1 + S is too strong to represent the logic of mathematics.
(f) PL1 + S is far more spendthrift in its ontological commitments than higher-order logic.

Ad (a): In view of the Löwenheim-Skolem theorems, no theory which has infinite models is categorical, for example Peano arithmetic, real analysis, Zermelo-Fraenkel set theory itself. There are non-standard models for all if we are not allowed to replace axiom schemata by axioms (e.g. the axiom schema of induction in Peano arithmetic): the Peano structure of the natural numbers is not definable in PL1 with the successor relation as additional constant. The condition of Dedekind-continuity, the defining characteristic of the real numbers, is likewise not first-order definable. There is no definition in PL1 of finitude or infinitude. (Because of the compactness of PL1 the following set of formulas is ω-consistent: $\{F(0), F(1), F(2), ..., $ for some number n, $\sim F(n)\}$.)

Ad (b): This is perhaps the most damaging criticism since it is internal to logic and metalogic. The following concepts are obviously logical constants: identity, self-identity, predicate conjunction, the relative product, the universal relation, the ancestral of a relation, the transitivity of a relation, and many more. These are all concepts

258

of the kind abundantly defined in *Principia Mathematica*. One may employ a method of Tarski,[2] and show that these are all logical constants, instead of doing what is normally done, simply presupposing that certain constant symbols are logical constants and treating them specially (as in the standard semantics for PL1). But to express some of the constants and to carry out all of the proofs of their logicality, it is essential that one employ higher-order bound variables.

Ad (c): The following quantifier expressions (among others) cannot be defined by means of PL1: *more, most, finitely many, exactly as many, more frequently than, at least 30% of.* Clearly this reason is connected with (b). The linguistic theory of determiners, to which such expressions belong, intersects with higher-order logic in the theory of generalized quantifiers. However, whereas most of those who treat generalized quantifiers do so within a set-theoretic metalanguage, I do not consider generalized quantifiers to require set theory or a metalanguage: they can be adequately and even advantageously expressed using the means of Leśniewski's Ontology.

Ad (d): Putnam's arguments in favour of anti-realism make essential use of the Löwenheim-Skolem theorems. Those who are not prepared to go beyond PL1 have no satisfactory answer to Putnam's argument.[3] Set theory cannot help out here: since as a first-order theory it also succumbs to Löwenheim-Skolem (as the latter stated to his own satisfaction in the 1920s): ZF1 has non-standard models too.

Ad (e): Many of these defects can be alleviated by adding a judicious dash of set theory. But not all. Consider 2nd order Peano arithmetic. This is categorical: all models have the same cardinality. Mathematicians claim to know what structure the natural numbers have, which is what enables them to clearly distinguish standard models from non-standard models in the case of the first-order theory. They also claim to be able to communicate about such things without radically misunderstanding one another. If they are right, they must

2. Cf. Tarski 1986; for applications, cf. Simons 1987.
3. Cf. Forbes 1986.

be able to understand the expressions they use without ambiguity, for instance *natural number, property of a natural number, relation between natural numbers*. This understanding is based on nothing more than what is discussed in a *standard* model of first-order Peano arithmetic. Standard set theory on the other hand assumes that the power set axiom can be iterated infinitely many times, in order to provide a model for Peano arithmetic in set theory. This is taking a nuclear-fusion-powered hammer to crack a nut. Similar consideration apply to real and complex analysis. The categoricity of mathematical theories is only restrictedly guaranteed within set theory. Any two models of Peano arithmetic within first-order set theory have the same cardinality, but only *relative to* a fixed model of set theory. As soon as the background theory is allowed to have different models, Skolem-relativity reasserts itself. Second-order Peano arithmetic is on the other hand categorical without any ifs and buts, and requires only one second-order quantifier ranging over properties of natural numbers to be so.

Ad (f): Every set theory with sufficient power to help out in overcoming the shortfalls of PL1 is committed to the existence of infinitely many individuals.[4] Standard models of ZF are of inaccessible cardinality. PL2 (second order predicate logic) is on the other hand usually committed only to the existence of at least one individual, which commits one to the existence of at least two property extensions, four binary relation extensions etc. Even in simple type theory the existence of no more than finitely many items of each type is required by the logic (which is the basis of Gentzen's proof of consistency). If the case is admitted (as in Leśniewski's Ontology or in some brands of predicate logic, called universal logics), that the domain of quantification may be empty, then these commitments shrink still further. Higher-order logic *enables* one to define different transfinite numbers (in contrast to PL1), but does not of itself commit one to the existence of numbers of objects sufficient for these definitions to be satisfied. As Whitehead and Russell recognized, assumptions about the number of individuals that exist are not

---

4. By an individual I mean not an *Urelement* but an object of lowest logical type. Sets are individuals in standard set theory.

260

logical but empirical in nature. At this stage I have to concede to giving logicism up. What is attractive about logicism, at least with regard to simpler theories like Peano arithmetic and real number theory, is that the concepts involved are logical. Also for example one may derive the second-order Peano axioms from second-order logic together with the Whitehead-Russell axiom of infinity (the assumption that there are infinitely many individuals, not the set-theoretic axiom of infinity, which allows infinite iterations of set formation) in a much more natural manner than in set theory. Many of the connections, cultivated by Frege, between pure and applied mathematics can be retained. What has to be jettisoned from logicism is the thesis that all the *truths* of arithmetic etc. are derivable from pure logic alone.

In order to further support the argument, it is important to show that all logics intermediate in strength between PL1 and PL2 have their problems. This task has been undertaken by Stewart Shapiro.[5] The perhaps surprising result of the investigation is that for mathematical purposes the full expressive power of PL2 is not used. Nevertheless, the main issue, the quantification of higher-order variables, remains a problem.

4. *Alternatives in the Semantics of Quantifiers*

Those with no fear of a Platonistic ontology can quite happily accept Quine's referential interpretation of quantifiers for variables of higher order. Nominalistically inclined philosophers on the other hand frequently prefer a substitutional interpretation of quantifiers, since even Quine admits that substitutional quantification does not commit one to the existence of entities quantified over like referential quantification does.

I find this way out unsatisfactory. For a start, I consider that first-order or nominal quantification and its equivalents in natural language are quite evidently objectual or referential. An astrophysicist who declares "There are black holes" or a palaeobiologist who

5. Cf. Shapiro 1991.

261

says "There were no endothermic dinosaurs" are clearly saying nothing about actual or possible names of such objects. That their statements have consequences for the semantics of various kinds of expression follows from the existence or non-existence of the objects in question and so also presupposes the referential interpretation. For another thing, for many purposes the substitutional interpretation is forced to assume the existence of infinitely many names or other expressions, which is uncongenial in view of **FIN**, and in any case, if we are talking about actual token expressions rather than potential ones, is simply false. If we are talking about expression types, then we have bought back into Platonism by the metalinguistic back door. The existence of uncountably many real numbers is, when all is said and done, much more plausible, natural and maybe even probable as an assumption than the existence of uncountably infinitely many expressions for such numbers. It would be preferable in such circumstances to welcome Platonism of numbers by the front door and have done with it.

There is the possibility of following a mixed strategy: referential quantification for individual variables, substitutional quantification for higher-order variables. In that way the objection of unnaturalness is avoided, but not the objection of back-door Platonism.

This mixed strategy suggestion has at least the merit of distinguishing higher-order quantification sharply from first-order quantification. But the reason for the distinction is bad: it is assumed, following Quine, that quantification binding higher-order variables is quantification binding nominal variables for higher-order entities. In this respect I agree with Arthur Prior, who considered that Quine's view involves an eccentric view of namehood, and proposed by contrast that higher-order quantification does not involve nominalization. On the other hand, Prior's attempt to sell higher-order quantification as ontologically harmless by pointing out that its natural language counterparts are natural and so harmless[6] is unsatisfying as long as no explicit semantics for such quantification is in view.

A more concerted account of higher-order quantification without ontological commitment is probably latent in the ontology of Leś-

6. In Prior 1971, ch. 3.

262

niewski, but it is a difficult problem to disentangle the point from the ramifications of Leśniewski's other logical views.

5. *Ways to Reduce the Tension*

There are I think at least the following three ways to learn to live with higher-order logic and still not be committed to Platonism. All of them involve fiddling with semantics. One way, which comes in effect closest to the method adopted by Tarski in his long truth essay, is to simply accept a Tarskian semantics for expressions with bound higher-order variables and buy the extensions over which such higher-order variables are to range, while denying that we have Platonic abstract individuals in such a case. In this way one follows the lead of medieval anti-realists such as Abelard, who accepted that for instance two things may have something in common, but denied that this something can be *named*. This view is, if we like, a kind of realism which is not as extreme as the sort which claims abstracta can be named. But in some sense "there are" such entities. The palette of positions which can be strung out between an extreme realism and an extreme nominalism has been extensively investiga- ted by Nino Cocchiarella.[7] Cocchiarella's work remains however within the bounds of Tarskian semantics, so that stronger ontological commitments and greater expressive power of the language go hand in hand.

A second possibility is to reject the Tarskian way of doing semantics, in particular in the case of mathematical discourse. The language of mathematics may be *prima facie* Platonistic, but this appearance may be compatible with for instance a fictionalistic interpretation of mathematics, according to which there are no mathematical objects and sentences implying that there are must be literally false. Field's programme has its difficulties and critics, and I should prefer a less sweeping attack on the problem, one which examined different kinds of discourse in a more piecemeal fashion.

A third possibility is that one should look for an interpretation of quantification and ontological commitment which steers between

7. Most copiously in Cocchiarella 1986.

263

the Scylla of referential Platonism and the Charybdis of substitutional Platonism. One suggestion, already made, is to follow up the extreme nominalistic approach to logic of Leśniewski, whose quantifiers are apparently neither referential nor substitutional. That is certainly a project for the future: the adepts of Leśniewski who strongly believe he has the right answer often have some difficulty in communicating their conviction to those who are not convinced but would like to be. Another attempt, which started out by considering how one would provide some recognizable semantics for a Leśniewskian style of language, is my combinatorial semantics,[8] about whose avoidance of Platonism I am however less than wholly certain. Although this method does not incur ontological commitments directly at the level of higher-order quantification, it passes commitments down to a next lower level via the complications of what I call ways of meaning. Whether such ways of meaning are themselves somehow hypostatized is the crucial problem.

So in the end I think we have to confront an as yet unsolved problem for those, like myself, who find themselves in the difficult cognitive situation portrayed at the outset. It would be nice to have one's cake (no Platonism) and eat it too (higher-order logic), and in the spirit of Arthur Prior I sincerely hope it is possible.

REFERENCES

Barwise, J. 1977. "An Introduction to First-Order Logic". In J. Barwise, ed., *Handbook of Mathematical Logic*. Amsterdam: North-Holland, pp. 54-6.

Cocchiarella, N.B. 1986. *Logical Investigations of Predication Theory and the Problem of Universals*. Naples: Bibliopolis.

Forbes, G. 1986. "Truth, Correspondence and Redundancy". In G. Macdonald & C. Wright, eds., *Fact, Science and Morality. Essays on A. J. Ayer's Language, Truth and Logic*. Oxford: Blackwell, pp. 27-54.

Prior, A.N. 1971. *Objects of Thought*. Oxford: Clarendon Press.

Shapiro, S. 1991. *Foundations without Foundationalism. A Case for*

8. Cf. Simons 1985.

264

Second-Order Logic. Oxford: Clarendon Press.

Simons, P.M. 1985. "A Semantics for Ontology". *Dialectica* 39, 193-216. Reprinted in Simons 1992, pp. 295-318.

— 1987. "Bolzano, Tarski, and the Limits of Logic". *Philosophia Naturalis* 24, 378405. Reprinted in Simons 1992, pp. 13-40.

— 1992. *Philosophy and Logic in Central Europe from Bolzano to Tarski*. Dordrecht: Kluwer.

Van Benthem, J. & Doets, K. 1983. "Higher-order Logic". In D.M. Gabbay & F. Guenthner, eds., *Handbook of Philosophical Logic*, I. Dordrecht: Reidel, pp. 275-330.

Part II
Ontological Reduction, Intended Interpretations, and the Löwenheim-Skolem Theorems

[10]

ONTOLOGICAL REDUCTION *

MY purpose is to investigate the notions of "ontological reduction" and "ontological economy." The discussion will
be centered around certain arguments and proposals
having to do with ontological matters that have been put forth in
two recent papers by W. V. Quine.** It is a complex issue with many
interesting and illuminating technical considerations. But in summary, I am going to argue that the primary technical proposals advanced by Quine are unsatisfactory, and, what is important, that an
analysis of their failure will lead to the conclusion that the over-all
program of ontological reduction and economy is misdirected.

I

There are natural examples that appear to exhibit some sort of economy, where certain kinds of entities, perhaps of a putative nature,
are eliminated. They are eliminated by showing that, in some sense,
other—perhaps more acceptable—entities will do the same work.
Some of the examples cited by Quine in the first paper are: Frege's
definition of the natural numbers, as well as those of Von Neumann
and Zermelo; various definitions of the real numbers using Dedekind
cuts; and Carnap's reduction of impure numbers to pure numbers.
The various versions of the natural numbers, as well as the various
versions of the real numbers, appear to be equally serviceable for ap-

⁹ See N. Goodman, 2d ed., *Fact, Fiction, and Forecast* (Indianapolis & New
York: Bobbs-Merrill, 1965), ch. III.

¹⁰ Essentially the same view is expressed by F. P. Ramsey, "Philosophy," in
The Foundations of Mathematics and Other Logical Essays (London: Routledge &
Kegan Paul, 1931), pp. 267–268.

* I would like to thank Hao Wang for much encouragement and many stimulat-
ing discussions. Thanks are also due W. V. Quine, Burton Dreben, and Charles
Parsons for a number of helpful clarifications and suggestions.

** "Ontological Reduction and the World of Numbers," in *The Ways of Paradox
and Other Essays* (New York: Random House, 1966), hereafter referred to as WP;
and "Ontological Relativity," in *Ontological Relativity and Other Essays* (New
York: Columbia, 1969), hereafter OR.

plication to the physical world. Quine infers that this is because the different versions are "structure-preserving" (WP 201). This could be interpreted in various ways, but Quine explicitly means by this that what counts is that they are all models of the same theory: "The problem of construing comes to no more, again, than modeling" (*loc. cit.*).

But there is an initial problem, because the models of the natural numbers, and the models of the real numbers that we are considering, are, respectively, isomorphic. That is, they have exactly the same structure in the strongest possible sense. There is a function that maps the domain of one model one-to-one onto the domain of the other model, preserving all relations. To infer from these examples that any two models of the same theory are equally acceptable would be unjustified. Because of my present restricted aims, I do not wish to argue for either the truth or falsity of the proposal that any two models are equivalent with regard to physical applications. But more justification is needed for the apparent claim that this is so, and undoubtedly interesting and subtle issues would emerge from an examination of this point. (Does one apply *models* or *theories?*)

It is clear, however, that, for mathematical applications, it is not the case that any two models of the theory of the real numbers are equivalent. Although it is commonplace to identify two isomorphic structures in mathematics, it would not do to identify the real numbers with some countable model. One frequently wishes to apply the theory of reals to other parts of mathematics, and one may need to know (say) that a set of reals, not a set in this countable model, has a least upper bound. Although Quine only suggests in the first paper that any two models are equivalent for physical purposes, in the second paper it is claimed that any two domains fulfilling arithmetic are equally good in general (OR 44, 45). However, here he says *progression* as well as model. A progression is of the standard isomorphism type and cannot be characterized by first-order axioms. Although Quine always seems to take theories and models in a first-order sense, he is tacitly taking this one in a second-order sense.

As it happens, both arithmetic and the theory of real numbers have a finite set of second-order axioms that characterize models up to isomorphism. But this fails for set theory, suggesting that, even if one greatly strengthens the definition of model, it is not to be expected that all models of a theory are equivalent for all purposes.

For first-order theories, there is a theorem, the Löwenheim-Skolem theorem, which asserts that all consistent theories have models whose domain is the numbers. This result is regarded by

ONTOLOGICAL REDUCTION   I53

Quine as a major threat to the subject of ontology. But if models are taken in a second-order sense (which we claimed was a necessary if not sufficient condition) then it is not clear that the Löwenheim-Skolem theorem applies.

There is confusion here caused by the fact that the introductory examples suggested that one is going to reduce models individually. One can think of reduction in some global sense so that all of the theoretical structure fits together and then our objections no longer apply. This seems to be the only possibility, and, if one looks at the question in this way, there is no completely obvious reason why one cannot maintain that numbers are all there is.

To see how Quine deals with the wholesale Pythagoreanism provided by the Löwenheim-Skolem theorem, it is helpful to introduce some terminology. We are dealing with first-order languages unless otherwise specified, and by a *theory* we shall usually mean a set of sentences closed under logical deduction. Quine uses *theory* to mean an interpreted theory; so we shall use the word ambiguously, and when it is not clear from context, *theory form* will mean a theory in the syntactical sense. The version of the Löwenheim-Skolem theorem which Quine uses in the first paper (and which we simply call *the* Löwenheim-Skolem theorem) asserts that any theory that has a model (or, equivalently, is consistent) has a model whose domain is the set of integers.

We shall consider both axiomatizable theories, i.e., those whose axioms are effectively given, and complete theories, i.e., those which contain either α or $\sim\alpha$ for any given sentence α. (Recall that some theories are in neither category, whereas others are in both.) These distinctions are quite important for part of the discussion. In the first paper, Quine is explicitly concerned with theories consisting of all truths of an interpretation, which are therefore complete in the above sense.[1]

Quine establishes (WP 202) the point that a reduction in ontology may be accomplished, typically, only at the cost of an increase in "ideology." More explicitly, when a given consistent theory is shown to have a model in the integers, it is not true that the relations over the integers need be arithmetical; in general, they are more complicated than arithmetical relations. On the other hand, for an *axiomatizable* theory T one can prove in arithmetic that $\mathrm{Con}(T) \to \alpha'$ whenever $T \vdash \alpha$, where α' is the result of replacing the predicates of α by certain arithmetical predicates. (This result, which we may call the

---

[1] Completeness must be allowed for the argument that follows. But in the second paper, while still dealing with interpreted theories, Quine only requires them to be logically closed. See p. 51 of "Ontological Relativity."

arithmetical Löwenheim-Skolem theorem, is got by carrying out a proof
of the theorem within arithmetic.) Therefore, we have a uniform reduction of all axiomatic theories to true arithmetic. Complete theories, as we have said, are generally too complicated to have an inner
model in true arithmetic. To prove this claim, note that if the starting theory is second-order arithmetic[2] with the natural model consisting of all integers and all sets of integers, then the existence of
arithmetical predicates corresponding to the original predicates
would enable one to define the truth of sentences of second-order
arithmetic in terms of the truths of arithmetic. But this definition
could then be carried out in second-order arithmetic itself, contradicting Tarski's theorem.

One might very well regard this latter demonstration as showing
that no general reduction of theories "to numbers" is possible, since
on the one hand one wants to preserve all truths of a theory, while
on the other hand, a reduction "to numbers" must mean a reduction
to arithmetic; that is, the new relations must be definable in terms
of the established arithmetical relations. The significance of the various reductions of numbers to sets resides in the fact that the arithmetical relations are defined purely in terms of the relations of an
antecedently accepted theory of sets. Similar remarks hold for the
other examples.

For Quine, an ontological reduction of a theory T to numbers may
very well entail new predicates over the numbers, predicates perhaps definable only in terms of the old predicates. This, we emphasize, is an important point in understanding Quine's position. The
Löwenheim-Skolem construction certainly seems to trivialize this
concept of reduction; so the notion of a proxy function is introduced
to narrow the concept of ontological reduction so that the serious
cases (e.g., Von Neumann's definition of number) qualify, but also
so that "there ceases to be any evident way of arguing, from the
Löwenheim-Skolem theorem, that ontologies are generally reducible
to the natural numbers" (WP 205). The conditions are: "We specify
a function, not necessarily in the notation of θ or θ', which admits as
arguments all objects in the universe of θ and takes values in the
universe of θ'. This is the proxy function. Then to each n-place
primitive predicate of θ, for each n, we effectively associate an open
sentence of θ' in n free variables, in such a way that the predicate is
fulfilled by an n-tuple of arguments of the proxy function always
and only when the open sentence is fulfilled by the corresponding
n-tuple of values" (*loc. cit.*).

[2] We mean the usual first-order theory of integers and sets of integers, or
"analysis."

A given sentence ⍺ in the old theory is to correspond to the sentence ⍺′ in the new theory got by replacing the old predicate letters by their corresponding open sentences. It is interesting to ask whether, given a proxy function satisfying the quoted conditions, each ⍺ true in θ goes into a true sentence ⍺′ in θ′. It turns out that this does not hold in general; but it is not difficult to establish that it does hold if the proxy function is *onto*. Therefore, one should add to this definition the condition that there exist an open sentence true of just the range of the proxy function, and that quantifiers of ⍺′ are to be restricted to this predicate. This ensures that truths go into truths.

We note that a related concept much used in the literature is that of *inner model*, where one does not require that objects be mapped to objects, but merely that the new predicates and domain be definable in the new theory in such a manner that truths go into truths.

The restriction to proxy functions does outlaw the Löwenheim-Skolem construction, since it generally bars any reduction of cardinality, at least in theories with identity.[3] That is, if f is the proxy function and the cardinality of θ is greater than that of θ′, then $fx = fy$ for some distinct x and y in the domain of θ. But if I plays the role of identity in θ′, $I(fx, fx)$ since $x = x$. So $I(fx, fy)$ while $x \neq y$, violating a condition the proxy function must satisfy. An analogous argument holds if identity is not a primitive of θ, but is nevertheless definable in θ.

There is a certain type of reduction, similar in some ways to Quine's, which is associated with another variant of the Löwenheim-Skolem theorem. Let us call f an *elementary imbedding* of one model into another of the same type if it maps the individuals of the first one-to-one into the individuals of the second, in such a manner that '⍺$(x_1, \ldots, x_n)$' is true in the former if and only if '⍺$(fx_1, \ldots, fx_n)$' is true in the latter, for any open ⍺ and any parameters $x_1, \ldots, x_n$ from the domain of the first model. In particular, f preserves all relations, and both models have the same true sentences. If the identity function on a model V is an elementary imbedding of V into U, then V (which is a submodel of U) is called an *elementary submodel* of U. The version of the Löwenheim-Skolem theorem that we shall call, along with Quine, the *strong* Löwenheim-Skolem theorem (and whose proof uses the axiom of choice), says that every model U has a countable elementary submodel V.

This is the version used in "Ontological Relativity." From the footnote on page 68 of that paper it is clear that Quine is willing to

---

[3] This point is made by Michael Jubien in "Two Kinds of Reduction," this JOURNAL, LXVI, 17 (Sept. 4, 1969): 533–541.

accept the type of reduction given by the strong Löwenheim-Skolem construction, and presumably also that given by imbedding functions. Ontological trivialization would result were it not for what apparently is a definability condition: "we must, of course, be able to say what class of objects is dropped, just as in other cases we had to be able to specify the proxy function."

Arguments are not advanced for the necessity of this condition, and it is not at all evident why it should assume such a critical role. There are familiar theories, such as set theory with the axiom of constructibility, in which it is provable that there is a definable well-ordering of the universe. In such a theory, if a model is definable then the submodel given by the strong Löwenheim-Skolem construction is likewise definable. (The model given by the ordinary Löwenheim-Skolem construction would in any case be definable in terms of the initial model.) Similarly, any definable model is reducible by means of a definable proxy function to any other definable ontology (i.e., set) if the cardinality of the latter is no smaller than that of the model's domain. Since the extent to which the Löwenheim-Skolem construction gives economy, does not, as an intuitive matter, seem to depend on such considerations as whether there is a universal definable well-ordering, doubt would seem to be reflected on the relevancy of the definability restriction.

One can go further. The strong Löwenheim-Skolem construction has the property that the submodel provided has the same relations as the original, restricted of course to a smaller domain. There is no reason to suppose this is important from Quine's viewpoint, and R. E. Grandy[4] has argued that, taking the usual Löwenheim-Skolem construction, one can always inflate the model, then define the class of objects dropped, and finally get a numerical model.

II

Technical obstacles have failed to block the Löwenheim-Skolem trivialization. It is disturbing that nowhere in this first paper is it argued why this general concept of ontological reduction should not be trivialized, that is, why the reductions given by the Löwenheim-Skolem construction are less satisfactory than others. Likewise, it is not demonstrated that the existence of a proxy function is a necessary or a sufficient condition for a reduction to fall under some independently explained concept of ontological reduction. It is noted that all the examples have proxy functions, but, for that matter, they are all isomorphisms.

[4] This JOURNAL LXVI, 22 (Nov. 20, 1969): 806–812. In certain ways Grandy's over-all position is similar to my own.

At this point it is perhaps appropriate to examine Quine's view of the general status and meaning of ontological reduction and economy. This is best described in the second paper. Quine views the reduction of U to V as given by specifying, relative to a background theory containing U and V, a proxy function mapping U to V; it need not exist as an object, but only as an open sentence. Even given that such a map exists, one might wonder how to formulate the reduction so that economy can be claimed. Quine answers this by:

> If the new objects happen to be among the old, so that V is a subclass of U, then the old theory with universe U can itself sometimes qualify as the background theory in which to describe its own ontological reduction. But we cannot do better than that; we cannot declare our new ontological economies without having recourse to the uneconomical old ontology.
>
> This sounds, perhaps, like a predicament: as if no ontological economy is justifiable unless it is a false economy and the repudiated objects really exist after all. But actually this is wrong; there is no more cause for worry here than there is in *reductio ad absurdum*, where we assume a falsehood that we are out to disprove. If what we want to show is that the universe U is excessive and that only a part exists, or need exist, then we are quite within our rights to assume all of U for the space of the argument. We show thereby that if all of U were needed then not all of U would be needed; and so our ontological reduction is sealed by *reductio ad absurdum* (OR 58).

These paragraphs give rise to a number of interesting questions. It is first of all implied, and further confirmed by later remarks, that the *reductio ad absurdum* argument is meant to apply to self-reducing theories and that these cases give economy in some satisfactory sense in which the other cases don't.

There is a certain prima facie plausibility to the idea that, if one can do a given theory with fewer objects and without making further commitments or assumptions, then there is a clear advantage in doing so. Indeed, I find it impossible to see that Quine's motivation could be other than this.

It is therefore rather startling to reflect that there are many self-reducing theories, and in fact the most important theories have this property. For example, the set of numbers larger than 16 can be proved in arithmetic to form the domain of a model of arithmetic.[6] A similar situation results in set theory if one lets $\{\phi\}$ take the role of ϕ and extends the correspondence in the obvious way; here the same relation is used in the submodel as in the starting model. No one could maintain that these are interesting examples of reduction; so they are instructive in that they tend to illustrate what reduction cannot be about.

---

[6] This is mentioned in another context on p. 54 of "Ontological Relativity."

It is not entirely clear how to fit the standard examples into the self-reducing framework, although this apparently is meant to be done. One normally views Von Neumann's definition of number as a reduction of numbers (or number theory) to sets (or set theory). It could be viewed perhaps as a reduction of a set theory that takes numbers and arithmetical relations as primitive (call it S), to a pure set theory S'. The natural way to describe this is to say, e.g., that $\{\phi\}$ is assigned to 1, and, of course, $\{\phi\}$ is also assigned to $\{\phi\}$.

But S would normally have identity; so the proxy function would have to be one-to-one. Thus, since Von Neumann lets the number 1 be replaced by $\{\phi\}$, we must rather arbitrarily assign $\{\phi\}$ to some other set. One can define such a map, but, besides the arbitrariness here, it seems unnatural to formulate the matter in such a manner, since the theory of S is not what we wish to preserve—with its *Urelemente* it is in fact somewhat inelegant.

Let us assume, though, that we have accepted a formulation of our major examples in terms of self-reduction, using some criterion other than proxy functions. What does the *reductio* argument demonstrate? There are three key terms used in this argument: 'exist', 'need exist', and 'needed', where the latter is evidently short for 'needed as a model for (a certain theory)'.

Certainly the argument does not show that only part of U exists. From the standpoint of the background theory—which is U—all of U exists. Perhaps from the standpoint of V one would say that not all of U exists. (Note that this is exactly the same situation as for an ordinary case of reduction.) One might also wonder about the claim that one has shown that U doesn't need to exist, since V depends on U.

What clearly is demonstrated is that, if one is considering the adoption of the universe U, one can without deeper commitment assume V and get an interpretation of the theory form of U in V.

Since we may fairly assume that in general V is defined and known only in terms of U, what is the point of accepting V? Surely it is no *safer* to accept V than U. V is no *clearer* than U. Economy in the sense rejected above would seem to be the only motive.

As further evidence against this kind of economy, note that there is in fact an *infinite regress*. If U is excessive and $U - V$ is nonempty, then $(\exists x) \sim \alpha(x)$ is true in U, where α defines V in U. All truths of U hold in V; so α defines a proper subuniverse V_1 of V. It is easy to conclude that one gets an infinite sequence $U, V, V_1, V_2, \ldots$ each a proper subuniverse of the preceding, and each satisfying all truths of U. Which is the most economical?

ONTOLOGICAL REDUCTION   159

It may be illuminating to compare the Quine self-reducing theories to a similar phenomenon which occurs in inner-model proofs of consistency. One shows in Gödel's relative-consistency proof of the continuum hypothesis that a certain universe $L(U)$ satisfies ZF plus the continuum hypothesis, given that U satisfies ZF. Here one does not get an infinite regress because it is quite possible that $U = L(U)$ to start with—in fact $L(L(U)) = L(U)$.

One does not show that *every* truth of U is a truth of $L(U)$, but only the axioms.[6] If one could show that all truths are preserved one would have proved the continuum hypothesis outright.

This proof might be viewed as giving economy in a different sense, because, although one hasn't shown that $L(U)$ is different from U, one known $L(U)$ is well-behaved. Since one cannot show that all truths are preserved, one presumably will restrict attention to axiomatizable theories. But as we remarked very early, for such theories one can get arithmetical models at once. So again economy proves elusive.

Consider again our example of S and S'. Now I think that, provided one is a little loose about one's meaning, it is reasonable to claim that one shows that the theory of S can be done in S'. In what sense was the move to S' motivated by one's wishes to show the universe S "excessive"? One did want to show that the numbers as primitive weren't needed. But is seems quite unfortunate to say that one wants to show the universe "excessive." One specifically wanted S'.

It is a self-sufficient, intuitively appealing universe in its own right; moreover its usual axiomatization is neater than that of S. Also note that one didn't introduce some new relations over S', but defined the arithmetical operations over S' in terms of membership.

It might be objected that there is no harm in construing ontological reduction very broadly. Just as not every inner model is interesting, one should not expect to find every ontological reduction interesting. The reply of course is that to construe it so broadly is to ignore the features that made the initial examples significant.

III

An obvious question is, To what extent do the reductions provided by self-reducing theories differ from the kind of reduction in which one shows that U is reducible to V on the basis of some background theory not identical with the theory of U. Typically, Löwenheim-Skolem constructions where one is reducing uncountable models are of this latter kind, and this is taken to be a distinction "which dis-

---

[6] As it happens, various other truths are preserved, e.g., all truths of arithmetic; however not all truths of the reals need be preserved.

courages any general argument for Pythagoreanism from the Löwenheim-Skolem theorem" (OR 62).

Certain differences between this type of reduction and the type considered earlier are evident: in Löwenheim-Skolem constructions one must assume that U is a set, and then perform certain set-theoretical constructions on U. Where U is the universe of all sets, and thus not a set, there may in fact be certain qualms about the sense and legitimacy of this move. But I think in general one need not feel discouraged by the fact that a "stronger" background theory is required.

It is important to note that these proofs are not like a proof of consistency given in a theory whose acceptance would entail acceptance of the theory whose consistency is in question. Rather we are concerned with convincing ourselves that, if there is an interpretation, then a countable model exists. Suppose that someone was considering adopting a universe for set theory. Lacking a cogent argument that would discredit it as an ontological reduction, he could simply take the form of the Löwenheim-Skolem theorem that says that a consistent theory has a model over the integers. The mathematical principles used to prove that if there is no countable model then there is an outright formal contradiction would hardly be questioned by him, since they are implied by his theory. Moreover he believes the theory is not formally inconsistent; so he must be willing to accept a numerical model.

It is perhaps interesting to note that, even in a theory whose universe is uncountable (seen from inside), one can sometimes have self-reduction to the integers.

The set theory that is able to prove its own ontological reduction to the integers is due to Myhill.[7] It extends the Von Neumann-Bernays-Gödel set-class theory (which is an inessential extension of ZF) by adding a sentence, '$(\forall X)(\exists n)D(X,n)$', where '$D(X,n)$' asserts that n is (the Gödel number of) a formula defining X. If we define $F(X,n)$ as $D(X,n) \wedge (\forall i)(i < n \rightarrow \sim D(X,i))$, then it is provable that $(\forall X)(\exists n)F(X,n)$ and that, if $F(X,m)$ and $F(X,n)$, then $m = n$. Since, moreover, $F(X,n)$ and $F(Y,n)$ imply $X = Y$, it follows that F is a virtual function mapping all classes one-to-one into the integers. The obvious contradictions are avoided since we may not infer that the range of F is a class.

---

[7] "The Hypothesis that All Classes Are Nameable," *Proceedings of the National Academy of Sciences*, XXXVIII (1951): 979. The brevity of the presentation has given rise to some question as to the validity of the proof of relative consistency; but it can be completed in detail. In any case, the consistency is not in doubt, since a model is $M(\gamma + 1)$, where $M(\gamma)$ is the minimal model for ZF.

The procedure is actually quite general. If T is a standard set theory consistent with a sentence giving a universal definable well-ordering, one can extend T consistently to a self-reducing theory T'. A subtle application of the Löwenheim-Skolem construction is used.

IV

The set theorist must agree that I, the Pythagorean, can add a new predicate letter to arithmetic, and assume, relative to some interpretation of it, all set-theoretical sentences he has accepted, and even all such sentences as are true. It would be quite reasonable for me to claim that I am taking everything to be a number, since I retain the interpretation of the arithmetical relations over my entire domain.

As we remarked earlier, Pythagoreanism must be understood in a global sense if things are to fit together. There might be advanced some argument against this global Pythagoreanism in that again things don't quite fit together. For example, there are two different kinds of numbers, namely, those which encompass everything, and then the more artificial kind of which the reals are an extension. But I don't think a Pythagorean need be too bothered by this duality. From a purely ontological standpoint such Pythagoreanism seems perfectly all right.

The serious objections to this type of Pythagoreanism are not of an entirely clear-cut nature, although they are overwhelming. Consider present-day set theory. The set of axioms one has accepted at any given time form an effectively specifiable set, say we take it to be ZF. Such a set of axioms, if consistent, has a numerical interpretation, with an arithmetical relation representing membership. But such a model is utterly different from the standard model. This is especially clear from the point of view of finding new truths about set theory. Our number-theoretical model gives a completion of ZF, but an essentially arbitrary one; that is, sentences other than the consequences of ZF come out true or false depending on certain syntactical properties. A new axiom accepted in set theory will, with probability $1/2$, be inconsistent with the completion in the numerical model. Nor need one look far for acceptable new axioms; Con-(ZF) is one.

Of course, there does exist a numerical (but nonarithmetical) interpretation satisfying, at once, all truths of set theory. Although surprising, this seems a quite unimportant fact. What one is really interested in is learning more about sets, and arithmetical or numerical models contribute nothing. All of one's knowledge of things true in these deviant models for set theory (or whatever) is derived from one's knowledge of sets (or whatever). In this respect the discovery that countable models exist is no more informative than the observation that the truths form, via Gödel numbering, a countable set of integers.

Progress in set theory or in the discovery of new mathematical principles comes slowly. The situation would be more striking if we had cited a physical science, for there not only are new principles added more rapidly, but there can be no question that a natural model provides a unique source of insight into new truths. I do not mean to imply that numbers (say) are particular and definite objects. Perhaps the natural "model" here is not a specific model, but the standard isomorphism type. But even if something like this is the case, the same principle holds true.

We are treading on very slippery ground indeed. I of course cannot pretend to give an account of the "perception" of mathematical truths. Yet one feels that the heart of the matter lies here, and, for the question at issue one need not provide such an account.

The attitude toward reductions that we have been contemplating is not very different from some sort of formalist position. If particular models don't matter, what matters is their theory form. One's objections to blanket Pythagoreanism should be similar to one's objections to formalism.

Quine, in the second paper, agrees that such Pythagoreanism only obscures matters. But he tries to block it by formal criteria (proxy functions and a definability condition), and certain informal arguments (the Löwenheim-Skolem construction takes place in a stronger background theory). One should have concluded that epistemological considerations are the source, and a valid source, of one's discomfort.

V

If ontological economy seems pointless, one might wonder what exactly is exemplified by the examples taken as a starting point for the analysis of ontological reduction. The examples involve many different considerations. The basic step seems to have been to single out, within a theory independently accepted as significant, an inner model satisfying the old theory. Even at this stage subtle considerations obtrude. It would not do to choose just any model for the first-order theory. Theories such as Peano arithmetic or the theory of real numbers have a categorical set of second-order axioms. Given the notion of an arbitrary set, one knows exactly what one wants, and it is unlikely that one would accept a candidate not of the standard isomorphism type. Of course this is not a problem in practice—it takes logical ingenuity to construct first-order models not of this type.

The example of the definition of the numbers in set theory may be claimed to be of purely ontological significance, since number theory is more secure than set theory. But even here there is some epistemological content—merely adding new primitive predicates to set theory and doing arithmetic relative to them would not have been taken seriously. Also it must be emphasized that the reduction of number theory to set theory was initially regarded as more important than it is now, just because the set theory was originally supposed to be logic, and thus provide a great deal of epistemological reduction.

Some of the other examples do not involve economy in any sense. If one has isolated one of the standard definitions of the reals, then one may be interested in a reduction of one model to another (by proving an isomorphism exists) even though both are accepted to exist, just because the new one has certain features desirable for the purpose at hand: e.g., it may be easier to develop formally or be more intuitive.

A more serious consideration is that, if one's knowledge of the reals consisted of an unclear mixture of some algebraic rules and geometrical intuitions, then a definition of the reals in terms (say) of sets of rationals would be an important advance.

I think this issue, suggested (if not exemplified) by our examples, is the most important.

It may be instructive to consider an example where one emphatically does not understand the old objects or their theory, but where one does achieve reduction, economy, and explication. Abraham Robinson's definition of a system of infinitesimals,[8] which we take to be in set theory, may for the present purposes be considered to be such an example. Here a model is defined, using, of course, only set-theoretical concepts. The language and theory form of infinitesimals are as obscure as the objects themselves; yet it is reasonable to claim that this model is a reduction, because the new infinitesimals do the sorts of things infinitesimals were supposed to do. For example, the derivative of f at a is infinitely close to $(1/z)[f(a+z) - f(a)]$ when z is any nonzero infinitesimal.

It might be debated whether such elimination and explication should be considered in the same context as ontological reduction. Quine does, though, intend to fit such cases into his general framework:

> Ontological reduction of a muddy theory, as of old infinitesimals, makes sense on my terms only relative to a correspondingly muddy background language. The discomfort that one might feel from this

[8] *Introduction of Model Theory and to the Mathematics of Algebra* (Amsterdam: North-Holland, 1963).

...is for me allayed by reflections akin to [the second paragraph quoted in section II above].[9]

Now it would seem that on this view one gets (under reasonable assumptions) an easy solution of the mind-body problem. Simply let correspond to each mind an ordinal number, and carry all relations over to the ordinals, etc.

Apparently the most that can be said in general about the kind of explication one wants is that one must provide an inner model in a theory agreed to be acceptable, and show that this model satisfies the theory, or bits and pieces of theory, of the problematical universe; or to put it even more vaguely, it must exhibit behavior that would justify us in calling the new objects infinitesimals, or whatever. One can make sense of such reduction from the standpoint of quite a weak metalanguage, since it is typically a matter of showing how certain sentences are provable in the theory to which we are reducing. To the extent one understands the objects of the acceptable theory, one understands the new infinitesimals (or minds, or whatever) as objects. To the extent one is confident of the truth of the accepted theory, one is confident of the truth of the (reinterpreted) old theory.

Perhaps this merely shows that explication is different from ontological reduction. Very well. Then we have evidence that, with regard to genuinely problematical theories, ontological reduction is again unimportant.

LESLIE H. THARP

Rockefeller University

[11]

VIRGINIA KLENK

INTENDED MODELS AND THE
LÖWENHEIM–SKOLEM THEOREM*

Formally the Löwenheim–Skolem theorem is unproblematical: any first-order theory which has an infinite model has a denumerable model. But recent controversy over the interpretation of this result indicates that its philosophical significance remains unclear.[1] The Skolemite sees it as onto-logically significant, as showing that there is no such thing as "absolute" uncountability, but only uncountability relative to a formal system, and hence that there are only finite or denumerable sets in the universe. For to claim uncountability for a set, the Skolemite argues, is just to assert that there is no enumerating function for the set, but since this non-existence claim is proved within a particular formalization, all it shows is that the formal theory itself is not powerful enough to generate the enumerating function. It does not exclude the possibility of such a function outside the formal theory, and that the required function does exist, according to the Skolemite, is shown by the Löwenheim–Skolem theorem, which provides a denumerable model for any first-order theory. For in the Skolem model, the term which designates the supposedly uncountable set is provided with a referent which is at most denumerable. Thus the set, though uncountable within the formal theory, is countable outside the theory, in the meta-language, and hence no term of a formal theory can be taken as designating a set which is anything more than relatively uncountable.

The Platonist, on the other hand, discounts the ontological significance of the theorem, claiming that the Skolem models, though denumerable, do not provide either an enumeration of sets (but only map *statements* about sets onto statements about numbers)[2] or an acceptable rendering of the concept of membership (but only a particular unintuitive fragment of number theory), and that furthermore, the Skolemite has overlooked the other side of the coin – the "upward" Löwenheim–Skolem theorem, which tells us that any theory with an infinite model has models of any infinite cardinality, including vastly nondenumerable models. And there is absol-utely no reason, according to the Platonist, to allow the countable models greater significance than their uncountable cousins, except for a stubborn

Journal of Philosophical Logic 5 (1976) 475–489. *All Rights Reserved*
Copyright © 1976 by D. Reidel Publishing Company, Dordrecht-Holland

476   VIRGINIA KLENK

preference for countability which has nothing to do with the proof itself.
Thus the theorem rules out nothing, and ontological questions must be
decided on other grounds.[3]

Much of the controversy centers around the question of what constitutes
an intuitively acceptable model for set theory. The Platonist rejects the
Skolem model as the intended interpretation of set theory on the grounds
that it provides only an unintuitive number-theoretic relation which cannot
be taken as a true representation of the concept of membership. The
Skolemite replies (not entirely consistently, perhaps) that it is not surprising
that the models are unintuitive, since the whole notion of an intended model
for set theory is unclear,[4] and that anyway, it is possible to obtain intuitively acceptable denumerable models by a proof of the Skolem theorem
which generates an elementary submodel of whatever model the Platonist
finds acceptable to begin with.[5]

I shall argue in the following pages that neither the Platonist nor the
Skolemite can make a convincing case: that the Skolemite cannot really
establish that there are only denumerable sets, but that on the other hand,
neither can the Platonist establish that there are nondenumerable sets. My
own non-conclusion is that the question is undecided, and perhaps undecidable except by fiat, but this does not mean that the discussion is philosophically unrewarding. There may be significance in the lack of ontological
significance of the Löwenheim—Skolem theorem: evidence for some sort of
Formalist view of mathematics.[6]

I shall first examine certain aspects of the Skolemite's claim that the
Skolem theorem shows that there are only denumerable sets.[7] One common
objection to the Skolemite position is that the denumerable models do not
give us an intuitive interpretation of the concept of membership at all, but
merely of a complex number-theoretic predicate which would never have
seen the light of day except for its isomorphism with the membership predicate as it appears in formalized set theory. As Resnik puts it, "How does it
make sense to speak of sets here? The membership predicate has been given
a numerical reinterpretation, and set-theoretic statements have been
absorbed into number theory. Thus we are no longer dealing with sets . . ."[8]
In other words, under the Skolem interpretation the theorems of formalized set theory must be taken as theorems about the natural numbers, and
there is a serious question whether the notion of set has not entirely disappeared.

The Skolemite might reply that even if this were a legitimate objection to the original Skolem model, it is hardly a definitive refutation of the Skolemite position, since this particular model is an accident of this particular proof, and there are several different methods of proving the Skolem theorem. Indeed, in what is now perhaps a more familiar proof the domain consists of a subset of the set of terms of the formalized theory, and the predicates are modeled nicely by just those sets of n-tuples of terms between which the predicate is said to hold. It is hard to see how this Henkin model could be considered less than intuitive, since it simply mirrors the sentences of the formal theory itself.

But the Platonist would no doubt rejoin at this point that such a mirroring is not what he has in mind by "intuitively acceptable", and that the Henkin model is no more acceptable than the number-theoretic interpretation. If our statements are now taken to be about terms rather than numbers, little has been gained from the Platonist's point of view. It is plausible to assume that the only sort of model the Platonist will find acceptable is one in which the domain is a set of sets and the membership predicate is given its "normal" interpretation.

It is possible, however, to find a denumerable model which fulfills these conditions. Through a proof of the Skolem theorem which utilizes the Axiom of Choice, we may obtain a denumerable submodel of the Platonist's original interpretation,[9] and since this submodel is just a restriction of the original model to a denumerable domain, even the Platonist is willing to admit that it is intuitively adequate,[10] and it seems the Skolemite is home free.

The Skolemite, however, wants to show that there are *only* denumerable sets, and the mere existence of a denumerable submodel does little to shore up this claim. For by hypothesis, the denumerable domain of the Skolemite's model is a subset of a larger domain, the existence of which must be assumed in the first place. One reply which has been given to this argument is that we no more need accept the existence of the original model than we need accept the truth of a statement which we assume for a Reductio proof. We simply "assume all of U for the space of the argument. We show thereby that if all of U were needed then not all of U would be needed; . . .".[11] There are really two questions being confused here, however: whether the domain exists, and whether it is needed for the model. No contradiction is generated by supposing that the original set *exists* (indeed, if this could be done there would be no need for these endless discussions about the significance of the Skolem theorem); thus the conclusion cannot be that the set does not exist. Rather, all we show is that a model can be constructed without the entire domain, i.e., the original set is not all *needed* for the model. Since the existence of the original set has not been refuted the only question is the cardinality of this set, and the Skolemite has not claimed that *this* set is denumerable; presumably he grants its nondenumerability at this juncture. The Skolemite can get his intuitive denumerable model, then, but in the process he seems to be cutting the ground from under his own feet.

But even if this dispute could somehow be resolved in favor of the Skolemite, there is a more serious problem with his argument. The Skolemite, and the Platonist as well, have been simply assuming that the submodel, which is just a restriction of the original interpretation to a denumerable domain, will retain the intuitive properties of sets and membership, but unfortunately for the Skolemite, there is some reason to doubt that this is the case. A simple example from number theory may illustrate the problem. If we consider the relation of properly less than, and restrict it to the first ten natural numbers, the resulting sets of pairs $\{\langle 1, 2\rangle, \langle 2, 3\rangle, \ldots \langle 4, 6\rangle \ldots \langle 9, 10\rangle\}$ is certainly still intuitively the less than relation. But if we restrict the original relation to a different domain, say $\{2, 4, 8, 16 \ldots\}$ the resulting set of pairs will look like $\{\langle 2, 4\rangle, \langle 2, 8\rangle, \langle 4, 8\rangle, \langle 2, 16\rangle \ldots\}$ and in this case we would likely say that the predicate represented by the relation was "divides evenly by $2x$", rather than "properly less than". There are similar situations in set theory. By an appropriate restriction we might end up with a domain consisting of just those sets which are both members of and subsets of each other, e.g., the natural numbers in the Zermelo construction. And in this case we would have no reason to take '∈' as membership rather than subset. Of course *formally* the relation can still be said to be that of membership, but that is not the point. The point was whether such a restriction would yield a relation which was *intuitively* the same as the original, and it is clear that this is not always the case. The existence of such submodels, then, should be of no solace to the Skolemite; if denumerable models are unintuitive without the Axiom of Choice, there is little reason to think the situation is much improved by applying it in this way.

The Skolemite has one last resort. He can simply deny, as Thomas does,

that there is any need for an intuitive interpretation of '∈', on the grounds
that the notion of an intended model for set theory is none too clear in the
first place. Thus we should not be surprised, or unduly alarmed, this argument goes, to find the membership predicate modeled by a complex, unintuitive number-theoretic relation.[12] This is a little odd, however, in view of
the fact that there are unintuitive Skolem models for *any* formalizable
theory, e.g., Newtonian mechanics or the kinetic theory of gases. No one
argues in these cases, however, that lack of clarity in the idea of a standard
model forces us back upon non-standard models. It is only in set theory that
the Skolem models are given any particular importance, and to carry his
point that all we can reasonably expect in set theory are unintended models,
the Skolemite will have to come up with some explanation of why set
theory is different from other theories in this regard.

Where does this leave us? The Skolemite has failed to demonstrate that
his denumerable model conforms to our ordinary notion of membership,
and his defense that it need not do so is, so far, unconvincing. We are forced
to the view, it seems, that while there are indeed denumerable models for
set theory, these are not the intended models, and since we must then turn
to other interpretations, the Löwenheim–Skolem theorem remains without
ontological force. This conclusion is only reinforced by a consideration of
the "upward" version of the Skolem theorem, which tells us that set theory
has models of any arbitrary cardinality. To conclude that there are only
denumerable sets because any set-theoretic term *can* be interpreted in a
denumerable domain is simply to beg the question, since the term can also
be interpreted in a non-denumerable domain.[13] The most sensible conclusion seems to be that urged by both Resnik and Myhill: formalized set
theory admits of models of any infinite cardinality, and what the
Löwenheim–Skolem theorem shows is not that nondenumerable sets do not
exist, but only that they cannot be completely characterized by the structure of first-order logic, since any first-order proposition can be construed
as a statement about a denumerable collection.

Of course, as Thomas and others have pointed out,[14] this conclusion, and
the objection to the Skolemite position on which it is based — that the
Skolem models are not intuitively adequate representations of the concept
of set — rest upon the supposition that there *is* an intuitively adequate
model, and that it can somehow or other be adequately characterized.
Unless the Platonist can establish this claim, there is little point to his

480   V I R G I N I A   K L E N K

objection. In the following paragraphs I should like to examine the basis of
this claim.

Both Myhill and Resnik, in arguing for the existence of nondenumerable
standard models for set theory, make an essential appeal to the distinction
between formal and informal, or intuitive, mathematics. Myhill claims, for
instance, that "there is evidently only one standard model of set theory",
and that the existence of nonstandard models shows us only that we must
never "forget completely our intuitions. . . . a formalism . . . only remains
set-theory as long as the intuition of membership has not slipped away from
us".[15] And Resnik argues that "the primary purpose of most formal systems
is the formalization of informal mathematical theories. It is *via* these
theories that the intended models of the system are furnished, . . .".[16] Thus
"the intended model is already at hand when the formalism is introduced".[17]

This distinction between formal and informal mathematics is essential
to the Platonists' claims, since it is obviously impossible to characterize the
standard model by formal means. Any formal description would again be
open to the Skolem interpretation and thus could just as well be taken as a
description of a denumerable nonstandard model. The question at this point,
then, is what this distinction amounts to, and what the relationship between
formal and informal mathematics might be.

Myhill is merely suggestive, noting that *formal* mathematics depends
upon an "informal community of understanding" among mathematicians.[18]
He says that we can no more assure formally that our interpretation of set-
theoretic concepts (and nondenumerability in particular) will conform to
other mathematicians' than we can assure that our "internal" perception of
green is the same as others'. But this is a particularly unfortunate analogy
for the Platonists' point, since for this very reason a great many philosophers
have concluded that it is futile even to discuss the question of whether your
green is the same as mine. It is not as if there is some other, informal , level
at which the question could be decided. Perhaps by the same token we
should conclude that it is futile to try to decide, even on informal grounds,
which is the intuitive, intended model for set theory. Or perhaps this
"informal community of understanding", as in the case of the word "green",
can be taken as something like community use of the word, and indeed, this
is how Resnik tries to spell out what is meant by an informal understanding
of set theory. But as we shall see, if this is all that informal understanding
amounts to, it does little to support the Platonist's claim that there is an
intended, informally specifiable model for set theory.

According to Resnik, set theory as formalized can be considered just a part of number theory, but on the *in*formal level number theory "can be distinguished from set theory and its terms by appealing to the *role* each has in the totality of mathematics. We learn the role of set theory when we study it. We learn more than its quantificational structure; we learn to *use* its terms and statements."[19] (My emphasis.) He goes on to say that we clearly understand the meanings of set-theoretic terms and statements, and do distinguish them from number-theoretic propositions, and that these meanings are probably best elucidated in terms of use.[20]

However, it is difficult to see how this appeal to the role of set-theoretic statements, or their use in mathematics, is going to help in making the distinction between standard set-theoretic interpretations and non-standard number-theoretic interpretations. Certainly the role or use of statements seems to have very little to do with what sorts of models they have. No matter what the use, no matter where a proposition appears in the formal theory, it is still open to various semantic interpretations. And to say that we learn the *role* of set theory when we study it, learn to *use* its formulas, sounds very much like saying that we learn to put the right formula in the right place — which is very much like saying we learn to manipulate the formulas — which is very much like saying we learn a purely formal procedure. The Platonist cannot get out of this bind by claiming that we learn something *more* than manipulation, e.g., truths about sets, because the problem in the first place was to show we were doing set theory and not number theory. We cannot appeal to a supposed distinction between numbers and sets (the objects of our discourse) in trying to explain what it is we are learning, since we were appealing to the learning process to explain the difference between sets and numbers. The attempt to elucidate the notion of informal mathematics in terms of use, then, is of no help to the Platonist in his search for an intended model.

Of course, there are pictures associated with the learning process, and the pictures of sets are certainly different from the "pictures" of numbers, and perhaps it is something like this that the Platonist has in mind when he talks about the informal difference between set theory and number theory. We are aided by graphics; we are given certain representations of finite sets, perhaps, and the results of the union and intersection operations upon them. But the Platonist should not put too much emphasis on this visual aspect, or suppose that the difference between formal and informal mathematics can be described in these terms, for such a move could lead to a conceptual-

ism akin to Intuitionism, a position very much opposed to the spirit of
Platonism.

In fact, it is not at all clear how we should characterize the difference
between formal and informal mathematics. According to Resnik we have a
great deal of informal knowledge about sets prior to any axiomatization, and
in fact, the point of formalization is simply to codify, and perhaps in the
process make more rigorous and precise, an already existing domain of
mathematics. Our concept of "set" is to be found at the informal level, and
can never be fully captured by the formalization. But it is not clear to what
extent this informal set concept is independent of the formalization, for by
Resnik's own admission the formalization may lead to changes in the concept by pointing up inconsistencies, or simply by increasing our precision.[21]
Thus the formal system is not just a repository for previously acquired concepts, but is in some sense creative, and it is difficult to see how the intended model could already be "at hand" when the formalism is introduced.

Perhaps a more adequate view of the relationship between formal and
informal mathematics would be the following: informal mathematics consists of sentences which may be vague, incomplete, and even inconsistent.
Formal mathematics is a more precise (and, one hopes, consistent) language
into which the sentences of the informal theory may be transcribed. The
relationship, above all, is between two *languages*, one far more precise and
comprehensive than the other. (In fact, it would no doubt be more accurate
to say that what we have is a whole series of languages, with a fragment of
our ordinary language on one end and the fully formalized theory of sets on
the other.) Our intuitive mathematics should be seen as something less, not
more, than our formal mathematics: less precise, less consistent, perhaps,
and less complete (unless, of course, it is inconsistent, in which case the
completeness hardly counts as an advantage.)

But on this view there is little reason to think that pre-formal mathematics has any special power to express the intuitive sense of mathematical concepts such as "nondenumerability". Intuitive mathematics must,
after all, be couched in language, and there is no reason to think that the
ordinary language of informal mathematics is in any way superior to the
formal language of first-order predicate logic. Aside from the fact that informal mathematics is *temporally* prior to formalization, it is difficult to see
what the evidence might be for Resnik's claim that the formalization provides only a *partial* specification of the informal set concept. Once we have

formalized the notion, moved from our ordinary language to the formal language, what *more* could there be in the informal theory, unless one supposes that vagueness is a hallmark of comprehensiveness. Where is this "real" concept of set to be found? "Nods, and Becks, and wanton Wiles" (to mangle a metaphor) won't do in mathematics.

There is one other possibility, of course, and that is that our ordinary language is in fact a second-order language, in which case it would be expressively superior. But aside from the inherent difficulties with this idea, the problem of the intuitive concept of "set" would still be more satisfactorily discussed on the formal level, as a question of the relative expressive powers of first and second-order languages.

It seems, then, that the attempt to make clear the notion of an intended model for set theory in terms of informal mathematics is a failure. Nevertheless, Resnik declares that there is no more difficulty in interpreting set theory than in interpreting elementary logic, arithmetic, or physics.[22] The analogies do not hold up, however, except in the case of arithmetic, and there the similarity points to a conclusion quite different from Resnik's own. In the first place, it seems a little odd to speak of an interpretation of elementary logic in this context, since absolutely anything will count as an interpretation. This can hardly help us with the issue of distinguishing standard from nonstandard models. Nor is the case of physics analogous to set theory. There are obvious physical interpretations for macroscopic terms, and strong, if not absolute, definitional links between macroscopic and theoretical terms. But there is no sort of definitional link between even the most abstract physical term and the terms of set theory.

This leaves arithmetic, and it is not surprising that the problems of interpreting set theory should be similar to those of interpreting arithmetic, since on the one hand numbers can be taken as sets, and on the other, sets can be taken as numbers (the fact which motivated this whole discussion). But as we shall see, the analogy undermines rather than supports the Platonist's claim that there is an intended model for set theory.

In the first place, there are also nonstandard models for arithmetic, and the same difficulties in distinguishing standard from nonstandard models. But even if we could make this distinction clearly, there are many different *sorts* of things which could serve as a basis for standard models. We might take numbers just as indefinable abstract entities, or we could construe them as sets. In the latter case, of course, there are various possibilities: we could

take the number two, for instance, either as {{0}} or as {0, {0}}. We might
even take numbers as collections of inscriptions. This does not worry
Resnik in the least, and in fact he even suggests that the same thing might
be said about sets. It is here that he gives away the whole game.

He wants to claim "ontological neutrality" for his thesis that there is an
intended model for set theory, and thus he asserts that it is possible to maintain that an ontology of sets is "actually an ontology of properties or even
that an ontology of sets is not even an ontology of abstract entities."[23] But
now what has become of our informal notion of set? If we had anything in
mind before we reached the point of formalization it certainly included the
idea that sets are abstract. And even apart from the difficulties noted
earlier in distinguishing between formal and informal mathematics, how can
it make sense now to say, as Resnik does, that formalized set theory is
derived from our informal concept of set, when that concept is so ill-defined
as to include non-abstract entities? How can he claim that the intended
model is "at hand" when so many different sorts of things could serve as
intended models? What on earth is an intended model supposed to be? And
finally, what is now the justification for rejecting the Skolem models? The
original objection was that they did not provide an intuitively adequate
representation of the concept of a set, but surely numerical models are
closer in spirit to sets than any collection of non-abstract entities. It almost
looks like a *Reductio* of the Platonist's position, and at the least he seems to
have given up any serious attempt to say what the intended model for set
theory is like.

Perhaps with good reason. We have seen what the difficulties are in
saying, apart from the formalization, what sets are like. Because of the
Löwenheim–Skolem theorem it is impossible to describe them formally,
and because of the problems cited above, we cannot succeed at the informal
level either. Perhaps the moral of the story is that sets (and numbers) are
just what the axioms tell us they are, and any set of objects which satisfies
these axioms will do as a model for set theory.

But now aren't we jumping too quickly to a Formalist conclusion?
Perhaps all Resnik has in mind by "ontological neutrality" is that we take
our set theory at face value, as Quine suggests, and simply accept the proposition that there are nondenumerable sets. And perhaps what he has in
mind by "informal mathematics" is what Quine calls the "background
theory" in which we discuss the formalization. This is no help, however.

For what can it mean to take a theory "at face value" when the theory has so many faces? As Quine points out, there is no ontological hay to be made from accepting the sentence of the formal language; it is only through an interpretation that we can say what model is *meant* by a theory.[24] But interpretations are made in the metalanguage, and in this case our metalanguage is precisely the language which tells us a denumerable model will do for set theory. Of course, it also tells us about other models, but how, in the midst of this abundance, are we to make a choice? Quine's suggestion is that we get the *intended* reference "by paraphrase in some antecedently familiar vocabulary",[25] but again, there seems to be a choice of vocabularies at this point. Perhaps it is the "background theory", presumably Resnik's informal mathematics, which is the vocabulary meant here; Quine does claim that the predicates "set" and "number" can be distinguished against such a background theory. *But*: he claims that the *way* they are to be distinguished is in terms of "the roles they play in the laws of that theory."[26] And as we saw earlier, such an appeal to the role of a term or sentence looks very much like a Formalist interpretation.

Leslie Tharp has suggested, in arguments directed as much against Formalism as Pythagoreanism, that standard models can be distinguished from non-standard counterfeits in terms of their epistemological roles. We want to learn about sets, and "arithmetical or numerical models contribute nothing."[27] This sounds persuasive, and looks like a promising approach to the distinction between standard and nonstandard models that the Platonists have been trying to make, until we ask ourselves what an *intended* model would contribute. A great deal, according to Tharp. "There can be no question that a natural model provides a unique source of insight into new truths."[28] But what is it that provides insight? Is it really the abstract model, the universe of sets? And if so, what is the mechanism by which this insight is obtained? Do we have, as Plato suggested, a special faculty, as yet unexplored by scientists, for apprehending abstract objects? I myself am not aware that I have ever learned a set-theoretic truth by being confronted with an army of sets. I do remember confronting new notation, drawing pictures, and working problems, and I suspect that my experience is typical Of course, there can be no definitive answer to this question until we learn more about how we learn, but I doubt very much that the learning process in set theory will be found to be analogous to that of zoology.

Yet there is clearly something in Tharp's suggestion; the number-theoretic predicates which would correspond to the set-theoretic vocabulary are highly unintuitive, and would be extremely difficult to work with. I submit, however, that this plays into the hands of the Formalist, rather than providing an answer for the Platonist. The problem is that if we spelled out in detail these numerical predicates we would have a formal system that was completely unwieldy; I think what bothers us is our vision of the complexity of the resulting *formulae*.

Having disposed of these arguments, perhaps we are now entitled to jump to our Formalist conclusions. The attempts, and failures, to distinguish between standard and nonstandard models illustrate the problems inherent in taking mathematics as a descriptive science — descriptive of either sets or numbers. In the end, we find that we are describing almost anything — in other words, nothing. What we are left with is a formal structure which may fit any number of different domains, and the only reasonable conclusion to be drawn at this point is that it is the formal structure, the formal language, and not any particular set of objects, which is the important thing in mathematics. But this should be neither surprising nor disturbing; mathematics, after all, like logic, has traditionally, and correctly, been seen as a discipline which is applicable in almost every situation. It is a formal theory which we *use* to talk about objects of all sorts, not a theory *about* objects. (And this, by the way, answers the question which was earlier put to the Skolemite, of why set theory differs from other theories in not requiring a standard model.)

A Formalist account of mathematics is often rejected out of hand, on the grounds that it makes mathematics nothing more than the manipulation of meaningless marks, a view which we all know to be mistaken. But this objection is based upon the supposition that meaning is reference, and there is no reason to suppose that this is an accurate account of meaning in mathematics. Indeed, in view of the ambiguity of reference for formal systems, we could hardly maintain that meaning is a function of reference for in that case we should have to admit that π means something different when we construct mathematics according to Zermelo's definitions than when we use von Neumann's, a conclusion few mathematicians would be willing to draw. It is likely, I think, that the most promising account of meaning in mathematics is to be found in something like the use.

Another objection to the Formalist view might be that if we take it seriously a whole field of mathematical endeavor may go out the window

Formalism may reasonably be taken to be the view that what counts in mathematics is the derivation of theorems from axioms and rules, and that truth is to be defined in terms of provability rather than satisfaction. It may seem, then, that there is little need for the investigation of various sorts of models. I think, however, that there is little danger that model theory will be abandoned, even if a Formalist view is generally accepted. What does need to be done, perhaps, is to reexamine the assumptions of model theory, and reassess the relationship between model theory and proof theory. Hilary Putnam has even attempted to elucidate the notion of "standard model" in purely formal terms,[29] and perhaps more research along these lines would be rewarding.

Finally, it might be objected (has been objected, in fact) that if the formal system is all that counts, then there is no place left for one very important aspect of mathematics: the analysis of intuitive concepts, which plays such a large role for people like Gödel and Kreisel. But I see nothing inconsistent in combining a Formalist view with this sort of enterprise. It is not necessary that the meaning of intuitive concepts be a function of objects which they represent, and indeed, both Gödel and Kreisel are careful not to insist on this connection. Intuitive analysis is important, but so is the investigation into the *source* of our intuitive concepts, upon which little effort has been expended. Perhaps greater effort would yield a satisfactory account of meaning in terms of something other than reference. In any case, the point we have tried to make here is that intuitive mathematics has nothing to do with *ontology*, not that there is no such thing as intuitive mathematics.

It remains to tie up a few loose ends. Has not the Skolemite won his case after all? We have concluded that since it is impossible to specify, even informally, an intended model for set theory, any model will do, including the nonstandard "unintuitive" models of the Skolemite. And we have noted that set theory is not analogous to physics, e.g., or to any other substantive theory, which helps the Skolemite make his point that there is no need for an intended model in set theory. These points may be conceded, but the most important claim, that there are only finite or denumerable sets in the universe, remains open to the same sort of objection raised against the Platonist. We saw that the Platonist cannot establish the existence of non-denumerable sets because of the difficulty in saying what they are like, in either the object or meta-language. By the same token, the Skolemite must

488   VIRGINIA KLENK

fail in trying to establish that there are only denumerable sets, since the
formalism leaves it wide open what sorts of models there are.

West Virginia University

                              NOTES

* This is a revised version of a paper which I read at the University of Colorado in
  July 1975; I gained a great deal from the lively discussion there. I would like to
  thank especially William Reinhardt, from the Department of Mathematics, who read
  the paper and made extensive comments. Many of the changes I made have been
  the direct result of his very useful remarks. The changes I did not make are, of
  course, no fault of his.
1 See, for example, the exchange between Michael Resnik and William J. Thomas, in
   Journal of Philosophy, LXIII, 15 (1966); *Analysis*, 28.6 (1968) and 31.6 (1971);
   Nous III, 2 (1969).
2 Resnik, "On Skolem's Paradox", *Journal of Philosophy*, LXIII, 15, p. 428.
3 *Ibid.*, p. 433.
4 Thomas, "On Behalf of the Skolemite", *Analysis*, 31.6, p. 184.
5 *Ibid.*, pp. 184–5.
6 At this point I must apologize to Resnik for over-simplifying his position for the
   sake of a more coherent discussion. He himself discounts the ontological significance of his arguments and thus would not want to be labeled a Platonist in the
   traditional sense of the word. His arguments, however, are so well-adapted to those
   a real Platonist might use that I have taken the liberty of construing the debate as
   one between a Platonist and a Skolemite.
7 I am omitting a great many points which may seem crucial to a discussion of the
   Löwenheim–Skolem theorem, on the grounds that they have already been covered
   in sufficient detail in the Resnik and Thomas articles. I shall here raise only those
   issues which I believe have not been given sufficient attention.
8 Resnik, "On Skolem's Paradox", p. 428. Quine makes much the same point in
   "Ontological Reduction and the World of Numbers" (in *The Ways of Paradox*)
   where he denies that set theory can be reduced to number theory. Myhill also
   argues along these lines, claiming that in the Skolem models none of the relations
   can be taken as membership. ("On the Ontological Significance of the Löwenheim–
   Skolem Theorem", in *Contemporary Readings in Logical Theory*, eds. Copi and
   Gould.)
9 William Reinhardt has pointed out to me that it is possible to obtain an elementary
   submodel without appealing to the Axiom of Choice, but I believe this does not
   materially affect my argument.
10 Resnik, "More on Skolem's Paradox", *Nous*, III, 2, p. 195.
11 W. V. Quine, "Ontological Relativity", *Journal of Philosophy*, LXV, 7 (1968),
   p. 206.
12 Thomas, *op. cit.*, p. 184.
13 Arthur Fine has suggested that the "upward" version doesn't really make much

sense, since it seems to rely on some absolute notion of "uncountability", to which
he pleads "lack of understanding". I am in sympathy with this point of view, but
in the absense of a clear account of meaning in mathematics I prefer not to base
my argument on these grounds. (See "Quantification over the Real Numbers",
Philosophical Studies, XIX, 1–2 (1968), p. 31.)

[14] Thomas, "Platonism and the Skolem Paradox", *Analysis*, 28.6, p. 195. Michael
Jubien makes much the same point, though from a different philosophical perspective, when he argues that we reject nonstandard models because we have "an already
accepted ontology". (See "Two Kinds of Reduction", *Journal of Philosophy*, LXVI,
17 (1969), p. 540.) And Arthur Fine point out that the objection that the Skolem
models only map statements onto statements, not sets onto numbers, rests upon the
supposition that there *is* a set of reals to be mapped, i.e., that there is a sense of
mapping other than just interpreting the theory in a countable domain. ("Quantification over the Real Numbers", p. 30).

[15] Myhill, *op. cit.*, p. 50.

[16] Resnik, "On Skolem's Paradox", p. 191.

[17] *Ibid.*

[18] Myhill, *op. cit.*, p. 50.

[19] Resnik, "On Skolem's Paradox", p. 436.

[20] *Ibid.*, p. 437.

[21] Resnik, "More", p. 190.

[22] *Ibid.*

[23] *Ibid.*, pp. 191–2.

[24] Quine, "Ontological Relativity", p. 204.

[25] *Ibid.*

[26] *Ibid.*, p. 207.

[27] Leslie Tharp, "Ontological Reduction", *Journal of Philosophy*, LXVIII, 6, (1971),
p. 161.

[28] *Ibid.*, p. 162.

[29] Hilary Putnam, "Mathematics Without Foundations", *Journal of Philosophy*, LXIV,
1 (1967), especially pp. 20–22.

[12]

History and Philosophy of Logic, 1 (1980), 187–207

Categoricity

JOHN CORCORAN

Department of Philosophy, State University of New York at Buffalo, Buffalo, N.Y. 14260, U.S.A.

Received 14 July 1979

After a short preface, the first of the three sections of this paper is devoted to historical and philosophic aspects of categoricity. The second section is a self-contained exposition, including detailed definitions, of a proof that every mathematical system whose domain is the closure of its set of distinguished individuals under its distinguished functions is categorically characterized by its induction principle together with its true atoms (atomic sentences and negations of atomic sentences). The third section deals with applications especially those involving the distinction between characterizing a system and axiomatizing the truths of a system.

0. PREFACE

Aside from the analysis of the logical structure of mathematical propositions and the formalization of mathematical reasoning, perhaps the most striking achievement of pre-Gödelian mathematical logic was the categorical characterization of traditional mathematical systems (Euclidean geometry, the natural numbers, the rational numbers, etc.) viewed as interpretations of formal languages. Section 1 treats historical and philosophical aspects of the notion of categoricity (and, thus, also isomorphism) within the broader context of a discussion of characterization of a mathematical system by means of a set of sentences which hold in it. Section 2 considers mathematical systems which are categorically characterized by means of one (second order) induction principle supplemented only with atomic sentences and negations of atomic sentences (i.e., using no properly first order sentences). In particular it is shown that a system can be categorically characterized by such means provided only that it is inductive, i.e. that its domain is the closure of a finite number of its individuals under a finite number of its operations (any number of relations may also be present). This theorem leads immediately to a very weak test of categoricity. Section 3 shows that the test has useful applications in axiomatizing inductive systems. The weakness of the test, especially when viewed in relation to examples given, suggests that the importance of categoricity may have been exaggerated and that the relationship between characterizing a model and axiomatizing its truths, is not as close as had been thought. In particular, the test is used to establish a categorical characterization of the natural number system from which it can not be deduced that zero is not a successor. Other examples of deductively very weak theories which are nevertheless categorical are also given.

1. CATEGORICAL CHARACTERIZATIONS OF MATHEMATICAL SYSTEMS

By the turn of the century mathematicians had distinguished mathematical systems from axiomatizations. A *mathematical system* was thought of, in effect, as a class of mathematical objects together with a finite family of distinguished relations, functions and elements.[1] An *axiomatization* of a system was often thought of as a set of *propositions* about the system. Some mathematicians took the notion of a proposition about a system so literally that they could not conceive of a reinterpretation of a set of axioms (Frege *1906*, 79).

At this time, one must recall, there was no such thing as a formal grammar. Nevertheless, certain mathematicians (e.g. Hilbert *1899*; Veblen *1904*) conceived of the axioms for a mathematical system as propositional forms interpreted in the given system but admitting of other interpretations as well.[2]

Today we *can* speak of mathematical systems without reference to particular formal languages interpreted in them, but we often do not. For example, when we speak of the system of natural numbers we often mean the intended interpretation of one of the formal languages commonly used for number theory. And we have to be reminded of the fact that we can refer to the system of natural numbers in itself, so to speak. We also have to be reminded of the fact that an interpretation of a formal language is not merely a mathematical system but it also involves (among other things) a precise specification of which formal symbols get assigned to which distinguished relations, functions, and elements. In general, a set of formal axioms can be interpreted in a given system in more than one way. Thus, strictly speaking, a

1. The term 'mathematical system' was and is widely used in just this sense (compare Huntington (*1917*, 8) and Birkhoff and MacLane (*1944*, *1953*)). From a philosophical and historical point of view it is unfortunate that the term 'mathematical structure' is coming to be used as a synonym for 'mathematical system'. In the earlier useage, which we follow here, two mathematical system having totally distinct elements can have the same structure. Thus in this sense a structure is not a mathematical system, rather a structure is a 'property' that can be shared by individual mathematical systems. At any rate a structure is a higher order entity. The relation between a given structure and a system having that structure is analogous to the relation between a quality and an object having that quality. For mathematical purposes it would be possible to 'identify' a structure with the class of mathematical systems having that structure, but such 'identification' may tend to distort one's conceptual grasp of the ideas involved.

A referee suggested that some readers could confuse 'mathematical system' in the above sense with 'axiom system' in the sense of an axiomatization. This would be analogous to confusing an event with a description of the event or to confusing the set of solutions to an equation with the equation. All three cases confuse subject-matter with discourse 'about' that subject-matter.

2. Resnik (*1974*, esp. pp. 390–392) discusses this particular aspect of Hilbert's thought as reflected in the so-called 'Frege–Hilbert controversy'. The interpretation of Hilbert advanced here is in full agreement with Resnik.

set of formal axioms does not delimit a class of mathematical systems but rather it delimits a class of interpretations of its language. (An exact definition of 'interpretation' is given in sub-section 2.2 below.)

In the rest of this paper certain confusions can be avoided by keeping the above distinctions in mind. Especially important is the fact that, strictly speaking, a set of formal sentences true in a given interpretation should be regarded as an axiomatization of *the interpretation* rather than an axiomatization of *the underlying mathematical system*.

Let K be a set of non-logical constants and let LK be a formal language having K as its set of primitives. Let i be an interpretation of LK and let $T(i)$ be the set of sentences of LK true in i. A complete axiomatization of i requires the choice of a subset A of $T(i)$ which logically implies the rest.

An axiomatization of a given interpretation provides a *description* of the interpretation. Without meaning to suggest that one can *uniquely* describe an interpretation by means of a set of sentences, Hilbert (e.g. in *1899*) permitted himself remarks to the effect that a set of sentences can '*define*' an interpretation. Frege's criticism of Hilbert shows that Frege *mis*understood Hilbert's remarks as implying the possibility of unique axiomatic characterizations (Frege *1899*, 6–10). But Hilbert's reply shows that Hilbert was fully aware of the impossibility of such characterizations. Hilbert (*1899L*, 13–14) wrote: '. . . each and every [satisfiable] theory can always be applied to infinitely many systems of basic elements'. (See also footnote 2 above.)

Even today one occasionally finds a passage which admits of the same misconstrual. For example, Kac and Ulam (*1968*, 171) write: 'The axioms are meant to describe simple properties of the objects under consideration; one hopes that in these properties the essence of the objects will be captured completely'.

Nevertheless, by the turn of the century, at least, it had become clear that truth in a formal language has nothing whatever to do with the 'essence' of the objects in an interpretation, but rather depends solely on the form of the interpretation or, as it is sometimes put, on the formal interrelations among the objects. The notion of isomorphism between two interpretations was adopted as a mathematical formulation of the idea of two interpretations having the same form.[3]

3. A mathematically precise definition of isomorphism is given in sub-section 2.3 below. It is important to realize that the concept comes into play in a context where the language is fixed and the interpretations are changed (not when the interpretation is fixed and the language is changed). This confusion is rather widespread in the informal parts of the literature of the recent past. For example, in introducing the proof that any two mathematical systems satisfying the integral domain postulates are isomorphic, Birkhoff and MacLane say that the postulates 'are true of the integers not only as expressed in the usual decimal notation; they are also true of the integers expressed in the binary, ternary or any other scale!' (*1944*, 37). Notice that the Birkhoff and MacLane remark is true and that it would still be true regardless of whether the postulate set in question were categorical. Their remark is totally beside the point and could only be made in this context by persons confusing change of system with change of notation.

190   J. CORCORAN

One occasionally reads that if *two* interpretations are isomorphic they are 'identical except for the names' of the elements and relations (Fraleigh *1967*, 55) or that isomorphic interpretations 'differ only in the notation for their elements' (Birkhoff and MacLane *1953*, 33). Such remarks can be misleading because generally two isomorphic interpretations have *different* elements and relations. The whole point is that the two have the same 'form', and what sets of objects the two each involve, be they identical or different, is beside the point. (Compare footnote 3 above.)

For example, let *LK* be the algebraic language based on one binary operation symbol *. One familiar interpretation takes as domain the four so-called complex units, 1, −1, *i* and −*i* and as interpretation of * it takes multiplication. Another interpretation takes as its universe the four social classes of the Kariera society and as interpretation of * it takes the function which yields the class of a child when applied to the class of its father and the class of its mother. Levi-Strauss has discovered that such a function exists and, indeed, that the interpretation just mentioned is isomorphic to the interpretation in the complex units (compare Barbut *1966*). Surely one would not want to say in this context that a complex unit and a social class differ only in notation. The form which is common to these two interpretations is, of course, the so-called 'the Klein group'.

The insight that truth in a formal language depends solely on the form of the interpretation (and is independent of content or matter) is partly reflected in the fact that isomorphic interpretations have the same set of truths, i.e. if *i* and *j* are isomorphic then $T(i) = T(j)$. Moreover, it has been clear at least since the turn of the century (Hilbert *1899L*, 14) that given any interpretation *i*, there are other interpretations isomorphic with *i* but having no content in common with *i*. The existence of such isomorphic 'images' implies, of course, the impossibility of uniquely characterizing an interpretation by means of a set of sentences in a formal language.[4] Accordingly, it is sometimes said that the best possible characterization of an interpretation would be a 'characterization up to isomorphism', where a set *A* of sentences is said to *characterize i up to isomorphism* if every interpretation which satisfies *A* is isomorphic to *i*.

Thus instead of an ideal of exact characterization, mathematicians adopted the ideal of characterization up to isomorphism, and terminology was introduced to indicate the property of sets of sentences which characterize up to isomorphism the interpretations they characterize (Veblen *1904*, 346). More precisely, a set of sentences *A* is said to be *categorical* if every two interpreta-

---

4. As late as 1944 some writers were still not clear about this point. For example, Birkhoff and MacLane *1944* are clear that, within the class of formal languages that they were using, no postulate set could distinguish between isomorphic systems; but they did not see that this is a feature of all classes of formal languages. They wrote: '. . . no postulate system for the integers (or the type which we have used) could distinguish between two isomorphic systems' (Birkhoff and MacLane *1944*, 37).

tions which satisfy A are isomorphic.[5] (Because of the peculiarity of 'every', a contradictory set A is vacuously categorical.) Incidentally, Veblen noted that the term 'categorical' was suggested by the philosopher John Dewey (*ibid.*).

By the middle of the first quarter of this century categorical characterizations of several important interpretations had been established. Many of these results are reported in Huntington *1905*. It became common to 'identify' the intended interpretation of a formal language used to discuss a standard mathematical system with the system itself. For example, the phrase 'the system of natural numbers' sometimes indicates the intended interpretation of a language LK, where K is a set of 'arithmetic primitive symbols'. Using this terminology it can be said that the following systems had been categorically characterized: the natural numbers, the integers, the rationals, the reals, the complex numbers, and Euclidean space.[6]

Investigation of formal languages led to the further insight that whether a categorical characterization is possible depends not only on the form of the interpretation in question but also on the *logical* devices (variables and logical constants) available in the language chosen. For example, if LK is a first-order

---

5. Calvin Jongsma (University of Toronto) pointed out that some early mathematicians and logicians understood Veblen to be applying the term 'categorical' to systems that would now be called 'semantically complete' or, to use Church's phrase, 'complete as to consequences', as opposed to deductively complete (Church *1956*, 329; compare Skolem *1928*, 523). Categoricity, of course, implies semantic completeness but the converse does not in general hold, as can be seen from Skolem's paper (*ibid.*). Later I noticed that Birkhoff and MacLane treat the categoricity of the axiom set for integral domains in a section called 'Completeness of the postulates for the integers' (*1944*, 36). It is worth noting that, in that section, there is not a word about completeness in either the semantic or deductive senses. Incidentally, the mathematical use of the term 'categorical' is certainly due to Veblen (*1904*, 346). However, it is not at all clear that Veblen uses it in its modern sense. In fact, he seems to be using 'categorical' to mean 'semantically complete'. The earliest use of 'categorical' in the modern sense is no later than Young (*1911*, 49).

6. The fact that categorical characterizations of the traditional mathematical systems were self-consciously obtained by early mathematical logic suggests that the discovery of such characterizations may have been a stated goal of the field (see footnote 7 below). Developments leading up to research aimed at categoricity results are not well-known. An idea similar to isomorphism is attributed to Galois (1811–1832) and the term is found in an 1870 paper of Camille Jordan (Kline *1972*, 765, 767). Related ideas are found in Cantor's *Grundlagen* of 1883 (Jourdain *1915*, 76, 112) and in Dedekind (*1887*, 93). Cantor and Dedekind each have theorems which can easily be applied to yield categoricity results, but neither seemed to have the idea of characterizing a class of systems by means of sentences (or propositional functions). The earliest genuine categoricity result I know of is due to Huntington (*1902*). Kline finds that 'this notion [categoricity] was first clearly stated and used by ... Huntington in a paper devoted to the real number system' (*1972*, 1014). Kline is referring to Huntington *1902*, which proves the categoricity of a set of 'axioms' for 'absolute continuous magnitude'. It was alleged, falsely and without justification, by Young (*1911*, 154) that Hilbert's *Foundations of geometry* (1899) contains a proof that Hilbert's axiomatization of geometry is categorical. Ironically, Hilbert *1899* does not even show awareness of semantic completeness, despite Veblen's apparent comment (*1904*, 346) to the contrary.

192   J. CORCORAN

language (i.e. one having only individual variables) without identity then *no* categorical characterizations are possible, and if *LK* is a first-order language with identity, then only finite interpretations can be categorically characterized.

Kreisel (*1965*, 148) points out that this limitation of first-order languages came as a surprise to logicians,[7] and he also makes the interesting observation that all finite interpretations are categorically characterizable in first order languages so that being finite and being first-order categorically characterizable are equivalent properties of interpretations. Many writers, including Kreisel (*ibid.*) and Montague (*1965*, 136), have noted that the many known categorical characterizations of the familiar classical systems all involve languages of second order, at least.

However, if one moves beyond a first-order language with identity by the smallest possible amount, i.e. by allowing *one* one-place predicate variable, then not only are some infinite interpretations categorically characterizable but many important infinite interpretations are so characterizable. For example, if mathematical induction is written

$$(Po \,\&\, \forall x(Px \supset Psx)) \supset \forall y\, Py$$

then the natural number system (relative to a primitive 0 for zero and *s* for successor) is categorically characterizable. Likewise, the integers, the reals and other important systems are also categorically characterizable using these 'slightly augmented first-order languages' (Montague *1965*).

To be more precise, define the formulas of the *slightly-augmented language with non-logical primitives K*, abbreviated '*saLK*', to be exactly the formulas of the first-order language with identity but based on $K + \{P\}$, where *P* is a one-placed predicate symbol not in *K*. The *sentences* of *saLK* are the formulas which lack free occurrences of the individual variables. The truth conditions for sentences of *saLK* are exactly these of the first-order sentences involving $K + \{P\}$ except that a sentence $S(P)$ involving *P* is true under an interpretation *i* iff it is satisfied by every assignment of a subset of the universe of *i* to *P*. Thus every sentence $S(P)$ is understood to be universally quantified with respect to *P* taken as a variable. Henceforth, *P* is called 'a one-placed predicate variable'. Note that $S(P)$ and $\sim S(P)$ are contraries, *not* contradictories.

Note that *saLK* is not equivalent to the language in which there is universal quantification of *P*, unless one requires (1) that only one occurrence of the universal quantification of *P* is allowed per sentence, and (2) that the single universal quantification must occur at the front. Thus for example, $\forall P S(P)$ is

7. How much of a surprise this was is another matter. Ellentuck says: 'One of the earliest goals of modern logic was to characterize familiar mathematical structures up to isomorphism . . . in a first order language' (*1976*, 639). In the opinion of this author it is doubtful whether any logicians held this as a goal, at least for very long. By the time of Skolem *1920*, it was clear that no uncountable systems (e.g., geometry, the reals, or the complex numbers) could be categorically characterized in first order, and there appears to have been very little interest in first order languages before that.

equivalent to the *saLK* sentence $S(P)$, but $\sim\forall P S(P)$ is not in general equivalent to a *saLK* sentence. In particular, the existential predicate-variable quantifier is not definable in *saLK*.

If somehow required to classify a slightly-augmented language as first-order or as second-order, many mathematical logicians would probably not hesitate to call it second-order. However, if slightly augmented languages are so classified it must be noticed that they are weaker than the usual second-order languages (Enderton *1972*, 268–269) in four ways. In the first place, *saLK* has no function variables. Second, it has no *n*-ary predicate variables for *n* greater than one. Third, instead of infinitely many one-placed predicate variables *saLK* has but one. Fourth, instead of formulas with arbitrarily many universal predicate-variable quantifications arbitrarily deeply imbedded, *saLK* contains only formulas with at most one such quantification occurring at the front, i.e. not imbedded at all. Thus *saLK* would be an extremely weak second-order language. In fact, Church has implicitly classified *saLK* as an applied first-order language with identity (*1956*, §48).

In the opinion of the author, (1) classification of *saLK* as a first-order language would introduce confusion because there are many properties usually thought of as intrinsically first-order but which do not hold of *saLK*, and (2) classification of it as second-order would tend to mask its expressive weakness and its simplicity. It would seem best simply to refer to *saLK* by the name given above, or by the name 'slightly augmented first-order language'.

The rest of the paper treats categorical characterization in the context of slightly augmented first-order languages.[8] Section 2 establishes an extremely weak sufficient condition for categoricity which is nevertheless useful in constructing categorical sets of sentences. The main theorem is that an interpretation which satisfies an induction principle is categorically characterized by its induction principle together with its true atomic sentences and the negations of its false atomic sentences. In effect, we show that the form of an inductive interpretation is determined by its atoms. The proof of the theorem involves no reference to truth-functional combinations or to quantifications (except, of course, to those involved in induction principles).

Section 3 applies the result of section 2. The paper is intended to be largely self-contained. Moreover, since terminology in logic has not yet been

---

8. The idea of categoricity is attractive even to logicians who want to avoid quantification over 'higher-order objects'. For example, in order to save categoricity in contexts devoid of such quantificiation Ellentuck *1976* goes to infintary languages, and Grzegorczyk *1962* restricts the interpretations to what he calls 'constructive models'. Other writers 'weaken' the idea of being categorical to being 'categorical in a power'. An axiomatization A is said to be categorical in k, where k is a cardinal number or power, if any two models of A whose universes have cardinality k are isomorphic. For further discussion see Enderton (*1972*, 147). This writer, however, has no interest in avoiding quantification over high-order objects, something he regards as a fundamental aspect of mathematical language. The motivation for considering slightly augmented languages is to isolate an idea which could have served as the core idea in many known categoricity proofs.

194   J. CORCORAN

standardized, it was thought worthwhile to repeat some rather elementary definitions. However, when this is done it is only done to the extent necessary for the immediate purposes at hand.

2. CATEGORICITY IN saLK

Sub-sections 2.1 and 2.2 below deal with 'grammatical' and semantic preliminaries. Since categoricity is a purely semantic concept having no *intrinsic* dependence on object-language deductions, no system of formal proofs is provided. If the reader wishes to have a system of formal proofs for *saLK* it is sufficient to take any standard system for first-order with identity, e.g. Mendelson (*1964*, 57, 75) and for the monadic predicate variable take the 'rule of substitution' (which amounts to regarding a sentence involving P as a scheme). Sub-section 2.3 repeats the standard, exact notions of isomorphism and categoricity. Sub-section 2.4 associates with each *interpretation* a 'bar interpretation'. Sub-section 2.5 proves the main theorem.

2.1. *Syntax for atoms and induction formulas*

Let K be a finite set of non-logical constants containing at least one individual constant and at least one function symbol. Besides these K can contain any number of individual constants and, for each $n \geqslant 1$, any number of n-ary functions symbols and any number of n-ary relation symbols. For each such K, TK is the set of *constant terms* of K, i.e. TK is the closure of the set of individual constants of K under the operations of attaching an n-ary function symbol f in K to a string $t_1 \ldots t_n$ of n constant terms (forming $ft_1 \ldots t_n$).

An *atomic sentence* of K is an *identity* '$t_1 = t_2$' or a string '$RT_1 \ldots t_n$' where R is an n-ary relation symbol in K and $t_1, \ldots,$ and t_n are all constant terms. The negation of an identity is written '$t_1 \neq t_2$', and the negation of '$Rt_1 \ldots t_n$' is written '$\sim Rt_1 \ldots t_n$'. Atomic sentences and their negations are called *atoms*.

Let P be the monadic predicate variable. If $K = \{0, s\}$ where 0 is an individual constant and s is a monadic function symbol, then the *induction formula for K* is the following:

$$I\{0, s\} \quad (P0 \;\&\; \forall x(Px \supset Psx)) \supset \forall y\, Py.$$

If $K = \{0, 1, s, +\}$ where 0 and s are as above, 1 is an individual constant, and $+$ is a binary function symbol then the *induction formula for K* is as follows:

$$I\{0, 1, s, +\} \quad ((P0 \;\&\; P1) \;\&\; \forall x_1 x_2 ((Px_1 \;\&\; Px_2) \supset (Psx_1 \;\&\; P + x_1 x_2))) \supset \forall y\, Py.$$

In general an induction principle has the form

$$I \quad (B \;\&\; \forall x_1 \ldots x_n (IH \supset IC)) \supset \forall y\, Py,$$

where B is the so-called 'basis', IH is the 'induction hypothesis', and IC is the 'induction conclusion'. Thus in order to define the induction formula IK for an arbitrary K it is sufficient to define each of the parts, BK, IHK and ICK. The *basis BK* is the conjunction of all the formulas Pc, where c is an individual constant of K. Let m be the maximum of the degrees ('arities') of the function symbols in K. Then the *induction hypothesis IHK* is the conjunction Px_1 & ... & Px_m. For each n-ary function symbol f in K, form $Pfx_1 \ldots x_n$. The conjunction of all such formulas involves only the m variables $x_1, \ldots, x_m$ and is called the *induction conclusion, ICK*. Thus IK, the *induction formula for K*, is the following:

$$IK \quad (BK \ \& \ \forall x_1 \ldots x_m (IHK \supset ICK)) \supset \forall y \, Py.$$

The fact that IK is not uniquely determined is not important.

2.2. *Semantics for atoms and induction formulas*

An *interpretation i of LK* is an ordered pair $\langle u, d \rangle$ where u is a non-empty set and d is a function defined on K and such that dc is in u if c is an individual constant, df is an n-ary function defined on u and taking values in u if f is an n-ary function symbol, and dR is set of n-tuples of members of u if R is an n-ary relation symbol.[9] As usual, the *denotation $d^i t$ of a term t under an interpretation i* is defined on TK as follows:

$$d^i[c] = dc \quad \text{and} \quad d^i[ft_1 \ldots t_n] = (df)d^i[t_1] \ldots d^i[t_n],$$

i.e. the denotation of an individual constant is its interpretation and the denotation of a function symbol attached to terms is the interpretation of the function symbol applied to the denotations of the terms.

As a result of the way that TK, the set of terms, is defined it is obvious that d^i is defined on all of TK. The range of d^i is then the set of objects in u which are denoted by constant terms. For example, if $K = \{1, s\}$, u is taken to be the set of natural numbers including zero, $d1$ is taken to be the number one and ds is taken to be the successor function, the range of d^i is only the set of positive numbers; the range of d^i need not be all of u.

---

9. It is true that, in the strict sense of the term 'set', each element "occurs" only once. Thus when one speaks of 'the set of solutions to an algebraic equation' the word 'set' is not being used in the strict sense. Sometimes the word "family" is used to indicate a 'set' wherein an element can have multiple occurrences. If repetitions of the same element are distinguished by indices, one speaks of an 'indexed family' (Halmos *1960*, 34). Thus a system, in the sense of section 1, is a class of mathematical objects together with a finite family of distinguished elements, functions and relations. An interpretation can then be seen as a certain kind of system, namely as a system wherein the family is indexed by a set of symbolic characters (namely, by K). It is important, however, that the indexing respect semantico/syntactic distinctions, i.e. that elements are indexed by individual constants, that n-ary functions are indexed by n-ary function symbols and that n-ary relations are indexed by n-ary symbols. Bridge (*1977*, 6, 16) treats the topic explicitly in this way. For a treatment of semantic categories, see Tarski *1934*, 215–236.

Let $d^i TK$ be the range of d^i. Notice that because of the way TK is defined, $d^i TK$ is the closure of the set of objects in u denoted by individual constants under the functions denoted by function symbols, i.e. $d^i TK$ is a subset of any subset of u containing the said objects and closed under the said functions. Where no confusion results 'd^i' is sometimes written 'd'.

As usual an identity '$t_2 = t_2$' is *true under i* if dt_1 is the same object as dt_2 (i.e. $dt_1 = dt_2$) and *false under i* if dt_1 and dt_2 are different (i.e. $dt_1 \neq dt_2$). '$Rt_1 \ldots t_n$' is *true under i* if the n-tuple of objects denoted by the terms is in the relation denoted by $R(\langle dt_1, dt_2, \ldots dt_n \rangle \in dR)$, and '$Rt_1 t_2 \ldots t_n$' is *false under i*, otherwise. Negation, of course, reverses truth-values.

For a given interpretation $i = \langle u, d \rangle$ an *assignment aP* to the monadic predicate variable P is simply a subset of u. When aP is assigned to P, (1) BK is *true under i* if aP contains the objects denoted by individual constants; (2) $\forall x_1 \ldots x_m (IHK \supset ICK)$ is *true under i* if aP is closed under the functions denoted by function symbols; and (3) $\forall y Py$ is true if $aP = u$. More particularly, to say that the induction formula *IK is true of aP under i* is to say that *if* aP both contains the objects denoted by individual constants and is closed under the functions denoted by function symbols, *then* aP is u. And *IK* is *true under i* if *IK* is true of *each* aP under i. More particularly, *IK* is *true under i* iff every subset of u which contains the objects named by individual constants and which is closed under the functions denoted by function symbols is u. Since the range of d^i, $d^i TK$, is such a subset, if *IK* is true under i then $d^i TK = u$. Conversely, if $d^i TK = u$ then *IK* is true under i. Thus, to say that *IK* is true under i is to say nothing but that $d^i TK = u$. In other words, the induction formula 'says' that every object is denoted by some constant term. When induction holds in i, we say that i is *inductive*.

2.3. *Isomorphism and categoricity*

Let $i = \langle u, d \rangle$ and $j = \langle v, e \rangle$ be two interpretations of LK. Then i is said to be *isomorphic* to j if there is a one-one function h from u onto v which 'preserves the structure' in the sense that: if c is an individual constant then $h[dc] = ec$; if f is an n-ary function symbol then for every $b_1, \ldots, b_n$ in u,

$$h[(df)b_1 \ldots b_n] = (ef)[hb_1] \ldots [hb_n];$$

and if R is an n-ary predicte symbol then for all $b_1, \ldots, b_n$ in u,

$$u, \langle b_1, \ldots, b_n \rangle \in dR$$

if and only if

$$\langle [hb_1], \ldots, [hb_n] \rangle \in eR.$$

Let $K = \{0, s\}$ and take i and j as follows. The universes u and v are both the set I of integers, ds and es are both the successor function, $d0$ is zero and $e0$ is one hundred. The function $hx = x + 100$ is an isomorphism between i and j.

Here (and below) we exploit the set theoretic notion of a function being a set of ordered pairs by writing $i = \langle I, \langle 0, \text{zero} \rangle, \langle s, \text{successor} \rangle \rangle$ and $j = \langle I, \langle 0, 100 \rangle, \langle s, \text{successor} \rangle \rangle$.

If i is isomorphic to j then any sentence of $saLK$ true in one is true in the other, i.e. $T(i) = T(j)$. The proof of this is straight-forward but it requires a set of definitions rather more complete than is otherwise required in this paper.

Let A and B be two sets of sentences. Then i is a *model* (or *true interpretation*) of A if $A \subset T(i)$ and A *logically implies* B if every model of A is a model of B. If A logically implies B then, for purposes of smooth expression, B is said to be a *logical consequence* of A. If A has no models, A is said to be *contradictory*. Because of the peculiarity of 'every', a contradictory set implies every set. If A implies B and $B = \{p\}$ then A is said to imply p.

A set A of sentences is *categorical* if all models of A are isomorphic to each other. Because of the peculiarity of 'all', contradictory sets are vacuously categorical. Notice that if A is categorical then for every sentence p not involving P, A implies p or A implies $\sim p$. It fails in the case of sentences involving P only because for them negation does not reverse truth-values: $S(P)$ and $\sim S(P)$ are contraries, not contradictories.

Examples. Let $K = \{0, s\}$. Let $A0$ be the set of sentences true in $iN = \langle N, d \rangle$ where N is the natural numbers, $d0$ is zero, and ds is the successor function. Let $S^n 0$ indicate n occurrences of S followed by 0. Notice that the true atoms are simply the logical identities ($S^n 0 = S^n 0$) and the negations of the other identities ($S^n 0 \neq S^m 0$, for $n \neq m$). Let A be the Peano Postulates for zero and successor, i.e.

$$A = \{IK, \forall xy(sx = sy \supset x = y), \forall x(sx \neq 0)\}.$$

Reasoning which shows that A is a categorical characterization of iN is familiar (compare Birkhoff and MacLane *1953*, 54–56). Let

$$A1 = \{IK\}.$$

It is obvious that $A1$ is not categorical because it is satisfied by the unit model $i1 = \langle \{0\}, d \rangle$ where $d0$ is zero and ds is the identity function. More generally it is clear that for any K, IK and any set of positive atoms is satisfied by the unit model, and thus, if IK and a set of atoms is categorical, a negative atom must be present. Let

$$A2 = \{IK, S0 \neq 0, \ldots, S^n 0 \neq 0, \ldots\}.$$

It is worth noting that $A2$ implies $\forall x(Sx \neq 0)$. (IK 'says' that every object is named by a term $S^n 0$. The joint effect of the rest of the sentences in $A2$ is to say that if an object is named by a term its successor is not zero.) Nevertheless, it is clear that $A2$ is not categorical because it has as a model $i2 = \langle \{0, 1\}, d \rangle$ where $d0$ is zero and ds is the identity function. Let $A3$ be the result of adding to $A2$ the rest of the negative atoms, $S^n0 \neq S^m0$, for $n \neq m$ and $m \neq 0$:

$$A3 = \{IK\} + \{S^n0 \neq S^m0, n \neq m\}.$$

Consider a sentence $SS^n0 = SS^m0 \supset S^n0 = S^m0$. If $m = n$ then the sentence is logically true. If $m \neq n$ then, since the negations of the antecedents are atoms in $A3$, the sentence itself is implied by the atoms in $A3$. By the reasoning of the previous paragraph, then, $A3$ implies $\forall xy(sx = sy \supset x = y)$. Thus $A3$ implies all three of the Peano postulates and is thus categorical.

From the perspective of the next sub-section, the main feature of $A3$ is that it is *atom-complete* in the sense that for every atomic sentence p, either $A3$ implies p or $A3$ implies $\sim p$. The main theorem is that every atom-complete set containing induction is categorical.

2.4. *Bar cointerpretations*

The denotation function d^i maps TK into u. Thus d^i can be used to define an equivalence relation Ei on TK in the usual way, i.e. let $t_1 Eit_2$ iff $d^it_1 = d^it_2$. Let $\bar{t}^i$ be the equivalence class of t, and let $\overline{TK}^i$ be the set of equivalence classes. Let $\bar{d}^i$ be the function from $\overline{TK}^i$ into u such that $\bar{d}^i\bar{t}^i = d^it$. Notice that $\bar{d}^i$ is one–one from $\overline{TK}^i$ onto d^iTK.

Let $i = \langle u, d \rangle$ be inductive. The mapping $\bar{d}^i$ is therefore a one–one onto function from $\overline{TK}^i$, the set of equivalence classes of terms, to the universe of i. Now we define a denotation function ^ib on K in such a way that $\bar{d}^i$ is an isomorphism *from* $\{\overline{TK}^i, {}^ib\}$ to $i = \{u, d\}$. The interpretation $\{\overline{TK}^i, {}^ib\}$ thus induced by $\bar{d}^i$ is called the *bar cointerpretation.*

One *could*, of course, define $\langle \overline{TK}^i, {}^ib \rangle$ as 'the isomorphic image of i under the inverse of $\bar{d}^i$'. *However, the information that we need to highlight is brought out better by defining* $\langle \overline{TK}^i, {}^ib \rangle$ explicitly and *then* verifying that i is its isomorphic image under $\bar{d}^i$.

(1) The universe of $\langle \overline{TK}^i, {}^ib \rangle$ is, of course, $\overline{TK}^i$. The interpreting function ^ib is defined as follows. (2) $^ibc = \bar{c}^i$. Now notice that $\overline{ft_1 \ldots t_n^i} = \overline{ft_{n+1} \ldots t_{2n}^i}$ is implied by the following taken together $\bar{t}_1^i = \bar{t}_{n+1}^i, \ldots, \bar{t}_n^i = \bar{t}_{2n}^i$. This means that the equivalence class of a term $ft_1 \ldots t_n$ is a "function" of the equivalence classes of its components $t_1, \ldots, t_n$. (3) Thus we can define ^ibf as follows:

$$^ibf[\bar{t}_1^i, \ldots, \bar{t}_n^i] = \overline{ft_1 \ldots t_n^i}.$$

Finally notice that

$$\bar{t}_1^{li} = \bar{t}_{n+1}^i, \ldots, \bar{t}_n^i = \bar{t}_{2n}^i,$$

taken together, imply that '$Rt_1 \ldots t_n$' and '$Rt_{n+1} \ldots t_{2n}$' have the same truth-value in i. (4) Thus we can define the relation ^ibR to hold of $\{\bar{t}_1^i, \ldots, \bar{t}_n^i\}$ iff '$Rt_1 \ldots t_n$' is true in i.

To check that $\bar{d}^i$ is an isomorphism, one need only check:

(1) that $\bar{d}^i$ is one–one and onto from $\overline{TK}^i$ to u.
(2) that $\bar{d}^i\{^ibc\} = dc$,
(3) that $\bar{d}^i\{(^ibf)(\bar{\iota}_1^i,\ldots,\bar{\iota}_n^i)\} = (df)(\bar{d}^i\bar{\iota}_1^i,\ldots,\bar{d}^i\bar{\iota}_n^i)$, and
(4) that $\langle \bar{\iota}_1,\ldots,\bar{\iota}_n\rangle \in {}^ibR$ iff $\langle \bar{d}^i\bar{\iota}_1^i,\ldots,\bar{d}^i\bar{\iota}_n^i\rangle \in dR$.

The above reasoning establishes the following:

Lemma. Each inductive interpretation is isomorphic to its bar cointerpretation.

Now we establish the main lemma of the paper.

Lemma. Two inductive interpretations which satisfy the same atoms have the same bar cointerpretation.

Proof: Let $\langle \overline{TK}^i, {}^ib\rangle$ and $\langle \overline{TK}^j, {}^jb\rangle$ be the two bar cointerpretations. To show:
(1) $\overline{TK}^i = \overline{TK}^j$, (2) ${}^ibc = {}^jbc$, (3) ${}^ibf = {}^jbf$ and (4) ${}^ibR = {}^jbR$.

To see (1) notice that t_1Eit_2 holds iff '$t_1 = t_2$' is true in i. By hypothesis the latter holds iff '$t_1 = t_2$' is true in j. Again the latter part holds iff t_1Ejt_2. Therefore $\overline{TK}^i = \overline{TK}^j$.

It follows then that $\bar{c}^i = \bar{c}^j$. Thus ${}^ibc = {}^jbc$. (2) is established.

It also follows that for all t, $\bar{t}^i = \bar{t}^j$. and, in particular, that $\bar{f}t_1\ldots t_n^i = \bar{f}t_1\ldots t_n^j$. Thus ${}^ibf = {}^jbf$, (3) is established.

To see (4) notice first that $\langle \bar{\iota}_1^i,\ldots,\bar{\iota}_n^i\rangle \in {}^ibR$ iff '$Rt_1\ldots t_n$' is true in i. By hypothesis the latter holds iff '$Rt_1\ldots t_n$' is true in j. Again using the definition of bar interpretation, now for j, it follows that '$Rt_1\ldots t_n$' is true in j iff

$$\langle \bar{\iota}_1^j,\ldots,\bar{\iota}_n^j\rangle \in {}^jbR.$$

But since

$$\overline{TK}^i = \overline{TK}^j, \bar{t}^i = \bar{t}^j,$$

so

$${}^ibR = {}^jbR. \qquad \text{Q.E.D.}$$

2.5. *Main theorem*

Any atom-complete set of sentences which includes induction is categorical.

Proof: Let S be such a set of sentences. If S is not satisfiable then S is vacuously categorical.

Assume that S is satisfiable. Let i and j be models of S. Since S is atom-complete, i and j satisfy the same atoms. Since S includes induction, i and j are inductive. By the lemmas, i and j are isomorphic.   Q.E.D.

In order to put this theorem in perspective one can note that its import is simply the following. The set of true atoms of an inductive interpretation, taken together with induction, characterize the interpretation up to isomorphism. Thus *if* one is characterizing an inductive interpretation, and *if* one can tell that the axioms already set down are sufficient to imply induction all of the true atomic sentences and the negations of the false ones, *then* the goal of categoricity is achieved.

It is not hard to see that the same theorem holds in all stronger languages, so it holds for second order languages and for higher order languages. Thus one might wonder whether *saLK* can be weakened without losing the theorem. The immediate answer is affirmative because in the entire *proof* not a word was said about any sentences besides (constant) atomic sentences, their negations, and induction. Thus the result holds for *inductive atomic languages which are defined as follows. Let K* be a finite set of non-logical constants as above. The logical constants of the *inductive atomic language based on K, iaLK* are $=$, $\sim$ and a new symbol '*I*' to be explained presently. The sentences are all of the atomic sentences of *saLK*, their negations, and *I*. The interpretations are the same as for *saLK* and the truth conditions for the atoms are the same. But *I* is true in *i* iff *i* is inductive.

The question also arises whether the condition of atom-completeness plus induction can be weakened without losing categoricity. It is clear that one cannot 'throw out' the negations of the atomic sentences because the example A2 in sub-section 2.3 above shows that the true atomic sentences of an interpretation (plus induction) do not characterize the interpretation up to isomorphism. One can also easily show that if *S* contains induction and is categorical then *S* is equivalent to an atom-complete set. Thus the above theorem is the strongest possible in the sense that atom completeness is the weakest possible condition sufficient to guarantee that a set of sentences containing induction is categorical.

3. APPLICATIONS

3.1. *Repetition theory*

Imagine that one is dealing with 'sets' of 'repeatable' objects where 'multiplicities' are counted. For example $[3, 3]$ is the 'set of roots of $x^2 - 6x + 9 = 0$' but $[3]$ is 'the set of roots of $x - 3 = 0$'. Such 'sets' are called iterates (or heaps or multiplicities: Hailperin *1976*, 88). At any rate if *I* is a set of objects being repeated then with each iterate, *r*, one can associate a unique function *f* from *I* into *N* such that for each repeatable object *x*, *fx* is the number of times *x* is repeated in *r*. The functions *f* are called *repetitions* of *I*.[10]

We consider the case where $I = \{a, b\}$. Let *u* be the set of repetitions of *I*, i.e. the set of functions from $\{a, b\}$ into *N*. Let 0, the null repetition, be the function:

10. Compare footnote 9 above.

$fa = 0, fb = 0$. The a-successor function s_a is the function which 'jacks up' the a-component of f by one, i.e. $s_a f(a) = f(a) + 1$ and $s_a f(b) = f(b)$. Likewise s_b is the b-successor function. Thus we are considering an interpretation $i = \langle u, d \rangle$ of $saLK$ with $K = \{0, s_a, s_b\}$.

A little thought suffices to see that the following axioms are true:

A1 $\forall x (s_a x \neq 0 \ \& \ s_b x \neq 0)$,

A2 $\forall xy ((s_a x = s_a y \supset x = y) \ \& \ (s_b x = s_b y \supset x = y))$,

A3 $\forall x (s_a s_b x = s_b s_a x)$,

A4 $\forall x (s_a x \neq s_b x)$,

A5 IK.

By looking at what the constant terms denote one can see that an identity '$t_1 = t_2$' is true in i iff the repetition of $\{s_a, s_b\}$ in t_1 is the same as the repetition of $\{s_a, s_b\}$ in t_2. For example the repetition of $\{s_a, s_b\}$ in the terms

$$s_a s_b s_a s_a s_b 0, \quad s_b s_a s_b s_a s_a 0, \quad s_b s_b s_a s_a s_a 0$$

is

$$\{\langle s_a, 3 \rangle, \langle s_b, 2 \rangle\}.$$

Using this idea one can show that the above axiom set, $A = \{A1, A2, A3, A4, A5\}$, implies each true identity and the negation of each false identity. Thus it implies an atom-complete set including induction, which is categorical by the theorem, and so it is categorical itself.

3.2. *Other possibilities*

It is already clear that various versions of Peano arithmetic (or number theory) admit of this treatment. Kleene (*1952*, 246) has discussed an infinite class of interpretations which he calls 'generalized arithmetics'. It is certainly possible to categorically characterize any one of them and perhaps to give a general formula for treating all of them – of course, using the above method. There is an infinite class of theories of strings based on Tarski *1934*, 172 and categorically axiomatized in Corcoran, Frank and Maloney *1974*. In the same work one finds another infinite class of theories dealing differently with strings based on an idea of Hermes. Both of these classes admit of treatment by this method. In addition it is possible to deal with finitely branching trees and hereditarily finite sets in this way.

For purposes of discussion assume that K has no relation symbols. This restriction does not matter in principle but it holds in all the familiar examples which will come to mind. Define an *equation in K* to be any constant identity (as above) or any sentence of the form $\forall x_1 \ldots x_n t_1 = t_2$ where t_1 and t_2 are terms in K and all variables occurring in t_1 or t_2 or both are among $x_1 \ldots x_n$. Let A be a set of equations. A model of A is called an *A-algebra*. Let $I(A)$ be the set of constant identities implied by A. Let $NI(A)$ be the set of negations of the constant identities *not implied* by A. If i is a model of $A + IK + NI(A)$ then i is what has been called a 'free' A-algebra. For example if $A = \{\forall xyz(x + (y + z) = (x + y) + z)\}$ and $K = \{a_1, \ldots, a_n, +\}$ then the models of $A + IK + NI(A)$ are the free semigroups on n generators.

It is clear that the above methods constitute *one* approach to getting categorical axiomatizations for the theories of free A-algebras. If A or $NI(A)$ is not recursively enumerable then the above approach may not work at all, even given a maximum of ingenuity.

3.3. *Strong induction*

The induction principle IK treated above is the weakest possible induction principle for LK, a language whose set of non-logical constants is K. We saw that IK is true in i iff every object of u is denoted by a term in TK. The weak induction principles IK are essentially unique, but for each K there are many stronger induction principles and, in fact, there are generally several which are maximally strong. To consider the first class of stronger induction principles for LK consider a proper subset $K1$ of K which still contains at least one individual constant and one function symbol. The induction principle $IK1$ used in LK is stronger than IK because it holds in i iff every object in u is denoted by a term in $TK1$ (a proper subset of TK). Clearly, $IK1$ implies IK but not vice versa. Thus $IK1$ is 'stronger' than IK.

For the general definition of a strong induction principle, let T be a proper subset of TK. Let $S(K)$ be a sentence possibly involving the monadic predicate variable. If $S(K)$ has the truth condition that it is true in i *iff* every object in u is denoted by a term in T, then $S(K)$ is a *strong induction principle* for LK. For example take $K = \{1, +\}$. Then IK is

$$(P1 \ \& \ \forall xy(Px \ \& \ Py \supset P(x + y))) \supset \forall zPz.$$

The following, however, is a strong induction principle:

$$(P1 \ \& \ \forall x(Px \supset P(x + 1))) \supset \forall zPz.$$

The truth condition for this sentence is that every object in u is denoted by a term of the following class: 1, $(1 + 1)$, $((1 + 1) + 1)$, ..., which leaves out $(1 + (1 + 1))$, $((1 + 1) + (1 + 1))$, etc. It is obvious that IK + associativity implies this strong induction principle.[11] At any rate, since every strong induc-

---

11. One of the referees wanted to know whether there are 'maximally strong' induction principles, i.e. whether there are strong induction principles which are not weaker than any others. One strengthens a strong induction principle by restricting the class T of terms which it 'forces to cover the universe'. For example, the principle cited above can be strengthened by replacing the (non-constant) term '$x + 1$' by '$(x + 1) + 1$'. The most restrictive class of terms is, of course, the null set which corresponds to a contradictory 'induction principle', e.g. $\forall yPy$. Short of this the strongest induction principles would 'force a unit set to cover the universe', e.g. $Pc \supset \forall yPy$.

tion principle implies *IK*, the above results hold when a strong induction principle is substituted for the induction principle.

3.4. *'Negative' applications*

In this sub-section let *LK* be any second-order language with identity and let *DK* be any sound and recursive system of deductions for *LK*. For example, let *DK* be the system of Church *1956* suitably extended with rules and axioms dealing with function symbols.

In this sub-section it is important to recall the distinction between (1) characterizing an interpretation *i* by means of a subset of *T(i)* (the truths of *i*) and (2) axiomatizing the truths of *i*. The point of characterizing *i* is descriptive and criterial; one aims at distinguishing *i* from other interpretations. The point of axiomatizing is to form the basis for a deductive development of the truths of *i*. From the standpoint of characterization the best that can be done (when it can be done) is a categorical characterization. With a categorical characterization an interpretation is distinguished from every other interpretation from which it can be distinguished by formal means. From the standpoint of axiomatization it is clear that the best that can be done (when it can be done) is a deductively complete axiomatization, i.e. a recursive subset of *T(i)* from which every member of *T(i)* is deducible by a (finite) deduction. It is obvious that the set of theorems deducible from a set of axioms is necessarily a recursively enumerable subset of the truths no matter which sound, recursive system of deductions is used but that, in general, the set of theorems is sensitive to choice of deductive system. Below we assume a fixed deductive system *DK*.

Some early postulate theories (e.g., Veblen *1904*, 346) were clear about the *conceptual* distinction between characterization and axiomatization *and* about the possibility of an axiomatically inadequate categorical characterization at least to the extent of explicitly mentioning the possibility that a categorical characterization need not be a (deductively) complete axiomatization. This possibility, of course, entails the possibility of 'logically' incomplete underlying logics (wherein semantic consequences of a given set of axioms are not deducible as theorems).

At that time, however, there was no suspicion of the idea of recursiveness, nor, *a fortiori*, of the relevance of recursiveness and recursive enumerability to problems of axiomatizability. Now we can see that if the set of truths of an interpretation is not recursively enumerable then there is no way to give a complete axiomatization even if the logic is complete. It follows immediately from the Gödel incompleteness result that a (recursive) set of sentences which provides a categorical characterization need not provide a complete axiomatization. Moreover, in such cases, it follows that there are infinitely many other categorical characterizations each of which provides a better axiomatization in the sense of providing the basis for the deduction of additional theorems not deducible from the first characterization.

Separate from the recursiveness considerations which lead to mismatches between characterization and axiomatization are the so-called 'compactness' considerations which lead to additional mismatches (compare Corcoran *1972*, 37*ff.*). Since every deduction is finite and therefore involves only finitely many axioms, no consequence of an infinite axiom set which depends on infinitely many of the axioms can be deducible from those axioms. It is compactness considerations rather than recursiveness considerations which are operative in the rest of this discussion.

It might be thought that *any* categorical characterization of an interpretation provides the basis (given a suitable deductive system) for the deduction of the 'obvious' truths of the interpretation. That is, one might expect that any truth not deducible from a categorical characterization must be a 'pathological' or 'complicated' proposition such as a so-called Gödel sentence or a statement of consistency. Admittedly this point has not been discussed much in the literature (see Paris *1978*). Nevertheless, the test of categoricity given above permits the establishment of categorical characterizations from which the most elementary general truths are not deducible, no matter what sound deductive system is used (regardless of the criterion of recursiveness).

Take $K = \{0, s\}$ and take S as the set of all true arithmetic identities ($s^n0 = s^n0$) and the negations of all of the false ones. By the above theorem $S + IK$ is a categorical characterization of $i = \langle N, \langle 0, \text{zero} \rangle, \langle s, \text{successor} \rangle \rangle$. Thus $S + IK$ implies $\forall x(sx \neq 0)$. However, it is impossible to deduce $p = \forall x(sx \neq 0)$ from $S + IK$ using DK or any other sound deductive system $DK1$ because if, say, $p_1 \ldots p_n p$ is a deduction of p from $S + IK$ and $DK1$ is sound then p is implied by the finite number of premises in $p_1 \ldots p_n$. But it is easy to see that no finite number of sentences in $S + IK$ implies $\forall x(sx \neq 0)$.

Examples of this sort can be multiplied. Take $K = \{0, s, +\}$. Take $i = \langle N, \langle 0, \text{zero} \rangle, \langle s, \text{successor} \rangle, \langle +, \text{addition} \rangle \rangle$.

For S take

$$\forall x(sx \neq 0), \forall xy(sx = sy \supset x = y),$$

and the true identities ($s^n0 + s^m0 = s^{n+m}0$) and the negations of the false ones. $S + IK$ is categorical but it is impossible to deduce $\forall xy((x + y) = (y + x))$ from $S + IK$ using a sound deductive system.

These examples are but other illustrations of the vast difference between characterizing an interpretation and axiomatizing its set of truths. The examples point to the conclusion that the connection between the two is weak. In particular, it is now clear that a 'best possible' characterization can be a very poor axiomatization.[12] The class of categorical characterizations of a given inductive system includes many which are virtually useless as axiomatizations.

---

12. Since semantic completeness is implied by but does not imply categoricity, it follows that semantic completeness is not sufficient for an axiomatization to be 'good'. In fact, as far as axiomatization is concerned, semantic completeness seems to be beside the point unless supplemented by other conditions formulated in accordance with goals arising in particular cases. These observations are due to George Weaver.

CATEGORICITY   205

It is clear from a survey of the relevant literature not only that the early postulate theorists were unaware of the recursiveness considerations (as mentioned above) but also that they were unaware of compactness considerations as well. One can *not* automatically conclude, however, that they were misguided in using categoricity as an index of worth of an axiomatization. One must realize that the above counterexamples all involve *infinite* sets of axioms whereas the earlier logicians, occasionally explicitly (Veblen *1904*, 343), conceived of an axiomatization as inherently finite. And in the opinion of this writer, the philosophical wisdom of abandoning the finiteness condition should be questioned despite the undeniable advances that came as a result of considering the mathematical consequences of relaxing that condition.

3.5. *Heuristics*

The above test of categoricity requires, for its application to a *given set of axioms* for an inductive system, that one first establish that the axiom set implies each of the true atoms of the system. It is entirely possible that this preliminary step is more demanding in a given case than a straight-forward categoricity proof. However, *if* one is given *the system* (interpretation) alone and the problem is to find a manageable categorical set of axioms *then* the goal of deducibility of the true atoms is often an effective heuristic which leads to the discovery of the required axiom set.

ACKNOWLEDGEMENTS

Sincere thanks to Nicolas Goodman (Mathematics, SUNY/Buffalo), William Frank (Philosophy, Oregon State University), George Weaver (Philosophy, Bryn Mawr College) and Michael Scanlan (Philosophy, SUNY/Buffalo) for useful criticisms and for pleasant discussions. This paper was presented to the Buffalo Logic Colloquium in November 1977 and to the Association for Symbolic Logic in December 1977. Several improvements in the final draft were made on the advice of Michael Resnik (Philosophy, University of North Carolina) of the Editorial Board of this publication. I would like to emphasize the fact that each of the persons here acknowledged made extensive contributions to this research.

BIBLIOGRAPHY

Barbut, Marc *1966* 'On the meaning of the word "structure" in mathematics', orig. publ. *Les temps modernes*, no. 246; translated in Lane *1970* by Susan Gray.

Birkhoff, G., and S. MacLane *1944* *A survey of modern algebra* (revised edition 1953), Macmillan, New York.

Bridge, Jane *1977* *Beginning model theory*, Oxford University Press, Oxford.

Cantor, Georg *1895/7* *Contributions to transfinite numbers* (translated and provided with introduction and notes by P. E. B. Jourdain), Open Court, Chicago, 1915; repr. Dover, New York, 1955.

206

J. CORCORAN

Church, A. *1956 Introduction to mathematical logic*, Princeton University Press, Princeton.

Corcoran, J. *1972* 'Conceptual structure of classical logic', *Philosophy and phenomenological research*, **33**, 25–47.

Corcoran, J. *1977* 'Categoricity' [abstract to appear in *J. symb. logic*].

Corcoran, J., W. Frank and M. Maloney *1974* 'String theory', *Journal of symbolic logic*, **39**, 625–637.

Dedekind, Richard *1887* 'The nature and meaning of numbers', in Dedekind, R. *Essays on the theory of numbers* (trans. W. W. Beman: Open Court, Chicago, 1901; repr. Dover, New York, 1963).

Ellentuck, Erik *1976* 'Categoricity regained', *Journal of symbolic logic*, **41**, 639–643.

Enderton. H. *1972 A mathematical introduction to logic*, Academic Press, New York and London.

Fraleigh. John *1976 A first course in abstract algebra*, Addison-Wesley Publishing Company, Reading, Massachusetts.

Frege. Gottlob *1899* Leter to Hilbert. dated 27 December 1899. in Frege *1976*. 60–64; cited from the English translation in Kluge *1971*, 6–10.

Frege. Gottlob *1906* 'Über die Grundlagen der Geometrie', *Jber. Dtsch. Math.-Ver.*, **15**. 293–309, 377–403, 423–430; cited from the English translation in Kluge *1971*, 49–112.

Frege, Gottlob *1976 Wissenschaftlicher Briefwechsel* (edited by G. Gabriel *et al.*), Felix Meiner Verlag, Hamburg, 1976.

Grzegorczyk, Andrzej *1962* 'On the concept of categoricity', *Studia logica*, **13**, 39–65.

Halmos, Paul R. *1960 Naive set theory*, D. Van Nostrand Co., Princeton.

Hailperin, T. *1976 Boole's logic and probability*, North-Holland Publishing Company, Amsterdam. New York and Oxford.

Hilbert. David *1889 The foundation of geometry* (authorized translation by E. J. Townsend), Open Court, LaSalle, Illinois, 1902.

Hilbert. David *1899L* Letter to Frege dated 29 December 1899, in Frege *1976*, 65–68; cited from the English translation in Kluge *1971*, 10–14.

Huntington, Edward V. *1902* 'A complete set of postulates for the theory of absolute continuous magnitude', *Transactions of the American Mathematical Society*, **3**, 264–279.

Huntington, Edward V. *1905* 'The continuum as a type of order', *Annals of mathematics*, (2) **6**, 151–184, (2) **7** (1906), 15–43; citations from next reference.

Huntington, Edward V. *1917 The continuum* (second edition), Harvard University Press, Cambridge, Massachusetts; repr. by Dover, New York, 1955.

Jourdain, P. E. B. *1915* 'Introduction', in Cantor *1895/7*, 1–82.

Kac. Mark and St. Ulam *1968 Mathematics and logic*, Praeger Publishers, New York; pagination from reprint edition, New American Library, New York. 1969.

Kline. Morris *1972 Mathematical thought from ancient to modern times*, Oxford University Press. New York.

Kleene. S. C. *1952 Introduction to metamathematics*. D. Van Nostrand Company, Princeton. Toronto and New York.

Kluge. Eike-Henner (ed.) *1971 Gottlob Frege on the foundations of geometry and formal theories of arithmetic*. Yale University Press, New Haven.

Kreisel, Georg *1965* 'Informal rigour and completeness proofs', in Lakatos *1967*, 138–186.

Lakatos, Imre (ed.) *1967 Problems in the philosophy of mathematics*, North-Holland Publishing Company, Amsterdam.

Lane, Michael (ed.) *1970 Introduction to structuralism*, Harper Torchbook, Basic Books, New York.

Mendelson, E. *1964 Introduction to mathematical logic*, D. Van Nostrand Company, Princeton, Toronto, New York and London.

Montague, R. *1965* 'Set theory and higher order logic', in Dummett and J. Crossley (eds.) *Formal systems and recursive functions*, North-Holland Publishing Company, Amsterdam, New York and Oxford, pp. 131–143.

Paris, J. B. *1978* 'Some independence results for Peano arithmetic', *J. symb. logic*, **43**, 725–731.

Resnik, Michael *1974* 'The Frege–Hilbert controversy', *Philosophy and phenomenological research*, **34**, 386–403.

Skolem, Thoralf *1920* 'Logico-combinatorial investigations, . . .'; cited from English translation in van Heijenoort *1967*, 252–263.

Skolem, Thoralf *1928* 'On mathematical logic'; cited from English translation in van Heijenoort *1967*, 508–529.

Tarski, Alfred *1934* 'The concept of truth in formalised languages', in *Logic, semantics, metamathematics* (Oxford, Clarendon Press, 1956), 152–278. [Revised edition of book to appear in 1980.]

van Heijenoort, Jean (ed.) *1967 From Frege to Gödel*, Harvard University Press, Cambridge, Massachusetts.

Veblen, Oswald *1904* 'A system of axioms for geometry', *Transactions of the American Mathematical Society*, **5**, 343–384.

Young, John Wesley *1911 Lectures on fundamental concepts of algebra and geometry*, Macmillan Company, New York.

[13]

CHARLES McCARTY AND NEIL TENNANT

SKOLEM'S PARADOX AND CONSTRUCTIVISM

INTRODUCTION

Skolem's paradox has not been shown to arise for constructivism.
Indeed, considerations that we shall advance below indicate that the
main result cited to produce the paradox could be obtained only by
methods not in mainstream intuitionistic practice. It strikes us as an
important fact for the philosophy of mathematics. The intuitionistic
conception of the mathematical universe appears, as far as we know,
to be free from Skolemite stress. If one could discover reasons for
believing this to be no accident, it would be an important new consideration, in addition to the meaning-theoretic ones advanced by
Dummett (1978, 'Concluding philosophical remarks'), that ought to
be assessed when trying to reach a view on the question whether
intuitionism is the correct philosophy of mathematics.

We give below the detailed reasons why we believe that intuitionistic
mathematics is indeed free of the Skolem paradox. They culminate in
a strong independence result: even a very powerful version of intuitionistic set theory does not yield any of the usual forms of a countable
downward Löwenheim–Skolem theorem. The proof draws on the
general equivalence described in McCarty (1984) between intuitionistic
mathematics and classical recursive mathematics. But first we set the
stage by explaining the history of the (classical) paradox, and the
philosophical reflections on the foundations of set theory that it
has provoked. The recent symposium between Paul Benacerraf and
Crispin Wright provides a focus for these considerations. Then we
inspect the known proofs of the Löwenheim–Skolem theorem, and
reveal them all to be constructively unacceptable.

Finally we set out the independence results. They yield, we believe,
the deep reasons for the localised constructive failures. Besides showing

Note: 'ε' is used both as the epsilon of set membership and for the existential quantifier. Context will always make clear which use is intended.

Journal of Philosophical Logic **16** (1987) 165–202.

166   CHARLES McCARTY AND NEIL TENNANT

that weak versions of the Löwenheim — Skolem Theorem cannot be
proved in extensions of the intuitionistic set theory IZF, we prove
that the Theorem entails principles which many constructivists would
reject (e.g., Kripke's Schema) and is falsified outright by principles
(Church's Thesis and Markov's Scheme) which a number of constructivists would accept. Let us also emphasize at the outset that the metamathematics which we adopt in giving our proofs is itself constructive.

                                   I

Skolem's paradox was thought by Skolem, and has been thought by
his skeptical successors since, to show that the notion of absolute
non-denumerability is ineffable. In Skolem's own words (1922):

> ... auf axiomatischer Grundlage sind höhere Unendlichkeiten nur in relativem Sinne
> vorhanden.
> (on axiomatic foundations higher infinities occur only in a relative sense.)

We say that the paradox has been thought to *show* this precisely because,
unlike Russell's paradox, it does not *say* anything inconsistent in the
viciously circular way we have come to associate with the logical and
set-theoretical paradoxes.[1] Skolem's paradox, unlike Russell's, is not
crisply expressed by any one sentence of the object language. Instead,
it involves a traversing of levels between object and metalanguage. It
arises because Cantor's theorem, provable in the object language of
set theory, says that there is no one-one correlation of the set of natural numbers onto the set of all its subsets; while a model existence
theorem, provable (using informal set theory) in the metalanguage,
says that the axiomatic set theory of the object language has a countable model.

   Two theorems therefore produce the paradoxical tension. Let $M[t]$
be the denotation, in model M, of the term t. Let '$P(\omega)$' be the term
for the power set of ω, the set of natural numbers. Suppose M is a
countable model of set theory. The tension is this: $M[P(\omega)]$ appears
'within' the countable model M (on pain of contradicting Cantor's
theorem) not to be the domain of any one-one mapping, *within M*,
onto the set ω; the model fails to contain an element in its own domain
serving as the set of ordered pairs that would establish such a one — one
correlation. But from 'outside' the model M the set $M[P(\omega)]$ does

appear to be the domain of a one-one mapping onto the set ω. Cantor's theorem, however, prevents this mapping from being in the model M. From 'inside' the countable model M, as it were, one fails to appreciate just how 'small' $M(P(\omega)]$ is.

That, in a nutshell, is the paradox. The Skolemite Skeptic invokes it to undermine our confidence that one can communicate a conception of the mathematical universe. If the mathematical universe is rich enough to contain the real numbers (in the form of the power set of ω), the paradox says it can be impoverished. Or, rather, the account of it can be so devalued by interpretation in a countable (sub-) universe that one cannot gain a purchase on its contents.

We shall argue that the devalued linguistic currency may be strictly classical. The *intuitionist* or *constructivist* mathematician appears not to be affected by the problem. For one of the two theorems involved in the classical case is not intuitionistically true. So, ironically, although being able to 'say' much more by virtue of his stronger logical methods, the classicist appears to speak to less pointed effect than the intuitionist.

Our method of exposition will be as follows. First we shall examine the recent exchange on the problem of Skolem and the Skeptic by Benacerraf and Wright. Then we shall look more closely at the method of proof of Cantor's theorem, and bring out its rich constructive import. Then we shall inspect the various methods of proof of the countable downward Löwenheim—Skolem theorem, this being the other result (available classically) that produces the paradox, as explained above. These methods include Skolem's original one, using Skolem normal forms, another using prenex normal forms, and the one by Henkin expansion. We shall show how each method fails in the intuitionistic case. Finally we shall give a general result behind all these failures: the independence, within a strong version of intuitionistic set theory, of forms of the arbitrary countable models theorem. This is an interesting and important phenomenon that deserves further investigation.

II

In his searching examination of the source of Skolem's paradox, Benacerraf (1985) reflects on the inadequacy of the formal axiomatic method as follows (p. 111):

168   CHARLES McCARTY AND NEIL TENNANT

Despite the imagined possible misunderstandings, mathematical practice reflects our intentions and controls our use of mathematical language in ways of which we may not be aware at any given moment, but which transcend what we may explicitly set down in any given *account* — or may ever be able to set down.

The account in question is, of course, an axiomatic one at first order, involving only countably many sentences. In his paper Benacerraf does attempt to locate more precisely where the trouble lies than merely the countable character of our sayings. He suggests that the problem lies not so much with the interpretation of 'ε' (the membership predicate), but rather with the interpretation of the universal quantifier:

> ... whether *T says* that a set is non-denumerable depends on *more* than whether the interpretation is over a domain of sets, 'ε' of the interpretation coincides with membership among *those* sets, and every element of any set in the model is also in the model. The universal quantifier has to mean *all*, or at least *all sets* — or at least it must range over a domain wide enough to include 'enough' of the subsets of (the set of natural numbers). (*loc. cit.* p. 103)

In his reply, Wright (1985) is sympathetic to this diagnosis of the classical set theorist's problem, even if not sympathetic to the classicism that generates it. He claims that the diagonal argument establishes a result about uncountability only on a nonconstructive interpretation. Otherwise, he thinks, all it can be taken to show is that there is no effective enumeration of all constructive (that is, decidable) sets of natural numbers. As Wright puts it (pp. 134 – 135):

> ... before the ... informal proof of the power set theorem can lead us to a conception of the intended range of the individual variables in set theory which will allow us to regard any countable set model as a non-standard truncation ... we need to grasp the notion of a *non-effectively enumerable denumerably infinite subset of natural numbers*. This is what, if he is in the business of giving explanations, the Cantorian needs to explain.

We shall now argue that, in this fascinating encounter between a classicist and a constructivist, the real problem has been mislocated and the proper solution has been missed. The points to be made are historical, logical and conceptual. But first, in order to set the stage properly for discussion, let us survey the many potential sources of the difficulty, and see how far we can agree with Benacerraf and Wright in putting them aside. This will enable us to focus more sharply on the problem they neglect and on the shape of its solution.

SKOLEM'S PARADOX AND CONSTRUCTIVISM 169

III

The following is a digest of points worth considering in this regard.

1. What *is* the classical notion of set?
   a. Can it be communicated only informally? Does it elude axiomatic characterization
      i.   at first order
      ii. at second or higher order
      iii. in finitary languages
      iv. in infinitary languages
      v.   using only countably many sentences
      vi. using uncountably many sentences?[2]
   b. Does it have *built into it* the existence of:
      i. an infinite set
      ii. an uncountably infinite set?
2. How are the two theorems involved in Skolem's paradox proved? Constructively or strictly classically?
3. What is the significance of having a countable *sub*model theorem as opposed to one which merely guarantees the existence of some countable model (whether or not it be a submodel of an intended model presumed given)?

Having posed these questions, let us now sketch what we take to be common ground between Benacerraf and Wright. (a) Both take the classical notion of set, if communicable, to be communicable only informally. (b) Both see no point in considering higher order or infinitary languages, or uncountable theories in the axiomatic characterization of the notion. For with all these (it could be argued) the notions of set or of non-denumerability are being presupposed in the very project of communicating or imparting an understanding of them. (c) Both consider worthwhile and admissible only countable axiomatic characterizations in finitary first order languages.

With this much we are in sympathy. Skolem himself remarked on the circularity of the presupposition just mentioned (1922, p. 144). But Skolem would also have accorded no particular significance to the distinction mentioned in (3) above (as historical considerations will in due course show); and in this regard we too would be unwilling

170   CHARLES McCARTY AND NEIL TENNANT

to move on to the common ground between Benacerraf and Wright. For both of them regard the countable *sub*model theorem as contributing an important part of the discomfiture produced by Skolem's paradox. Their thought is that a countable submodel extracted from the intended model arguably preserves the interpretation of 'ε', and thereby shifts the locus of the difficulty to the interpretation of the universal quantifier (at least as it applies to sets). Benacerraf himself states as one version of the Löwenheim—Skolem theorem the following:

SMT (a transitive submodel version): Any transitive model for ZF has a transitive countable submodel. (A model is *transitive* if and only if each element of each set in the model belongs to the domain of the model.) (p. 101).

And Wright endorses his preference for this version of the countable model theorem as follows:

In order to get a line worth considering we must, I think, go for something like the more sophisticated reconstruction of the argument which Benacerraf builds on the transitive countable sub-model version of LST (= SMT). The intended interpretation for ZF involves, I take it, a transitive model: that is, every member of every set which the intended interpretation would include in the subject matter of set theory is likewise part of that subject matter. According to SMT, then, if ZF can sustain its intended interpretation at all, it may be interpreted in a countable sub-domain of the sets involved in the intended interpretation, *in such a way that 'ε' continues to mean set-membership* and every set in the domain of the new interpretation is itself at most countably infinite . . . (p. 118; our emphasis).

In his eagerness to concentrate on the countable *sub*model version of the theorem, Benacerraf even goes so far (p. 94) as mistakenly to claim that it was this version that Skolem proved in his 1922 paper. In the next section we shall return to the distinction between the two versions of the model existence theorem, and argue that it is philosophically irrelevant. We shall argue also that both Benacerraf and Wright have followed a red herring, despite clear contextual clues in Skolem's own writings that the distinction *is* irrelevant, and that the problem cannot be shifted away from the interpretation of 'ε' and onto the interpretation of 'all sets' or 'all subsets of . . . '. But first let us complete our summary responses to the lists of questions above.

As to the method of proof of the theorems involved in the paradox — Cantor's theorem and the theorem on the existence of countable models of countable theories — Benacerraf says nothing, apart from an inaccurate historical remark (p. 91, n. 3) that Löwenheim employed the axiom of choice in his original proof. In fact, it was Skolem who introduced the use of choice in order to simplify Löwenheim's proof. More to the point would be the question whether the countable models theorem is constructively provable. Wright enters constructivist considerations, and sees the embarrassment of Skolem's paradox as affecting, first and foremost, the Cantorian. He makes the Skeptic sound thoroughly constructivist; he suggestively asserts (p. 124) that

the Skeptic will urge (that) a full and complete explanation of the concept of set is neutral with respect to the existence of uncountable sets. But if there really were uncountable sets, their existence would surely have to flow from the concept of set as intuitively satisfactorily explained.

And later, on p. 126 he writes:

. . . if the ZF-axioms, with 'ε' interpreted as set membership, did constitute a satisfactory explication of the extension of the intuitive concept of set, the fact that they do not, so interpreted, entail the existence of uncountable sets would force the conclusion that there is no such entailment from the intuitive concept of set either.

It would have been natural, in registering constructivist doubts about the *existence* of (uncountably) infinite sets as completed totalities, to have mentioned also the relevance of the more narrowly restricted *logical* methods of which the constructivist may avail himself compared with the classicist. It would have been appropriate then to ask whether indeed the countable models theorem is *constructively* provable; whether, that is, the Cantorian embarrassment arises from paradisial *logical* manoeuvres as much as it does from ontological excess. Now Wright did not himself pursue this question; for he was allowing the classicist the full use of classical methods in order to bring out the inadequacies internal to realism. But it is only natural to ask whether, once the problem has been posed by the proofs of the two main theorems − Cantor's theorem, and the countable models theorem − the same problem will continue to bedevil the constructivist who cuts back on permissible methods of proof.

IV

Let us suppose, with all parties to the debate, that what is sought and what is being assessed for explicatory adequacy, is some formal characterization of the notion of set using a finitary first order language

172   CHARLES McCARTY AND NEIL TENNANT

and a countable formulation using rules and or axioms. Let us for the
moment register, but set aside, the problem of the correct choice of
logical rules of inference. Note that there are two aspects to the 'notion
of set'. One may be called *structural-theoretical*, the other *ontological*.
The *structural-theoretical* aspect is addressed by such principles as the
principle of extensionality and Church's conversion schema. To wit,
sets with the same members are identifiable; and the members of the
set of F's are precisely the F's.

The *ontological* aspect is addressed by such questions as:
     is there a null set?
     is there an infinite set?
     is there an uncountably infinite set?
     is there a universal set?
and it is not at all clear that (the first three at least of) *these* questions
should be answerable by anyone who has mastered the *concept* of set
as governed by the principles mentioned earlier. At least, it is not at
all clear without further argument that this is indeed so. Take, for
example, a mastery of the concept 'tiger'. Does that mastery entail
ability to decide *a priori* whether there are only finitely, or infinitely
many, tigers? We should think not. But, it may be objected, this is
simply because the concept in question applies to what the set-theorist
calls *urelements*. Were we to take instead any sensible concept applying only to *pure sets*, then (so this reply goes) answers to such questions would be entailed *a priori* merely by adequate grasp of the concept involved.

   But *would* they? Certainly the recent history of mathematics tells
against such an assertion. Many a writer has denied the existence of
infinite sets as completed totalities, even though not denying that
there are infinitely many things of different kinds, such as natural
numbers. And even if one concedes *a priori* that there be infinitely —
indeed, even uncountably — many pure sets, that *still* falls short of
securing a pure set with infinitely (or uncountably) many members.
Are constructivists with reservations such as these to be accused of
deficient grasp of the concept of set, even though they are perfectly
well acquainted with the agreed principles governing the interrelation-
ships among predication, set formation and membership?

SKOLEM'S PARADOX AND CONSTRUCTIVISM   173

The point we are making may best be put as follows. There is what may be called the *logic of sets*: a collection of rules governing these interconnections between set formation and membership, and predication and existence; and it is this logic alone which underlies proper grasp of the *notion* or *concept* of set. There is then what may be called the *theory of sets*, formulated according to one's ontological convictions. One may or may not postulate the existence of ω, the set of natural numbers. But whether we do or do not, we are all (classicists and intuitionists alike) agreed that, *if* we do, we shall be able to show that ω has strictly more subsets than it has members. This is because Cantor's proof, using separation most importantly among the ZF axioms, is thoroughly constructive (cf. Greenleaf, 1981). It is not even necessary to have the power set of ω in the picture itself as a completed set.

To see this, let us look more closely at how Cantor's theorem is proved. Cantor's reasoning shows that it absurd to assume that one can correlate subsets of ω one-one with members of ω, in such a way that every subset is dealt with. Given merely that ω exists, the diagonal argument requires only that one be able to 'cull' from ω a 'diagonal subset' which, on pain on contradiction, cannot be dealt with by the method of correlation presumed given. If R is the method in question, so that xRy means 'the subset x of ω is correlated with the member y of ω, then the diagonal subset is simply defined as the set of all z in ω such that z is not a member of the y such that yRz.

Since ω is assumed to exist, and since R is assumed to serve up, for each z in ω, the unique subset y of ω such that yRz, it follows, by separation, that this diagonal set will exist. Call it d. Consider now the member e of ω such that dRe. Is e a member of D? It is if and only if it isn't. No appeal is made to the existence of $P(\omega)$, the power set of ω. Given this *reductio* of the assumption that we could have any such method R, the constructivist is able to assert that *there are strictly more* subsets of ω than there are members of ω. He cannot be denied this cardinality reading of his result, since he agrees with the Cantorian analysis of equinumerosity in terms of one-one correlations. It is not a justifiable move even from the *classicist's* vantage point to re-interpret the Cantor proof (as Wright does on pp. 133–134) as establishing, not a result about uncountability (of the subsets of ω),

174   CHARLES McCARTY AND NEIL TENNANT

but rather one to the effect that there can be no effective enumeration
of all decidable subsets of ω. And certainly the constructivist would
refuse to be read that way. For the constructivist takes himself to be
talking about correlations and subsets *tout court*. Wright himself does
not do justice to the implicit power, from the constructive standpoint,
of Cantor's result. Not only is Cantor's proof constructively acceptable, but, given the constructive interpretations available for the term
'countable', its conclusion can be made even stronger. Cantor's argument shows not only that $P(\omega)$ is uncountable, but that it is *not subcountable*, and that is ω-*productive*.

A *subcountable* set is one which is the range of some function
whose domain is a *sub*set of ω. An ω-*productive* set is one in which,
should any of its subsets be the range of a partial or total function
defined on the natural numbers, one can find an element that lies outside that range. The notions *countable*, *subcountable* and *non-ω-
productive* coincide classically. Intuitionistically, however, they come
apart. Countable implies subcountable; subcountable implies non-ω-
productive. To each converse, however, there is a constructive counterexample. It is ω-productivity which yields the strongest intuitionistic
reading of the conclusion of Cantor's proof; and the method of proof
directly justifies that reading. Detailed analyses of the constructive
content of Cantor's proof with reference to the notions 'countable',
'subcountable' and 'ω-productive' have appeared in Grayson (1978)
and Greenleaf (1981).

Note that in proving Cantor's theorem, the constructivist does not
have to appeal to the power set axiom. Depending therefore on one's
view of separation — is it a 'logical' axiom governing sets, or a
'mathematical' one? — one might regard Cantor's result as embedded
in the very concept of set. One might even go so far as to question
whether Wright is entitled to say (pp. 123—124)

Let somebody have as rich an informal set-theoretic education as you like — which,
however, is to stop short of a demonstration of Cantor's theorem, or any comparable
result, since these findings are, after all, supposed to be available by way of discovery
to someone who has mastered the intuitive concept of set.

But to pursue this point here would be to digress, since what we have
to say below is independent of any decision one might reach concerning the precise status of separation.

The logic of sets is formulated in Tennant (1978).[3] It consists of rules for the introduction and elimination of the set term-forming operator in contexts of identity. The introduction rule codes extentionality; the elimation rules code the conversion schema. The logic is a free logic, so that, for example, the reasoning behind Russell's paradox furnishes a proof that the Russell set does not exist. The logic is proved sound and complete with respect to the obvious semantics. Properly *mathematical* assumptions may then be made about the existence of sets: in particular, the null set and the set of natural numbers. But this, on our account, is to go strictly *beyond* what is involved in the correct analysis of the notion of set, as enshrined in the logical rules alone.

The null set axiom and the infinity axiom both make *outright* claims about existence. In so doing they are the clearest possible cases of what we take to be strictly *mathematical* claims about sets. But there is a penumbral family of axioms falling between the outright existence claims and the introduction and elimination rules mentioned earlier. These are the axioms of *conditional existence*: power set, pairs, unions, replacement, and choice. They all say that *if* such-and-such sets exist, *then* so too does one of a certain kind where the latter kind gives the axiom its special character. Now the interesting thing about separation is that it too makes a conditional existence claim, yet makes it so generally that it is difficult to regard it as doing anything more than merely contributing to the explication of the notion of set itself. Separation is an axiom schema, with instances obtained by choosing a particular formula $F(x)$. An instance will say 'for all sets y' there exists a set whose members are exactly those members x of y such that $F(x)$'. If we take separation as part of the logic of sets, then we have also as a purely 'logical' result that the null set exists if *any* set does: simply apply separation with '$\neg x = x$' for $F(x)$. But there will be no similarly quick way to the set of natural numbers. The existence of *that* set remains, on this analysis, a strictly mathematical postulate.

Having more or less clearly separated the ontological from the conceptual aspects of 'set', one can then go on to exercise further choice as to the logic appropriate for developing the consequences of whatever existential theoretical commitments one might care to make. For

176   CHARLES McCARTY AND NEIL TENNANT

the so-called logic of sets concerns itself thus far only with the curly
brackets and epsilon, and identity. Nothing has yet been laid down
for the logical connectives and quantifiers. A range of positions thus
becomes available, each founded upon but properly extending the
common core of analytical agreement over the conceptual aspect of
'set':

   (i)   postulate the existence of the null set
        and
       work with intuitionistic logic in the object language
  (ii) postulate the existence of the null set
        and
       work with classical logic in the object language
 (iii) postulate the existence of the set of natural numbers
        and
       work with intuitionistic logic in the object language
 (iv) postulate the existence of the set of natural numbers
        and
       work with classical logic in the object language.

(iv) is the position of classical ZF. (ii) gives the classical theory of the
hereditarily finite sets. (i) represents an extreme Ockhamite constructivism. (iii) is a natural and inviting alternative, for the intuitionist, to
classical ZF. What we want to investigate is whether, had Wright but
made such a clear and explicit choice as (iii), and had he been
prepared to concede that (iii) was all that was available *informally in
the metalanguage* as well, he might have seen the Skolemite problem
in a different light, and offered a different constructive resolution of
the paradox.

V

Skolem gave two proofs of his theorem. The first was in the paper of
1920. There he used the axiom of choice to construct a countable *sub*-
model of any given model of a first order theory. The method was as
follows: first one replaces each sentence of the theory, without loss of
generality, by its Skolem normal form. This would be a sentence

SKOLEM'S PARADOX AND CONSTRUCTIVISM   177

beginning with universal quantifiers, followed by existentials, all appended to a matrix of a certain form. Next one chooses an element from the domain of the given model, and uses it multiply to instantiate the universals of the first chosen sentence. One then chooses (using the axiom of choice) at most finitely many existential satisfiers of the following existential quantifiers, and puts these alongside the original chosen element. Then one extends one's attention to the second sentence of the theory as well. (The sentences of the theory are assumed given in some countable enumeration.) One tries every possible way, using the finitely many elements so far in the picture, of instantiating the universal quantifier strings of both the first and the second sentence. For each way at most finitely many new existential satisfiers for the following existential quantifiers have to be chosen (again using the axiom of choice). One proceeds in this way, progressively taking into account more and more sentences of the theory, and recruiting new satisfiers for the existentials in each sentence with respect to each of the increasingly numerous ways of instantiating their universal prefixes. The countable model being extracted is in an obvious sense the product of this process in the limit. It arises, one might say, by *interated existential closure*, via the axiom of choice, from the Skolemised surrogates of the original sentences of the theory.

There is another proof using choice, which Skolem could have used, given that Zermelo had established in 1904 that choice is equivalent to the well-ordering principle. This proof applies choice globally at the outset, by taking the domain to be well-ordered. Iterated existential closure then simply trawls the ordering for its countable catch.

In his second proof of the theorem, in his paper of 1922, Skolem drops the appeal to the axiom of choice and thereby proves a slightly different result. No longer is it a *sub*model of a given model that is being constructed; rather, it is a model erected on the natural numbers. The assumption that Skolem normal forms are available is, as before, an absolutely crucial feature of his method of proof for the classically understood object language.

But there is the possibility also of using *prenex* normal forms, as is done in the classical case by Quine (1959) and Grandy (1977). We note this as an alternative to Skolem's method, and return to it below.

Now what is remarkable, in the light of Benacerraf's discussion, is that it is only in the 1922 paper that Skolem formulates the set theoretic paradox that now bears his name. He saw it as quite sufficient simply to produce *some* countable model of the theory, rather than a countable submodel of some given intended model. And his philosophical conclusion about higher infinities cited at the beginning of this paper came but one paragraph after a much more general conclusion that he drew concerning the relativity of the (classical) *notion of set* itself:

Die axiomatische Begründung der Mengenlehre führt zu einer Relativität *der Mengenbegriffe*, und diese ist mit jeder konsequenten Axiomatik untrennbar verknüpft. (Our emphasis; in the original the whole sentence is italicised.)

One might even speculate that Skolem himself would have been aware of the significance, if any, of the difference between a countable model extracted from an intended model of set theory and a countable model erected directly upon the natural numbers, insofar as philosophical conclusions about conceptual relativity were in the offing. For he, after all, was the author of both kinds of theorem. The first used choice to rummage within a given model and pare it down. The second eschewed choice by starting with a (set theoretically) phoney line-up.

It might be maintained in response to this, and on Benacerraf's behalf, that Skolem could well have been blind to the difference, since he had not bothered to reflect further on the possibility of apportioning blame between our interpretation of the universal quantifier and grasp of the notion of the uncountable. But what precisely is the extra significance accorded by Benacerraf to the version of the countable model theorem which has one *extract* it from an intended model?

Benacerraf sees the extraction as somehow preserving the interpretation (assumed correctly given in the intended model) of the membership relation. This is because, according to him, the countable model of set theory extracted from the intended model will, like its parent model, be *transitive*. Transitivity, an important global feature of the membership relation, must not be lost if the constructed model is to have any claim at all to be a model of *set* theory. But what is transitivity, exactly? Benacerraf defines it as follows:

A model is *transitive* if and only if each element of each set in the model belongs to the domain of the model. (p. 101)

Let *b* be a set in the model. That is, let *b* be a member of the domain of the model. What is it for *a* to be an element of *b*? There are two answers to this question. First, the *internal* one: *a* is an element of *b* just in case *a* is, like *b*, in the domain of the model, and the ordered pair (*a*, *b*) is in the extension of 'ε' within the model. That is, *a* is, *according to the model*, an element of *b*. Secondly, the *external* answer: both *a* and *b* might have genuine properties, or genuine internal structure, not registered within the model. They are recruited as members of the domain of the model and assigned, within the model, various skeletal relations to each other (perhaps) and to other members of the domain. The assignment can ignore their genuine properties and internal structure. The web of relations within the model can fail to unpack the metaphysical richness which they intrinsically bring with them 'from outside' the model, so to speak. It is a little like treating a Royal procession as a model for a strict linear discrete ordering with first and last elements. The kinship relations and the character traits go unheeded. So too with sets — according to Benacerraf. We may put the genuine power set of ω into the domain of a countable model of set theory, but this model will be a transitive model only if every one of that set's members — that is, every set of natural numbers — is also in the model. The external reading has it that what is really the case with membership must be properly reported within the model itself — otherwise the model won't be transitive.

There are problems with both the internal reading and the external reading of transitivity. First, on the internal reading every model will be transitive; so the requirement of transitivity is trivial. For what the model *says* bears epsilon to *b* will obviously have to be in the domain of the model! Secondly, on the external reading no countable model of ZF that contains the genuine power set of the naturals can possibly be transitive! For, if the model contains $P(\omega)$ then, in order to be transitive, it would also, as just observed, have to contain every member of $P(\omega)$. But there are uncountably many such members — so

180   CHARLES McCARTY AND NEIL TENNANT

some of them would have to be missing from the domain of the model in order for it to be countable, as supposed.

Thus, for the model to be transitive, the genuine $P(\omega)$ itself would have to be missing. One might put the familiar response as follows: the set term '$P(\omega)$' is not *rigid* — it does not denote the *same* set as one passes from one transitive model of set theory to another. And the failure of *sameness* is a radical one, involving (from an external perspective) not just substitution of an isomorph, but also collapse of cardinality at times. Any countable *transitive* submodel of the intended model therefore cannot contain the genuine $P(\omega)$ as its own denotation of the set term '$P(\omega)$'. With this symptom of Skolemite non-standardness, it is difficult to see how in a countable model the further requirement of transitivity could yield assurance that the model be any better behaved on epsilon than any other model would be. The intension of 'ε', in allowing an extension to be determined for 'ε' in any countable domain for ZF, is irretrievably parableptic.

There is a further, and in our view clinching, reason not to be persuaded of the alleged philosophical relevance of Benacerraf's distinction, accepted by Wright, between 'set models' delivered by SMT and the 'numerical models' delivered by Skolem's 1922 proof of the countable models existence theorem. Benacerraf's version SMT of the Löwenheim–Skolem theorem is no more telling in posing Skolem's paradox than would be any version merely guaranteeing the existence of a countable model of set theory. For his own philosophical purposes in his paper, Benacerraf could just as well have stated his version of the theory as follows (a version which follows from Mostowski's contraction lemma):

Any *standard* model for ZF has a countable submodel ε-isomorphic to the *minimal model*; where the latter is the sole model which is a countable standard transitive model of ZF + V = L and also a submodel of every standard transitive model of ZF.

A *standard* model is one in which 'ε' is interpreted as set-membership. The *minimal model* helps, as it were, to uniformize the Skolemite's ministrations.[5] Here is the best behaved countable fragment of the *real* epsilon relation that one can get. But the *real* power set of the naturals must, by our foregoing considerations, *not* be caught up in this fragment. Something else in the fragment is playing the role of that power set. $P(\omega)$ is therefore quite unlike the Benacerrafian 3 (cf. Benacerraf, 1965). For while Benacerraf was able to conclude that the number 3 was no particular set on any set-theoretic construal of numbers, but rather a structural locus — the role played by whatever 'is' 3 in any standard recursive progression — he is robbed of a similar thought concerning $P(\omega)$. He cannot say that $P(\omega)$ is the role played by whatever 'is' $P(\omega)$ (that is, whatever is denoted by '$P(\omega)$') in any standard transitive model of set theory. For the most crucial feature of $P(\omega)$ is that it has (by Cantor's reasoning) uncountably many members. Yet here, in the countable model called the minimal model (which is both standard and transitive), whatever it is that stands as the denotation of '$P(\omega)$' does *not* have uncountably many members! — neither within itself, 'genuinely' (for the model is both standard and transitive) nor by model-relative alliance via 'ε' (for the model is countable). By contrast, the Benacerrafian 3 always has *three* predecessors (0, 1 and 2) in any progression.

VI

Thus far we have made the classicist's predicament more pointed. And we take the challenge to be not so much how to get out of that predicament, but rather how to avoid getting into it in the first place. We have not yet investigated the consequences of position (iii) above: the one that postulates the existence of the null set and the set of natural numbers, but restricts one to intuitionistic logic (both in the object language and, let us also assume, in the metalanguage). Restriction to intuitionistic metalogic is important. We shall now see how all the familiar proofs of the countable models theorem are intuitionistically objectionable. These include Skolem's proof discussed above; the proof using prenex normal forms, which we mentioned in passing, and the proof by Henkin's method. Once we have seen how all these methods of proof fail, we shall advance perfectly general reasons for their doing so: we shall show that a general form of the countable models theorem is independent of a strong version of intuitionistic Zermelo Fraenkel set theory (IZF).

First, the use of the axiom of choice in Skolem's proof of his theorem is intuitionistically unacceptable. Choice is constructively correct

182   CHARLES McCARTY AND NEIL TENNANT

for certain sets, such as the natural numbers, but not for arbitrary
sets. Yet it is in the more general setting that Skolem needs choice. It
is simply mistaken to think that on an intuitionistic construal of
operators in the object language, the truth of a sentence of the form

$$\forall x_1 \ldots \forall x_n E y_1 \ldots E y_m P(s_1, \ldots, x_n, y_1, \ldots, y_m)$$

requires that there be a uniform effective method for choosing, for
each $a_1, \ldots, a_n$, appropriate members $b_1, \ldots, b_m$ of the domain so
that $P(a_1, \ldots, a_n, b_1, \ldots b_m)$. A sentence of that form could be
intuitionistically provable for reals without there being any *uniform*
effective method as described; although over the naturals it is always
possible to uniformize the respective methods for each instantiation of
the universal prefix (cf. Diaconescu, 1975). (Thus Dummett is in error
when he claims (1977, p. 64) "As *always*, a form of the axiom of
choice holds good. . . ." (our emphasis).)

Intuitionistically incorrect also is the over-swift assumption that
each sentence of the theory to be Skolemized can be replaced by an
intuitionistically equivalent sentence of an appropriate syntactic form
in order for the construction of the model to go through, even with
liberal use of choice (which, however, as we have noted, the intuitionist cannot permit in the general case). So if we are in the business of
looking for an intuitionistic analogue of the well known proof, due to
Skolem, of the existence of countable submodels, we have to enquire
more closely about the first step in his proof for the projected intuitionistic version of his result.

What results are there concerning Skolem normal forms for sentences of first order languages? The answer is that, for the would-be
constructivist Skolemite, they are distressingly meagre. First, there
appears to be little prospect of furnishing a countable model using
choice in the light of the limited extent to which sentences could be
replaced by intuitionistically equivalent Skolem normal forms, as
revealed in a proof-theoretical study by Minc (1972). Smorynski
(1978) has since established the same negative result by simpler
model-theoretic methods: Skolemization cannot in general be obtained
within the bounds set by constructive logic. (It is worth remarking
here that the axiom schema of separation in IZF will have arbitrarily
complex instances. Thus it would be futile to look for "Skolemisability

within limits" on logical complexity of one's set of axioms.) Skolem himself started, it is now clear, from an unregenerately classical vantage point. At this point one could ask (as Michael Resnik did in correspondence) whether the intuitionist might not be able to mimic the prenex method of proof, even if not the Skolem method. But here too (with the most obvious mimicking) he would be frustrated. The cut elimination theorem for intuitionistic logic (cf. Dummett, 1977, p. 150) has as a corollary that theoremhood of prenex forms is decidable. But then this would contradict the undecidability of theoremhood in the full language (which holds even in the monadic case for intuitionistic logic), *if* the prenex normal form theorem held for intuitionistic logic. For, given a sentence A, one could find an equivalent prenex form A' simply by enumerating proofs; and then apply to A' the decision method for theoremhood provided by the cut-elimination theorem. This would yield a decision as to the theoremhood of A.

To re-inforce this point, note what can happen on the intuitionistic front if one tries to apply the standard prenex normal form algorithm from the classical camp. Consistent theories can then be converted into inconsistent ones. For example, $\neg \forall x(FX \vee \neg Fx)$ is intuitionistically *consistent*. But the standard algorithm for prenexing converts this to $Ex\neg(Fx \vee \neg Fx)$, which is intuitionistically *inconsistent*.

Another way of proving the classical Löwenheim – Skolem theorem is by Henkin's method. Could *this* way possibly be adapted so as to meet constructivist requirements? On this approach, one starts with a consistent set of sentences and expands it to a maximal consistent set with witnesses. Then one defines a canonical model autonomously on the language of the expanded set and shows that its theory is precisely the expanded set of sentences. And since the model, by construction, is countable, we have the desired result. This method, however, is non-constructive at the point where one expands the original set of sentences. One does so by contemplating both sentences and formulae in one free variable drawn from two assumed denumerable lists. One adds a sentence when it is consistent to do so, and one adds a fresh instance of a formula should it be consistent to assume its existential quantification. This requires an infinite sequence of choices, each made after deciding a question of consistency. But we know by Church's theorem that there is no general recursive method for making such decisions; hence the Henkin method is not available. Therefore, the intuitionist, as long as his mathematics is consistent with Church's Thesis, cannot use Henkin's method to establish a general countable model theorem and so to produce a Skolem paradox. (This criticism applies just as forcefully to the classical proof of the completeness of intuitionistic logic with respect to Kripke models, as given in Tennant (1978).)

But what about *intuitionistic* proofs of the completeness of intuitionistic logic? There are two of these in the literature, by Veldman (1976) and de Swart (1976). Veldman's proof uses a single 'universal' model, whereas de Swart's uses a certain class (more precisely: *fan*) of (countable) models. The models under discussion here are neither the familiar structures of classical model theory nor standard topological models but generalized Beth and Kripke structures. By 'generalized', we mean that allowance is made for the possibility that both a sentence and its negation might be forced at some node in the underlying frame.

It suffices to consider de Swart's method more closely in order to see that, even here, there is no *constructive* countable models theorem in the offing. Let the fan be B. We define 'the set X of sentences B-implies the sentence P' in the obvious way: for every structure M in B, if every member of X is valid (i.e. intuitionistically true) in M, then so is P. De Swart's completeness proof provides a constructive method for producing an intuitionistic proof of P from some finite subset of X, on the assumption that X B-implies P. Unlike Henkin, he does *not* provide a method (let alone a constructive one) for producing, for any consistent set Y of sentences (that is, a set Y which cannot be proved inconsistent using intuitionistic logic), a countable structure in B making every member of Y true. Nor is this a construction in their completeness proofs.

A result of Gödel, as reported by Kreisel (in Kreisel, 1962) shows that there can be no direct analogue of Henkin's method in the intuitionistic case. Let $T(A, M)$ be the sentence in set theoretic notation that expresses "A is true in the model M"; let $\mathrm{Con}(A)$ and $\mathrm{Prov}(A)$ be as usual (with reference to intuitionistic logic). Let CM (for "consistency implies model existence") be the claim

$$\text{for all } A, \text{ if } \mathrm{Con}(A) \text{ then, for some } M, T(A, M)$$

and let VP (for "validity implies provability") be the claim

for all A, if for all M, $T(A, M)$, then Prov(A).

Each of CM and VP constructively implies (in IZF) arithmetic Markov's principle. But at least infinitely many instances of Markov's principle are independent of IZF. Therefore neither CM nor VP is provable in IZF.

There is a fairly extensive literature on the completeness problem for intuitionistic predicate logic, a representative sample of which would include Kreisel (1962), Leivant (1972) and van Dalen (1973). Surveys of results on completeness appear in Dummett (1977) and Troelstra (1977). Relatively little of this material bears directly on the constructivity of the general Löwenheim – Skolem theorem. Those results that do apply, e.g., theorems of the form

if A is not a theorem, then there is a subcountable model for not-A

hold at most for restricted classes of formulae such as the class of negative formulae. It would not be possible, therefore, to apply these results, without further ado, to the axioms of set theory and of arithmetic. (See the final section for more detailed argument on this last point.)

Note that the results of Gödel and Kreisel, as well as those to be obtained below, do not require the creation of any arcane version of "intuitionistic model theory". The model theory that we shall do in (informal) IZF and its extensions involves simple duplications of the existing definitions from classical model theory for such notions as "sentence A holds in model M". All that differs in our treatment is the underlying (meta)logic, which of course is intuitionistic. In particular, we can help ourselves to the normal Tarskian clauses in the definition of model relative satisfaction.

VII

We have found no evidence so far that the intuitionist can visit upon himself, on the mere assumption that his set theory is consistent, that Skolemite embarrassment that now may be the peculiar and dubious

186   CHARLES McCARTY AND NEIL TENNANT

privilege of the classicist. Indeed, this is no mere appearance. We can advance rather general considerations in support of the claim that no countable models theorem of any of the usual straightforward forms is intuitionistically provable. These general negative results consolidate and extend all the frustrations so far of attempts to devise constructive analogues of the countable downward Löwenheim – Skolem theorem. Let us now explain how the results are obtained. Note first that one obviously cannot *refute* the downward claim in IZF, since it holds in a classical, consistent extension of IZF. So what we have so show is that we cannot *prove* the downward claim in IZF.

A corollary to a general theorem in McCarty (1984) is the well known more specific fact that;

a set of natural numbers is recursively enumerable in the classical sense if and only if that set is countable in the Kleene realizability model V(Kl) for IZF +.

IZF + is intuitionistic Zermelo – Fraenkel set theory and other strong principles. These include Church's Thesis; Markov's Principle; Brouwer's theorem; various forms of choice (strong natural forms of which are relativised dependent choice and the Blass – Aczel presentation axiom) and a panoply of other axioms including the uniform reflection principle.

Moreover, is it a simple (constructively provable) recursion-theoretic fact that:

immune sets exist.

These are sets that have no infinite recursively enumerable subsets (cf. Rogers, 1967, p. 106). Putting these two results together, it is easy to see that the countable downward Lowenheim – Skolem claim is independent of IZF plus the other principles that hold in V(Kl). One version of this claim which we shall now show cannot be proved in IZF + is the following:

(C₁)   for every set X of sentences, for every model M of X, there is a countable submodel of M satisfying X.

In what follows the turnstile represents intuitionistic deducibility.

THEOREM 1. $IZF+ \nvdash C_1$.

Proof. To see why C_1 cannot be proved in IZF +, take any immune set I (such sets exist). Consider I^*, its analogue in V(Kl). In V(Kl), I^*

SKOLEM'S PARADOX AND CONSTRUCTIVISM 187

is a subset of ω and has no infinite *countable* subset. (This is because countability, as noted above, is the realizability analogue of recursive enumerability.) It follows that there is no countable submodel of $(I^*, =)$ for the theory of identity over I^*. For this theory contains the sentences E_n (n any integer) saying "There are not not at least n individuals". If there *were* a *countable* submodel of the theory, it would accordingly have to be infinite, contradicting what we know about I^*. ∎

Note that this proof actually establishes a stronger independence result than the one already stated. For it shows the independence of the following weaker claim:

(C_2)   For every *recursively enumerable* (r.e.) theory X, for every model M of X, there is a countable submodel of M satisfying X.

THEOREM 2. *IZF+* ⊬ C_2.
Proof. See above. ∎

This observation deals with the possible complaint from the Skolemite that he is concerned to visit the paradox on set theory, which is axiomatisable.

The reader should be reminded that the realizability methods used in the proofs do not adversely affect the generality of the independence claims just made. Granted, the proofs themselves rely upon phenomena which are at present of very restricted mathematical application: immune sets and their theories of identity. The statements shown to be independent of IZF do not partake of any correlative restriction; on the contrary, they remain perfectly general versions of the downward Löwenheim–Skolem Theorem. Assertions such as C_2 are not to be understood as restricted to those models whose domains are immune sets or to theories which are theories of immune sets. The statements shown to be independent do not refer to or suffer restriction from immune sets in any way. (It might be well to compare this situation with the more familiar one of Cohen forcing. The Continuum Hypothesis neither refers to nor suffers restriction from forcing

188   CHARLES McCARTY AND NEIL TENNANT

conditions and generic sets, even though these constructs enter into
the proof of its independence.)

We should also point out that an examination of the details of
the proof of Theorem 1 will license two further strengthenings of
Theorem 2. Consider the statement:

($C_{2.5}$)   For every r.e. theory X, for every model M of X, it is
            not not the case that there is a countable submodel of
            M satisfying X.

THEOREM 2.5. ($C_{2.5}$) *is not provable in IZF+*
Proof. this is immediate from the proof of Theorem 1.   ■

Next, recall that a sentence is negative whenever it is equivalent,
within a constructive theory, to a sentence devoid of disjunctions and
existential quantifications and in which every atomic sentence appears
doubly negated. Our proof of Theorem 1 also shows that we can take
X to be a set of negative sentences such as "There are not not at least
n individuals" for arbitrary n.

What $C_{2.5}$ and our remark on negative sentences show is that no
superficial 'negativization' or application of a Gödel – Gentzen negative translation suffices to bring back the strong Löwenheim – Skolem
Theorem in its standard form.

Our proofs of Theorems 1 and 2 do not, in themselves, require that
the existence of immune sets be provable constructively – for example,
within IZF or IZF plus MP. We can suppose that the metatheory in
which we define V(Kl) and work with it is classical ZF. So in the
metatheory we can avail ourselves of all the benefits of classical
mathematics, including the ready assurance that immune sets exist.
On the other hand, we can prove, even constructively, that immune
sets exist – Post's original argument (Rogers, p. 106) is readily con-
structivized. And we can define the realizability structure and prove
the fundamental results about it in a constructive metamathematics.

So the prospects of an intuitionistic analogue of the full countable
downward Löwenheim – Skolem theorem are bleak indeed. In fact, it
can already be seen that it would even be a slight understatement of
our result to point out that *the assumption of a strong counterexample*

SKOLEM'S PARADOX AND CONSTRUCTIVISM 189

(using immune sets) *to the countable downward claim is consistent with all of Bishop's constructive mathematics.* This understatement is true because all the latter can be done in IZF plus relativised dependent choice.

So far the Skolemite appears to be intuitionistically empty-handed as far as countable *sub*models are concerned. But might he clutch at a surviving straw: the existence of countable models *überhaupt*, be they submodels or not of the original model? Once again, the answer is no. For the following claim is independent of IZF+:

(C_3) for every set X of sentences, for every model M of X, there is some countable model M' (not necessarily a submodel of M) satisfying X.

Bear in mind that on a constructive interpretation, if we have X and M then the countable model M' whose existence is guaranteed by the claim depends parametrically on X and M, and so too does the counting function (with domain ω) that makes it countable. Once again we concentrate only on the need to show that the claim (C_3) does not *follow* from IZF+.

THEOREM 3. *IZF+ $\nvdash C_3$.*

Proof. Here first is a summary of the argument:

Consider once more the realizability model V(KI). Let $[i]$ be the i-th partial recursive function under the standard enumeration. Were (C_3) to be provable, and (by the soundness of the realizability semantics) true in V(KI), the predicate "$[i]$ is a total recursive function" would be recursively enumerable. But the predicate in question is known not to be recursively enumerable (cf. Rogers, 1967, p. 264). Hence (C_3) is not provable.

Let us now expand this summary of the result with more argumentative detail.

Assume for *reductio* that (C_3) is true in V(KI). For each natural number i, consider the set

$$\mathbf{i} =_{\mathrm{df}} \{0\} \cup \{1 : [i] \text{ is total}\}$$

Let M be the model with $\mathbf{i}$ as its domain, and identity as its only relation. We consider only the language of identity. Let X be the

190   CHARLES McCARTY AND NEIL TENNANT

theory in this language for M. It follows from the assumed truth of (C_3) in $V(K1)$ that X has a countable model M', with enumerating function f_i. (Here we are following our recent advice, by bearing in mind that this counting function depends parametrically on i.) This is because the recursion-theoretic properties of i help at least in part to determine the original model M of the claim (C_3) above, which is the focus of our *reductio*.

Since the domain of M is a subset of the integers, M satisfies the claim that identity is decidable:

$$\forall x \forall y (x = y \lor \neg x = y)$$

Thus this sentence is in X. Hence, by assumption, it holds in M' as well. Now (c.f. Minio, 1974) countable sets with decidable equality are isomorphic (with respect to $=$) to subsets of ω. Thus the domain of M' can without loss of generality be taken to be a subset of ω. Now, since f_i maps ω into ω, we may assume that f_i is total recursive, with index e_i depending *effectively* on i. (That is, f_i is $[e_i]$.) This is because $V(K1)$ satisfies *Church's Thesis*: that every number-theoretic function is total recursive.

Now the statement that says that the recursive function $[i]$ is total has the form

$$\forall n Em P(i, m, n)$$

where the three-place predicate P is primitive recursive.

We shall now prove the equivalence

$$\forall n Em P(i, n, m) \text{ if and only if } En Em \neg f_i(n) = f_i(m)$$

which contradicts well-known results of ordinary recursion theory. For the right hand side is a recursively enumerable predicate of i. But the left hand side, expressing the claim "$[i]$ is total", is not recursively enumerable (cf. Rogers, p. 264). This will complete the *reductio* of the assumption that (C_3) is true in $V(K1)$. Hence (C_3) cannot be proved in $IZF+$.

We establish the equivalence above by arguing first in the direction from left to right:

SKOLEM'S PARADOX AND CONSTRUCTIVISM   191

$$\cfrac{\cfrac{\cfrac{\cfrac{\cfrac{\forall n EmP(i,\,n,\,m) \text{ (that is, } [i] \text{ is total)}}{i \,=\, \{0,\,1\}}\text{ by definition of } \mathbf{i}}{M \vDash ExEy\neg x = y}\text{ by Tarski clauses}}{M' \vDash ExEy\neg x = y}\text{ by choice of } M'}{Ex \text{ in } M' \quad Ey \text{ in } M' \quad \neg x = y}\text{ by Tarski clauses}}{EmEn \, \neg f_i m \,=\, f_i n}\text{ since } f \text{ is onto}$$

and then from right to left:

the proof schema

$$\cfrac{\cfrac{\cfrac{\cfrac{\quad\quad\quad\quad\quad\quad\quad}{t = 1 \,\&\, [i] \text{ is total}}(1)}{[i] \text{ is total}}}{\text{i.e.} \quad \forall n EmP\,(i,\,n,\,m)}}{\;}$$

$$\cfrac{\quad\cfrac{t \text{ in } i}{t = 0 \,\vee\, (t = 1 \,\&\, [i] \text{ is total)}} \quad\quad \cfrac{\cfrac{EmP\,(i,\,a,\,m)\quad\neg EmP(i,\,a,\,m)}{\quad}(1)}{t = 0} \quad\quad \cfrac{\wedge}{t = 0}\quad}{t = 0}(1)$$

establishes the inference $\cfrac{t \text{ in } i \quad \neg EmP(i,\,a,\,m)}{t = 0}$ for arbitrary t.

This is now applied twice over in the following proof, along with Markov's principle (for primitive recursive F) in its inferential form

$$\cfrac{\cfrac{\overline{\quad\quad}^{(1)}}{\neg EmFm}}{\;}$$
$$\vdots$$
$$\cfrac{\wedge}{EmFm}(1), \quad \text{and the axiom scheme} \quad \overline{f_i t \text{ in } M'}.$$

192   CHARLES McCARTY AND NEIL TENNANT

$$\cfrac{\cfrac{\overset{(2)}{\overline{\quad}}}{f_i c \text{ in } M' \quad \neg EmP(i, a, m)}{f_i c = 0} \qquad \cfrac{\overset{(2)}{\overline{\quad}}}{f_i d \text{ in } M' \quad \neg EmP(i, a, m)}{f_i d = 0}}{\cfrac{f_i c = f_i d}{\qquad\qquad \neg f_i c = f_i d}} \quad (1)$$

$$\cfrac{\cfrac{\cfrac{EmEn \,\neg f_i m = f_i n \qquad\qquad\qquad\qquad\qquad \wedge}{ \wedge} \; (1)}{} \;\; (2) \quad \text{Markov}}{\cfrac{EmP(i, a, m)}{\forall n EmP(i, n, m)}}$$

i.e. $[i]$ is total

This completes our proof of the equivalence, and also the proof of Theorem 3. Again, the reader is cautioned not to confuse the method by which we prove the theorems with the statements whose independence the theorems establish. We have shown that a general countable Löwenheim–Skolem theorem is not constructively provable by giving a formalization of the theorem its realizability interpretation. On the interpretation, the statement of the general Löwenheim–Skolem theorem comes to imply that a certain uniform effective method exists. By marshalling other considerations, one shows that the required method cannot exist and, hence, that the Löwenheim–Skolem theorem is not constructively provable. It is essential to note that neither the statement of the Löwenheim–Skolem theorem, nor its formalization nor its standard constructive interpretation asserts, or even implies, that there is such an effective method. It is not the case that our proof shows no more than that an "effectivization' of the Löwenheim–Skolem theorem is independent of IZF+. Rather, it shows just what it purports to show — that the general theorem itself is not constructively provable.

There now arises the following possible objection:

"You have proved the independence of (C_3) only from IZF+. But IZF+ does not contain the Fan Theorem or Bar Induction. So how do you know that there is not proof, using these stronger principles, of a form of the downward Löwenheim–Skolem theorem?"

The answer to this objection is as follows. One can show (as we shall below) that the strong downward countable Löwenheim–Skolem theorem is constructively inconsistent with Markov's Principle for arbitrary natural number functions. In the terminology of Brouwerian intuitionism, the conjunction of the two claims has a "weak counter-example". That is, *modulo* IZF (or even second-order Heyting arithmetic), they imply the law of excluded middle for arbitrary sentences. For the formal statement of the result, let us introduce some abbreviations:

(C_4) is the claim: for every model M there is a countable model M' elementarily equivalent to M.

MPF is Markov's Principle for arbitrary natural number functions:

$$\text{if } \neg \forall x \forall y (fx = fy) \quad \text{then} \quad \text{Ex} \text{Ey} \neg fx = fy.$$

THEOREM 4. *Let A be an arbitrary formula.*
         Then IZF, MPF, $C_4 \vdash A \vee \neg A$.

COROLLARY 1. *Since $V(\text{K}1)$ satisfies MPF but not all formulae of the form $A \vee \neg A$, it follows that C_4 is independent of IZF+.*

COROLLARY 2. *Since there is a sheaf model (cf. Fourman and Hyland 1979) for the fan theorem, bar induction and Markov's Principle that does not satisfy all formulae of the form $A \vee \neg A$, it follows that C_4 is independent of IZF plus the fan theorem plus bar induction.*

Proof. Since $\neg\neg(A \vee \neg A)$ is an intuitionistic theorem, it suffices to show, for any sentence B, that IZF, MPF, C_4, $\neg\neg B \vdash B$. So let $\mathbf{B} =_{df} \{0\} \cup \{1 : B\}$. In other words, x is in $\mathbf{B}$ if and only if $(x = 0 \vee (x = 1 \ \& \ B))$. We shall identify $\mathbf{B}$ with the model whose domain is $\mathbf{B}$ and whose only relation is the identity relation.

Assume B holds, Then $\mathbf{B}$ by definition contains 0 and 1. Thus were we to assume C: $\mathbf{B} \vDash \forall x \forall y (x = y)$, a contradiction would ensue by virtue of the Tarskian clause for the universal quantifier, and the standard interpretation of $\vDash$.

By the intuitionistic proof schema

$$
\begin{array}{cc}
(2)\ \overline{\quad C\quad} & \overline{\quad B\quad}\ (1) \\
\end{array}
$$

$$
\vdots
$$

$$
\dfrac{\wedge}{\neg B}\ (1) \qquad\qquad \neg\neg B
$$

$$
\dfrac{\wedge}{\neg C}\ (2)
$$

We have a proof of $\neg C$ from the assumption $\neg\neg B$. We shall now continue with a proof of B from $\neg C$. This will establish overall that B follows from $\neg\neg B$.

So assume $\neg C$, that is, $\neg(\mathbf{B} \vDash \forall x \forall y(x = y))$. By the Tarskian clause for negation,

$$\mathbf{B} \vDash \neg \forall x \forall y(x = y).$$

Now, by C_4, let N be a countable model elementarily equivalent to $\mathbf{B}$ with respect to the language of identity. Let f be the function from ω to (the domain of) N that enumerates it. Since $\mathbf{B}$ is a subset of ω, identity is decidable on $\mathbf{B}$; that is, $\mathbf{B} \vDash \forall x \forall y(x = y \ \vee \ \neg x = y)$. By elementary equivalence,

$$N \vDash \neg \forall x \forall y(x = y).$$

By elementary equivalence, we also have that identity is decidable on N. By Minio (1974) as before, we can assume without loss of generality that N is a subset of ω and, hence, that f is a number-theoretic function. Since

$$N \vDash \neg \forall x \forall y(x = y),$$

we have

$$\neg \forall m \forall n(fm = fn)$$

By MPF we obtain

$$EmEn \ \neg fm = fn$$

Thus $N \vDash ExEy \ \neg x = y$. Hence by elementary equivalence, $\mathbf{B} \vDash ExEy \ \neg x = y$. It follows by the Tarskian clauses for the

SKOLEM'S PARADOX AND CONSTRUCTIVISM   195

universal quantifier and negation that

$$\text{E}x \text{ in } \mathbf{B} \ \text{E}y \text{ in } \mathbf{B} \ \neg x = y$$

Now note that by the definition of **B**, the following schematic inference is valid:

$$\frac{t \text{ in } B}{t = 0 \ \vee \ (t = 1 \ \& \ B)}$$

We proceed intuitionistically as follows:

$$
\cfrac{
  \cfrac{
    \cfrac{(3)}{b \text{ in } B}
  }{b = 0 \ \vee \ (b = 1 \ \& \ B)}
  \quad
  \cfrac{
    \cfrac{(3)}{a \text{ in } B}
  }{a = 0 \ \vee \ (a = 1 \ \& \ B)}
  \quad
  \cfrac{
    \cfrac{(1)}{a=0} \quad \cfrac{(2)}{b=0} \quad \cfrac{(3)}{a=b} \quad \cfrac{(3)}{-a=b}
  }{\wedge}
  \quad
  \cfrac{
    \cfrac{\cfrac{(1)}{a = 1 \& B}}{B} \quad \cfrac{(2)}{b = 1 \& B}
  }{B}
}{\text{E}x \text{ in } \mathbf{B} \ \text{E}y \text{ in } \mathbf{B} \ \neg x = y}
$$

This completes our proof of B from $\neg\neg B$, and also the proof of Theorem 4.

Insofar as *countable* models are concerned, then, our four theorems appear to block the most obvious routes to the result, in any of its usual straightforward forms, that the Skolemite needs.[6]

VIII

But what about *subcountable* models — whether or not they are *sub*models of the original model? So far we have shown the independence of the downward claim involving the cosntructively strong notion of countability, thereby apparently weakening the independence result. It remains to be seen whether independence survives upon substitution of 'subcountable' for 'countable'. But whether it does or not, we may have already drawn the Skolemite's sting. For it is consistent with IZF+ to assume that extremely capacious sets are subcountable. For example, in V(K1) every metric space — including the reals — is subcountable! The intuition of the capaciousness of the reals holds firm, backed by Cantor's argument. Indeed, it would be one of the standards against which one would judge the appropriateness or adequacy of any attempted explication, in mathematical terms, of the notion of cardinality.

It is clear from McCarty (1984) and Grayson (1978) that subcountability cannot serve as a fully satisfactory constructive measure of the size of a set. For subcountability is incapable of sustaining distinctions of cardinality which the constructivist wishes to make. Insofar as there is a constructive theory of cardinality, it uses the notion of countability. This is the notion based on *total* counting functions, which is already so familiar to the classicist.

Finally, what about negative translations? On the basis of these translations, it may appear, at first sight, as though the paradox could be re-instated. The line of thought would run as follows: As is well known, there are negative translations (Friedman, 1973; Powell, 1975; Beeson, 1985; Leivant, 1985) of classical set theory into intuitionistic 'correlates'. Cannot now the 'negative' version of LST induce Skolemite stress? Our reasons for disagreeing are as follows. The general pattern is this: a translation f is defined so that the following holds:

$$\text{If } ZF \vdash_C \phi \text{ then } IZF \vdash_I f\phi.$$

But the translation f has to be defined with some care, in order to overcome certain difficulties presented by set theory, Friedman, in his choice of f, went beyond the usual double-negation treatment of atomic formulae, disjunctions and existential quantification by also replacing any atomic formulae $a \in b$ with an extremely complicated formula in a and b. And Powell, in his choice of f, went beyond the Gödel – Gentzen dualization of $\vee$ and $\exists$ in terms of $\neg$, & and $\forall$ by further restricting all quantifiers to range over *stable sets* (sets in which not not being a member implies being a member).

Now, in the light of these remarks, we have to consider the putative objection based on the observation that (for some such choice of f)

$$\text{If } ZF \vdash_C LST \text{ then } IZF \vdash_I f(LST).$$

Is this an adequate objection to our pessimism over the prospects for a constructively acceptable analogue of the downward Löwenheim –

Skolem theorem? We think not. For the statement f(LST) is not a statement of constructive model theory. The tampering with $\in$ or with the range of the quantifiers give the lie to the objector's reading of f(LST) as a version of LST in any serious sense. But quite apart from those features of f, the logical rewriting via the negative part of the translation totally obstructs such a construal. For even f(LST) would not deal with theories (sets of sentences closed under derivability) but with collections X of sentences such that, if it is not not the case that B is derivable from X, then B is not not a member of X. In the same way, the "double negative translation" of the predicate "is a model" is not "is a model".

One would want to say very much the same thing about the negative form of the classical mean value theorem (MVT), which is also not constructively provable. The negative translation of the MVT does not afford a counterexample to the claim that the MVT is independent of constructive set theory because the negative form of the theorem is, quite simply, not a statement of real analysis. It deals neither with real-valued functions nor with real values.

Secondly, the negative translation of the Löwenheim – Skolem theorem is not likely to give rise to worries of the Skolemite sort because it does not assert the existence of countable non-standard models. All it would assert (even if it *did* concern theories, models and the like) is that there *not not* exists a countable model (or rather: a countable$^{\prime}$ model$^{\prime}$), a claim which is much weaker than the claim that gives rise to the Skolem paradox. When a constructivist claims that there *not not* exists a certain structure, he is claiming that one can rule out on mathematical grounds the assumption that no such structure exists. But this is less than what is needed to get the paradox off the ground: that countable$^{\prime}$ nonstandard model$^{\prime}$s fail to be prohibited is not tantamount to *countable* nonstandard *model*s receiving, as they do in classical mathematics, a general licence.

There remains one further possible worry about whether we have been successful in blocking the application of a downward Löwenheim – Skolem theorem to set theory. The worry takes this form: might we not have left open the possibility that one could (constructively) derive *specific* completeness or countable models theorems for formal set theories or more 'ordinary' theories such as arithmetic? We believe

the following results go some way toward warding off such worries.
We will limit ourselves to statements of results; complete proofs will
appear in McCarty (forthcoming). On the basis of the first two theorems below, one sees that there is no hope of arriving at a countable
downward Löwenheim–Skolem Theorem by way of a model existence theorem for simple extensions of arithmetic and set theory.

THEOREM 5. *If IZF is consistent, then one cannot prove, in IZF +
A, the statement*

> *If T is consistent, then there not not exists a model of T
> where T ranges over extensions (even finite or r.e.) of
> Heyting (intuitionistic) arithmetic.*

*A can be Church's Thesis, Markov's Principle or the Uniform Reflec-
tion Principle.*

THEOREM 6. *Let S be any formal set theory (e.g., a suitable sub-
theory of IZF) which contains arithmetic separation and which is at
least as strong as Heyting arithmetic. Then, if IZF is consistent, one
cannot prove in IZF + A the statement*

> *for all sentences B, if S + B is consistent then it is not
> not the case that there is a model of S + B.*

*A should be such that IZF + A proves that A is true under realizability
and such that ZF + A does not prove that ZF is inconsistent.*

Consequently, one cannot provide various weakened forms of
model existence theorems for theories representing a reasonable
amount of constructive mathematics.

Next, one can prove outright, using Church's Thesis in IZF, that
strong set theories have absolutely no models which are either very
small in cardinality or have "well structured sets" as their carriers.

THEOREM 7. *In IZF plus Church's Thesis, there is a proof that,
if a set theory T has arithmetic separation and is at least as strong as
Heyting arithmetic, then T has no models of the same cardinality as
some subset of the natural numbers. In fact, T will have no models
which support stable equality.*

The equality relation on a set will be stable when it is invariant with respect to double negation. Theorem 7 implies that set theory will have no true interpretations of the same cardinality as a metric space; this will include the reals, Baire space and Cantor space.

Finally it is consistent with IZF to assume that there are theories in the language of first-order arithmetic for which none of the standard problems of "ontological relativity" can arise. It will follow from Church's Thesis plus Markov's Principle that Heyting arithmetic determines its models up to isomorphism.

THEOREM 8. *In IZF, Church's Thesis plus Markov's Principle proves that Heyting arithmetic is categorical.*

Obviously, these results, even taken together with those given earlier, do not absolutely rule out the possibility that a version of the Skolem Paradox might be applicable to some strong constructive theory. Nothing that one could do in the way of independence results would show definitively that all statements which could conceivably be thought 'versions' of the Löwenheim — Skolem Theorem are independent of all theories which could conceivably be thought 'constructive'. Besides, it would be foolish to attempt to draw a formal circle around just those mathematical claims which might pose metaphysical problems. But this is not to say that nothing has been accomplished; the theorems of the paper do suffice to show that, if 'Skolemism' can arise for some extension of constructive set theory, then it must be a relatively 'local' phenomenon. As we have seen, the countable models theorem is not, at least in constructive mathematics, an ineliminable feature of the study of any countable consistent first-order theory. If the theorem is constructively available, it will only be so in virtue of the fine details of the theory under consideration and of the assumptions in the attendant metamathematics.

What we have shown is that even very weak forms of the Löwenheim — Skolem theorem are independent of the strongest intuitionistic set theories commonly considered. It follows that none of the theorems of Bishop-style constructivism, none of the work of the members of the Markov — Sanin 'School' and none of the axioms of standard Brouwerian constructivism will prove the

200   CHARLES McCARTY AND NEIL TENNANT

Löwenheim—Skolem theorem in anything approaching its ordinary general form. Consequently, the possibility of a Skolem-style paradox is definitely not the concommitant of any attempt to formalize a sufficiently large part of constructive mathematics. Or, if some refined form of the paradox were to be resurrected, it could only be on the basis of axioms (such as those of the creative subject) which lie outside the 'core' area of constructive mathematics or by employing metamathematical methods which are truly novel. If there is some way of infecting constructive mathematics with 'Skolemism', we have yet to see what it is and from whence it could come.

ACKNOWLEDGEMENTS

We are grateful to Michael Resnik, Timothy Smiley and Crispin Wright for comments on an earlier draft on this paper by Neil Tennant. That draft did not contain Charles McCarty's results (Sections 7 and 8) on the independence of versions of the countable models theorem from extensions of intuitionistic set theory. It was these results that led to joint authorship of the present paper. We are grateful for comments on the joint paper from W. V. Quine and from the referees for the *Journal of Philosophical Logic*. The results have been presented to colloquia at Michigan and Western Ontario.

NOTES

[1] For a proof-theoretic analysis of this vicious circularity, see Tennant (1982).
[2] We have not raised here the question of characterising the uncountable directly by means of the quantifier 'There exist at least uncountably many x such that . . . '. As Timothy Smiley has observed, Keisler's completeness proof for a (classical) logic based on a simple set of axioms and rules for this quantifier gives an intriguingly quick answer to the question whether there is *any* way at all of characterizing the uncountable. (Vaught had earlier established that the logical truths in this language were recursively enumerable; the interest of Keisler's result is that he shows four simple schemata using the new quantifier to be sufficient for its axiomatisation.) This victory is made to look somewhat Pyrrhic, however, by the impossibility of recursively axiomatising the logic of 'there exist at least infinitely many x such that . . . '. (This impossibility follows from Vaught's test: see Bell and Slomson, 1971, p. 266). Thus, direct expression by means of quantifiers has a freakish pattern of success and failure. The interest in how *set theory* fares in characterizing both the infinite and the uncountable derives, we believe, from the thought that both these notions either reduce to, or can *some*how be conveyed by, the use of some more basic notion, such as that of set. And there, says the Skolemite, lies the rub.
[3] The interested reader should compare this with Quine's 'virtual set theory'.
[4] Mostowski (1949).
[5] We are indebted to Kit Fine for this point.

SKOLEM'S PARADOX AND CONSTRUCTIVISM 201

[6] Those familiar with constructive mathematics will see that our proof of theorem 4 actually supports a much stronger (and more interesting) conclusion. We have shown that C_4, even restricted to models of the pure theory of identity having at most two elements, implies Kripke's Scheme. The latter is commonly taken to axiomatise Brouwer's theory of the creative subject.

REFERENCES

Beeson, M., *Foundations of Constructive Mathematics* (Springer Verlag, Berlin, 1985).

Bell, J. and Slomson, A., *Models and Ultraproducts* (North Holland, Amsterdam, 1971).

Benacerraf, P., 'What numbers could not be', *Philosophical Review* **74** (1965).

Benacerraf, P., 'Skolem and the Skeptic', *Proceedings of the Aristotelian Society, Supplementary Volume LIX* (1985), pp. 85. 115.

Diaconescu, R., 'Axiom of choice and complementation', *Proceedings of the American Mathematical Society* **51** (1975), 175 – 8.

Dummett, M., (with the assistance of R. Minio), *Elements of Intuitionism* (Oxford University Press, 1977).

Fourman, M. P. and Hyland, J. M. E., 'Sheaf models for analysis', in eds. M. P. Fourman, C. J. Mulvey and D. S. Scott, *Applications of Sheaves* (Springer Lecture Notes in Mathematics, no. 753, 1979).

Friedman, H., 'The consistency of classical set theory relative to a set theory with intuitionistic logic', *Journal of Symbolic Logic* **38** (1973), pp. 315 – 319.

Grandy, R. E., *Advanced Logic for Applications* (D. Reidel, 1977).

Grayson, R. J., *Intuitionistic Set Theory* (D. Phil. Thesis, University of Oxford, 1978).

Greenleaf, N., 'Liberal constructive set theory', ed. F. Richman: *Constructive Mathematics, Proceedings, New Mexico, 1980*, Springer Lecture Notes in Mathematics No. 873 (Berlin, 1981), pp. 213 – 240.

Keisler, H. J., 'On the quantifier "there exist uncountably many"', *Notices of the American Mathematical Society* **15** (1968), p. 654.

Kreisel, G., 'Weak completeness of intuitionistic predicate logic', *Journal of Symbolic Logic* **27** (1962), 139 – 158.

Kreisel, G. and Dyson, V., 'Analysis of Beth's semantic construction of intuitionistic logic', Technical report no. 3, Applied Mathematics and Statistics laboratories, Stanford University, Part II (1961).

Leivant, D., 'Notes on the completeness of the intuitionistic predicate calculus', Report ZW 40/72, Mathematisch Centrum, Amsterdam, 1972.

Leivant, D., 'Syntactic translations and provably recursive functions', *Journal of Symbolic Logic* **50** (1985), 682 – 8.

McCarty, D. C., *Realizability and Recursive Mathematics* (D. Phil. Thesis, University of Oxford, 1984). Released as a technical report by the Department of Computer Science, Carnegie-Mellon University, Pittsburgh, No. CMU-C2-84-131; revised version forthcoming as *Computation and Construction* (Oxford University Press).

McCarty, D. C., 'Constructive validity is nonarithmetic', (1986) (submitted to *Journal of Symbolic Logic*).

Minc, G. E., 'The Skolem method in intuitionistic calculi', *Proceedings of the Steklov Institute of Mathematics 121 (1972): Logical and Logico-Mathematical Calculi, 2*. (American Mathematical Society, 1974), pp. 73 – 109.

202   CHARLES McCARTY AND NEIL TENNANT

Minio, R., *Finite and Countable Sets in Intuitionistic Analysis* (M.Sc. Dissertation, University of Oxford, 1974), 35 pp.

Mostowski, A., 'An undecidable arithmetical statement', *Fundamenta Mathematicae* **36** (1949), pp. 143 – 164.

Powell, W. C., 'Extending Gödel's negative translation to ZF', *Journal of Symbolic Logic* **40** (1975), 221 – 229.

Quine, W. V. O., *Methods of Logic*, 2nd edition (Holt, Rinehart and Winston, 1959).

Quine, W. V. O., *Set Theory and Its Logic* (Belnap Press, Harvard, 1963).

Rogers, H., *Theory of Recursive Functions and Effective Computability* (McGraw-Hill, 1967).

Skolem, T., 'Logische-kombinatorische Untersuchungen über die Erfüllbarkeit und Beweisbarkeit mathematischen (sic) Sätze nebst einem Theoreme über dichte Mengen', 1920; pp. 103 – 135, in Fenstad, J. (ed.), *Selected Works in Logic*, Universitetsforlaget, Oslo, 1970.

Skolem, T., 'Einige Bemerkungen zur axiomatischen Begründung der Mengenlehre', 1922; *ibid*, pp. 137 – 152.

Smorynski, C., 'The axiomatization problem for fragments', *Annals of Mathematical Logic* **14** (1978), pp. 193 – 227.

de Swart, H. 'Another intuitionistic completeness proof', *Journal of Symbolic Logic* **41** (1976), pp. 644 – 662.

Tennant, N., *Natural Logic* (Edinburgh University Press, 1978).

Tennant, N., 'Proof and Paradox', *Dialectica* **36** (1982), pp. 265 – 296.

Troelstra, A. S., 'Completeness and validity for intuitionistic predicate logic', Proceedings of the "Séminair internationale d'été et Colloque international de Logique à Clermont-Ferrand", July 1975 (1977).

van Dalen, J., 'Lectures on intuitionism', in *Cambridge Summer School in Mathematical Logic*, edited by A. Mathias and H. Rogers, pp. 1 – 94 (Springer 1973).

Vaught, R., 'The completeness of the logic with the added quantifier "there are uncountably many"', *Fundamenta Mathematicae* **54** (1964), pp. 303 – 304.

Veldman, W., 'An intuitionistic completeness theorem for intuitionistic predicate logic', *Journal of Symbolic Logic* **41** (1976), pp. 159 – 166.

Wright, C., 'Skolem and the Skeptic', *Proceedings of the Aristotelian Society, Supplementary Volume LIX* (1985), pp. 117 – 137.

Department of Computer Science and the
Centre for Cognitive Science,
University of Edinburgh,
2, Buccleuch Place,
Edinburgh EH8 9LW, Scotland

and

Department of Philosophy,
The Faculties,
Australian National University,
Canberra, ACT 2600, Australia

[14]

THE JOURNAL OF PHILOSOPHY

SECOND-ORDER LOGIC, FOUNDATIONS, AND RULES*

> Not empiricism and yet realism in philosophy, that is
> the hardest thing.
>
> And the picture that might occur to someone here is
> that of a short bit of handrail, by means of which I
> am to let myself be guided further than the rail
> reaches. [But there *is* nothing there; but there isn't
> *nothing* there!]
>
> The difficult thing here is not, to dig down to the
> ground; no, it is to recognize the ground that lies
> before us as the ground.
>
> Ludwig Wittgenstein[1]

AN interpreted, or partially interpreted, language is called
"second-order" or "higher-order" if it has variables that
range over relations, propositional functions, properties,
classes, or sets of whatever is in the range of the ordinary, or first-
order variables. Many studies have paid careful attention to the differences among these items. It is said, for example, that propositional functions are intensional, while sets and classes are extensional; sets are "constituted by their elements," while classes are
extensions of properties.[2] Here I gloss over these differences and use
the terms interchangeably. If an identity relation on such items is
needed, I take them to be extensional, but few, if any, of my remarks
depend on this. I apologize to sensitive readers.

In any case, the distinguishing feature of second- or higher-order
languages is not so much the nature of the individual items that fall
in the range of the extra variables, but rather their extension or
totality. It is often noted along these lines that the central feature of
second-order logic is its semantics, as opposed to the language itself
and its deductive system. Indeed, one may go so far as to assert that

* I would like to thank the participants of the Ohio University Conference, especially John Corcoran, Richard Butrick, and George Boolos, for many hours of
fruitful discussion. I am also indebted to the members of a discussion group on the
Skolem paradox—Beth Cohen, Rick DeWitt, Robert Kraut, Ronald Laymon, Barbara Scholz, and George Schumm—and to my colleagues David McCarty and Allan
Silverman.
[1] *Remarks on the Foundations of Mathematics* (Cambridge: MIT, 1978, Rev.
ed.), pp. 325, 430, 333.
[2] The topic is treated and illuminated throughout Charles Parsons, *Mathematics
in Philosophy* (Ithaca: Cornell, 1983); see esp. Essay 8, "Sets and Classes."

0022-362X/90/8705/234-261   © 1990 The Journal of Philosophy, Inc.

languages and deductive systems, by themselves, are neither first-order nor higher-order. I have said this myself, for example.[3] I still think it is correct, especially when one focuses on formal or semi-formal Tarskian model theory, but it can be, for this reason, misleading. It depends on what one takes "semantics" to be.

I show here that the major issues surrounding second-order languages re-emerge when one compares theories of formal proof with informal or preformal mathematical practice. In a sense, this is "semantics" also, since issues of "meaning" are involved, but it is not traditional model theory.

I

Let L2 be a second-order language, in which, for simplicity, all second-order variables are monadic. Let L1 be the first-order restriction of L2. The range of the first-order variables, or the *domain*, may be unspecified, as in formal logic, or there may be a structure, such as the natural numbers, the real numbers, or the set-theoretic hierarchy, that serves as *intended domain*.

As usual, first-order variables are lower-case letters and predicate variables are upper case letters. A deductive system for L2 can be obtained from a deductive system for L1 by adding straightforward extensions of the quantifier axioms (e.g., $\forall X(\Phi(X)) \rightarrow \Phi(Y)$), and the axiom scheme of comprehension:

$$\exists X \forall x(Xx \equiv \Phi(x))$$

for each formula Φ not containing X free.[4]

A brief sketch of the model theory will help focus attention on the central issues. There are (at least) three semantics that have been introduced for languages like L2, but two of these are "equivalent" in a straightforward sense.

In *standard semantics*, which makes the logic properly second-order, the predicate variables range over the collection of all subsets of the domain. With a little more detail, a *standard interpretation* of

[3] See my "Second-order Languages and Mathematical Practice," *The Journal of Symbolic Logic*, 1. (1985): 714–742; also my *Foundations without Foundationalism: A Case for Second-order Logic* (New York: Oxford, forthcoming).

[4] See David Hilbert and Wilhelm Ackermann, *Grundzüge der theoritischen Logik* (Berlin: Springer, 1928). Notice that the formula Φ in the comprehension scheme may contain bound-predicate variables. Thus, some instances are impredicative. To avoid this, restricted versions of the scheme have been considered, resulting in ramified type theory. On the other hand, if function and relation variables are introduced into the language, the axiom of choice may be included:
$$\forall R(\forall x \exists yRxy \rightarrow \exists f \forall xRxfx)$$
A symbol for identity-between-predicates can also be introduced, together with the axiom of extensionality:
$$\forall X \forall Y[\forall x(Xx \equiv Yx) \rightarrow X = Y]$$

L2 is a structure $\langle d, I \rangle$, in which d is the domain and I an appropriate assignment on d to the common nonlogical terminology. A *variable assignment* consists of a function from the collection of first-order variables to d and a function from the collection of predicate variables to the powerset of d. The relation of *satisfaction*, between interpretations, variable assignments, and formulas is defined in the straightforward way.

Although this semantics is sound for all common deductive systems, the completeness, compactness, and Löwenheim-Skolem theorems all fail; and categorical characterizations of infinite structures are possible.[5]

Notice that a standard interpretation for L2 is exactly the same as an interpretation for the first-order L1, namely, a domain and an assignment to the (common) nonlogical terminology. That is, in standard semantics, by fixing a domain, one thereby fixes the range of both the first-order and the second-order variables. There is no further "interpreting" to be done. This is not the case with the next two candidate semantics. In both cases, one must separately determine a range for the first-order variables and a range for the second-order variables. That is the crucial difference.

In *Henkin semantics,* the predicate variables range over a fixed collection of subsets of the domain (which may not include all of the subsets). That is, a *Henkin interpretation* of L2 is a structure $\langle d, D, I \rangle$, in which d is a domain, D a subset of the powerset of d, and I an appropriate assignment on d to the nonlogical terminology. Informally, D is the range of the predicate variables. A *Henkin variable assignment* consists of a function from the collection of first-order variables to d and a function from the collection of predicate variables to D. Again, the relevant satisfaction relation is straightforward.

Ironically, this semantics is not sound for many deductive systems. The reason is that some Henkin interpretations do not satisfy the comprehension scheme (and other axioms that may have been added). Define an interpretation to be *faithful* to a given deductive system for L2 if it satisfies its second-order axioms. Leon Henkin[6] established a completeness theorem for L2 with this semantics: a sentence Φ is provable in a given deductive system for L2 if and only if Φ is satisfied by every Henkin interpretation faithful to that deductive system. The compactness and Löwenheim-Skolem theorems

[5] See George Boolos and Richard C. Jeffrey, *Computability and Logic* (New York: Cambridge, 1980, 2nd ed.), ch. 18; and my *Foundations without Foundationalism.*

[6] "Completeness in the Theory of Types," *Journal of Symbolic Logic,* xv (1950): 81–91.

are also provable for L2 with (faithful) Henkin semantics. Consequently, there are no categorical characterizations of infinite structures.

It is straightforward to see that a standard interpretation is equivalent to the Henkin interpretation in which the designated range of the predicate variables is the entire powerset of the domain. These "full interpretations" are faithful to all common deductive systems.

In first-order semantics, L2 is regarded as a *two-sorted* first-order language, and the "predication" or "membership" relation between properties (or classes) and objects, is regarded as nonlogical. In particular, a first-order interpretation of L2 is a structure $\langle d_1, d_2, \langle I, p \rangle \rangle$, in which d_1 and d_2 are sets, p is a subset of $d_1 x d_2$, and I is an appropriate assignment (on d_1) to the (other) nonlogical terminology. Informally, d_1 is the range of the first-order variables, d_2 is the range of the predicate variables, and p is the interpretation of the "predication" relation. A *first-order variable assignment s* consists of a function from the collection of first-order variables to d_1 and a function from the collection of predicate variables to d_2. Once again, the notion of satisfaction is straightforward. One clause is:

> Let s assign u (an element of d_1) to the variable x and v (an element of d_2) to X. Then $\langle d_1, d_2, \langle p, I \rangle \rangle$ and s satisfy the atomic formula Xx if and only if the pair $\langle u, v \rangle$ is in p.

As above, define a structure to be *faithful* to a deductive system if it satisfies each second-order axiom thereof. Of course, the completeness, compactness, and Löwenheim-Skolem theorems all hold for this semantics (restricted to faithful interpretations).

Every Henkin interpretation $\langle d, D, I \rangle$ is equivalent to the first-order interpretation in which the "second" domain is D and p is the "real" membership or predication relation. A converse (of sorts) holds as well: for every first-order interpretation M^1, there is a Henkin interpretation M^{II}, whose domain is the first domain of M^1, such that for each sentence Φ of L2, $M^{II} \models \Phi$ iff $M^1 \models \Phi$.[7]

Let us return to the difference between first-order logic and second-order logic. Clearly, there is nothing to prevent a first-order language from being "about" predicates, properties, sets, or any-

---

[7] See Paul Gilmore, "The Monadic Theory of Types in the Lower-predicate Calculus" (*Summaries of Talks Presented at the Summer Institute of Symbolic Logic at Cornell,* Institute for Defense Analysis, 1957). For each $A \in d_2$ let $S(A)$ be the set $\{a \in d_1 \mid \langle a, A \rangle \in p\}$. Let $D^1 = \{S(A) \mid A \in d_2\}$. Then the Henkin interpretation equivalent to M^1 is $\langle d_1, D^1, I \rangle$. To take a frivolous example, suppose that in M^1, d_1 is a collection of apples and d_2 is a collection of oranges. Then in the Henkin model equivalent to M^1, each orange is "replaced" by the set of apples that bear the (interpretation of the) "predication" relation to it.

thing else for that matter, provided only that such items exist and that one can make coherent statements about them. Similarly, there is nothing to prevent the introduction of a two-sorted first-order language, with one sort of variable intended to range over "objects" and another sort intended to range over "properties" of these objects. The relation of "predication" between properties and objects would then be nonlogical, but I would suggest that determining the distinction between logical and nonlogical terminology is not separate from determining the intended "semantics" of the language.

In a recent conference proceedings, John Corcoran[8] argues that discourse about properties (and relations) is a natural (and inevitable) extension of discourse about "objects," as indicated by the presence of common nouns and straightforward grammatical constructions. In present terminology, the thesis is that languages with variables ranging over properties (and relations) are natural extensions of languages with variables ranging over objects. This is particularly clear, I believe, in the development of formal languages to codify the underlying logics of mathematical systems.

There is a striking affinity between Corcoran's remarks and W. O. Quine's analysis of ontological commitment in language. Both focus on the development of linguistic techniques for acknowledging and referring to objects. Quine, of course, is a steadfast opponent of second-order logic *as logic,* but he is not an opponent of set theory, as mathematics. This indicates that the analysis of language that shows that talk of sets is natural does not thereby determine the totality of the range of the property variables. In the present framework, the Quine-Corcoran analysis does not, by itself, decide among the semantics for the extended languages. Many of Corcoran's arguments in favor of second-order *languages* equally support standard semantics and Henkin, or first-order, semantics.[9]

The proponents of second-order logic *as logic,* then, hold that

---

[8] "Second-order Logic," in *Proceedings Inference OUIC 86,* D. Moates and R. Butrick, eds. (Athens: Ohio UP, 1987), pp. 7–31.

[9] See Quine, *Word and Object* (Cambridge: MIT, 1960), and, for the attack on second-order logic, Quine, *Philosophy of Logic* (Englewood Cliffs, NJ: Prentice-Hall, 1970). Corcoran's presentation of second-order languages as (natural) *extensions* of first-order ones does not reflect the history of logic (nor was it intended to). First-order languages were first studied [by Leopold Löwenheim, "Über Möglichkeiten im Relativkalkül," *Mathematische Annalen,* LXXVI (1915): 447–479; and Hilbert and Ackermann, *op. cit.*] as distinct *subsystems* of previously developed higher-order logics [see also G. Moore, "Beyond First-order Logic: The Historical Interplay between Logic and Set Theory," *History and Philosophy of Logic,* I (1980): 95–137; and "The Emergence of First-order Logic," in W. Aspray and P. Kitcher, eds., *Essays in the History and Philosophy of Mathematics* (Minneapolis: Minnesota Studies in the Philosophy of Science, 1988), pp. 95–135].

SECOND-ORDER LOGIC, FOUNDATIONS, AND RULES   239

second-order terminology with standard semantics is sufficiently clear, intuitive, or unproblematic to serve the underlying framework for axiomatization and foundations of mathematics. The claim is that, once a domain (for the first-order variables) is fixed, there is a clear and unambiguous understanding of such locutions as 'all properties' or 'all subsets' thereof.[10] Such locutions play central roles in foundations of mathematics.

The first-order "opposition," on the other hand, rejects this. Most do not claim that sets do not exist, nor that discourse about sets or properties is somehow illegitimate. Indeed, many of the major proponents of first-order logic, such as Skolem, Quine, and Gödel, have explicitly acknowledged set theory and contributed to its development. Moreover, Quine and Gödel adopted and defended attitudes of realism toward set theory (each in his own way). The claim seems to be that variables ranging over all of the properties of a fixed domain are not sufficiently clear to serve the foundation of mathematics, and that this terminology itself stands in need of further "foundation." That is, the relevant theory of sets ought to be formulated as an axiomatic theory in its own right, presumably a *first-order* theory. Quine calls second-order logic "set theory in disguise."[11] It almost goes without saying that the relevant notion of "sufficient clarity" could use some clarification. The issue here is a difficult one and it seems impossible to avoid begging at least some of the questions.

It might be noted in passing that *foundationalism,* the view that there is, or can be, a unique, secure "foundation" for all knowledge, is almost universally rejected today. Variants that restrict the foundation to mathematical knowledge may have some contemporary advocates, but if these too are rejected, then one should expect

[10] Notice that the foregoing claim is markedly weaker than the assertion that the entire set-theoretic hierarchy is clear and intuitive. Zermelo set theory, for example, has a powerset operator that can be iterated—even into the transfinite. The second-order thesis is only that, once a domain is fixed, the range of the second-order variables is also fixed. Let d be a domain, regarded as unproblematic. The thesis at hand then accepts the notion of *all* subsets of d—one powerset of d. If *that* is now taken as another "domain," the second-order thesis sanctions its powerset—the collection of all sets of subsets of d. But even this "construction" cannot be iterated into the transfinite. At most, the "commitment" is to "n^{th} order" sets (of d) for any natural number n.

[11] Quine, *Philosophy of Logic.* As far as I can determine, neither Skolem nor Gödel ever *argued* in print for the restriction to first-order logic. Gödel's correspondence with Zermelo illustrates this well. See I. Grattan-Guinness, "In Memoriam Kurt Gödel: His 1931 Correspondence with Zermelo on His Incompletability Theorem," *Historia Mathematica,* VI (1979): 294–304; J. Dawson, "Completing the Gödel-Zermelo Correspondence," *Historia Mathematica,* XII (1985): 66–70; and Moore, *op. cit.*

neither a unique nor an absolutely secure foundation for mathematics. There may be different foundations, perhaps serving different purposes, but, in any case, the goal of absolute certainty is not among them. The present contribution is made in this spirit.[12]

II

Imagine a dialogue between an advocate of second-order logic, call him Second, and an advocate of first-order logic, call her First. Second begins by developing a formal language with variables ranging over properties, and he explains (in informal language) the intuitive, standard "semantics," the idea that the property variables range over all properties, or all sets, of numbers. He then proves that arithmetic and analysis, as formulated in the second-order language, are categorical, and he shows that the real numbers are not countable. At this point, First raises a question concerning the range of the second-order variables. She asserts that the meaning of the second-order terminology is not very clear and she requests that it be developed more fully. Second could retort that First knows perfectly well what locutions like 'all subsets' mean, and he may accuse her of making trouble for the sake of making trouble. They would then be at a standoff.

On the other hand, Second may regard the request for clarification as legitimate. One can, indeed, do formal semantics and, in effect, provide a foundation for the background theory. So Second proceeds to develop a (weak) version of axiomatic set theory sufficient to formulate the standard semantics of the second-order theories. Call the formal metatheory MT. Second shows how the proofs of the categoricity and cardinality theorems can be rendered in MT.

---

[12] The relationship between foundationalism and foundations of mathematics would make a fascinating study. Skolem's "Some Remarks on the Foundation of Set Theory" [*Proceedings of the International Congress of Mathematicians, Cambridge, Massachusetts* (Providence: American Mathematical Society, 1952): 695–704] presents three goals for foundations of mathematics. One is to "obtain a way of reasoning which is logically correct so that it is clear and certain in advance that . . . what we prove are truths in some sense." This is called a "natural" standpoint, and is identified with logicism. Another goal is to "have a foundation which makes it possible to develop present day mathematics, and which is consistent so far as is known yet." It is lamented that this "practical" or "opportunistic" standpoint has the "unpleasant feature that we can never know when we have finished the foundation of mathematics. We are not only adding new floors at the top of our building, but from time to time it may be necessary to make changes in the basis." The third outlook is the Hilbert program, which is "giving up of the logicistic standpoint and not being content with the opportunistic one." Skolem's preference for foundationalism is clear, but he realized its difficulties. See S. Wagner, "The Rationalist Conception of Logic," *Notre Dame Journal of Formal Logic*, XXVIII (1987): 3–35, for a contemporary discussion of rationalism and foundationalism in logic.

At this point, things get interesting. First applauds the effort. She agrees that the development of MT is a significant event and clarifies things considerably. She even contributes to the development of MT by proving some theorems and suggesting new axioms. But she takes MT as a first-order theory. In particular, she claims that the semantical theory is itself subject to the Löwenheim-Skolem theorems, and the like. Thus, MT has many interpretations. First argues that the categoricity theorems only show that, within each interpretation of the semantics MT, all models of arithmetic are isomorphic and, similarly, within each interpretation all models of analysis are isomorphic. Also, within each model of the semantics, the interpretation of the real numbers is uncountable *in that model*. Such is relativism.

Second retorts that MT is not to be regarded as just another uninterpreted theory with various models of various cardinalities. On the contrary, MT has an *intended interpretation*—the intuitive semantics of the original languages. In short, the categoricity results should be taken at face value, referring to the possible domains for the original languages, not to this or that model of a formal theory like MT. The results are that arithmetic and analysis are really categorical, not just categorical in each model of MT, and Second insists that the real numbers are really uncountable. Moreover, he points out that the argument that First makes depends in part on the Löwenheim-Skolem theorems, which are demonstrably false for the second-order languages under study. Remember the categoricity theorems.

The reply at this point is predictable. First claims that the "refutations" of the Löwenheim-Skolem theorems only show that, within each interpretation M of MT, each model *m* of, say, analysis, is uncountable in M. It does not rule out the possibility that this same model *m* is countable in some more encompassing model of MT (but of course *m* would not be a model of analysis in the more encompassing structures). Indeed, we cannot rule out the possibility of countable models of MT itself. The theorems only show that the same structure cannot be both uncountable and a model of analysis from the same perspective. As for the assertion that MT is already interpreted, First requests clarification of this "intended semantics for the original languages." Second reminds First that he has already accomplished this. MT is the characterization, and First accepted it. But she seems to insist that MT must itself be interpreted. We seem to have reached another standoff.

Upon further examination (since neither Second nor First are stubborn), it is seen that the dispute centers on the axiom of separation (in MT), which states that, if *x* is a set and *P* a property that is well-defined on the members of x, then there is a subset y of x that contains all and only the members of x that have the property P. Its straightforward formulation is:

$$\forall x \, \exists y \, \forall z [z \in y \equiv (z \in x \ \& \ Pz)]$$

This, of course, is a second-order sentence. Second insists that it be given a *standard* reading, with the variable P ranging over all properties of the universe (i.e., all classes), while First insists on a first-order (or Henkin) interpretation. At this point, Second can respond to the challenges by developing a semantics for MT, a metasemantics. Call it MMT. But then First would understand MMT as a first-order theory and Second would not. Our desire to avoid (or postpone) the standoff between First and Second has led to a regress.

The skepticism, or relativism, that First advocates is not limited to second-order logic. There is a remarkably similar regress or standoff concerning the range of the first-order variables in rather basic theories. They can be regarded as uninterpreted and subject to a wide variety of standard and nonstandard modeling, or else one can insist that there is an unambiguous, intuitive understanding of the range of the variables. Suppose, for example, that First asks what one means by 'the natural numbers', the intended domain of arithmetic. Second responds with an axiomatization of Peano arithmetic, the crucial item of which is the induction axiom:

$$(P0 \ \& \ \forall x(Px \rightarrow Psx)) \rightarrow \forall x Px$$

First raises the (by now) usual question concerning the range of the property variable P. She takes induction to be a first-order axiom scheme, and points out that the theory, thus construed, has lots of models of various cardinalities. Of course, Second rejects this construal and replies that the understanding he has in mind—the intended model—is the "smallest" of the models. It is the structure that is an initial segment of all of the first-order models. First asks for a characterization of 'smallest model' or 'initial segment of all models', claiming that the quantifiers in these locutions are too problematic to serve foundations without further ado. Perhaps in frustration, Second asserts that by 'natural number', he means "member of the sequence 0, 1, 2, 3, 4, ..." First asks about the '. . .'.

In sum, our advocate of second-order logic holds that reference to, say, the natural numbers is clear and unambiguous, at least now. Any two mathematicians who discuss arithmetic are talking about the same (or isomorphic) structures. Suppose that First and Second agree, at least for the sake of argument, that formal arithmetic adequately codifies pretheoretical discourse—as they agreed that MT codifies informal semantics. Second takes the categoricity results to confirm his prior belief that arithmetic is unambiguous. But the results involve locutions like 'all properties' or 'all subsets', and the "confirmation" depends on these locutions themselves being unambiguous. First insists that there is no reason to hold this for the second-order variables and, thus, no reason to accept the confirmation of the original belief that arithmetic is unequivocal.

I conclude that the skepticism or relativism concerning second-order languages applies (or not) just as well to mathematical theories whose intended domains are infinite. Relativism concerning second-order variables goes with what may be called "object language relativism." On the other hand, one who is willing to accept classical mathematics, more or less as it stands, should also accept second-order languages. No more is involved. Alonzo Church[13] makes a similar point:

> Objections may indeed be made to this new point of view, on the basis of the sort of *absolutism* it presupposes . . . But it should be pointed out that this . . . is already inherent in classical mathematics generally, and it is not made more acute or more doubtful, but only more conspicuous, by its application to [logic]. For our definition of the consequences of a system of postulates . . . can be seen to be not essentially different from [that] required for the . . . treatment of classical mathematics . . . It is true that the non-effective notion of consequence, as we have introduced it . . . presupposes a certain absolute notion of ALL propositional functions of individuals. But this is presupposed also in classical mathematics, especially classical analysis, . . . (*ibid.*, p. 326N).

Parts of these dialogues can be seen in exchanges between historical figures early this century. Ernst Zermelo,[14] who corresponds to Second, formulated axiomatic set theory in 1908. He defined a propositional function $P(x)$ to be *definit* for a domain d if, for each element x of d, "if the fundamental relations of the domain, by means of the axioms and the universally valid laws of logic, determine without arbitrariness whether [P(x)] holds or not." The axiom of separation was stated in these terms:

> Whenever the propositional function [P(x)] is *definit* for all elements of a set M, M possesses a subset containing as elements precisely the elements x of M for which [P(x)] is true.

---

[13] *Introduction to Mathematical Logic* (Princeton: University Press, 1956).
[14] "Neuer Beweis für die Möglichkeit einer Wohlordnung," *Mathematische Annalen*, LXV (1908): 107–128; and "Untersuchungen über die Grundlagen der Mengenlehre. I," *Mathematische Annalen*, LXV (1908): 261–281. Both of these are translated in J. van Heijenoort, *From Frege to Gödel* (Cambridge: Harvard, 1967).

Much of the extensive criticism that Zermelo's work engendered was directed at the axiom of choice, but separation was not ignored. Among the attempted improvements, the most successful was that of Thoralf Skolem,[15] in 1922. Like our character First, he formulated separation as a scheme, one instance for each formula of the (first-order) language. This, of course, is current practice.

When Zermelo returned to the axiomatization in 1929, he agreed that the original notion of *definit* was not formulated with sufficient precision, and he seemed to acknowledge that the explication depends on the logical resources available:

> At the time, there did not exist a generally recognized "mathematical logic" to which I could appeal, any more than it does today, where every . . . researcher has his own system.[16]

So Zermelo proposed to give the notion of *definit* an axiomatic treatment. The result was essentially second-order, as indicated by the clause that, if $P(g)$ is *definit* for each propositional function g (with only individual variables), then so are $\forall f(P(f))$ and $\exists f(P(f))$.

In his almost immediate response, Skolem[17] noted that Zermelo's new formulation was quite similar to his own (of 1922), the only major difference being the (uncritical) use of 'propositional functions'. Skolem wondered whether Zermelo intended to provide still more axioms to characterize those. Presumably, Zermelo did not think that a further formulation was needed—this is the material of which foundations are made.

Of course, Zermelo could have provided axioms to characterize the propositional functions. Indeed, around that time, Hilbert and Ackermann's *Grundzüge der theoritischen Logik* did this in their treatment of higher-order logic. But Skolem stands ready to interpret anything Zermelo might say as thoroughly first-order and thus subject to reinterpretation via the Löwenheim-Skolem theorem. The dialogue ended with this standoff.

---

[15] "Einige Bemerkungen zur axiomatischen Begründung der Mengenlehre," in *Matematikerkongressen i Helsingfors den 4–7 Juli 1922* (Helsinki: Akademiska Bokhandeln), pp. 217–232; translated in Van Heijenoort, *op. cit.*, pp. 291–301). For an extensive discussion of Zermelo's axiomatization, see Moore, *Zermelo's Axiom of Choice: Its Origins, Development, and Influence* (New York: Springer, 1982).

[16] "Über den Begriff der Definitheit in der Axiomatik," *Fundamenta Mathematicae*, XIV (1929): 339–344. The quoted passage is from p. 340. See Moore, "The Emergence of First-order Logic," *op. cit.* Of course, today there *is* a "generally recognized" mathematical logic, the classical first-order predicate calculus. The irony is that Zermelo would not "appeal" to it, because he thought that it is too weak. In elaborating his view, Zermelo proposed a massive infinitary language.

[17] "Einige Bemerkungen zu der Abhandlung von E. Zermelo: 'Über die Definitheit in der Axiomatik'," *Fundamenta Mathematicae*, XV (1930): 337–341.

Skolem's "Über die mathematische Logik,"[18] an expository article, contains a few remarks on second-order logic. After developing a first-order language, he showed how variables ranging over "propositional functions" could be introduced; and he raised the possibility of quantifiers over these second-order variables:

> If "all" and "there exists" are applied to variable propositional functions, the question arises: what is the totality of all propositional functions (*ibid.*).

The latter question, of course, is the sort of thing Zermelo (and Second) did not think had to be raised. Once a domain is fixed, the range of the locution 'all subsets' is determined. After all, similar questions are not raised about the first-order variables and the other logical terminology, such as the connectives. Moreover, from this perspective, the only way to answer such questions is to *use* the corresponding terminology in the metalanguage. We say, for example, that *A&B* is true if *A* is true and *B* is true.

But for Skolem (and First) the question is legitimate and pressing. He asserted that only two conceptions for the range of the higher-order variables are "scientifically tenable." The first is, in effect, ramified type theory, in which propositional functions are associated with formulas of an expanding language. The second is to introduce the notion of propositional function axiomatically:

> The axioms will then become first-order propositions, since the . . . "propositional functions" . . . will assume the role of individuals. The relation between arguments and functions will then appear . . . as primitive (*ibid.*).

The debate is over the material of which foundations are made.

Of course, neither propositional functions nor sets are the first controversial entities to be studied by mathematicians. Negative, irrational, and complex numbers readily come to mind.[19] There are, to speak (very) roughly, three different stances that have been taken toward such entities by their proponents. Here, however, each has a troubling circularity. The first is simply to *postulate* the existence of the entities. Any axioms that are given are taken to *describe* the postulated entities. One insists on a distinction between the axioms and the entities themselves, and, thus, alternate interpretations are regarded as irrelevant. It need hardly be mentioned that postulation,

---

[18] *Norsk matematisk tidsskrift*, x (1928): 125–142; translated in Van Heijenoort, *op. cit.*, pp. 508–524.

[19] See E. Nagel, "Impossible Numbers: A Chapter in the History of Modern Logic," in *Teleology Revisited and Other Essays in the Philosophy and History of Science* (New York: Columbia, 1979), pp. 166–194.

by itself, is not going to convince the wary. It begs the question if anything does. Often, postulation is accompanied by arguments concerning the usefulness of the entities, or the fruitfulness of the resulting theory. Much of Zermelo's and Cantor's writings fit this mold. The fruitfulness would apply to *any* structure that satisfies the axioms, however. Thus, the second possibility is *implicit definition*. One first gives axioms and then holds that the defined entities "are" or "can be" anything that satisfies them. In arithmetic, for example, Richard Dedekind's celebrated "The Nature and Meaning of Numbers"[20] defines a "simply infinite system" to be any collection of objects that has an operation with certain properties. He then gave a categoricity proof and defined 'the natural numbers' to be one such system. Skolem's second "scientifically tenable" attitude toward propositional functions, the axiomatic treatment, is also an implicit definition. The problem here is that the very issue at hand concerns just which systems satisfy which axioms. Sets, properties, propositional functions, or the like, are central items in that very enterprise. If the axioms used in the definitions are taken to be first-order, as Skolem explicitly intended, then many (nonisomorphic) systems are so defined, and there is no way to distinguish a preferred interpretation. This is relativism, and its confirmation. If the axioms are second-order, then the implicit definitions are circular. One "defines" the entities in question by using those same entities. Again, the understanding of the (object language) first-order variables is tied to the items in the logic. The third outlook toward problematic entities is *construction*. One shows how the objects in question can be taken as combinations of less problematic entities. Examples include the Dedekind construction of real numbers as sets of rationals, and the "definition" of complex numbers as pairs of reals. Skolem's first conception of propositional functions along the lines of ramified type theory is of this form. A series of languages is constructed and each propositional function is associated with formulas thereof. From the perspective of the advocates of higher-order logic, this construction is inadequate. Only countably many propositional functions are constructed this way. More importantly, virtually every construction appeals to the intuitive notion of set and, as Skolem acknowledged, natural number.[21] Again, these are the "problematic" entities under study.

[20] In *Essays on the Theory of Numbers*, W. W. Berman, ed. (New York: Dover, 1963), pp. 31–115 (originally published in 1888). See P. Kitcher, "Frege, Dedekind, and the Philosophy of Mathematics," in *Frege Synthesized*, L. Haaparanta and J. Hintikka, eds. (Dordrecht: Reidel, 1986), pp. 299–343.
[21] Recall that set theory is supposed to be the theory in which all of mathematics —including arithmetic—is (re)formulated. It would thus violate the envisioned

III

In philosophy, regresses like the ones encountered here are familiar. One cannot forever continue to define one's terms using other terms; one cannot forever continue to prove one's premises from still more premises; and, in the cases at hand, one cannot forever continue to codify the meaning of one's discourse by providing more metasemantics. The hope of foundationalism was that such regresses would end at the bedrock of self-evidence, but this was not to be. In the present situation, it looks like questions must be begged. The first-order camp rejects or, at any rate, challenges the very framework that the higher-order advocates use in explicating the meaning of informal mathematical discourse.

Several options are available at this point. One is to maintain that most of the discourse of informal mathematics can be taken at face value (i.e., not re-interpreted), and it successfully refers to structures that are unique (up to isomorphism). Such structures include those of the natural numbers, the real and complex numbers, and, perhaps, the set-theoretic hierarchy. I call this outlook *neutral realism,* because no position is taken concerning the nature of the indicated structures. Questions can be raised at this point concerning how the structures are *apprehended* or understood by individual mathematicians, how they are *characterized* or described in practice, and how descriptions and inferences about them are *communicated.* That is, how do we know that, say, all number theorists are talking about the same structures? Of course, it is not clear how far one can go in answering these questions while remaining philosophically neutral. But it is a constraint on any view called "realism" that such answers be possible.

My "Second-order Logic and Mathematical Practice" assumes neutral realism and argues from that perspective that first-order languages are inadequate. If the language of informal mathematics is

---

"foundation" to appeal to natural numbers in the formulation of set theory. Some of the early attempts to codify Zermelo's notion of *definit* function, such as those of H. Weyl ["Über die Definitionen der mathematischen Grundbegriffe," *Mathematisch-naturwissenschaftliche Blätter,* VII (1910): 93–95, 109–113); and Fraenkel ("Zu den Grundlagen der Mengenlehre," *Jahresbericht der Deut. Math.-Verein.* XXXI (1922): 101–2], used phrases like 'finitely many' which, of course, appeal to natural numbers. This was taken to be problematic. Zermelo ("Über den Begriff der Defintheit in der Axiomatik") claimed that his second-order formulation was free of such appeals. In the aforementioned expository article ("Über die mathematische Logik"), Skolem conceded that both of the "scientifically tenable" conceptions of propositional functions rely on numbers. He suggested that neither sets nor numbers be regarded as prior, but that the foundations for both be laid simultaneously "in an interrelated way."

understood as an uninterpreted first-order system, then the characterization and communication of the presumed structures is impossible. One "solution" is to maintain that the underlying languages of informal discourse are (at least) second-order, with standard semantics. This is to assume or presuppose that the second-order terminology is itself understood unambiguously, and that this is no more problematic than neutral realism in informal mathematics. As we have seen, Church held a view like this (and so do I). The presuppositions of second-order logic are those of classical mathematics, taken at face value.

Another option, still within the rubric of neutral realism, is to hold that mathematical structures are successfully characterized (up to isomorphism) and communicated, but to maintain that this occurs only intuitively, or only in informal practice. That is, the communication of mathematics is inherently informal.[22] This view thus denies a significant role to any formal semantics in explicating mathematical description and communication. One may then hold that, to the limited extent that mathematics can be codified, the underlying logic is first-order. Against Church, this combination rejects the connection between the presuppositions of classical mathematics, as practiced, and second-order logic. The former is accepted; the latter is not.

A third option is to reject (neutral) realism, and to embrace a version of Skolemite relativism. One maintains that mathematical theories do not have unequivocal interpretations—or even interpretations unique up to isomorphism. With a moderate amount of set theory, of course, one can characterize models of various theories, and, in some cases, it is possible to use the resources of set theory to specify an "intended" interpretation for a theory. Thus, one shows that the standard model of arithmetic is isomorphic to the set of finite ordinals. But, and here is the rub, this view invariably insists that the mathematical theories and metatheories have first-order semantics. Thus, any theory at any level, if satisfiable in an infinite domain, has models of every infinite cardinality, and ultimately there is no principled way to identify an intended one. In other words, the underlying model theory is yet another theory with lots of "interpretations" (presumably in its own metatheory). There

---

[22] See John Myhill, "On the Ontological Significance of the Löwenheim-Skolem Theorem," [*Academic Freedom, Logic and Religion*, M. White, ed. (Philadelphia: American Philosophical Society, 1951), pp. 57–70; also in *Contemporary Readings in Logical Theory*, I. Copi and J. Gould, eds. (New York: MacMillan, 1967), pp. 40–54] for a lucid presentation of this option. Similar views have been suggested (in correspondence or conversation) by several prominent logicians.

is no unequivocal understanding for such terms as 'set', 'subset', 'finite', and quantifiers like 'all models'.

Paul Benacerraf [23] shows that Skolem shifted from the "mathematics is inherently informal" view, which he held in 1922 (*op. cit.*), to such a relativism. Another illustration of the dilemma is J. von Neumann's "Eine Axiomatisierung der Mengenlehre." [24] In the concluding section, he noted that it is easy to prove the categoricity of, say, Euclidean geometry, but von Neumann adds a parenthetical remark that every version of the theorem makes use of set-theoretic or, in present terms, higher-order terminology. [25] The theorems thus presuppose that such terminology is itself unequivocal and, perhaps, unquestioned. In examining this presupposition, von Neumann considers the possibility of a categorical set theory. One attempt at this was Abraham Fraenkel's [26] "axiom of restriction" that states, in effect, that no model of the theory is to have a proper subclass that is a model of the (other) axioms. Von Neumann objected that this axiom uses "the notions of naive set theory" and, thus, does not belong in the formulation of set theory itself—especially if it is to provide assurance that the antinomies have been avoided. To overcome this, a noncircular codification of the axiom of restriction must be provided. He proposed, first, that a "higher" set theory be developed axiomatically, one that contains the sets and proper classes of ordinary set theory as elements. [27] One can then use the relation of

---

[23] "Skolem and the Skeptic," *Proceedings of the Aristotelian Society, Supplementary Volume*, LIX (1985): 85–115.

[24] In *Journal für die reine und angewandte Mathematik*, CLIV (1925): 219–244; translated in van Heijenoort, *op. cit.*, pp. 393–413.

[25] Categorical axiomatizations of geometry include an axiom of continuity. One version is that every bounded, infinite *set* of points has a cluster point. Another is Hilbert's axiom of "completeness" that no *model* of the theory is to be a proper subclass of another structure that satisfies the (other) axioms [Hilbert, *Les principes fondamentaux de la géometrie* (Paris: Gauthier-Villars, 1982); see Moore, "The Emergence of First-order Logic"].

[26] "Über die Zermelosche Begründung der Mengenlehre," *Jahresbericht der Deut. Math.-Verein*, XXX (1921): 97–8. If suitably interpreted, the axiom of restriction "rules out" urelements, non-well-founded sets, and inaccessible cardinals. Notice the similarity between this axiom and Hilbert's axiom of completeness in geometry (*op. cit.*). Carnap and Bachmann's "Über Extremalaxiome" [*Erkenntnis*, VI (1936): 166–188; translated by H. G. Bohnert, *History and Philosophy of Logic*, II (1981): 67–85] is an insightful discussion of such axioms [see also Fraenkel, Bar-Hillel, and Levy, *Foundations of Set Theory* (Amsterdam: North-Holland, 1973, 2nd rev. ed.), pp. 113–9].

[27] A model of von Neumann's set theory would be a collection of sets (i.e., a class) together with the collection of its subclasses. But (proper) classes are not supposed to be elements of collections. Thus, talk of models requires a more encompassing system that has collections whose elements are the classes of ordinary set theory.

"subset in the higher theory" to formulate the notion of a class satisfying various statements. He notes that the intersection (within the higher system) of all models of the (other) axioms might turn out to be itself a model of the axioms. If so, it satisfies the axiom of restriction, and no other model does. It is emphasized, however, that this construction is relative to the choice of the "higher" set theory and the system satisfying it. Other (perhaps still "higher") systems may have different classes of "models of the axioms" and, thus, different intersections and different models that "satisfy" the axiom of restriction.

This, of course, leads to the sort of regress or relativism that we encountered above, and there seems to be no (non-question-begging) reason to think it stops. Von Neumann concludes that

> . . . no categorical axiomatization of set theory seems to exist at all; for probably no axiomatization will be able to avoid the difficulties connected with the axiom of restriction and the "higher" systems. And since there is no axiom system for mathematics, geometry, and so forth that does not presuppose set theory, there probably cannot be any categorically axiomatized infinite systems at all (*op. cit.*, p. 412).

It is then noted that a similar relativism seems to apply to basic notions of cardinality, since even these concepts are formulated with variables which occur "with reference to the entire [set-theoretic] system." An example is the concept of "Dedekind-finite," which refers to "all subsets" of a given set. If this relativity is sustained, of the notion of finitude,

> . . . nothing but the shell of its formal characterization would . . . remain . . . It is difficult to say whether this would militate more strongly against its intuitive character, . . . or its foundation as given by set theory (*op. cit.*, p. 413).

The dilemma that von Neumann registers here corresponds to two parts of the present trilemma: (1) one can reject the assertion that the fundamental notions of number, cardinality, subset, etc., as formulated in practice, are unequivocal. The best grasp we have of these concepts is that codified in axiomatic set theory, taken to be uninterpreted. This is relativism. (2) One can maintain that we have an intuitive, unequivocal grasp of the notions and reject the adequacy of the axiomatic, set-theoretic foundation. I add another option: (3) one can maintain the intuitive character of the notion of subset and *use* this to provide second-order characterizations.

IV

In practice, of course, the present regresses do not go on very long. No one, for example, has bothered to formulate a meta-meta-meta-language, and then wonder how to understand, or interpret it. The regresses end, on all accounts, in the discourse of informal mathematics. At this level, there is no further formal codification, perhaps because it is realized that this would only push the problems further back. It is not that we cannot precisely formulate the next metalanguage, but we see little point in doing so.

There is also almost universal agreement among present-day mathematicians concerning the practice of informal mathematics and the use of informal discourse. With the exception of traditional constructivists, there is no widespread and systematic disagreement over examples of correct proof, at least not now. There is at most an occasional skirmish. The more philosophical disputes noted here do not concern the correctness of informal mathematics, but rather things like how the discourse should be described, what it means, what it refers to, and what its nonlogical terminology is. None of the real or fictional disputants here should be regarded as advocating the revision of current practice.

This explains why the proof theories of the logics under examination here are remarkably similar, and underscores my opening remark that the differences between first-order logic and higher-order logic lie primarily in the different views on the totality of the range of the extra variables—in the "semantics." Indeed, one cannot prove more theorems using a second-order logic than one can with a first-order logic with set variables (and appropriate axioms). Both have admirably succeeded in "saving the appearances" concerning the correct use of informal discourse.[28]

Thus, to borrow a Quinean phrase, the regress ends in practice when we "lapse into the mother tongue" of informal mathematical discourse, and there things proceed rather smoothly, at least for now. This observation is rather congenial with the view held by some

---

[28] The deductive systems of common second-order theories are a bit stronger than their first-order counterparts. For example, one can prove the consistency of first-order arithmetic in second-order arithmetic. The reason is that the comprehension axiom scheme of second-order logic has instances that contain bound predicate variables. These impredicative definitions yield classes that cannot be defined in the first-order languages and, thus, do not fall under the first-order axiom schemes. Second-order arithmetic is equivalent to first-order analysis, however, and first-order set theory is much stronger than second-order analysis. On the comparison of first-order set theory with second-order set theory, see my "Second-order Languages and Mathematical Practice" and *Foundations without Foundationalism*.

philosophers that there is no more to understanding concepts than knowing how to use the relevant language. Variations on this theme have been championed by Hilary Putnam, Michael Dummett, and the later Wittgenstein.[29]

A slogan often associated with such views is "meaning is use," but I believe this is misleading. For present purposes, it is not a thesis about meaning, but about understanding. The claim is that understanding should not be ineffable. One understands the concepts embodied in a language to the extent that she knows how to use the language correctly. Call this the *use thesis*.

In the philosophy of language, the most straightforward opponents of the use thesis are those who tie understanding to a grasp of something that transcends use, or is conceptually independent of use. One example, perhaps, is Gottlob Frege, who held that each meaningful expression of an interpreted language is associated with a timeless, eternal, objective, and mind-independent entity called its *sense*. The concepts associated with a part of language are understood only if the requisite senses are grasped. To be sure, a given person's ability to use a language, such as that of arithmetic, is evidence for his having grasped the senses, but use and understanding are two different things. A similar view seems to underlie Noam Chomsky's distinction between competence and performance, but a consideration of this would take us too far afield. Against such views, the use thesis is that the ability to use a language constitutes understanding and thus grasping the concepts.

As Dummett puts it:

> An individual cannot communicate what he cannot be observed to communicate: if an individual associated with a mathematical symbol or formula some mental content, where the association did not lie in the use he made of the symbol or formula, then he could not convey that content by means of the symbol or formula, for his audience would be unaware of the association and would have no means of becoming aware of it (*op. cit.*, p. 216).

Moreover, in learning a language, such as that of arithmetic, one does not directly learn the sense of each expression, much less the denotation of each term. One learns, in effect, how to use each expression: how to perform the computations that underlie simple

[29] Putnam, "Models and Reality," *Journal of Symbolic Logic*, XLV (1980): 464–482); Dummett, "The Philosophical Basis of Intuitionistic Logic," in *Truth and Other Enigmas* (Cambridge: Harvard, 1978), pp. 215–247; Wittgenstein, *Remarks on the Foundations of Mathematics* and *Philosophical Investigations*, G.E.M. Anscombe, trans. (New York: MacMillan, 1958).

equations, how to apply arithmetic to everyday problems, and, later, how to prove and refute arithmetic statements and reason hypothetically in the language:

> These things are all that we learn when we are learning the meanings of the expressions of a language of the mathematical theory in question, because they are all that we can be shown (*op. cit.*, p. 217).

Perhaps another way to characterize the present contrast is to note that, on Fregean views, it is possible, at least in principle, for someone (or some machine) to be able to use a language correctly and still not grasp the relevant concepts. One can "fake it." Dummett writes:

> To suppose that there is an ingredient of meaning which transcends the use that is made of that which carries the meaning is to suppose that someone might have learned all that is directly taught when the language of a mathematical theory is taught to him, and might then behave in every way like someone who understood the language, and yet not actually understand it, or understand it only incorrectly (*op. cit.*, pp. 217–8).

Conversely, on the views under attack, it seems to be possible to grasp concepts and yet not be able to say anything correct about them. Again, understanding is conceptually independent of use, even if the gap is usually not very large in practice. The use thesis rejects both of these possibilities out of hand.[30]

It is often thought that the use thesis stands opposed to realism, the view that the variables of the language range over a realm of objects that exist independent of the mathematician and her mental life. Indeed, the slogan, "meaning is use," is often taken as a hallmark of antirealism. Dummett, for example, argues that the use thesis suggests, or even demands, that Tarskian semantics, with domain of discourse and truth conditions, is inappropriate. He proposes that 'use' replace 'truth' or 'satisfaction' as the main ingredient of a compositional semantics. One provides "proof conditions" instead of "truth conditions."

[30] Admittedly, these possibilities-in-principle invoke a rather broad metaphysical modality. It is open to a Fregean to defend a thesis that a complete fake, or even a large fake, is impossible in some, perhaps epistemic, sense. Plato may have held such a view in that the Socratic method seems to presuppose that, if a person fails to grasp a concept, then there are situations in which he cannot give intelligent responses. As for the converse, it is surely reasonable to hold—again on epistemic grounds—that there is a limit to the gap between competence and performance. It might be noted that the use thesis can itself countenance a distinction between competence and performance, but both are related to a subject's ability to use the language. Competence refers to global use and the ability to relate the relevant aspects of language to the rest of it. Performance concerns particular utterances in restricted areas, which may or may not be consistent with global use. On this sketch, the distinction between competence and performance is one of degree, not of kind.

I do not believe that this broad conclusion is warranted. The use thesis is quite plausible, and places sensible constraints on any theory of meaning. But, by itself, it does not demand an antirealist ontology, nor a non-Tarskian semantics. It does not follow that 'use' must be the central item in semantics. It does follow that ontology and truth conditions must be compatible with the learnability, and understanding of a language through its use. Truth conditions should not be that far removed from the knowledge implicit in the correct use of a language, and from whatever it is that one learns when acquiring facility with a language. There should be a natural (if not inevitable) link between use and truth conditions. Ideally, it should be clear, or at least defensible, that an object-language theory, plus its semantics, is a further articulation of the object-language theory alone, not an abrupt shift away from it. To use a Wittgensteinian metaphor (probably out of place), it should be plausible that, in adding semantics to a theory, one has "gone on as before." Without this, Dummett is correct that meaning is ineffable, or at least the charge is left unrefuted.[31]

In a similar context, Putnam notes that an advocate of the use thesis does not have to reject notions like model and reference. It is a question of understanding the roles these play in philosophy of mathematics. The grasp of particular models cannot be the central criterion of understanding. The model theory is itself formulated (originally) in informal language, which is understood through its use. Once one

> . . . has succeeded in understanding a rich enough language to serve as a meta-language for some theory T, he can define 'true in T' . . . he can talk about "models" for T, etc. He can even define 'reference' . . . exactly as Tarski did (*op. cit.*, p. 479).

The model theory and the formal semantics extend the resources of the original language, and may even articulate it further, but both come after the language is understood, and the original understanding does not consist of grasping a model, intended or otherwise. Understanding the language—original and extended—is knowing how to use it.

---

[31] In my "Logic, Ontology, Mathematical Practice" [*Synthese*, LXXIX (1989): 13–50], I argue that structuralism is compatible with realism, Tarskian semantics, *and* the use thesis. See essay 3 of Parsons' *Mathematics in Philosophy* for an insightful discussion of the sense in which the arithmetic within the theory of arithmetic-plus-arithmetic-truth is the same or different from arithmetic *simpliciter*.

At the level of informal discourse, prior to semantics, one can ask about the "meanings" of certain words and their referents. Many of the answers turn out to be platitudes like '12 is a natural number' or '6 is the sixth natural number after zero'. Similarly, the informal metalanguage is capable of distinguishing standard from nonstandard models (or semantics): "Standard models of arithmetic are those which are isomorphic to the natural numbers. Nonstandard models are those which begin with a copy of the natural numbers, followed by other elements of the domain." We actually say things like this to students of logic, and some of them understand it, or seem to. From the perspective of the regress above, of course, it begs the question (like everything else). The "problems" concerning meaning and reference apply to those very statements. We cannot rule out unintended interpretations of the metalanguage, unless we can somehow fix its semantics, and we cannot do that without using more language, in which case we must worry about its interpretation. And on it goes. But the use thesis blocks this regress or, at any rate, decides to stop by refusing to carry it further. We use the languages without (further) interpretation, but, to echo Wittgenstein, it does not follow that we use them without right. Both (informal) metalanguage and object language are understood when we know how to use them. We make the statements, we understand them, and we mean what we say. It makes no sense to ask for the meaning and reference of the entire informal language and to expect the answers to be somehow independent of that language.

Recall that two of the options in the above trilemma involve what I call *neutral realism,* the view that mathematical discourse is to be taken at face value. Arithmetic is about natural numbers, analysis is about real numbers, set theory is about sets, etc. This much is compatible with the use thesis and, indeed, seems to follow from it. What else is the use theorist going to hold? To be explicit about it, the informal, or preformal language of mathematics has a predicate N for 'is a natural number'. It is a basic item of the lexicon. The mathematician says that four is a natural number, $N4$. It follows, by existential generalization, that $\exists xNx$, numbers exist. Period. I propose that, on the use thesis, Skolemite relativism is to be rejected out of hand. The possibility of reinterpretation of the language has nothing to do with original use and, thus, nothing to do with understanding. The informal language, as it is used, is not equivocal.

At some point, it seems, we asked for a formal semantics to be used in accounting for how structures are apprehended, characterized, and communicated. As Putnam notes, this is the fatal step. It is not that the desire for a formal semantics is somehow illegitimate. Quite the contrary. The problem is with the idea that formal semantics is needed, or even helpful, in giving an account of what we grasp when we learn the meaning of a language:

> To adopt a theory of meaning according to which a language whose whole use is specified still lacks something—viz. its "interpretation"—is to accept a problem which *can* only have crazy solutions. To speak as if *this* were my problem, "I know how to use my language, but, now, how should I single out an interpretation?" is to speak nonsense (*op. cit.*, pp. 481–2).

If this conception of understanding is to be sustained, there is an interesting problem analogous to that of reference and nonstandard models. Use theorists are often criticized for leaving the concept of "use" vague. Surely an account is needed, even if problems of interpreting that account inevitably arise. As Wittgenstein[32] puts it:

> It all depends [on] *what* settles the sense of a proposition . . . The use of the signs must settle it; but what do we count as the use? (*ibid.*, pp. 366–7)

It might appear that attention is shifting from semantics to proof theory, but this is misleading. The problem is to account for how the understanding of a language of mathematics, through its use, constrains the future use of the language. How do we "go on as before"? Specifically, what is the relationship between formal proof and broad, global "use," as it figures in understanding? I maintain that the use thesis and formalism make a very bad philosophical partnership.

There is an important aspect of mathematical practice—the *use* of mathematical discourse—that is not captured by first-order languages, namely, the sense in which the discourse outstrips, or even "transcends," its previous manifestations. I would also suggest that many of the considerations in favor of second-order languages can be straightforwardly understood in these terms.[33] The arguments that certain concepts and theories resist first-order treatment amount to claims that first-order languages are not adequate to capture important aspects of the use of the relevant languages. I take up an example presently. To anticipate the conclusion, even if the use thesis dissolves the above regress of semantics, it does not dissolve the issue of second-order logic. The problem of interpreting the interpretations is replaced with the problem of rules for following rules.

[32] *Remarks on the Foundations of Mathematics.*
[33] See my "Second-order Logic and Mathematical Practice" and *Foundations without Foundationalism.*

SECOND-ORDER LOGIC, FOUNDATIONS, AND RULES   257

There is an interesting affinity between the way the problem of second-order languages is posed above and the discussion of rule following in Crispin Wright's *Wittgenstein on the Foundations of Mathematics.*[34] In the above scenarios, the two characters Second and First seemed (at times) to be in agreement. They accepted the same verbal description of the interpretation of the language. But they later discovered that they were interpreting the description (of the interpretation) differently. Thus begins the regress. Wright presents a thought experiment involving two people who agree on a description of a rule, and yet go on to apply it differently. There is a hidden divergence on the interpretation of the description of the rule. Let us pursue the analogy.

Saul Kripke[37] formulates an interesting problem on Wittgenstein's behalf. Most of us were taught how to add in grade school, and we now employ the symbol '+'. The problem is to account for our meaning the addition function by '+'. The thesis at hand is that the meaning of the term is fixed by our use of it and other expressions. But "what do we count as the use?" The addition function has infinitely many arguments and values. We have not "used" all of them yet, nor will we ever. Let 'quus' denote a binary function that agrees with addition for all arguments less than some large number N, but differs thereafter. I suggest that the role of the quus function here is analogous to the role of nonstandard models in the above discussion. Kripke's version of the problem is to show how we can refute a "skeptic" who tries to claim (ad hominem) that, in the past, we really meant quus by '+'. The analogy is with a skeptic (i.e., Skolem) who tries to claim that our past use of mathematical discourse may have referred to a nonstandard model.

If the value of N is large enough, both the plus function and the quus function are consistent with all of our past activities—all the numbers we have added (or quadded) so far. Both are also consistent with our physical dispositions to add (assuming that there is some limit to the size of numbers we are disposed to consider). Moreover, anything we have said about our practice, such as the recursive definition of '+', is subject to alternate interpretations, supporting either the "plus" or the "quus" reading.

Putting aside the exegetical and philosophical merits of Kripke's proposed (antirealist) solution, the problem points to a difficulty for the use thesis, or at any rate, a presupposition behind the use of ordinary language. Virtually every (nonfictitious) person believes that we do mean plus, and not quus, by '+'. It follows that our ability

---

[34] (Cambridge: Harvard, 1980), ch. 2. I owe this observation to Barbara Scholz.
[35] *Wittgenstein on Rules and Private Language* (Cambridge: Harvard, 1982).

to add, and our knowledge of arithmetic, somehow go beyond our previous performance, our dispositions to behave, etc., or at least we assume that they do. When we encounter new instances of addition and handle them correctly, we have "gone on as before." Wittgenstein's problem is to maintain the use thesis and yet have the pattern of "use" of an expression at a given time go beyond its previous manifestations. We need a plausible account of what it is to "go on as before." I think this is the meaning of the first passage quoted at the top of this paper. The problem is to maintain realism in the sense that we do mean something by our terms—Skolemite relativism is surely absurd—and yet not be limited by the "observation" of our previous behavior, not "empiricism."

The same sort of difficulty arises in the present situation, but now on a grand scale. Here it is "realism but not formalism, that is the hardest thing." Consider, for example, real analysis and, to make the analogy better, think of the completeness axiom as sanctioning a rule:

> When one learns that a set S of real numbers is bounded from above, one can infer that S has a least upper bound.

Suppose that some mathematician, or some community of mathematicians, formulates analysis, learns the completeness rule, and uses it successfully for several decades. Call him (or them) "Karl." Suppose also that, during those decades, analysis is the most powerful theory Karl has. In particular, suppose that Zermelo-Fraenkel set theory has not been developed.

Now, a first-order formulation of real analysis takes the completeness rule as a scheme, one instance for each open formula of *Karl's language*. Like Kripke's quus function, first-order analysis characterizes (1) Karl's past (correct) use of the completeness rule and (2) his physical dispositions to apply the rule (correctly). But it captures much more: (3) every use of the rule he *can* make given his expressive resources: here the modal term 'can' is not limited to physical possibility—the first-order scheme includes every application of the rule in Karl's present language; (4) his ability to describe his possible uses; (5) our ability to describe his ability to apply the rule (correctly); etc.

It should come as no surprise, then, that, from Karl's perspective, there is no significant difference between the class of theorems he can prove in first-order analysis and the class he can prove in second-order analysis. The first-order scheme has instances covering just about any use he can make given his (or their) resources. That is to say, any use he can make *at this stage* in his development. Yet, as with Kripke's quus function, this is not enough. The problem is to maintain the use thesis and yet hold that the "use" of at least some expressions (like the completeness rule) goes beyond, say, (1) to (5) above.

Suppose that Karl now develops, or becomes aware of, new expressive resources. For example, he may formulate some set theory and model the real numbers therein. Suppose also that he uses his extended language to define sets of real numbers he could not define prior to the new developments. He then applies the completeness rule to these sets and thereby learns more analysis—he learns more about the real numbers. It seems evident, at least to me, that he (or they) has "gone on as before." Karl's reaction to the newly defined sets of real numbers is exactly analogous to ours when we encounter a particular sum for the first time. We apply the same rule to a new case.

Similarly, Parsons' forthcoming "The Uniqueness of the Natural Numbers" describes a scenario in which two (communities of) mathematicians meet and learn each other's language. Each proves that the other's natural and real numbers are isomorphic to his own. This is not accomplished by an esoteric appeal to intuition. Rather, each uses the resources of the combined language to define sets of numbers, and then applies the relevant axiom (of induction or completeness) to these sets. That is, each mathematician applies the relevant "rule" to cases he could not formulate before the meeting. As with Karl and set theory, the combined language expands the resources, and provides new cases to which the rule can be applied, but the rule itself is not changed.

Second-order analysis captures this intuition of "going on as before." For Karl, the terminology of set theory is involved in extending the comprehension scheme (of second-order logic) that is used to describe and establish the existence of the "new" sets of real numbers. Once this is accomplished, the completeness rule is applied to these sets—the same completeness rule that Karl used before. The rule itself is not extended, and not reinterpreted.

With first-order analysis, the situation is not as straightforward. To derive the new theorems, the set-theoretic terminology must be used to produce new instances of the completeness scheme itself. That is, Karl must produce new cases of the very defining statements of analysis. In effect, the new *language* entails that Karl is working in a new *theory*, with a new set of rules. It is an extension of the old one, perhaps, but that theory has lots of extensions (since it is incomplete). What is lacking is a principled account of how the chosen new theory is the "same"—how Karl went on as before. It is straightforward to prove that there are models of the original first-order theory in which some of the new theorems are false. Thus, if Karl's (prior) use of the discourse of analysis involves no more than what is captured in the first-order theory (with its perfect match between semantics and proof theory), then the "new" theorems are not "about" the real numbers.

In sum, the first-order view and the second-order view seem to diverge in practice when the expressive resources are expanded. The first-order theory is limited to the sets of real numbers that are definable in the original language. The second-order account applies the same rule to the newly defined sets. Perhaps we did not note the potential divergence in the above scenarios because it was assumed that the total expressive resources are held constant throughout the regress, or maybe we silently allowed our advocate of first-order logic to reconstruct the basic theories at each level.[36]

I do not claim that adopting higher-order logic somehow solves the twin problems of rule following and Skolemite relativism. It does not refute the skeptics. If our character First were reintroduced at this point, she would argue that the second-order language creates a false appearance of "going on as before." With the adoption of the set theory, the completeness rule is the same, perhaps, but the comprehension scheme changes. She would claim that there is a shift in the range of the second-order variables. Recall that she denies that this range is fixed. It changes from one model of the semantics to another.

My thesis here is that, for better or worse (well, for better), the attitude underlying the practice of mathematics is that skepticism is false. We do go on as before, both in following rules and in extending our theories. At any rate, we talk and act that way. It may be an illusion, I suppose, but when philosophy is said and done, we do not take skepticism seriously. Second-order languages echo this presupposition. It is in line with Church's statement that the assumptions of second-order logic may be more conspicuous than those behind the practice of classical mathematics, but they are not more troublesome.

[36] There is an interesting analogy between these considerations and a major argument in Michael Friedman's *Foundations of Space-time Theories* (Princeton: University Press, 1983) for the reality of space-time. Friedman concedes that, if we focus attention exclusively on the most powerful and complete physical theory we have, general relativity for example, the substantival versions that refer to space-time are equivalent to the relationalist versions that do not. But space-time is needed to see the smooth relationships between, say, gravitational theory alone and gravitational theory plus electro-dynamics. The relationalist theories do not go together that well, and the reality of space-time is confirmed in the historical advance from the weaker theory to its more comprehensive successor.

SECOND-ORDER LOGIC, FOUNDATIONS, AND RULES   261

Similar considerations rule out a formalistic account of the "use" of mathematical discourse. No single explicitly formulated algorithm can capture Karl's prior understanding and use of his language. If nothing else, the incompleteness theorem establishes this. No algorithm can account for Karl's "going on as before" when his resources are expanded. At best, an explicit algorithm can codify the applications of Karl's theories in their present state of articulation —in their present languages.

Benacerraf puts the situation well:

> We *do* need a metaphysically and epistemologically satisfactory account of the way mathematical practice determines or embodies the meaning of mathematical language. (We could also use a satisfactory account for areas other than mathematics.) We may even need to devise new concepts of meaning to forge such an account.
>
> . . . mathematical practice reflects our intuitions and controls our use of mathematical language in ways of which we may not be aware at any given moment, but which transcend what we have explicitly set down in any given *account*—or may ever be able to set down.
>
> With Gödel, I incline toward this view. But I am sufficiently aware of its vagueness and inadequacy not to be tempted into thinking it constitutes a *view*. It is merely a direction (*op. cit.*, pp. 110–1).

I also incline to this "view" and, as far as I have put it here, I think the later Wittgenstein did as well. The difference with Gödelian Platonism (and with formalism) is the extent to which we are able to fix the use of our terms completely. What is needed here is a synthesis between "relativism," which holds that the constraints on the use of our language do not extend much beyond our present ability to describe and apply our rules, and traditional Platonism, which holds that our minds can somehow grasp infinite structures, and that the concepts are thereby fixed completely, once and for all. The question concerns the extent to which our present practice in, say, analysis, determines the future use of the terms or, to use another Wittgensteinian phrase, the extent to which meaning and use are and are not fixed "throughout logical space."

However this is resolved, I make the modest proposal that we can shed some light on the requisite concepts of meaning and reference, and the sense in which these can outstrip prior manifestations, by studying the use of second-order concepts in mathematics.

STEWART SHAPIRO

Ohio State University

Part III
Plural Quantification

[15]

TO BE IS TO BE A VALUE OF A VARIABLE
(OR TO BE SOME VALUES OF SOME VARIABLES)*

ARE quantification and cross reference in English well represented by the devices of standard logic, i.e., variables x,y,z, . . ., the quantifiers $\forall$ and $\exists$, the usual propositional connectives, and the equals sign? It's my impression that many philosophers and logicians think that—on the whole—they are. In fact, I suspect that the following view of the relation between logic and quantificational and referential features of natural language is fairly widely held:

No one (the view begins) can think that the propositional calculus contains all there is to logic. Because of the presence in natural language of quantificational words like 'all' and 'some' and words used extensively in cross reference, like 'it', 'that', and 'who', there is a vast variety of forms of inference whose validity cannot be adequately treated without the introduction of variables and quantifiers, or other devices to do the same work. Thus everyone will concede that the predicate calculus is at least a part of logic.

Indispensable to cross reference, lacking distinctive content, and pervading thought and discourse, *identity* is without question a logical concept. Adding it to the predicate calculus significantly increases the number and variety of inferences susceptible of adequate logical treatment.

And now (the view continues), once identity is added to the predicate calculus, there would not appear to be all that many valid inferences whose validity has to do with cross reference, quantification, and generalization which cannot be treated in a satisfactory way by means of the resulting system. It may be granted that there are certain valid inferences, involving so-called "analytic" connections, which cannot be handled in the predicate calculus with identity. But the validity of these inferences has nothing to do with quantification in natural language, and it may thus be doubted whether a logic that does nothing to explain their validity is thereby deficient.

In any event (the view concludes), the variety of inferences that

* I am grateful to Richard Cartwright, Helen Cartwright, James Higginbotham, Judith Thomson, and the editors of the *Journal of Philosophy* for helpful comments, criticism, and discussion. Helen Cartwright's valuable unpublished Ph.D. dissertation, "Classes, Quantities, and Non-singular Reference" (University of Michigan, 1963) deals at length with many of the issues with which the present paper is concerned.

0022-362X/84/8108/0430$02.00

cannot be dealt with by first-order logic (with identity) is by no means as great or as interesting as the variety that can be handled by the predicate calculus, even without identity, but not by the propositional calculus.

It is the conclusion of this view that I want to take exception to. (At one time I thought the whole view was probably true.) It seems to me that we really do not know whether there is much or little in the province of logic that the first-order predicate calculus with identity cannot treat. In the first part of this paper I shall present and discuss some data which suggest that there may be rather more than might be supposed, that there may be an interesting variety both of quantificational and referential constructions in natural language that cannot be represented in standard logical notation and of valid inferences for whose validity these constructions are responsible. Whether quantification and cross reference in English are well represented by standard logic seems to me to be an open question, at present.

Several kinds of constructions, sentences, and inferences that cannot be symbolized in first-order logic are known. Perhaps the best-known of these involve numerical quantifiers such as 'more', 'most', and 'as many', e.g., the inference

> Most Democrats are left-of-center.
> Most Democrats dislike Reagan.
> Therefore, some who are left-of-center dislike Reagan.

Another is the construction "For every A there is a B," which, although it might appear to be symbolizable in first-order notation, cannot be so represented, for it is synonymous with "There are at least as many Bs as As."[1] The construction is not of recent date; it is exemplified in a couplet from 1583 by one T. Watson:[2]

> For every pleasure that in love is found,
> A thousand woes and more therein abound.

Jaakko Hintikka has offered a number of examples of sentences that cannot, he claims, be represented in first-order logic.[3] One of these is:

> Some relative of each villager and some relative of each townsman hate
> each other.

---

[1] Cf. my "For Every A There Is a B," *Linguistic Inquiry*, XII (1981): 465–467.
[2] See the entry for 'for' in the Oxford English Dictionary.
[3] "Quantifiers vs. Quantification Theory," *Linguistic Inquiry*, V (1974): 153–177.

There appears to be a consensus regarding this sentence, viz., that if it is O.K., then it can be symbolized in standard first-order logic as follows:

$$\forall x \forall y \exists z \exists w(Vx \,\&\, Ty \rightarrow Rzx \,\&\, Rwy \,\&\, Hzw \,\&\, Hwz \,\&\, z \neq w)$$

I find this sentence marginally acceptable at best and not acceptable if not symbolizable as above.

Jon Barwise has offered "The richer the country, the more powerful is one of its officials" as another example of a sentence that cannot be symbolized in first-order logic.[4] However, since the sentence seems to me, at any rate, to mean "Whenever x is a richer country than y, then x has (at least) one official who is more powerful than *any* official of y," it also seems to me to have a first-order symbolization:

$$\forall x \forall y([Cx \,\&\, Cy \,\&\, xRy] \rightarrow \exists w[wOx \,\&\, \forall z(zOy \rightarrow wPz)])$$

Are there better examples?

Perhaps the best-known example of a sentence whose quantificational structure cannot be captured by means of first-order logic is the Geach-Kaplan sentence, cited by W. V. Quine in *Methods of Logic*[5] and *The Roots of Reference*[6]:

(A) Some critics admire only one another.

(A) is supposed to mean that there is a collection of critics, each of whose members admires no one not in the collection, and none of whose members admires himself. If the domain of discourse is taken to consist of the critics and Axy to mean "x admires y," then (A) can be symbolized by means of the *second*-order sentence:

(B) $\exists X(\exists x\, Xx \,\&\, \forall x \forall y[Xx \,\&\, Axy \quad x \neq y \,\&\, Xy])$

And since (B) is not equivalent to any first-order sentence, (A) cannot be correctly symbolized in first-order logic.

The proof, due to David Kaplan, that (B) has no first-order equivalent is simple and exhibits an important technique in showing nonfirstorderizability: Substitute the formula $(x = 0 \lor x = y + 1)$ for Axy in (B), and observe that the result:

(C) $\exists X(\exists x\, Xx \,\&\, \forall x \forall y[Xx \,\&\, (x = 0 \lor x = y + 1) \rightarrow x \neq y \,\&\, Xy])$

---

[4]"On Branching Quantifiers in English," *Journal of Philosophical Logic*, VIII, 1 (February 1979): 47–80.

[5]4th ed. (Cambridge, Mass.: Harvard, 1982), p. 293, where "people" is substituted for "critics" in the example.

[6]LaSalle, Ill.: Open Court, 1973; p. 111.

is a sentence that is true in all nonstandard models of arithmetic but false in the standard model.[7]

I must confess to a certain ambivalence regarding the Geach-Kaplan sentence. Although it usually strikes me as a quite acceptable sentence of English, it doesn't invariably do so. (The "only" seems to want to precede the "admires" but the intended meaning of the sentence forces it to stay put.) I find that if the predicates in the example are changed in what one might have supposed to be an inessential way matters are improved slightly:

> Some computers communicate only with one another.
> Some Bostonians speak only to one another.
> Some critics are admired only by one another.

I don't have any idea why replacing the transitive verb 'admires' by a verb or verb phrase taking an accompanying prepositional phrase helps matters, but it does seem to me to do so.

I turn now from this brief survey of known examples of sentences not representable in first-order logic to examination of some other nonfirstorderizable sentences. Like the Geach-Kaplan sentence but unlike the sentences involving 'most', these sentences *look* as if they "ought to be" symbolizable in first-order logic. They contain plural forms such as 'are' and 'them', and it is in large measure because they contain these forms that they cannot be represented in first-order logic.

Consider first the following sentence, which, however, contains no plurals and which can be symbolized in first-order logic:

> (D) There is a horse that is faster than Zev and also faster than the sire of any horse that is slower than it.

Quantifying over horses, and using 0, s, $>$, and $<$ for 'Zev', 'the sire of', 'is faster than', and 'is slower than', respectively, we may symbolize (D) in first-order logic:

$$\text{(E)} \qquad \exists x(x > 0 \mathrel{\&} \forall y[y < x \rightarrow x > s(y)])$$

[7] To see that (C) is true in any nonstandard model, take as X the set of all nonstandard elements of the model. X is nonempty, does not contain 0, hence contains only successors, and contains the immediate predecessor of any of its members. To see that it is false in the standard model, suppose that there is some suitable set X of natural numbers. X must be nonempty: if its least member x is 0, let $y = 0$; otherwise $x = y + 1$ for some y. Since x is least, y is not in X, and 'Xy' is false. The nonfirstorderizability of "For every A there is a B" can be established in a similar way: Select variables x and y not found in any presumed first-order equivalent, substitute $[(1) < x + 5 \mathrel{\&} \sim\exists y3\cdot y = (1)]$ for $A(1)$, substitute $[(1) < x + 5 \mathrel{\&} \exists y3\cdot y = (1)]$ for $B(1)$, and existentially quantify the result with respect to x; the result would be true in all nonstandard models but false in the standard model.

Sentence (F), however, cannot be symbolized in first-order logic:

(F) There are some horses that are faster than Zev and also faster than the sire of any horse that is slower than them.

(F) differs from (D) only in that some occurrences in (D) of the words 'is', 'a', 'horse', and 'it' have been replaced by occurrences of their plural forms 'are', 'some', 'horses', and 'them'. The content of (F) is given slightly more explicitly in:

(G) There are some horses that are all faster than Zev and also faster than the sire of any horse that is slower than all of them.

I take it that (F) and its variant (G) can be paraphrased: there is a nonempty collection (class, totality) X of horses, such that all members of X are faster than Zev and such that, whenever any horse is slower than all members of X, then all members of X are faster than the sire of that horse.[8] (F) and (G) can be symbolized by means of the second-order sentence (domain and denotations as above):

(H) $\exists X(\exists x\, Xx \;\&\; \forall x(Xx \rightarrow x > 0)$
$\&\; \forall y[\forall x(Xx \rightarrow y < x) \rightarrow \forall x(Xx \rightarrow x > s(y))])$

(H) is equivalent to no first-order sentence; for it is false in the standard model of arithmetic (under the obvious reinterpretation) but true in any nonstandard model, since the set of nonstandard elements of the model will always be a suitable value for X. Thus (F) cannot be symbolized in first-order logic.[9]

(F) is not an especially pretty sentence. It is hard to understand, awkward, and contrived. But ugly or not, it is a perfectly grammatical sentence of English, which has, as far as I can see, the meaning given above and no other. Moreover, such faults as it has appear to be fully shared by (D).

Another example, shorter and perhaps more intelligible:

(I) There are some gunslingers each of whom has shot the right foot of at least one of the others.

---

[8] Zev won the Kentucky Derby in 1923.

[9] Cf. my "Nonfirstorderizability Again," to appear in *Linguistic Inquiry*, xv, 2 (1984). In an important unpublished manuscript entitled "Plural Quantification," Lauri Carlson has given "If some numbers all are natural numbers, one of them is the smallest of them," as an example of a sentence that cannot be symbolized in the first-order predicate calculus. I have heard it claimed that this is not a proper sentence of English. Perhaps it is not, but "If there are some numbers all of which are natural numbers, then there is one of them that is smaller than all the others," surely is. I am grateful to Irene Heim for calling this reference to my attention.

(I) may be rendered in second-order logic:

(J) $\exists X(\exists x\,Xx \mathbin{\&} \forall x[Xx \to \exists y(Xy \mathbin{\&} y \neq x \mathbin{\&} Bxy)])$

(Here we quantify over gunslingers and use B for "has shot the right foot of.") By substituting $x = y + 1$ for Bxy, we may easily see that (J) is equivalent to no first-order sentence. (Alternatively, we may note that if we negate (J), substitute $y \leq x$ for Bxy, and make some elementary transformations, we obtain:

$$\forall X(\exists x\,Xx \to \exists x[Xx \mathbin{\&} \forall y(Xy \mathbin{\&} y \leq x \to y = x)])$$

a formula that expresses the least-number principle, which is one version of the principle of mathematical induction.)

When used as a demonstrative pronoun, 'that' is marked for number, as singular, but when used as a relative pronoun, as in (F), it is unmarked for number, i.e., can be used in either the singular or plural. 'Who', 'whom', and 'whose', however, are unmarked for number when used either as relative or as interrogative pronouns. 'Which' is also unmarked for number as a relative pronoun, but 'which ones', when it can be used, is strongly preferred to 'which' as an interrogative plural form; it may well be that interrogative 'which', like demonstrative 'that', is marked as singular.

It is the plural forms in (F) and (I), as well as the unmarkedness of 'that' and 'whom', that are responsible for the nonfirstorderizability of these sentences. And by taking a cue from the well-known second-order definitions of "x is a standard natural number" and "x is an ancestor of y," we can use plurals to define these notions in English (in terms of "zero" and "successor of" and in terms of "parent of," respectively):

(K) If there are some numbers of which the successor of any one of them
     is also one, then if zero is one of them, x is one of them.

(L) If there are some persons of whom each parent of any one of them
     is also one, then if each parent of y is one of them, x is one of them;
     and someone is a parent of y.

There are some comments on (K) and (L) to be made: (a) 'which' and 'whom' are used in these sentences as we have noticed they can be used, in the plural. (b) Instead of saying "of which the successor of any one of them is also one," one could as well say "of which the successor of any one is also one of them": at least one "them" is needed to cross-refer to the "witnessing" values of 'which': this 'them' is sometimes called a *resumptive* pronoun, and appears to be needed to capture the force of $\forall y(Xy \to Xs(y))$, with its two occurrences of X. (c) Like (F) and (I), (K) and (L) cannot be given cor-

rect first-order symbolizations, and thus the following (valid) inference cannot be represented in first-order logic:

> If there are some persons of whom each parent of any one of them is
> also one, then if each parent of Yolanda is one of them, Xavier is one
> of them; and someone is a parent of Yolanda.
> Every parent of someone red is blue.
> Every parent of someone blue is red.
> Yolanda is blue.
> Therefore, Xavier is either red or blue.

(To see that this is a valid inference, consider the persons who are
either red or blue. By the second and third premises, every parent of
any one of these persons is also one of them; and since Yolanda is
blue, each of her parents is red, hence red or blue, and hence one of
these persons. Thus Xavier is also one of them and thus either red
or blue.) (d) The 'there are's in the antecedents of course express
universal quantification, as does the 'there is' in "If there is a logician present, he should leave." (e) Like (F), (K) and (L) are somewhat ungainly, in part because of the resumptive 'them' they contain, but principally because of the complexity of the thoughts they
express. However, they seem to be perfectly acceptable vehicles for
the expression of those very thoughts. And although they are indeed contrived—they have been contrived *to take advantage of referential devices that are available in English*—the fact that they are
so hardly begins to bear on the question whether they are un-
grammatical, unintelligible, or in some other way unacceptable.

The suggestion that it is the complexity of the thoughts expressed in (K) and (L) that is responsible for their ungainliness
rather than the presence of any construction not properly a part of
English draws support from the ease and naturalness with which
"x is identical to y" may be defined in the same style: if there are
some things of which x is one, then y is one of them too. (Or: it is
not the case that there are some things of which x is one, but of
which y is not one.)

Another example, of a different sort, is:

> (M) Each of the numbers in the sequence 1,2,4,8, . . . is greater than the
> sum of all the numbers in the sequence that precede it.

(M) states something true, which, using a mixture of logical and
arithmetical notation, we can express as follows:

$$(N) \quad \forall x \forall y (Px \mathbin{\&} y = \Sigma\{z : Pz \mathbin{\&} x > z\} \rightarrow x > y)$$

In (N), Σ is a sign for a function from sets of objects in a domain to objects in that domain and attaches to a variable and a formula to form a term in which that variable is bound. Signs for such functions are simply not part of the primitive vocabulary of first-order logic, although on occasion mention of functions of this type can be paraphrased away (e.g. "the least of the numbers z such that . . . z . . ."). No one function sign of the ordinary sort can do full justice to "the sum of the numbers z such that . . . z . . . ," as can be seen by considering:

> (O) Although every power of 2 is 1 greater than the sum of all the powers of 2 that are smaller than it, not every power of 3 is 1 greater than the sum of all the powers of 3 that are smaller than it.

We certainly cannot symbolize (O) as:

$$\forall x \forall y (Px \ \& \ y = f(x) \ \rightarrow \ x = y + 1)$$
$$\& \sim\forall x \forall y (Qx \ \& \ y = f(x) \ \rightarrow \ x = y + 1)$$

and were we to try to improve matters by changing the second occurrence of f to an occurrence of (say) g, we should fail to depict the recurrence of the semantic primitive 'the sum of . . .' in the second conjunct of (O). Nor could any ordinary function sign express the dependencies that may obtain between predicates contained in '. . . z . . .' and those found in the surrounding context.

A short and sweet example of the same type is:

No number is the sum of all numbers.

The last example for the moment of a sentence whose meanings cannot all be captured in first-order logic is one that is again found in Quine's *Methods of Logic*—but not, this time, in the final part of the book, "Glimpses Beyond." It is the sentence (P):

> (P) Some of Fiorecchio's men entered the building unaccompanied by anyone else.

On Quine's analysis of this sentence, it can be represented as $\exists x (Fx \ \& \ Ex \ \& \ \forall y [Axy \ \rightarrow \ Fy])$, where Fx, Ex, and Axy mean "x was one of Fiorecchio's men," "x entered the bulding," and "x was accompanied by y."[10] Quine states that "x was unaccompanied by anyone else" clearly has the intended meaning "Anyone accompanying x was one of Fiorecchio's men."

Quine's is certainly one reading this sentence bears: there are some Fiorecchians each of whom entered the building unaccom-

---

[10] Page, 197. Quine uses K, F, and H instead of F, E, and A, respectively.

panied by anyone who wasn't a Fiorecchian. But since (P) appears, at times, to mean something like:

> There were some men, see.
> They were all Fiorecchio's men.
> They entered the building.
> And they weren't accompanied by anyone else.

it can also be understood to mean: there are some Fiorecchians each of whom entered the building unaccompanied by anyone who wasn't one of *them*. On this stronger reading, there is no asymmetry between the predicates "*x* was one of Fiorecchio's men" and "*x* entered the building," 'else' means "not one of them," and the whole can be symbolized by:

$$\exists X (\exists x\, Xx \;\&\; \forall x(Xx \to Fx) \;\&\; \forall x(Xx \to Ex) \;\&\; \forall x \forall y(Xx \;\&\; Axy \to Xy))$$

whose nonfirstorderizability can be seen in the usual way, by substituting $x > 0$ for both Fx and Ex and $x = y + 1$ for Axy.

It is because of these examples that I think that the question whether the first-order predicate calculus with identity adequately represents quantification, generalization, and cross reference in natural language ought to be regarded as a question that hasn't yet been settled.

Changing the subject somewhat, I now want to look at a number of sentences whose most *natural* representations are given by second-order formulas, but second-order formulas that turn out to be equivalent to first-order formulas.

The sentence:

(Q) There are some monuments in Italy of which no one tourist has seen all.

might appear to require a second-order formula for its correct symbolization, e.g.,

(R) $\exists X(\exists x\, Xx \;\&\; \forall x[Xx \to Mx] \;\&\; \sim\!\exists y[Ty \;\&\; \forall x(Xx \to Syx)])$

Of course, (Q) can be paraphrased:

(S) No tourist has seen all the monuments in Italy.

and this can be symbolized in first-order logic as:

(T) $\exists x Mx \;\&\; \sim\!\exists y[Ty \;\&\; \forall x(Mx \to Syx)]$

which is equivalent to (R).[11] But just as $\sim\!\sim\! p$ can sometimes be a

----

[11] It take it that since (S) implies that there are some monuments in Italy, but does not imply that there are tourists, the conjunct $\exists x Mx$ is indispensable.

better symbolization than p of "It's not the case that John didn't go," e.g., if p were used to symbolize "John went," so (R) captures more of the quantificational structure of (Q) than does the equivalent (T). (Q) might appear to say that there is a (nonempty) collection of monuments in Italy and no tourist has seen every member of this collection; (S) doesn't begin to hint at collections of monuments. Nevertheless, (Q) and (S) say the same thing, if any two sentences do, and (R) and (T) are, predictably enough, equivalent.

Another example of the same "collapsing" phenomenon:

> (U) Mozart composed a number of works, and every tolerable opera
> with an Italian libretto is one of them.

has the second-order symbolization:

(V) $\qquad \exists X(\exists x Xx \ \& \ \forall x(Xx \ \to \ Mx) \ \& \ \forall x(Tx \ \to \ Xx))$

But as (U) says what (W) says:

> (W) Mozart composed a number of works, and every tolerable opera
> with an Italian libretto is a work that Mozart composed.

so (V) is equivalent to the first order

(X) $\qquad \exists x Mx \ \& \ \forall x(Tx \ \to \ Mx)$

The construction 'Every . . . is one of them' bears watching; suffice it for now to observe that it is a perfectly ordinary English phrase.

Collapses can also occur unexpectedly. (Through a publisher's error) the sentence:

> (Y) Some critics admire one another and no one else.

meaning (approximately), "There is a collection of critics, each of whom admires all and only the *other* members of the collection," and possessing the second-order symbolization:

(Z) $\exists X(\exists x \exists y[Xx \ \& \ Xy \ \& \ x \neq y]$
$\qquad\qquad \& \ \forall x[Xx \ \to \ \forall y(Axy \ \leftrightarrow \ \{Xy \ \& \ y \neq x\})])$

was claimed in the first American printing of the third edition of *Methods of Logic* to be a sentence incapable of first-order representation.[12] But although (Z) might appear to be susceptible to the same kind of treatment given out above, it was in fact observed by Kaplan to be equivalent to the first-order formula:

(a) $\exists z(\exists y Azy \ \& \ \forall x[(z = x \ \lor \ Azx)$
$\qquad\qquad \to \forall y(Axy \ \leftrightarrow \ \{(z = y \ \lor \ Azy) \ \& \ y \neq x\})])$

---

[12] *Methods of Logic*, 3d ed. (New York: Holt, Rinehart & Winston, 1972), p. 238/9.

Consider now sentence (b):

(b) There are some sets that are such that no one of them is a member
of itself and also such that every set that is not a member of itself is
one of them. (Alternatively: There are some sets, no one of which
is a member of itself, and of which every set that is not a member
of itself is one.)

By quantifying over sets and abbreviating "is a member of" by ϵ,
we may use a second-order formula to symbolize (b):

(c) $\exists X (\exists x\, Xx \,\&\, \forall x[Xx \rightarrow \sim x\epsilon x] \,\&\, \forall x[\sim x\epsilon x \rightarrow Xx])$

(c) is obviously equivalent to (d):

(d) $\exists X (\exists x\, Xx \,\&\, \forall x[Xx \leftrightarrow \sim x\epsilon x])$

Let us notice that (d) immediately implies $\exists x \sim x\epsilon x$. Conversely, if
$\exists x \sim x\epsilon x$ holds, then there is at least one set in the totality X of sets
that are not members of themselves, and X witnesses the truth of
(d). Thus (d) turns out to be equivalent to $\exists x \sim x\epsilon x$, the symbolization of an obvious truth concerning sets.

(The worry over Russell's paradox which the reader may be ex-
periencing at this point may be dispelled by the observation that
logical equivalence is a model-theoretic notion, the "sets" just referred to may be taken to be elements of the domain of an arbitrary
model, and the "totalities," subsets of the domain of the model.)

In view of the near-vacuity of (b) and the fact that instances of
the second-order comprehension schema $\exists X\, \forall x[Xx \leftrightarrow A(x)]$, including (e):

(e) $\exists X\, \forall x[Xx \leftrightarrow \sim x\epsilon x]$

are logically valid under the standard semantics for second-order
logic, the collapse of (d) is not at all surprising. The rendering (d)
of (b) is considerably more faithful to the semantic structure of (b)
than is $\exists x \sim x\epsilon x$, however, and (b) is more nearly synonymous with
(d) than with $\exists x \sim x\epsilon x$.

But can we use (c) or (d) to represent (b) at all? May we use second-order formulas like (c), (d), or (e) to make assertions about *all
sets?*

Let's consider (e), which is slightly simpler than (c) or (d). (e)
would appear to say that there is a totality or collection X containing all and only those sets x which are not members of themselves.
Are we not here on the brink of a well-known abyss? Does not acceptance of the valid (e), understood as quantifying over all sets
(with ε taken to have its usual meaning), commit us to the exis-

tence of a set whose members are all and only those sets which are not members of themselves?

There are a number of ways out of this difficulty. One way, which I no longer favor, is to regard it as illegitimate to use a second-order formula when the objects over which the individual variables in the formula range do not form a set (just as it is illegitimate to use a first-order formula when there are *no* objects over which they range).[13] This stipulation keeps all instances of the comprehension principle as logical truths; it also enables one always to read the formula Xx as meaning that x is a *member* of the set X.

The principal drawback of this way out is that there are certain assertions about sets that we wish to make, which certainly cannot be made by means of a first-order formula—perhaps to claim that there is a "totality" or "collection" containing all and only the sets that do not contain themselves is to attempt to make one of these assertions—but which, it appears, could be expressed by means of a second-order formula if only it were permissible so to express them. To declare it illegitimate to use second-order formulas in discourse about all sets deprives second-order logic of its utility in an area in which it might have been expected to be of considerable value.

For example, the principle of set-theoretic induction and the separation (Aussonderung) principle virtually cry out for second-order formulation, as:

(f) $$\forall X (\exists x \, Xx \rightarrow \exists x [Xx \,\&\, \forall y (y \epsilon x \rightarrow \sim Xy)])$$

and

(g) $$\forall X \, \forall z \, \exists y \, \forall x (x \epsilon y \leftrightarrow [x \epsilon z \,\&\, Xx])$$

respectively. It is, I think, clear that our decision to rest content with a set theory formulated in the first-order predicate calculus with identity, in which (f) and (g) are not even well-formed, must be regarded as a compromise, as falling short of saying all that we might hope to say. Whatever our reasons for adopting Zermelo-Fraenkel set theory in its usual formulation may be, we accept this theory because we accept a stronger theory consisting of a *finite* number of principles, among them some for whose complete expression second-order formulas are required.[14] We ought to be able to formulate a theory that reflects our beliefs.

[13] I took this view in "On Second-order Logic," this JOURNAL, LXXII, 16 (Sept. 18, 1975): 509–527.

[14] Cf. the remarks about "full expression" and "part of the content" of various notions in my "The Iterative Conception of Set," this JOURNAL, LXVIII, 8 (April 22, 1971): 215–231.

We of course also wish to *maintain* such second-claims as are made by e.g., $\exists X \forall x[Xx \leftrightarrow \sim x\epsilon x]$; if we are to utilize second-order logic in discourse about all sets, these comprehension principles must remain among the asserted statements. Nor do we want to take the second-order variables as ranging over some set-like objects, sometimes called "classes," which have members, but are not themselves members of other sets, supposedly because they are "too big" to be sets. Set theory is supposed to be a theory about *all* set-like objects.

How then can we legitimately maintain that such (closed) formulas as $\exists X \forall x[Xx \leftrightarrow \sim x\epsilon x]$, (f), and (g) express truths, without introducing classes (set-like non-sets) into set theory and without assuming that the individual variables do not in fact range over all the sets there really are?

There is a simple answer. Abandon, if one ever had it, the idea that use of plural forms must always be understood to commit one to the existence of sets (or "classes," "collections," or "totalities") of those things to which the corresponding singular forms apply. The idea is untenable in general in any event: There are some sets of which every set that is not a member of itself is one, but there is no set of which every set that is not a member of itself is a member, as the reader, understanding English and knowing some set theory, is doubtless prepared to agree. Then, using the plural forms that are available in one's mother tongue, translate the formulas into that tongue and see that the resulting English (or whatever) sentences express true statements. The sentences that arise in this way will lack the trenchancy of memorable aphorisms, but they will be proper sentences of English which, with a modicum of difficulty, can be understood and seen to say something true.

Applying this suggestion to:

(h) $\sim \exists X (\exists x\, Xx \,\&\, \forall x[Xx \to (x\epsilon x \lor \exists y[y\epsilon x \,\&\, Xy \,\&\, y \neq x])])$

which is equivalent to (f), we might obtain:

   (i) It is not the case that there are some sets each of which either contains itself or contains at least one of the others.

From Aussonderung we might perhaps get:

   (j) It is not the case that there are some sets that are such that it is not the case that for any set z there is a set y such that for any set x, x is a member of y if and only if x is a member of z and also one of them.

or, far more perspicuously,

(k) ~ there are some sets such that

$$\sim \forall z \exists y \forall x [x \epsilon y \leftrightarrow (x \epsilon z \ \& \ x \text{ is one of them})]$$

(k) is of course neither an English sentence nor a wff of any repu-
table formalism—for that matter neither is (j), which contains the
(non-English) variables x, y, and z—but is readily understood by
anyone who understands both English and the first-order language
of set theory. It would be somewhat laborious to produce a fully
Englished version of (g), but the labor involved would be mainly
due to the sequence $\forall x \exists y \forall x$ of *first-order* quantifiers that (g) contains. (j) and (k) are actually not quite right; properly they have the
meaning:

(l) $\sim \exists X (\exists x \, Xx \ \& \sim \forall z \exists y \forall x [x \epsilon y \leftrightarrow (x \epsilon z \ \& \ Xx)])$

whereas the full Aussonderung principle omits the nonemptiness
condition $\exists x \, Xx$; to get the full content in English of Aussonderung, however, we need only conjoin "and there is a set with no
members" to (j) and $\exists y \forall x \sim x \epsilon y$ to (k). This observation calls to
our attention two small matters connected with plurals which must
be taken up sooner or later.

Suppose that there is exactly one Cheerio in the bowl before me.
Is it true to say that there are some Cheerios in the bowl? My view
is no, not really, I guess not, but say what you like, it doesn't matter very much. Throughout this paper I have made the customary
logician's assumption, which eliminates needless verbiage, that the
use of plural forms does not commit one to the existence of two or
more things of the kind in question.

On the side of literalness, however, I have assumed that use of
such phrases as "some gunslingers" in "There are some gunslingers each of whom has either shot his own right foot or shot the
right foot of at least one of the others" does commit one to—as one
might say—a *nonempty* class of gunslingers, but not to a class containing two or more of them. Thus I suppose the sentence to be
true in case there is exactly one gunslinger, who has shot his own
right foot, but to be false if there are aren't any gunslingers. It is
this second assumption that is responsible for the ubiquitous
$\exists x \, Xx$ in the formulas above.

Translation will be difficult from any logical formalism into a
language such as English, which lacks a large set of devices for expressing cross reference. And since plural pronouns like 'them', although sometimes used as English analogues of second-order vari-

ables, much more frequently do the work of individual variables, translation from a second-order formalism containing infinitely many variables of both sorts into idiomatic, flowing, and easily understood English will be impossible nearly all of the time. My present point is that, in the cases of interest to us, the things we would like to say can be said, if not with Austinian or Austenian grace.

It is, moreover, clear that if English were augmented with various subscripted pronouns, such as 'it$_x$', 'that$_x$', 'it$_y$', . . . , 'them$_x$', 'that$_x$', 'them$_Y$', . . . , then any second-order formula[15] whose individual variables are understood to range over all sets could be translated into the augmented language, as follows: Translate Vv as 'it$_v$ is one of them$_V$', $v \varepsilon v'$ as 'it$_v$ is a member of it$_{v'}$', $v = v'$ as 'it$_v$ is identical with it$_{v'}$', $\&$ as 'and', $\sim$ as 'not', and, where F^* is the translation of F, translate $\exists v F$ as 'there is a set that$_v$ is such that F^*'.

The clause for formulas $\exists VF$ is not quite so straightforward, because of the difficulty about nonemptiness mentioned above. It runs as follows: Let F^* be the translation of F, and let F^{**} be the translation of the result of substituting an occurrence of $\sim v = v$ for each occurrence of Vv in F. Then translate $\exists VF$ as 'either there are some sets that$_V$ are such that F^*, or F^{**}'.

For example, $(Xx \leftrightarrow \sim x \varepsilon x)$ comes out as "It$_x$ is one of them$_x$ iff it$_x$ is not a member of itself"; $\forall x(Xx \leftrightarrow \sim x \varepsilon x)$, as "Every set is such that it is one of them$_x$ iff it is not a member of itself"; and $\exists X \forall x(Xx \leftrightarrow \sim x \varepsilon x)$, as "Either there are some sets that are such that every set is one of them iff it is not a member of itself or every set is a member of itself." (We have, of course, improved the translations as we went along.)

I want to emphasize that the addition to English of operators 'it$_0$', 'that$_0$', 'them$_0$', etc. or variables 'x', 'X', 'y', etc. is not contemplated here. The 'x' of 'it$_x$' is not a variable but an index, analogous to 'latter' in 'the latter', or 'seventeen' in 'party of the seventeenth part'; 'X' and 'x' in 'them$_X$' and 'it$_x$' no more have *ranges* or *domains* that does '17' in 'x_{17}'. We could just as well have translated the language of second-order set theory into an English augmented with pronouns such as 'it$_{17}$', 'them$_{1879}$', etc. or an elaboration of the "former"/"latter" usage. Note also that such augmentation will be needed for the translation into English of the language of *first*-order set theory as well.

---

[15] We assume that no quantifier in any formula occurs vacuously or in the scope of another quantifier with the same variable; every formula is equivalent to some formula satisfying this condition.

Charles Parsons has pointed out to me that although second-order existential quantifiers can be rendered in the manner we have described, it is curious that there appears to be no nonartificial way to translate second-order *universal* quantifiers, that the translation of $\forall X$ must be given indirectly, via its equivalence with $\sim \exists X \sim$. Because our translation "manual" relies so heavily on the phrases 'there is a [singular count noun] that is such that . . . it . . .' and 'there are some [plural count noun] that are such that . . . they . . .', the logical grammar of the construction these phrases exemplify is worth looking at.

Of course, in ordinary speech, the construction 'that is/are such that . . . it/they . . .' is almost certain to be eliminable: the content of a sentence containing it can nearly always be conveyed in a much shorter sentence. But the difference between the two 'that's bears notice. The second one, following 'such', is a 'that' like the one found in oblique contexts and may be—as Donald Davidson has suggested that the 'that' of indirect discourse is—a kind of demonstrative, used on an occasion to point to a subsequent utterance of an (open) sentence; the first 'that', following the count noun and more frequently elided than the second, is no demonstrative, but a relative pronoun used to bind the 'it' or 'they' in the open sentence after 'such that'. Thus the first but not the second 'that' works rather like the variable immediately following an $\exists$, binding occurrences of that same variable in a subsequent open formula. Whether the preceding count noun is singular or plural appears to make no difference to the quantificational role of the first 'that'; as we have observed, 'that' is not marked for number and can serve to bind either 'it' or 'they'.

Whether any such second-order formula of the sort we have been considering can be translated into intelligible *un*augmented English is not an interesting question, and I shall leave it unanswered. Since English augmented in the manner I have described is intelligible to any native speaker who understands the term of art 'party of the seventeenth part', I shall assume that devices like 'it$_x$' and 'them$_X$' are available in the language we use.

I take it, then, that there is a coherent and intelligible way of interpreting such second-order formulas as (e), (f), and (g) even when the first-order variables in these formulas are construed as ranging over all the sets or set-like objects there are. The interpretation is given by translating them into the language we speak; the translations of (e), (f), and (g) are sentences we understand; and we can see that they express statements that we regard as true: after all, we do think it false that there are some sets each of which either contains itself or contains one of the others, and, once we cut through the verbiage, we do find it trivial that there are some sets none of which is a member of itself and of which each set that is not a member of itself is one. It cannot seriously be maintained that we do not *understand* these statements (unless of course we really *don't* understand them, as we wouldn't if, e.g., we knew nothing at all about set theory) or that any lack of clarity that attaches to them has anything to do with the plural forms found in the sentences expressing them. The language in which we think and speak provides the constructions and turns of phrase by means of which the meanings of these formulas may be explained in a completely intelligible way.

It may be suggested that sentences like (i) are intelligible, but only because we antecedently understand statements about collections, totalities, or sets, and that these sentences are to be analyzed as claims about the existence of certain collections, etc. Thus "There are some gunslingers . . ." is to be analyzed as the claim that there is a collection of gunslingers. . . .[16] The suggestion may arise from the thought that any precise and adequate semantics for natural language must be interpretable in set theory (with individuals). How else, one may wonder, is one to give an account of the semantics of plurals?

One should not confuse the question whether certain sentences of our language containing plurals are intelligible with the question whether one can give a semantic theory for those sentences. In view of the work of Tarski, it should not automatically be expected that we can give an adequate semantics for English—whatever that might be—in English. Nothing whatever about the intelligibility of those sentences would follow from the fact that a systematic semantics for them cannot be given in set theory. After all, the semantics of the language of ZF itself cannot be given in ZF.

In any event, as we have noticed, there are certain sentences that cannot be analyzed as expressing statements about collections in the manner suggested, e.g., "There are some sets that are self-identical, and every set that is not a member of itself is one of them."

---

[16] In a similar vein, Lauri Carlson writes, "I take such observations as a sufficient motivation for construing *all* plural quantifier phrases as quantifiers over arbitrary *sets* [Italics Carlson's] of those objects which form the range of the corresponding singular quantifier phrases." His "Plural Quantifiers and Informational Independence," *Acta Philosophica Fennica*, xxxv (1982): 163–174, is a recent interesting article in which this claim is made once again. He is by no means the sole linguist with this belief. Carlson does not face the question of what is to be done when the corresponding singular quantifier phrase is 'some set'.

That sentence says something trivially true; but the sentence "There is a collection of sets that are self-identical, and every set that is not a member of itself is a member of this collection," which is supposed to make its meaning explicit, says something false.

I want now to consider the claim that a sentence of English like "There are some sets of which every set that is not a member of itself is one" is actually false, on the ground that this sentence *does* entail the existence of an overly large set, one that contains all sets that are not members of themselves.

The claim that this sentence entails the existence of this large set strikes me as most implausible: there may be a set containing all trucks, but that there is certainly doesn't seem to *follow* from the truth of "There are some trucks of which every truck is one." Moreover, and more importantly, the claim conflicts with a strong intuition, which I for one am loath to abandon, about the meaning of English sentences of the form "There are some *A*s of which every *B* is one," viz. that any sentence of this form means the same thing as the corresponding sentence of the form "There are some *A*s and every *B* is an *A*." If so, the sentence of the previous paragraph is simply synonymous with the trivial truth "There are some sets and every set that is not a member of itself is a set," and therefore does not entail the existence of an overly large set.

Two worries of a different kind are that the construction "there are some [plural count noun] that are such that . . . they . . .' is unintelligible if the individuals in question do not form a "surveyable" set and that our understanding of this construction does not justify acceptance of full comprehension. I cannot deal with these worries here; I shall only remark that it seems likely that not much of ordinary, first-order, set theory would survive should either worry prove correct.

We have now arrived at the following view: Second-order formulas in which the individual variables are taken as ranging over all sets can be intelligibly interpreted by means of constructions available to us in a language we already understand; these constructions do not themselves need to be understood as quantifying over any sort of "big" objects which have members and which "would be" sets "but for" their size. There can thus be no objection on the score of unintelligibility or of the introduction of unwanted objects to our regarding ZF as more suitably formulated as a finitely axiomatized second-order theory than as an infinitely axiomatized first-order theory, whose axioms are the instances of a finite number of schemata, as is usual. (Of course, in the presence of the usual other first-order axioms of ZF, i.e., the axioms of extensionality, foundation, pairing, power set, union, infinity, and choice, only the one second-order axiom, Replacement:

$$\forall X (\forall x \forall y \forall z [X \langle x,y \rangle \mathbin{\&} X \langle x,z \rangle \to y = z]$$
$$\to \forall u \exists v \forall y [y \epsilon v \leftrightarrow \exists x (x \epsilon u \mathbin{\&} X \langle x,y \rangle)])$$

would be needed.) The great virtue of such a second-order formulation of ZF is that it would permit us to express as single sentences and take as axioms of the theory certain general principles that we actually believe. The underlying logic of such a formulation would be any standard axiomatic system of second-order logic, e.g., the system indicated, if not given with perfect precision, in Frege's *Begriffsschrift*.[17] The logic would deliver the comprehension principles $\exists X \forall x [Xx \leftrightarrow A(x)]$ (which are needed for the derivation of the infinitely many axioms of the first-order version of ZF from the finitely many second-order axioms) either through explicit postulation of the comprehension schema, as in Joel Robbin's *Mathematical Logic*,[18] or via a rule of substitution, like the rule given in chapter 5 of Alonzo Church's *Introduction*[19] or the one implicit in the *Begriffsschrift*. The interpretation of this version of ZF would be given in a manner similar to that in which the interpretation of the usual formulation of ZF is given, by translation into English in the manner previously described.

Entities are not to be multiplied beyond necessity. One might doubt, for example, that there is such a thing as the set of Cheerios in the (other) bowl on the table. There are, of course, quite a lot of Cheerios in that bowl, well over two hundred of them. But is there, in addition to the Cheerios, also a set of them all? And what about the $>10^{60}$ subsets of that set? (And don't forget the sets of sets of Cheerios in the bowl.) It is haywire to think that when you have some Cheerios, you are eating a *set*—what you're doing is: eating THE CHEERIOS. Maybe there are some reasons for thinking there is such a set—there are, after all, $>10^{60}$ ways to divide the Cheerios into two portions—but it doesn't follow just from the fact that there are some Cheerios in the bowl that, as some who theorize

---

[17] Gottlob Frege (Hildesheim: Georg Olms Verlagsbuchhandlung, 1964).

[18] *Mathematical Logic: A First Course* (New York: W. A. Benjamin, 1969). Section 56 of Robbin's book contains a presentation of the version of set theory here advocated. It is noted there that this theory is "essentially the same as" Morse-Kelley set theory (MK), but the difficulties of interpretation faced either by MK or by a set theory in the ZF family for which the underlying logic is (axiomatic) second-order logic are not discussed.

[19] *Introduction to Mathematical Logic* (Princeton, N.J.: University Press, 1956).

about the semantics of plurals would have it, there is also a set of them all.

The lesson to be drawn from the foregoing reflections on plurals and second-order logic is that neither the use of plurals nor the employment of second-order logic commits us to the existence of extra items beyond those to which we are already committed. We need not construe second-order quantifiers as ranging over anything other than the objects over which our first-order quantifiers range, and, in the absence of other reasons for thinking so, we need not think that there are collections of (say) Cheerios, in addition to the Cheerios. Ontological commitment is carried by our *first*-order quantifiers; a second-order quantifier needn't be taken to be a kind of first-order quantifier in disguise, having items of a special kind, collections, in its range. It is not as though there were two sorts of things in the world, individuals, and collections of them, which our first- and second-order variables, respectively, range over and which our singular and plural forms, respectively, denote. There are, rather, two (at least) different ways of referring to the same things, among which there may well be many, many collections.

Leibniz once said, "Whatever is, is one."

Russell replied, "And whatever are, are many."[20]

GEORGE BOOLOS

Massachusetts Institute of Technology

[16]

The Philosophical Review, XCIV, No. 3 (July 1985)

NOMINALIST PLATONISM[1]

George Boolos

Frege's definition of "x is an ancestor of y" is: x is in every class that contains y's parents and also contains the parents of any member. A philosopher whom I shall call N. once asked me, "Do you mean to say that because I believe that Napoleon was not one of my ancestors, I am committed to such philosophically dubious entities as *classes?*" Although it is certain that Frege's definition, whose logical utility, fruitfulness, and interest have been established beyond doubt, cannot be dismissed for such an utterly crazy reason, it is not at all easy to see what a good answer to N.'s question might be.

The germ of an answer may lie in the observation that there are sentences containing *plural* forms such as "are" and "them" whose logical forms look as though they ought to be representable in first-order logic, but which cannot be so represented, because of the plural forms the sentences contain. An example is "There are some horses all of which are faster than Zev and all of which are faster than the sire of any horse that is slower than all of them."[2] In

---

[1] I want to thank Martin Davies, Michael Dummett, Harold Hodes, David Lewis, John McDowell, Robert Stalnaker, Linda Wetzel, and the referee for helpful comments. This paper was written while I was on a Fellowship for Independent Study and Research from the National Endowment for the Humanities.

[2] G. Boolos, "Nonfirstorderizability again," *Linguistic Inquiry* 15 (1984), p. 343. To see that this sentence cannot be symbolized by a first-order formula, first substitute "number," "greater," "zero," "successor," and "slower" for "horse," "faster," "Zev," "sire," and "smaller," obtaining "There are some numbers all of which are greater than zero and all of which are also greater than the successor of any number that is smaller than all of them," and then notice that any correct symbolization of the latter sentence (using a constant 0 for "zero," a function sign s for "successor," and relation letters > and < for "greater than" and "less than") is a sentence true in a model M of the set of all first-order truths of arithmetic if and only if M is non-standard, that is, not isomorphic to the standard model of arithmetic. Since non-standard models exist, and since (trivially) every first-order sentence has the same truth-value in any model of the set of all first-order truths of arithmetic as in any other, no such correct symbolization can be a first-order sentence. For an account of non-standard models of arithmetic, see for example, chapter 17 of G. Boolos and R. Jeffrey, *Computability and Logic*, second edition, Cambridge University Press, 1985.

GEORGE BOOLOS

contrast, the sentence "There is a horse that is faster than Zev and that is faster than the sire of any horse that is slower than it," which differs from the other in containing singular forms in place of plural, can be symbolized in the notation of first-order logic:

$$\exists x(x>0 \ \& \ \forall y(y<x \rightarrow x>sy)).$$

If "some" is understood to mean "one or more," this sentence is *stronger* than the former: the horses of whose existence the former informs us are guaranteed to be faster than the sire of y only if y is slower than *all* of them; unlike the latter, it does not imply that there is any one horse x that is faster than the sire of y whenever y is slower than x.

   Geach and Kaplan gave an earlier example of a nonfirstorderizable sentence, that is, a sentence not expressible in first-order notation, containing no numerical or quasi-numerical words like "more" or "most": some critics admire only one another. This sentence is supposed to mean: there is a non-empty class of critics, each of whose members admires someone only if that person is someone else in the class. The meaning can also be put: there are some critics each of whom admires a person only if that person is one of them and none of whom admires himself. If we explain the meaning of the Geach-Kaplan sentence in this second way, we do not, it appears, quantify over classes of critics. We can also put the meaning: there are some critics who are such that (a) each of them admires a person only if he is one of them and (b) none of them admires himself.

   Here then is an answer to N.'s question: in a similar vein, we may say that Napoleon is not an ancestor of N. because either no one is a parent of N. or there are some people who are such that (a) each of N.'s parents is one of them, (b) each parent of any one of them is also one of them, and (c) Napoleon is not one of them. The response that this definition commits one to such philosophically dubious entities as classes now seems wholly out of place; classes aren't mentioned anywhere in the paraphrase. By using plural forms in English quantifiers and employing the construction "one of them" we are able to define *ancestor of* in a way that preserves the essence of Frege's idea and, *at least at first blush,* avoids commitment to such "philosophically dubious entities" as classes: x is an ancestor

NOMINALIST PLATONISM

of y if and only if (I) someone is a parent of y and (II) it is not the case that there are some people who are such that (a) each parent of y is one of them, (b) each parent of any one of them is also one of them, and (c) x is not one of them.[3] This definition may be symbolized by means of a second-order formula:

$$[\exists w wPy \;\&\; -\exists X(\exists wXw \;\&\; \forall w(wPy \to Xw) \;\&$$
$$\forall w \forall z(wPz \;\&\; Xz \to Xw) \;\&\; -Xx)],$$

which is equivalent to the familiar:

$$\forall X(\forall w(wPy \to Xw) \;\&\; \forall w \forall z(wPz \;\&\; Xz \to Xw) \to Xx)$$

The first question I want to discuss is whether the use of expressions like "there are some people who are such that . . . they . . . " or monadic second-order quantifiers like "$\exists X$" does in fact commit one to classes, despite appearances. The Geach-Kaplan sentence suggests that we consider the following case. Suppose that I assert that there are some critics, none of whom admires himself, and each of whom admires someone only if that person is one of those critics. Suppose further that I write down, with assertive intent:

$$\exists X(\exists xXx \;\&\; \forall x \forall y(Xx \;\&\; x \text{ admires } y \to x \neq y \;\&\; Xy)).$$
(The first-order variables are intended to range over all critics.)

In doing either of these things have I committed myself to the existence of a class, of critics, none of whom etc.?

Let us deal first with the formula. On the usual treatment of second-order formulae, I *would* have committed myself to the existence of such a class. The formula is normally read and understood to mean, "There is a non-empty class X of critics each of whose members x admires a person y only if x is other than y and y is in the class X." If that is what the formula means, then in writing it down with assertive intent, I would, I suppose, have committed myself to the existence of a class as thoroughly as I would have

---

[3]Since (II) is true if (I) is false and y alone exists, (I) is indispensable to the definition.

GEORGE BOOLOS

committed myself had I simply said "There is a non-empty class X etc." But suppose that the formula isn't to be understood as meaning "There is a non-empty class X etc." Does my writing it down then commit me to the existence of a class? The answer, obviously, is that whether or not it does depends on what the formula is supposed to mean: if the formula means something that does not commit one to classes, then it doesn't; if something that does, then it does.

Suppose now that it is said that the formula means: there are some critics, none of whom admires himself, and each of whom admires someone only if that person is one of those critics. On this understanding, interpretation, or reading of it, does the formula commit one to classes?

Before attempting to answer this question, let us note that there is a systematic way to utilize plural forms to translate into English all formulae of second-order logic in which the second-order variables are monadic.[4] Some features of this translation scheme are that it is an extension of the usual scheme for translating first-order formulae into English, and thus respects the propositional connectives, first-order quantifiers $\exists x$, and the equals-sign; that atomic formulae Xy are translated (more or less) as "it is one of them" and that second-order quantifiers $\exists X$ are translated (roughly) as "there are some objects that are such that . . . ." As in the translation of first-order formulae such as $\exists x\exists y\exists z(xLy\ \&\ yLz\ \&\ zLx)$, some devices like "the former," "the latter," "party of the third part," must be employed to do the cross-referencing done in a formal language by the identity and difference of variables; one convenient way to accomplish this is to introduce into English pronouns such as "it_x," "$them_Y$," "$that_z$," and "$that_Z$," to which variables of the formal language have been attached as subscripts. On this scheme, the translation of any instance of the comprehension principle of second-order logic is a truism. Thus the translation of the notorious $\exists X\forall x(Xx \leftrightarrow x$ is not a member of $x)$, where the first-order variables are taken to range over absolutely all sets is "(If there is a set that is not a member of itself, then) there are some sets that are such that

---

[4]In many of the most important applications of second-order logic, a pairing function will be available and monadic variables can then be made to do the work of all second-order variables.

NOMINALIST PLATONISM

each set that is not a member of itself is one of them and each set that is one of them is not a member of itself," as vacuous an assertion about sets as can be made, as desired. I have set out the details of the translation scheme elsewhere and will not repeat them here.[5] On this scheme, "There are some critics, none of whom etc." turns out to be an abbreviation of the translation of the formula: $\exists X(\exists x Xx \ \& \ \forall x \forall y(Xx \ \& \ x \text{ admires } y \rightarrow x \neq y \ \& \ Xy))$.

We are thus thrown back to answering the first question: does asserting "There are some critics, none of whom etc." commit one to the existence of a non-empty class of critics? The difficulty with this question is that the ground rules for answering it appear to have been laid down, by Professor Quine.

According to Quine, to determine whether or not "There are some critics etc." commits us to classes, we translate it into logical notation, and then see whether the variables contained in the translation must be supposed to range over classes, to have classes as their values. Now the Geach-Kaplan sentence "There are some critics etc." can be translated into standard logical notation, that is, the notation of *first*-order logic, only if one introduces special variables ranging over classes (or properties or other "dubious" entities). And then the sentence will be translated:

$\exists a(a$ is a class $\& \ \exists x x \in a \ \& \ \forall x \forall y(x \in a \ \& \ x \text{ admires } y \rightarrow x \neq y \ \& \ y \in a))$.

Since this sentence cannot be true unless there is a suitable class to assign as value to the variable a, it follows, according to Quine, that assertively uttering the Geach-Kaplan sentence commits one to the existence of classes.

But this is a weird outcome; that was N.'s point. It shouldn't turn out that I'm committed to classes if I state that there are some critics etc. What I ought to be committed to is *some critics*, but not to a class of critics. Furthermore, I would have thought, I ought not to be committed, on any reasonable sense of the word "committed," to a class containing all infinite classes if I say, "There are some

---

[5]See my "To be is to be a value of a variable (or to be some values of some variables)," *The Journal of Philosophy* 81 (1984), pp. 430–449 or "Reading the *Begriffsschrift*," to appear in *Mind*.

GEORGE BOOLOS

classes such that every infinite class is one of them," which is, as I
suppose, only an awkward way of saying "There are some classes
and every infinite class is one of them" or "There are some classes
and every infinite class is a class."[6]

Our problem arises from the thought that if we wish to assess the
commitment of a theory, we must first put it into first-order notation as well as we can (sometimes this will not be possible) and then
determine what the variables must be assumed to range over.
There are two suggestions we should resist at this point: that if we
are concerned with the *ontological* commitment of a theory couched
in some natural language, we must first translate it into a *first-order*
formalism and that we must suppose that second-order variables in
a formula must range over, or have as values, classes of objects over
which the first-order variables of that formula range, or have as
values.

With regard to the first suggestion, we want to ask: what does
translation into a first-order language have to do with "ontological
commitment?" "There are some critics etc." doesn't, it seems, commit us, in any ordinary sense of the word "commit," to the existence of a class of critics; what it commits us to, one would have
thought, is, as we have noted, some critics none of whom etc. We
are forced by Quine's criterion to say that it commits us to a class;
but why, we should ask, should we accept the criterion? If it is
answered: because it's Quine's phrase, and he is at liberty to define
it as he pleases, then we should rejoin: if so, Quine is defining a
relation which holds between us and certain objects (in this case,
classes) whose existence the normal use of our words does not force
us to admit, and hence a relation that ought not to be called "on-

---

[6]It is sometimes alleged that there are certain set-like objects, which have
elements, but which are "too big" to be sets; the term "class" is used in set
theory to apply to each such gigantic element-container, as well as to each
set. A *proper class* is a class that is not a set. Every current theory admitting
the existence of proper classes denies that there is a class that contains all
infinite classes. If the existence of proper classes is denied, then "class" and
"set" become coextensive, but every current set theory denies that there is
a set containing all infinite sets. Of course, class theories typically imply the
claim that there is a class containing all infinite sets. On the distinction
between sets and classes, see Charles Parsons' "Sets and Classes," in *Nôus* 8
(1974), pp. 1–12, and reprinted in his *Mathematics in Philosophy: Selected
Essays*, Cornell University Press, 1983, pp. 209–220.

NOMINALIST PLATONISM

tological commitment." If it is said that by admitting that there are some critics etc., we are *ipso facto* committed to a class, we must ask how this is supposed to have been shown.

We ought to recall that logicians have devoted attention to quantifiers other than the usual *for all x* and *for some x*. Among these less familiar quantifiers, which cannot be defined by means of the apparatus of first-order logic, are *for most x, for infinitely many x, for uncountably many x*, and *for at least as many x as there are objects (in the domain)*. To claim that a statement to the effect that there are infinitely many objects of a certain kind, made with the aid of the quantifier *for infinitely many x, implies the existence* (on the customary acceptation of those words) of an (infinite) class solely on the ground that the only way to utilize more familiar logical vocabulary to eliminate the unfamiliar quantifier is to employ a quantifier ranging over classes is to invite the response: what makes first-order logic the touchstone by which the ontological or existential commitments of these statements are to be assessed? The statements do not appear to commit us to classes; why believe that it is their translation into the notation of first-order logic augmented with variables ranging over classes that determines what they are actually committed to?

In the case of quantifiers like *for infinitely many x* or *for most x*, it is comparatively easy to hold one's ground in maintaining that assertions involving them need not be taken as *committing* one to the existence of classes; the variable x is, after all, a first-order variable. To see that second-order quantifiers are analogous in this regard to the less familiar quantifiers containing first-order variables, we must rebut the second suggestion mentioned above, that in any formula, the second-order variables have to be understood as ranging over (or having as values) classes of objects over which its first-order variables range.

This suggestion is less easily rebutted. The difficulty was well put in a recent, highly interesting article by Harold Hodes.[7] Hodes writes, "Unless we posit such further entities [as Fregean concepts], second-order variables are without values, and quantificational expressions binding such variables can't be interpreted referen-

---

[7] Harold Hodes, "Logicism and the ontological commitments of arithmetic," *The Journal of Philosophy* 81 (1984), pp. 123–149.

GEORGE BOOLOS

tially."[8] It will become clear that I disagree with this claim. We needn't posit concepts, classes, Cantorian inconsistent totalities, etc. in order to interpret second-order quantification referentially.

The heart of the matter is this: it is only with respect to a truth-"definition" of the standard sort, a Tarski-style truth-theory for a first-order language, that the notion *value of a variable* is defined. In the case of a second-order language, such as the second-order language of set theory, there are at least two different sorts of truth-theory that can be given: on one of these, it would be quite natural to define "value" so that the second-order variables would turn out to have classes as values; on the other, it would not. Before we present the two theories, we need to review the usual truth-definition for the usual first-order language of set theory. (It is because the difficulty of interpreting second-order quantification is most acute when the underlying language is the language of set theory that we examine truth-theories for set-theoretic languages.)

A *sequence* is a function that assigns a set to each first-order variable of the language; we inductively define satisfaction of a formula F by a sequence s as follows:

> If F is u∈v, then s satisfies F iff s(u) ∈ s(v);
> if F is u=v, then s satisfies F iff s(u) = s(v);
> if F is −G, then s satisfies F iff −(s satisfies G);
> if F is (G&H), then s satisfies F iff (s satisfies G & s satisfies H);
> if F is ∃vG, then s satisfies G iff
>   ∃x ∃t(t is a sequence & t(v) = x &
>   ∀u(u is a variable & u≠v → t(u) = s(u)) & t satisfies G).

Having given this definition, we may prove a lemma stating that if s and t are sequences that assign the same sets to the free variables of a formula F, then s satisfies F if and only if t satisfies F. Since a sentence contains no free variables, it follows from the lemma that if one sequence satisfies a sentence, all do. We may thus define truth as satisfaction by all, or by some, sequences. Finally we

---

[8]Ibid. p. 130.

NOMINALIST PLATONISM

may demonstrate that the Tarski biconditionals are provable from this definition, with the aid of a small amount of set theory.

The truth-theory provides an obvious way to define the notion *value of a variable*. We may say, simply, that x is the value of the variable v relative to the sequence s if and only if s(v) = x. Notice that we wish to give a *relative* definition. And we may say that x is a *value* of v if x is a value of v relative to some sequence.

There is an obvious way to extend this development to the second-order case. The resulting theory is the first of the two theories mentioned earlier. We define a sequence to be a function from the set of first- and second-order variables whose value for each first-order variable v as argument is a set and whose value for each second-order variable V as argument is a class (the existence of suitable sequences will of course have to be guaranteed by principles not available to us in standard set theory). We then inductively define satisfaction of a formula F by a sequence s as follows:

> If F is u∈v, then s satisfies F iff s(u) ∈ s(v);
> if F is u=v, then s satisfies F iff s(u) = s(v);
> if F is Vv, then s satisfies F iff s(v) ∈ s(V);
> if F is −G, then s satisfies F iff −(s satisfies G);
> if F is (G&H), then s satisfies F iff (s satisfies G & s satisfies H);
> if F is ∃vG, then s satisfies F iff
>> ∃x[x is a set & ∃ι(ι is a sequence & ι(v) = x &
>> ∀u(u is a variable & u≠v → ι(u) = s(u)) & ι satisfies G)];
>
> if F is ∃VG, then s satisfies F iff
>> ∃x[x is a class & ∃ι(ι is a sequence & ι(V) = x &
>> ∀u(u is a variable & u≠V → ι(u) = s(u)) & ι satisfies G)].

As before we may define "x is the value of the (first- or second-) order variable v relative to the sequence s" as s(v) = x, and define "x is a value of v" as x is the value of v relative to some sequence. Thus if we have this sort of truth-definition in mind, we may say, speaking informally, that in the second-order language of set theory, classes are values of second-order variables.

But there is another sort of truth-theory that can be given for the second-order language of set theory in which no mention is made

335

GEORGE BOOLOS

of sequences any of whose values are (proper) classes. Unlike the previous theories, this theory is formulated in a second-order language, the second-order language of set theory together with a new predicate containing two first-order variables "s" and "F" and one second-order variable "R": R and the sequence s satisfy the formula F. In this new theory a sequence is what it was in the case of the first theory, a function from the set of first-order variables whose values are all sets. The key clauses of the theory are:

If F is u∈v, then R and s satisfy F iff $s(u) \in s(v)$;
if F is u=v, then R and s satisfy F iff $s(u) = s(v)$;
if F is Vv, then R and s satisfy F iff $R\langle V,s(v)\rangle$;
if F is −G, then R and s satisfy F iff −(R and s satisfy G);
if F is (G&H), then R and s satisfy F iff (R and s satisfy G &
                                                    R and s satisfy H);

if F is ∃vG, then R and s satisfy F iff
  ∃x∃ι (ι is a sequence & $ι(v) = x$ &
  ∀u(u is a first-order variable & $u{\neq}v \to ι(u) = s(u)$) &
  R and ι satisfy G);

if F is ∃VG, then R and s satisfy F iff
  ∃X∃T(∀x(Xx ↔ $T\langle V,x\rangle$) &
  ∀U(U is a second-order variable & $U{\neq}V \to \forall x(T\langle U,x\rangle \leftrightarrow R\langle U,x\rangle)$) & T and s satisfy G).

("$\langle,\rangle$" is the ordered-pair function sign.)

In this theory it is reasonable both to define "x is a value of the first-order variable v with respect to the sequence s" as $s(v) = x$ and to say that sets are values of the first-order variables of the second-order language of set theory, since the sequences s mentioned in the new predicate "R and s satisfy F" are functions whose values are all sets. The present theory, however, makes no explicit mention of sequences whose values are (proper) classes. It does not proceed by introducing functions that assign to each second-order variable a unique class, possibly proper. Instead it employs a new predicate which, as one may say, is true or false relative to an assignment of a formula to the first-order variable F, a sequence to the first-order variable s, and some (or perhaps no) ordered pairs of second-order

NOMINALIST PLATONISM

variables and sets to the second-order variable R. There is, however-
er, no need to take the theory as assigning classes, or collections, of
those sets, to the second-order variables. Of course one might attempt to argue for the claim that classes are values of the second-
order variables of the original language, even according to the
present truth-theory, by claiming that they are also values of the
second-order variables of that theory. That claim, however, is also
one that we *needn't* suppose to be true; we need't interpret either
our original second-order language of set theory or the new truth-
theory for this language in this manner. Friends of classes will insist
that our latest theory may be so reinterpreted and will (reasonably
enough) claim that so reinterpreted it is different from the second
theory only in an inessential way; but foes of (proper) classes—
those who believe that enough is enough, already—will reject the
second theory and accordingly resist the suggestion that the third
theory may be reinterpreted in the way that the friends suggest.
The point of our third truth-theory is to show that the foes of
classes have a satisfactory way to define truth for the second-order
language of set theory.

A foe of classes may also utilize the third theory to define "x is a
value of the second-order variable V with respect to R" as: R⟨V,x⟩.
If he does so, he may then say that sets are values of second-order
variables as well as of first-. On this way of speaking, it will not in
general be the case that if sets x and y are values of V with respect
to R, then x = y. But, the foe will then emphasize, if one adopts this
definition, to say that second-order variables have values is not at
all the same thing as to say that their values are classes (or concepts,
etc.) The foe will be at pains to reject the suggestion that in general
there is any one object whose members are all and only the sets x
such that R⟨V,x⟩. Alternatively, instead of defining "x is a value of
V with respect to R" at all, he could say that second-order variables
have no values, perhaps on the ground that the truth-theory makes
no mention of functions that assign objects to second-order
variables.

The liar paradox prevents us from explicitly defining truth, or
satisfaction, in either the first- or the second-order language itself.
But in the first-order case we can expand the language by adding a
single new primitive predicate for the notion of satisfaction and
then axiomatically characterize satisfaction for the old language by

GEORGE BOOLOS

adjoining to (a rather weak) set theory a finite number of axioms containing the new primitive. Truth of a sentence of the first-order language of set-theory may be defined in this theory, and important facts concerning satisfaction and truth can be deduced, including the Tarski biconditionals for satisfaction and truth and many "laws" of truth, for example, the statement that a conjunction is true if and only if both conjuncts are, etc. In setting up our third truth-theory we have proceeded in like manner: we have expanded the language for which we wish to define truth by adjoining to it a single new predicate, and laid down axioms containing this predicate, but have otherwise exceeded the resources of the language in no way. In particular, we have made no additional ontological assumptions not made in the original theory. And with the aid of standard axiomatic second-order logic (e.g., the system of the *Begriffsschrift*) we can prove in the third theory the usual lemmas about free variables, make the usual definition of "true sentence," and derive desired laws of truth.

One technical point deserves mention: in the first-order case, we need a guarantee that $\forall s \forall v \forall x (s$ is a sequence & v is a variable $\rightarrow \exists t(t$ is a sequence & $t(v) = x$ & $\forall u(u$ is a variable and $u \neq v \rightarrow t(u) = s(u))))$ in order to derive the consequences we desire from the axioms of the truth-theory. A small amount of set theory provides us with this guarantee. We need a like guarantee in the present case; we need to be able to show that $\forall R \forall V \forall X [V$ is a second-order variable $\rightarrow \exists T (\forall x [Xx \leftrightarrow T\langle V,x \rangle]$ & $\forall U[U$ is a second-order variable & $U \neq V \rightarrow \forall x (T\langle U,x \rangle \leftrightarrow R(U,x))])]$. This time the requisite guarantee is forthcoming, again with the aid of a small amount of set theory, from a comprehension principle which will be a theorem of any standard axiomatic system for second-order logic:

$$\exists T \forall x [Tz \leftrightarrow \exists x \exists U(U \text{ is a second-order variable } \& z = \langle U,x \rangle \&$$
$$(U = V \rightarrow Xx) \& (U \neq V \rightarrow Rz))].$$

A somewhat disconcerting conclusion emerges: it is not in general possible to tell by inspection of its asserted formulae alone whether or not classes are to be counted among the values of the variables of a theory formalized in a second-order language of the usual sort, even in the most favored case, in which the asserted formulae of the theory include instances of the comprehension

NOMINALIST PLATONISM

schema $\exists X \forall x(Xx \leftrightarrow A(x))$. Since one can neither presume that the formula 'Xx' must have the meaning 'x is a member of the class X' or that the quantifier '$\exists X$' is to be read 'there is a class X such that . . . ', one is not entitled to "read off" a commitment to classes from the asserted statements of such a theory. The reinterpretation of the quantifiers $\exists X$ and atomic formulae Xx given by the third truth-theory alters neither the interpretation of the apparatus of first-order logic nor the truth-values to be assigned to sentences of the original language, if the friend of classes is right after all. It is therefore no mere matter of a formalism put to a deviant use that one cannot for example, discern commitment to a universal class even when $\exists X \forall x Xx$ is one of the asserted statements of a theory. The possibility of so reinterpreting second-order notions shows that assessment of the ontological costs of a theory is rather less routine a matter than we may have supposed.

Having dealt at length with the truth-conditions of sentences of the second-order language of set theory, we should wish to give an account of the validity-conditions of these sentences.[9]

A sentence of the first-order predicate calculus is called (logically) *valid*, or a *logical truth*, if it is true in all models; a model M is an ordered pair of a non-empty set D and a function F from some set L of symbols to a set of relations and functions on D of appropriate degrees. D is the *domain* or *universe* of M, L is the *language* of M, and F assigns suitable denotations or references to the symbols of L. The aspect of this familiar definition of validity that here concerns us is the set-theoretic definability of the notion of validity, a consequence of the stipulation that D and L, and therefore M as well, be *sets*.

There is no universal set; there is no set of all pairs $\langle x,y \rangle$ such that x is in y; and there is no model $\langle D,F \rangle$ in which D is the universal set and F is a function from $\{\in\}$ to the set of all such pairs. At any rate, there are no such items if Zermelo-Fraenkel set theory is correct, as we shall henceforth assume.

---

[9]The classical discussion of validity and second-order logic is G. Kreisel's "Informal rigour and completeness proofs," in *Problems in the Philosophy of Mathematics*, ed. Imre Lakatos, North-Holland Publishing Company, Amsterdam, 1967, pp. 138–171. Stewart Shapiro's, "Principles of Logic and principles of reflection," *Journal of Symbolic Logic* 49 (1984), pp. 1446–1447 discusses some of the issues raised in Kreisel's paper.

GEORGE BOOLOS

From the nonexistence of such a model ⟨D,F⟩ there arises a certain difficulty: suppose that some sentence G of the language of set theory is logically valid, true in all models. What guarantee have we have G is *true*, that is, true when its variables are taken as ranging over all the sets there are and ∈ as applying to (arbitrary) x,y if and only if x is in y? If there were such a model ⟨D,F⟩, there would be no problem: G would then be true in ⟨D,F⟩ and therefore true period. It appears that in set theory at least, the truth of a statement does not immediately follow from its validity.

Set theory itself provides a way out of the difficulty. In fact, it provides two. The first is via the reflection principle: it is a theorem (-schema) of set theory that for each sentence G of the language of set theory there is a model M, indeed one of the form

$$\langle V_{\alpha}, \{\langle \in, \{\langle x,y \rangle : x,y \in V_{\alpha}\ \&\ x \in y\}\rangle\}\rangle$$

such that the sentence is true in M if and only if it is true. Thus if G is false, −G is true, hence −G is true in some model, and therefore G is not valid. Thus it cannot happen that a sentence of the language of set theory be valid and yet false.

The second way out is via the completeness theorem, according to which G has a proof in any standard axiomatic system of first-order logic if it is valid. Since (the universal closure of) any axiom of logic is true and the rules of inference preserve truth, any valid sentence of the language of set theory is true, since (the universal closure of) any sentence occurring in a proof is true.

But it is rather strange that appeal must apparently be made to one or another non-trivial result in order to establish what ought to be obvious: viz., that a sentence is true if it is valid.

I want to point out that in addition to the usual notion of validity, there is another notion of validity stronger than the usual one, susceptible, like the notions of truth and satisfaction, only of schematic definition, and on which it is obvious, as is fitting, that a valid sentence is true. Moreover, on this notion, the fact that whatever is valid is true is not much more than an effect of the rules of inference UI (universal instantiation) and substitution, as is also appropriate.

I shall call the notion to be defined *supervalidity*. (I do think that *it*

NOMINALIST PLATONISM

ought to be called validity and the usual notion ought to be called subvalidity. But never mind.)

The idea of supervalidity can be informally explained as follows: a sentence of the language of set theory is supervalid if it is true, no matter what sets its variables range over (as long as there is at least one set over which they range) and no matter what pairs of sets ∈ is taken to apply to.

I suspect that it is the mistaking of validity for supervalidity and the mathematical interest of validity, together with certain doubts about the intelligibility of supervalidity, that are responsible for the prominence of the notion of validity in logical theory.

Before defining supervalidity, I would like to mention a further odd feature of the concept of validity, or logical truth, viz., that a true sentence to the effect that another sentence is valid is not itself *valid*, but rather a true statement of set theory. Of course, in view of the celebrated "limitative" theorems of logic, the thought that we should *want* true assertions of validity to be valid may strike one as greedy, but one really should not lose the sense that it is somewhat peculiar that if G is a logical truth, then the statement that G is a logical truth does not count as a logical truth, but only as a set-theoretical truth.

The formal definition of supervalidity is this: let G be a sentence of the language of set theory. Select two monadic second-order variables X,Y. Replace all formulas u∈v in G by formulas Y⟨u,v⟩. Relativize all quantifiers ∀v and ∃v in the result to the formula Xv; that is, replace contexts ∀v(. . .) by ∀v(Xv → . . .) and contexts ∃v(. . .) by ∃v(Xv & . . .). Quantify universally with respect to Y. Take the result as the consequent of a conditional with antecedent ∃xXx. Finally, quantify this conditional universally with respect to X. The result is the formalization of the assertion that G is supervalid.

Thus we do not define "is supervalid" by constructing a formula with one free variable that applies to the (Gödel numbers of) supervalid sentences and only to these. It is instead defined schematically, by associating with each sentence G of the first-order language of set theory, another sentence, of the second-order language, that expresses the assertion that G is supervalid, as informally explained above. In a similar way, "is true" is schematically

341

GEORGE BOOLOS

defined by associating with each sentence G of the language the
sentence G itself, and not by constructing a single formula satisfied
by the (Gödel numbers of) true sentences and nothing else.

We have defined supervalidity only for sentences of the first-
order language {∈} of set theory, but it is clear how it may be done
for any formula of any first- *or second-* order language at all. We
shall confine attention to sentences in which the only non-logical
constant is ∈. We shall also suppose that the only non-individual
variables found in second-order formulae are *monadic*.

It is apparent that for any sentence G, the sentence expressing
the truth of G, that is, G itself, can be derived in axiomatic second-
order logic from the sentence G' asserting the supervalidity of G,
together with a suitable axiom governing the ordered pair opera-
tion $\lambda x,y \langle x,y \rangle$. (Mention of an ordered pair axiom could have been
omitted had we replaced ∈ with a *dyadic* variable Y.) One need only
instantiate ∀X with the abstract {x:x=x} in G', resolve the abstracts,
discharge the antecedent ∃x x = x via logic, instantiate ∀Y with
{z:∃x∃y(⟨x,y⟩ = z & x∈y)} and use the ordered pair axiom to resolve
the abstracts; the result is trivially equivalent to G. Instantiation
with abstracts is legitimated in second-order logic by the comprehension schema.

It is also apparent that the result of restricting all of the quantifiers of any supervalid sentence to some one formula (possibly
containing one or more parameters, for example, a parameter for
a model) satisfied (with respect to any assignment of objects to
those parameters) by at least one object (= set) is true (with respect
to that assignment). Consequently, any supervalid sentence is valid.

Finally, it should be apparent that the axioms of axiomatic
second-order logic, including the instances of the comprehension
schema ∃X∀x[Xx ↔ A], X not free in A, are all supervalid and that
the rules of inference of second-order logic, including the rule of
substitution (of formulas for free variables), preserve supervalidity.
Substitution is a rule of inference that was used by Frege, in the
Begriffsschrift; the deductive equivalence of the comprehension
schema and the rule of substitution is well known.[10] I have of
course been assuming the intelligiblity and legitimacy of second-

---

[10]Cf. "Reading the *Begriffsschrift,*" op. cit.

NOMINALIST PLATONISM

order quantification over all sets or over objects of unbounded set-theoretic rank.

Any provable first-order formula is supervalid; any supervalid first-order formula is valid; and, by the completeness theorem, any valid first-order formula is provable. Thus validity, supervalidity, and provability coincide for first-order formulae. In the absence of a completeness theorem for (real, full, standard) second-order logic, we cannot make the analogous claim for second-order formulae. We know how to produce counter-examples to the claim that any given (r.e.) axiom system yields as theorems all valid sentences of second-order logic. Since the counterexamples turn out to be not only valid but supervalid, the only question about inclusion among the three notions that remains is whether all valid sentences are supervalid. Otherwise put, is every (second-order) sentence of the language of set theory that is true in all models *true*?

Many set theorists find it probable or plausible that the answer is yes. They speculate that "there is no property of the universe of sets that is not reflected by some type V_α." Thus if they are right, there could be no second-order sentence that is false but nevertheless true in all models, or even true in all models of the form $\langle V_\alpha, \{\langle \in, \{\langle x,y \rangle: x,y \in V_\alpha \ \& \ x \in y\}\}\rangle\rangle$. Any such sentence would show, in technical parlance, that On is Π^1_n-describable for some n; and this has seemed extremely unlikely to most set theorists who have written on axioms of infinity and the structure of the set-theoretic universe. (The claim, however, would appear to be insusceptible of anything like proof from currently accepted axioms.) Thus the notion of validity has, after all, greater interest than the foregoing, belittling, line of thought might have inclined one to suppose it has: it is plausible and a reasonable "working hypothesis" that validity coincides with supervalidity, and hence that supervalidity *can* be defined by means of a single formula, and indeed by a formula of the first-order language of set theory.

In conclusion, let us mention an apparent defect of the account of supervalidity we have given: it would seem that there is no natural or obvious way to generalize the notion of supervalidity to a notion of "superconsequence" or "supersatisfiability." What we want is some way to explain what it is for some sentences (in the first instance, of the language of set theory) to be true under some one interpretation, that is, for there to be some sets and some pairs

GEORGE BOOLOS

(to assign to ∈) under which all of those sentences are true, without introducing classes, infinitely long sentences, or an unanalyzed notion of truth or satisfaction. We have shown above how this may be done for any given sentence. There seems no satisfactory way to do it for an infinite set of sentences, however. And although the sense of loss may be mitigated by the knowledge that many important theories such as Peano Arithmetic and Zermelo-Fraenkel Set Theory, which are not finitely axiomatizable, are axiomatizable by a finite number of schemata and have a natural second-order extension that is a finite extension of axiomatic second-order logic, there is no denying that there is a loss.

Massachusetts Institute of Technology

[17]

SECOND-ORDER LOGIC STILL WILD 75

SECOND-ORDER LOGIC STILL WILD*

IN *Philosophy of Logic,*[1] W. V. Quine summed up a popular opinion among mathematical logicians by referring to second-order logic as "set theory in sheep's clothing." A few years later George Boolos argued that Quine's claims were overblown. He conceded that second-order logic commits one to some classes—every subclass of one's individual domain, to be exact—but he also observed that in most cases this falls far short of the commitments of the usual set theories.

I.

In two very interesting recent papers,[2] Boolos has tried to tame second-order logic completely. Second-order logic *per se,* he now claims, does not commit one to sets, classes, Fregean concepts, or anything else. The gist of his argument is this:

1. We need not posit classes or collections in order to render second-order sentences intelligible. We can simply translate them into ordinary language using plural quantifiers. (For example, the second-order version of the least-number principle, "For any *F,* if there are numbers that *F,* then there is a least such" may be translated as "It is false that there are some numbers such that no one of them is the least.")

2. Using plural quantifiers does not commit one to classes or collections. Indeed, it does not commit one to anything that one is not already committed to by means of one's use of singular quantifiers.

3. Thus, the use of second-order logic need not commit one to collections or sets. Quine is wrong; second-order logic is not class theory in disguise.

If this is correct, Boolos has made a significant contribution to the philosophy of mathematics. He will have vindicated those who have held that conflating second-order logic with set theory has perverted Frege's logicism, and he will have strengthened their version of logi-

* I am grateful to Susan Hale, Bill Lycan, Mark Risjord, Jay Rosenberg, Geoffrey Sayre-McCord, Stewart Shapiro, and Steven Wagner for helpful comments.

[1] Englewood Cliffs, NJ: Prentice-Hall, 1970; reprinted, Cambridge, MA: Harvard, 1986.

[2] Boolos's initial criticism of Quine occurs in his "On Second-order Logic," this JOURNAL, LXXII, 16 (Sept. 18, 1975): 509–537. His two recent papers are "To Be Is to Be a Value of a Variable (or to Be Some Values of Some Variable)," this JOURNAL, LXXXI, 8 (August 1984): 430–449; and "Nominalist Platonism," *Philosophical Review,* XCIV, 3 (July 1985): 327–344. I will refer to these as TB and NP, respectively. The bulk of the argument that I here attribute to Boolos may be found in TB.

0022-362X/88/8502/0075$01.30 © 1988 The Journal of Philosophy, Inc.

76   THE JOURNAL OF PHILOSOPHY

cism.[3] He will also have helped Hartry Field out of a pitfall he fell
into while trying to nominalize Newtonian mechanics.[4] He will have
shown structuralists, such as Stewart Shapiro and me, how we can
use second-order characterizations of set-theoretic hierarchies without fear of circularity.[5] Finally, he will have shown set theorists how
they can embrace second-order axioms without abandoning the idea
that the first-order universe of sets contains all the setlike entities
that there are. (Shapiro tells me that this has been the principal
motive for Boolos's work on second-order logic.)

Although one might see Boolos's translation as a "nominalization"
of second-order logic, it does not significantly affect the central
ontological problem in the philosophy of mathematics. Unless we
adjoin it to a program such as Field's, which has other problems
besides its use of second-order logic, Boolos's methods do not dispense with sets and other abstract objects. At most, his technique
eliminates the use of abstract entities from several specific (but important) contexts.

Despite the attractiveness of Boolos's view for philosophers of
mathematics, and for me in particular, I am not convinced that it is
an advance over the standard view of second-order logic. Here are
my reservations. First, my logical/linguistic intuitions differ sharply
from Boolos's. As I see it, some plural quantifications do commit us

[3] Second-order logic has many friends, but, for someone who wants to combine it
with logicism, see J. M. B. Moss, "The Mathematical Philosophy of Charles Parsons," *British Journal for the Philosophy of Science*, xxxvi, 4 (December 1985):
437–455, p. 447; and his review of my *Frege and the Philosophy of Mathematics*,
this JOURNAL, LXXIX, 9 (September 1982): 497–512.
[4] Field abandoned his use of second-order logic in "On Conservativeness
and Incompleteness," this JOURNAL, LXXXII, 5 (May 1985): 239–260. During a
commentary at the 1984 Eastern Division meeting of the American Philosophical Association, however, he stated that Boolos's work had convinced him that
he could use second-order logic to achieve a nominalist formulation of Newtonian physics over which Platonist mathematics is conservative.
[5] The structuralist's dilemma is this. We maintain that mathematics is a
science of patterns or structures and that sets are positions in a particular type
of structure or structures—the (an) iterative hierarchy of sets. However, the
fullest accounts of such structures use second-order logic. Thus, if we construe second-order logic as quantifying over setlike entities, we seem to be
presupposing the notion of set to say what a set is. Of course, this is an old
difficulty and not unique to the structuralist approach. And there are ways
out: (1) One can maintain that the second-order variables range over sets qua
manys rather than sets qua ones or over Fregean concepts, or over properties,
where each of these are taken to be *logical* entities rather than elements of a
structure. (2) Or one can claim that we just cannot hope to get fully complete and
categorical descriptions of the richer mathematical structures. (This is the position
toward which I now incline.) Boolos's approach would wipe out the problem of
circularity in one quick blow. For further discussion, see Stewart Shapiro, "Second-order Languages and Mathematical Practice," *Journal of Symbolic Logic*, L, 3
(September 1985): 714–742.

to collections, sets, or some other sort of second-order entity, and I
find fault with Boolos's arguments against my position. I will expand
on this in section II. But there is more to deciding matters of ontic
commitment than brandishing logical and linguistic intuitions.
Boolos may have had this in mind when he constructed an alternative
semantics for second-order logic according to which no second-
order variable has a second-order entity as one of its values. In
section III, I argue that, although this semantics is necessary for the
consistency of Boolos's position, it fails to demonstrate that second-
order logic is not committed to collections. In section IV, I argue that
neither the intelligibility of plural quantifications nor their prima
facie lack of commitment to collections is sufficient to demonstrate
that they never commit us to collections. Determining whether they
do involves representing them in an acceptable and suitably interpreted logical notation. Thus, those of us who find plural quantification in need of logical analysis will not be enlightened by formal
explications such as Boolos's, which presuppose plural quantification for their interpretation.

II.

I will start with my dissenting logical/linguistic intuitions. I find that
I cannot process many sentences containing plural quantifications
without understanding them in terms of collections. Take one of
Boolos's leading examples, the Geach-Kaplan sentence,

(1) Some critics admire only one another.

On the face of it this makes no mention of classes or collections.
However, I am inclined to understand it as saying:

(2) There is a nonempty collection of critics each member of which
admires no one but another member.

Boolos urges me to see (1) as a second-order sentence, symbolized by

(3) $(\exists X)[(\exists y)Xy \ \& \ (x)(y)(Xx \ \& \ Axy \ \rightarrow \ x \neq y \ \& \ Xy)]$

and to translate this, using his general scheme, into English as

(4) There are some critics such that any *one of them* admires another
critic only if the latter is *one of them* distinct from the former.

But this sentence seems to me to refer to collections quite explicitly.
How else are we to understand the phrase 'one of them' other than
as referring to some collection and as saying that the referent of
'one' belongs to it?

Of course, if we render (3) and (4) as (1), we can avoid the problematic phrase 'one of them', and see (4) as just an awkward way of

putting (1). But I have two worries with this way out. First, we have
no assurance that we can always smooth out Boolos's translations in
this way. Second, I started out by worrying about the ontic commitment of (1). The buck was passed to (3) and thence to (4). If (4) is just
an awkward way of putting (1), we have come full circle.

Perhaps, I have been looking in the wrong place for a reference
for 'them' in 'one of them'. Could not Boolos reply that plural
quantifiers and the plural pronouns they govern do not refer singly
to collections but rather divide their reference over the members of
collections? I think he could. The notion of divided reference makes
good sense with respect to predicates: 'sleek' refers in the divided
sense to a locomotive just in case "sleek" is true of it. Of course,
neither quantifiers nor variables refer to specific individuals outright. We might, however, extend the notion of reference to quantifiers and variables as follows: a quantifier, with its associated variable, refers in a given sense just in case it generalizes over positions
that are open to terms that refer in that sense. Then second-order
quantifiers, by virtue of generalizing over predicate positions, would
count as referring in the divided sense. Of course, although this
fledgling theory of divided reference helps Boolos's case, it does not
show that my intuitions concerning 'one of them' are wrong.

But there is more: Boolos argues that it is often a mistake to
represent plural quantifications as referring to collections. He claims
that

(5) There are some sets that are self-identical, and every set that is not a
member of itself is one of them.

should not be rendered as

(6) There is a collection of sets that are self-identical, and every set that
is not a member of itself is a member of this collection.

For, he claims, the former is obviously true but the latter is false (TB,
446/7). Note, however, that (6) need not be false provided collections are ultimate classes or something similar. So we could put
Boolos's point in more neutral terms as follows: (5) is obviously true
but (6) is not; thus one should not paraphrase (5) as (6).

Is (5) obviously true? Yes, if we read it as

(7) There are some sets that are self-identical, and every set that is not a
member of itself is a set.

For this is paraphrasable as the *first-order* sentence:

(8) $(\exists x)(Sx \ \& \ x = x) \ \& \ (x)(Sx \ \& \ x$ is not a member of $x \rightarrow Sx)$.

Should we read (5) as (7)? Boolos thinks that the answer is an unqualified affirmative. In just a paragraph beyond the one last cited we read:

> . . . I am loath to abandon [my strong intuition] about the meaning of English sentences of the form "There are some *A*s of which every *B* is one," viz. that any sentence of this form means the same thing as the corresponding sentence of the form "There are some *A*s and every *B* is an *A*" (TB, 447).

From which Boolos concludes that (5) "is simply synonymous with" (7). Furthermore, in NP (331/2) he says that "There are some classes such that every infinite class is one of them" is only an awkward way of saying "There are some classes and every infinite class is a class," rather than a way of saying that there is a class (collection) of infinite classes. Again, taking this reading, we have no reason to use a second-order translation and arrive at a first-order triviality of ZF rather than a well-known ZF falsehood.

More generally, Boolos equates

(9) There are some *A*s such that every *B* is one of them.
(10) There are some *A*s and every *B* is one.
(11) There are some *A*s and every *B* is an *A*.

which are equivalent within second-order logic, if rendered as

(12) $(\exists X)[(\exists x)Xx \ \& \ (x)(Xx \rightarrow Ax) \ \& \ (x)(Bx \rightarrow Xx)]$
(13) $(\exists X)[(\exists x)Xx \ \& \ (x)(Xx \rightarrow Ax) \ \& \ (x)(Bx \rightarrow Ax)]$
(14) $(\exists x)Ax \ \& \ (x)(Bx \rightarrow Ax)$

But this surely should not count as substantial evidence that (9), (10), and (11) are "synonymous" forms. It is often possible to prove within a theory that some of its sentences are equivalent to sentences in one of its more elementary subtheories. Thus, it is well known that within counting theory we can prove that

(15) There are three wise men.
(16) There is a number that counts the wise men and equals 3.

are equivalent, and in this way reduce some counting theorems to (obvious?) truths of first-order logic with identity. However, to my knowledge, no one today wants to conclude that (15) and (16) are synonymous.[6] Furthermore, it is generally agreed that (16) but not (15) commits one to numbers. By parity of reasoning, the contrast

---

[6] Similarly, within set theory "nothing fails to be self-identical" is equivalent to "there is a set of all non-self-identical things and it has no members"; yet these are not counted as synonymous.

between the apparent commitment of (9) and (10) to collections and
(11)'s lack of it (as well as its logical transparency) should count as
strong evidence that they are *not* synonymous.

I do not think we should grant Boolos his examples, because they
depend upon the second-order equivalence of (12), (13), and (14),
and the status of second-order logic is at issue. But suppose that we
grant them anyway. Then they show that *some* plural quantifications
should not be construed as quantifying over collections. But this is
because these examples can be plausibly put into first-order terms.
Thus, the examples do not bear on the question of whether what one
might call "genuine" plural quantifications, such as,

> (17) It is false that there are some sets such that every one of them has a
>   member which is also one of them.

are wrongly construed as quantifying over collections. In (17) we
have an example—Boolos's translation of the second-order axiom of
foundation—which cannot be put in first-order terms. It is a substantial set-theoretic axiom and by no stretch of the imagination a
triviality. Until Boolos can show us that construing it as quantifying
over collections renders it plainly false or more controversial than it
already is, my intuitions and his concerning it and the Geach-Kaplan
sentence are at a standoff.

                                   III.

Since consulting intuitions concerning ordinary language has failed
to resolve the issue between Boolos and me, it might be useful to turn
to formal truth theories for second-order languages. If these require
second-order variables to range over collections, then Boolos's
translation would seem to show nothing. "Nominalist Platonism,"
Boolos's second paper, takes this possibility to heart. Truth theories
now come to the forefront. With Quine's "To be is to be a value of a
variable" in mind, Boolos writes:

> The heart of the matter is this: it is only with respect to a truth-"defini-
> tion" of the standard sort, a Tarski-style truth-theory for a first-order
> language, that the *notion of a value of a variable* is defined. In the case
> of a second-order language, such as the second-order language of set
> theory, there are at least two different sorts of truth-theory that can be
> given: on one of these, it would be quite natural to define "value" so that
> the second-order variables turn out to have classes as values; on the
> other it would not (NP, 324).

He then introduces a new truth theory which assigns no classes as
values to second-order variables. The key idea is this: Instead of
saying

> Sequence s (which assigns individuals to the individual variables and classes of them to the predicate variables) satisifes "*Xy*" if and only if $s(y)$ belongs to $s(X)$.

Boolos says

> Sequence s (which assigns values to only the individual variables) and relation R satisfy "*Xy*" if and only if $R\langle X, s(y)\rangle$.

Lapsing into class talk for a moment, we can see that $R\langle V, b\rangle$ holds if and only if (1) V is a second-order variable of the object language, (2) b is an element of the individual domain, and (3) b belongs to some class $C(V, r)$, dependent on V and R, of members of that domain. Thus, given any sequence of the type used in a standard truth theory, i.e., one which assigns classes to the second-order variables, there is an equivalent relation R in Boolos's truth theory. Conversely, for any such R in a Boolos-style truth theory, there is an equivalent sequence of the usual type in an associated standard truth theory.

Notice that Boolos's relation R now takes the place of the problematic phrase 'one of them'. Moreover, we can even use R to give content to the idea that a predicate variable has a divided reference by stipulating that, with respect to s and R, V refers in the divided sense to b just in case $R\langle V, b\rangle$. (Cf. NP, p. 337.)

On this way of construing Quine's criterion, Boolos can certainly claim that, on his truth theory, classes are not values of variables. But I think that putting the issue of ontic commitment in terms of *sequences* and Tarski-style semantics obscures it. Sequences are something of an engineering trick, which allow one to have uniform recursion clauses in the truth definition for a quantificational language. Since, on the sequence approach, a first-order existential quantification is true only if an appropriate *sequence* exists, the ontic commitment to an *individual* that this entails is obscured. The question of the ontic commitment of second-order existential quantifications is even harder to evaluate in Boolos's truth theory. Because of this, I am going to present a variant of Boolos's truth theory which displays the ontic commitment of existential quantifiers (of both types) more perspicuously. I will start with the first-order case.

> Definition. An interpretation I of a first-order language L in a nonempty domain D is an assignment of an element of D to each variable of L and an n-ary relation over D to each n-adic predicate letter of L. E^* is the assignment by I to the expression E.

Then we define truth for I in D as follows:

(a) *Fxyz* is T for I in D iff $\langle x^*, y^*, z^*\rangle$ belongs to F^*.

(b) $\sim S$ is T for I in D iff S is not T for I in D.

(c) $S \lor W$ is T for I in D iff either S or W is T for I in D.

(d) $(\exists v)S$ is T for I in D iff there is a d in D such that S is T for $I[v^*/d]$ in D.

Here $I[v^*/d]$ is the interpretation of L which is just like I except for assigning d (rather than v^*) to the variable v.

Clause (d) makes it clear that an existential quantification is true in a domain only if an appropriate element of the domain exists.

Let us try a similar approach to Boolos's truth theory. We cannot regard an interpretation as something that assigns entities to the second-order variables since Boolos carefully avoids doing that in his truth theory. Instead, let us regard an interpretation as a *relation* which relates variables and predicates to the appropriate elements of the domain. Of course, our interpretations for first-order languages are relations too, since all assignment functions are relations. But the relations that serve as Boolos-style interpretations need not be functions where the second-order variables are concerned.

More precisely, a Boolos-style interpretation I of a second-order language L in a nonempty domain D is a relation that relates each first-order variable v to a unique element v^* of D, each n-adic predicate letter F to a unique n-ary relation F^* over D, and each second-order variable to 0 or more elements of D. Then we define truth for I in D as follows:

(a) Xy is T for I in D iff $\langle X,y^* \rangle$ belongs to I.

(b) $Fxyz$ is T for I in D iff $\langle x^*, y^*, z^* \rangle$ belongs to F^*.

(c) $-S$ is T for I in D iff S is not T for I in D.

(d) $S \lor W$ is T for I in D iff either S or W is T for I in D.

(e) $(\exists v)S$ is T for I in D iff there is a d in D such that S is T for $I[v^*/d]$ in D.

(f) $(\exists V)S$ is T for I in D iff there is a relation R over D consisting solely of 0 or more ordered pairs of the form $\langle V,d \rangle$ (d in D) such that S is T for $I[V/R]$ in D.

As before, $I[v^*/d]$ is the interpretation of L which is exactly like I except for assigning d to v. $I[V/R]$ is the interpretation that is exactly like I except for containing the pairs in R in place of the pairs $\langle V,d \rangle$.[7]

Comparing clauses (e) and (f) makes it quite clear that the truth conditions for the two types of existential quantifiers are parallel. Both use metalinguistic existential quantifiers. The only difference between them is that the condition for the second-order quantifier uses a second-order metalinguistic quantifier. Thus, one who is already convinced that second-order quantification commits one to

---

[7] For simplicity I have restricted my account to languages without identity, constants, functional terms, or second-order variables beyond monadic ones. Also I have defined assignments over nonempty domains, whereas Boolos presented a truth theory for set theory, using the entire universe (a nonset) as his universe of discourse.

collections will find nothing to shake this conviction in Boolos's approach. To the firm believer this detour through Boolos's truth theory shows only that an *explicit* ontic commitment to classes need not emerge in a rigorous truth theory for a second-order language, but it fails to show that it is eliminable altogether.

Boolos admits as much. However, to those who are skeptical and to the "foes of classes" he offers the by now obvious way out: if you construe the second-order metalinguistic quantification used in his truth theories in terms of plural quantification, then you need not commit yourself to classes (NP, 337).

But now we are back to the issue of the last section. Whether we are a friend or foe of classes, we are likely to find it difficult to assess the ontic commitment of certain sentences containing plural quantifiers, such as the Geach-Kaplan sentence. Perhaps the most straightforward method for attacking this problem is to see it as a question of deciding whether such sentences should be represented in second-order logic or in class theory. But this suggestion—one which Boolos appears to endorse—will be of no avail unless we already know what ontic commitment second-order quantification carries. That is why telling us that it is the same as that of plural quantification hardly helps.

IV.

Boolos is involved in a circle: he uses second-order quantification to explain English plural quantification and uses this, in turn, to explain second-order quantification. Of course, the same is true of our standard explanations of English singular quantification in terms of first-order quantification and even of our account of English and truth-functional conjunction. We could not even begin the logical enterprise if we could not break some of these circles by taking some logical particles of ordinary language as reasonably clear starting points. Gottlob Frege was clearly aware of this. He objected to using ordinary language to settle logical matters, but he also argued that it is a suitable place at least to start to understand logical notations. I agree; to begin to understand the quantifiers, for instance, we need not have already grasped the Tarski semantics. Using ordinary language as a start, we can bootstrap ourselves into the language of logic and mathematics. In a related vein, Quine has emphasized the triviality of his criterion of ontic commitment: it simply reflects the idea that the existential quantifier explicates our ordinary concept of existence—that the quantifier '$(\exists x)$' is just another way of saying "there exists an x."[8]

[8] These points deflect the charge that Quine's criterion is applicable only if it's unnecessary, because, in order to interpret the existential quantifier, one must first specify its domain and, thus, one's ontology. They also defuse the

Reflections such as these seem to be in the background when Boolos emphasizes that his scheme translates second-order sentences as *intelligible* English sentences which carry no explicit commitment to collections. Just as Quine finds 'there is' sufficiently clear to found his doctrine of ontic commitment, Boolos seems to think that the intelligibility of plural quantifiers shows that they are sufficiently clear to ground his doctrine of noncommitment. I take the opposite view.

Before I discuss Boolos's choice of circle, I want to remind us that, again at least since Frege, we have known that ordinary language is unsuitable for settling questions of logic and ontology. We have learned from examples, such as "Mary did it for John's sake" and "There is a possibility that it will rain today," that ordinary language often introduces excessive ontic commitments. Ironically, the Geach-Kaplan sentence is usually seen as a case in which ordinary language *hides* its ontic commitment.

It will be useful for us to consider hidden ontic commitment more carefully. Since Boolos has turned the Geach-Kaplan on its head, here is another example for us to consider:

(18) Every time Saul visited Catherine he visited Matthew beforehand.

On a first pass, (18) appears to commit us to nothing more than times and people. But, if we try to symbolize it in first-order logic by, say,

(19) (t) $(Vsct \rightarrow (\exists t')(Bt't \ \& \ Vsmt'))$

problems begin to emerge. Let us suppose that Catherine is Saul's lover and Matthew is his boring cousin. Let us further suppose that Saul's mother has insisted that any time he visits Catherine he must first visit Matthew. Finally, interpret (18) as a report that Saul obeyed his mother. If (19) were an acceptable paraphrase of (18), Saul could have obeyed his mother by visiting Matthew just once before paying any visits to Catherine. I do not know about Saul's mother, but my mother would not put up with that. We cannot obtain an easy paraphrase of (18) by enlisting Boolos's "For every *A*, there is a *B*" either (TB, 431). For even this stronger sentence will not do:

(20) $(\exists f)[f$ is 1–1 & $(t)(Vsct \rightarrow (Vsmf(t) \ \& \ Bf(x)t)]$

(There is a 1–1 correlation between the times of Saul's visits to Catherine and those of his visits to Matthew which mates every one of the former with an earlier one of the latter.) Saul could satisfy this by

worry that we cannot quantify over all sets, since doing so presupposes a universal set to serve as the domain of quantification. See my "Note on Interpreting Theories," *Noûs*, VIII, 3 (September 1974): 289–294.

first visiting Matthew, then visiting Catherine and then visiting Matthew again (during the same trip) and thereafter visiting Matthew *after* visiting Catherine. As far as I can tell, whether we analyze (18) in terms of times, visits, or events, we must introduce some means for appropriately linking Saul's visits to Matthew with his visits to Catherine. One way of doing this is to use contextually determined time intervals. (For example, if Saul lived in New York and Catherine and Matthew both lived in Princeton, then the intervals in question might consist of Saul's entire visit to Princeton.) If we symbolize "*i* is an appropriate interval" by '*Ai*' and "*t* and *t′* are part of *i*" by '*Ptt′i*', then (21) seems to be a suitable explication of (18):

(21) $(t)(Vsct \rightarrow (\exists i)(\exists t')(Ai \ \& \ Ptt'i \ \& \ Bt't \ \& \ Vsmt'))$

(For any time at which Saul visits Catherine, there is an appropriate interval containing a prior time at which he visits Matthew.) Given that Saul did visit Catherine, (21) not only commits us to times and people, it also commits us to time intervals.

Of course, someone else may prefer a treatment that eliminates the use of intervals. For example, if we introduce the relation "appropriately before," symbolized by '*A*', then we can render (18) as

(22) $(t)(Vsct \rightarrow (\exists t')(At't \ \& \ Vsmt'))$

and apparently avoid the commitment to intervals. But then we are left with an adverbial problem, i.e., that of explaining how (22) implies that Saul's visits to Matthew occurred before his visits to Catherine.

I do not mean to settle the example here, but I do wish to draw some morals from it. First, it should remind us that communication generally presupposes shared background beliefs and theories (such as beliefs concerning which visits to Matthew are appropriately prior to visits to Catherine). Hence, by virtue of their place within such systems of belief, our utterances may carry additional ontic commitments with them, which we frequently do not suspect until we subject them to logical analysis. Thus, the mere intelligibility of plural quantification and its prima facie lack of commitment to collections fails to be decisive.

Furthermore, as the competing analyses of both Geach-Kaplan and (18) illustrate, formalization is not simply a matter of considering candidate symbolizations in the light of our linguistic intuitions, but rather more a matter of applying broad linguistic, logical, and metaphysical theories. When there are competing candidate symbolizations, deciding between them may involve choosing between pervasive theories having different advantages and disadvantages on a variety of dimensions. Meaningful comparisons of competing pro-

posals might prove elusive. In any case, a comparison between
Boolos's proposal and the standard view of plural quantification is in
order, although we have reason to doubt that this will produce decisive results.

Boolos's treatment has the obvious advantage of offering us a
theory of ontic commitment which forestalls some commitments to
classes.[9] This leads to many benefits, including those I canvassed in
my opening section. Of course, the standard treatment via classes
fails miserably on this comparison.

On the other hand, we should not forget that Boolos's approach
has significant disadvantages too. It entails embracing second-order
logic (over and above any set or class theory one might accept). If we
accept Boolos's package, then one of the main objections to second-
order logic—its ontological commitment—would dissolve. But the
other—its lack of proof procedure—would remain. I will not rehearse the reasons why this is a highly undesirable feature of second-order logic, since this has been very well done in a recent paper
by Steven Wagner.[10]

There are also some technical drawbacks to Boolos's proposal. So
far at least, no one has been able to find a direct translation of
second-order universal quantification in terms of plural quantification. [Even Boolos's rule for the existential case is not particularly
natural. The plural "There are some things F " implies that there is
at least one F, but the second-order existential quantifier has no
first-order existential import. Thus, Boolos's rule must render
"$(\exists F)Fa$" as "Either $a \neq a$ or there are some things such that a is one
of them." The translation in the other direction is simpler: "There
are some things F. . ." becomes '$(\exists F)((\exists x)Fx$ & . . .)'.] Also,
Boolos's procedure works only for monadic second-order quantification. To be sure, we can reduce the other cases to the monadic case
by enlisting ordered n-tuples. Thus, we can construe an assertion of
the form "$(\exists F)(\exists x)(\exists y)Fxy$" as "There are some things such that $\langle x,y \rangle$
is one of them." But that means that the ontology of the theory in
which the assertion is made must be closed under ordered pairing.
This is no problem for set theory or even number theory, but it
boosts the cost of applying Boolos's procedure to theories lacking
devices for forming n-tuples. In sum, under Boolos's treatment there

---

[9] It is important to realize that Boolos's criterion of ontic commitment
differs from Quine's, since the latter is embedded in a logical theory that holds
that ontic assessments make sense only with respect to referential first-order
quantification.
[10] See his "The Rationalist Conception of Logic," forthcoming in the *Notre
Dame Journal for Formal Logic*. This paper also contains a valuable discussion of
the issue of changing logic in order to achieve ontological economy.

are complications and ungainly asymmetries, which the treatment of plural quantification by means of collections avoids.

Let me return to the topic with which we began this section: Boolos's circle. Despite Boolos's careful analyses, plural quantification remains problematic. We know something about its logic, but the question of its ontic commitment remains open. By taking it as a primitive for his logical theory, Boolos has put himself on the poor methodological footing of assuming that this issue is already closed.

The problematic nature of plural quantification is accentuated by Boolos's claims (discussed in section II) concerning synonymies. For, if he is correct, the plural quantifier is significantly equivocal: some substitutions for the dots in "There are some things such that . . . them . . ." give rise to plural quantifications that are irreducibly so, whereas others just produce variants on singular quantifications. Given that the latter occur only if certain formulas are second-order equivalent, there is no complete set of complete rules for establishing, much less determining, whether a given quantifier is genuinely plural.

What is more, if there is such an equivocacy, then Boolos is worse off than the truth-functional logician is when dealing with the equivocacy attached to 'or'. For the latter can always define his disjunctions in terms of conjunction and negation, whereas Boolos still lacks a definition of plural existentials in terms of unequivocal plural universals. Furthermore, suppose that he tried to deal with the equivocacy as we usually deal with that attached to 'or'. He could start by pointing out that there are two senses to "There are some things such that" and that in one sense it just means "There is at least one thing such that." But where would he go from there? Obviously not to: In the other sense it means "There is a collection of things such that." But that seems to put him in the ironic position of saying that in the other sense it simply means what it says.

MICHAEL D. RESNIK

University of North Carolina/Chapel Hill

Part IV
Philosophy of Set Theory

[18]

THOMAS WESTON

KREISEL, THE CONTINUUM HYPOTHESIS AND SECOND ORDER SET THEORY

Summary. The major point of contention among the philosophers and mathematicians who have written about the independence results for the continuum hypothesis (CH) and related questions in set theory has been the question of whether these results give reason to doubt that the independent statements have definite truth values. This paper concerns the views of G. Kreisel, who gives arguments based on second order logic that the CH does have a truth value. The view defended here is that although Kreisel's conclusion is correct, his arguments are unsatisfactory. Later sections of the paper advance a different argument that the independence results do not show lack of truth values.

I

The continuum hypothesis says that there are only two possible "sizes" (cardinalities) for infinite sets of real numbers, the size of the set of natural numbers and that of the set of all real numbers.[2]

Since its proposal by Cantor in 1878, the CH has attracted the interest and efforts of several generations of mathematicians. Naturally enough, Paul Cohen caused something of a stir when he proved that the CH could neither be proved nor refuted in Zermelo—Frankel set theory (ZF), even if the axiom of choice is assumed. Not only is the continuum question itself an interesting one, but the methods used to show its independence are sufficiently profound that they have provided a basis for attacking a host of other problems. Cohen's ideas were quickly adapted to show that the CH could not be decided by axioms asserting the existence of enormously large cardinal numbers, and that various other hypotheses about cardinals are also independent of ZF. Considerable applications were also found in other areas of logic.[3]

This flurry of mathematical activity was not the only result of the independence discoveries, however. Some philosophers of science and philosophically-minded mathematicians declared that the CH is independent because the notion of set is too vague to allow a decision ([15], p. 89). It has even been claimed that the fact of the independence discoveries itself shows or tends to show that the notion of set is vague or ambiguous ([15], p. 94).

Journal of Philosophical Logic 5 (1976) 281–298. *All Rights Reserved*
Copyright © 1976 by D. Reidel Publishing Company, Dordrecht-Holland

282   THOMAS WESTON

Others have predicted that set theory will split into incompatible sub-theories, only the common part of which could serve as a working foundation for mathematics ([20], p. 115) or even claim a central place in mathematics ([15], p. 94). As one writer put it: "I guess that in the future we shall say as naturally 'Let us take a set theory S' as we now take a group G or a field F" ([8], p. 105). For convenient reference, I will call views such as these "alarmist."

Not everyone agrees with the alarmists that the good old days when we could speak of set *theory* (rather than *theories*), are gone. Godel ([5]) and a number of others ([9], [17]) still maintain that set theory is the theory of a definite mathematical structure. Kreisel has been particularly active in arguing that 'set' is neither vague nor ambiguous, that set theory has a unique intended interpretation and that under that interpretation, the CH is either true or false, although we do not know which ([9, 10, 11, 12, 13, 14]).

I am inclined to agree with Kreisel on all these points, but *not* with the arguments and claims he makes in their support. Specifically, he asserts that the continuum hypothesis "*is* decided by the second order axioms of Zermelo" ([10], p. 99). Kreisel repeats this assertion in various articles and regards the "second order decidability of CH" as "the main theme" of his article "Informal Rigour and Completeness Proofs" ([9], p. 152).

Kreisel's claim is certainly seductive. It invites us to dispose neatly of the apparent unprovable and unrefutable status of CH by keeping the same set theory and merely beefing up our logic. But the invitation is misleading; CH is still unprovable and unrefutable in second order set theory (we will see below how this is proved). All that Kreisel could hope to show via second order logic is that CH has a truth-value. I argue that he either doesn't do this or he doesn't need to. That is, I will show that his second order argument actually presupposes a unique intended interpretation for ZF. I argue that if ZF has such an interpretation, then that fact itself shows that CH has a definite truth-value, and if there is no unique intended interpretation then the second-order argument is whistling in the dark. Worse, in defending his position via second order logic (SOL), Kreisel at least suggests that he is making various concessions to the alarmists that are, in my view, entirely unnecessary.

Following out Kreisel's polemic against the alarmists as it concerns SOL requires a bit of technical exposition about SOL and – later – second order set theory. I will turn to this now and get back to Kreisel's argument shortly.

II

Roughly, second order logic is logic in which quantification over predicates is permitted. Depending on just what notion of possible model for second-order theories is employed, these theories can have very different model-theoretic properties from theories formulated in the usual first order logic. It is on these special properties that Kreisel's argument depends. Since these features are most easily illustrated in number theory, without the complexities of set theory, we will deal first with second-order logic in this simplified setting.

We formulate second order number theory, T^2, as follows: our language contains, besides logical constants (including identity), the constant '0', the one-place symbol 's' ("successor") and the two-place function symbols '+' and '×'. We include only one-place predicate variables P, Q, R, etc., and adopt the usual (first order) axioms of number theory except that the induction schema is replaced by a single axiom with predicate quantification:

(1) $\qquad (P)((P(0)\ \&\ (x)(P(x) \supset P(s(x)))) \supset (x)P(x))$.

Beside the usual first order logical axioms, we need two more logical schemata:

(2) $\qquad (P)(\phi \supset \psi) \supset (\phi \supset (P)\psi)$, where P is not free in ϕ,

and

(3) $\qquad (P)\phi \supset \check{S}^{P(x)}_{\psi(x)}\phi$

where $\check{S}$ is the substitution operation defined in [4]. (2) is simply a second order version of the familiar first order schema. (3) ensures the existence of values of predicate variables with the same extension as open sentences $\psi(x)$ which satisfy the substitution restrictions on $\check{S}$. In particular, (3) allows us to prove all the instances of the first order induction schema.

A possible model M of T^2 is a sextuple $\langle \mathbf{N}, \mathbf{C}, +, \times, \mathbf{s}, \mathbf{0} \rangle$, where $\mathbf{N}$ is the range of the first order variables, $\mathbf{C}$ is the range of the second order variables, and the remaining boldface symbols designate the interpretations of '+', '×', 's', and '0', respectively. $\mathbf{C}$ is a subset of $P(\mathbf{N})$, the set of all subsets of $\mathbf{N}$, so that values of second order variables are subsets of the first order universe. If $\mathbf{C} = P(\mathbf{N})$, we will call M a "$*$-structure." When we are considering only *-models of second order theories, we will indicate this by prefixing a '*' to the name of the theory, e.g., '*T^2'. If **C** contains sufficient subsets of **N** to satisfy the logical axioms of T^2, M will be called a "general structure." Plainly, every *-structure is a general structure, but not conversely.

The model theory of general structures for T^2 (or for second order theories generally) resembles that of first order theories. But for *-structures, the kind which interest Kreisel, quite different results are obtained. On account of the Skolem—Lowenheim Theorem, no first order theory with any infinite models can be categorical — i.e., have all its models isomorphic. On the other hand, it is easy to show that *T^2 is categorical. We sketch a proof:

Let $a_0, a_1, a_2, \ldots$ be the sequence of (necessarily distinct and without other predecessors) denotations of the numerals of T^2 ('0', '$s(0)$', '$s(s(0))$', etc.) with respect to a given *-model $M = \langle$ **N**, **C**, $\ldots \rangle$ of T^2. Let $A = \{a_i | i \in \omega\}$. Since **C** $= P($**N**$)$ and $A \subseteq$ **N**, we have $A \in$ **C**. So A is within the range of the universal quantifier (P) in the induction axiom (1). Since M is a model of (1), and both **0** $\in A$ and $x \in A \Rightarrow s(x) \in A$, all numbers are in A, i.e., **N** $= A$. Thus the first order universe of each *-model must be isomorphic to the series of numerals, and hence they must all be isomorphic to each other.[4]

It is an easy consequence of the categoricity of *T^2 that the model theory of SOL for *-structures is not semantically complete. That is, there are sets of sentences Γ of T^2 such that some sentence σ of T^2 is true in all *-models of Γ, but cannot be proved from Γ. For example, if we make the modest assumption that T^2 is consistent (i.e. that no contradiction can be proved in it) then the Godel—Rosser Theorem yields a sentence true in the standard model of T^2 which cannot be proved from the axioms of T^2. Since *T^2 is categorical, however, this sentence is true of *all* *-models of the axioms of T^2.

This result may be generalized as follows: Given any effective set of axioms true in all *-structures (of some particular language type) and any effective set of sound and effective rules for SOL, *SOL is not semantically complete w.r.t. these rules and axioms. This remarkable result, that no set of rules and axioms meeting these eminently reasonable conditions can prove all the "semantic consequences" of every *SOL theory has important application in the case of the CH. Kreisel argues that either CH or its negation is a semantic consequence of second-order ZF. If so, however, CH (or not-CH) runs afoul of *SOL's lack of a completeness theorem, since it isn't provable. We return to this point shortly.

The Godel—Rosser result can also be used to show another remarkable property of $*T^2$: if T^2 is consistent, then it has a consistent extension with no *-models at all.[5] As one might guess, none of these "remarkable" properties are found when general structures are considered.

To anyone familiar with the ordinary (first order) model theory, these properties of $*T^2$ will seem bizarre. As L. Kalmar put it, "One can say humorously, while first order reasonings are convenient for proving true mathematical theorems, second order reasonings are convenient for proving false metamathematical theorems" ([8], p. 104). One of these properties in particular, that second order theories can be categorical, is of great importance to Kreisel. In fact, he attaches such importance to categoricity that he sees the fact that no infinite mathematical structure can be characterized by a categorical first order theory as "establishing the *inadequacy* of first order foundations" of mathematics ([11], p. 238). Below, we will try to figure out why categoricity could possibly be this important.

For the present, we note that if categoricity is important, it is obtained for T^2 only by the special restriction to *-models. But if the restriction is justifiably imposed for second order theories, it could perfectly well be imposed on *first* order formulations.

To illustrate this, we formulate the *first order* theory $T^{2'}$: The language of $T^{2'}$ contains the non-logical constants '0', 's', '+', and '×' as before, and two new one-place predicate symbols 'N' and 'C', read "is a number" and "is a set", respectively. Finally, we add a two-place predicate symbol '∈', interpreted as usual as "is a member of". In addition to the other usual Peano axioms, we have an induction axiom stated as follows:

(4) $(x)((C(x) \& 0 \in x \& (y)((N(y) \& y \in x) \supset s(y) \in x)) \supset$

            $(y)(N(y) \supset y \in x)).$

Of course, these axioms alone will not yield a viable formulation of number theory. We will need some axioms of set existence, an extensionality axiom for '∈', axioms which state that everything is either a number or a set, but not both, and axioms to settle similar technical matters.[6] A possible model of $T^{2'}$ will be of the form $\langle U, N, C, +, x, s, 0, \in \rangle$ which is obtained from a *-structure $\langle N, C, \ldots \rangle$ for T^2 as follows: let $U = N \cup C$, and $\in = \{\langle x, y \rangle | x \in N \& y \in C \& x \in y\}$. It is clear that the categoricity proof

286   THOMAS WESTON

given above for $*T^2$ also works for $T^{2'}$ if we consider only possible models
of this sort. Technically, of course, this is simply a trick; we have quite
arbitrarily restricted the possible models so that if they are models at all,
they are isomorphic to the standard one. It remains to be seen whether the
restriction to $*$-structures is any less arbitrary.

III

Our second-order formulation of ZF, ZF^2, is the following theory: The
language of ZF^2 contains only one-place predicate variables, and the only
non-logical constant is '$\in$'. For axioms, we have the usual Extensionality,
Pairing, Sumset, Powerset and Infinity. In addition, we have two second
order axioms, Foundation:

(5) $\qquad (P)((\exists x)P(x) \supset (\exists x)(P(x) \& (y)(y \in x \supset \sim P(y))))$

and Replacement:

(6) $\qquad (P)((x)(y)(z)((P(\langle x, y \rangle) \& P(\langle x, z \rangle)) \supset y = z) \supset$

$\qquad\qquad (x)(\exists y)(z)(z \in y \equiv (\exists u)(u \in x \& P(\langle u, z \rangle))))$

We also add the new logical axioms for second-order formulas, as in T^2.

A possible model for ZF^2 is a triple $\langle U, C, R \rangle$ where $C \subseteq P(U)$ and
$R \subseteq U \times U$. As before, if $C = P(U)$, we call $\langle U, C, R \rangle$ a $*$-structure, and if C
is such that $\langle U, C, R \rangle$ satisfies the logical axioms of ZF^2, it is a general structure. We can now examine Kreisel's argument.

To over-simplify slightly, we can put Kreisel's argument as follows:
(a) ZF^2 is categorical, so
(b) All models of ZF^2 are isomorphic, so
(c) The CH has the same truth value in all models, so
(d) The CH has a truth value, period.

The oversimplification in this presentation of Kreisel's case is that (a) is
probably not true, even for $*$-models, but a similar result is. To describe
this result, we need some preliminary definitions.

We define the function of ordinals R by induction:

$\qquad R(0) = $ the empty set,

$\qquad R(\alpha) = $ the powerset of $R(\alpha)$.

If λ is a limit ordinal, $R(\lambda)$ is just the union of all $R(\alpha)$'s where α is smaller than λ. This inductive definition cannot quite be carried out (i.e., made explicit) in ZF, because the collection of ordered pairs R is not a set. However, each initial portion which is a set can be defined in ZF, and R itself can be defined in relatives of ZF, such as von-Neumann–Bernays–Godel set theory (VBG), which permit quantification over such "large collections" (proper classes) as well as sets. In such theories, it is a simple matter to prove that R is one-one, that every set lies in the range of R and that if $\alpha \leq \beta, R(\alpha) \subseteq R(\beta)$. Thus R divides the "universe" of sets into a cumulative hierarchy of levels, called *ranks*. The least α such that $x \in R(\alpha + 1)$ is called the rank of x. It is convenient to think of the rank of a set as a measure of its complexity — the higher the rank, the more complex.

Since it is a theorem (of, say, VBG) that every set has a rank, every interpretation under which the axioms of set theory are true "carries with it" a notion of rank since every such interpretation satisfies the theorem. Let us suppose, as Kreisel does, that there is a *unique* intended interpretation of those axioms. Then there is a unique structure of cumulative ranks (call it the CRS) which contains all sets. Unfortunately, this "structure" cannot be a model of ZF, at least not in any ordinary sense of 'model', since any model is a set, and the "structure" of all sets cannot be a set. Kreisel does not regard this as any special problem; he is content with an "intuitive" notion of truth in the CRS ([19], p. 144). Perhaps he is right: in any case, we are already dealing with structures which are not sets, such as the function R. We will assume, therefore, that some notion of truth encompassing structures like the CRS is available to us, and agree to include such structures as models.

Let us say that two ∗-models of ZF^2 are *almost isomorphic* if they are isomorphic or if one is isomorphic to some initial portion of the rank structure of the other. If every pair of ∗-models of ZF^2 are almost isomorphic, we will say that that theory is *almost categorical*. In this terminology, we can now state the result on which Kreisel relies:

PROPOSITION: ∗ZF^2 is almost categorical.

A ∗-model for ZF^2 can also be shown to contain a large portion of the CRS, and it is possible (although unlikely) that ∗ZF^2 is in fact categorical. In any case, the continuum hypothesis can be so phrased that it involves quantification only over fairly simple sets — i.e. sets of low rank. In fact we only need the initial portion of the CRS out to $R(\omega + 3)$.[8] Since $R(\omega + 1)$

288   THOMAS WESTON

is the lowest place in the CRS at which any infinite sets appear, and
$R(\omega + 3)$ is only two iterations of the operation "set of all subsets of"
beyond that, we are not talking about anything like the most complex sets
here.

Since CH need only concern $R(\omega + 3)$, and since almost categoricity
guarantees that $R(\omega + 3)$ or an isomorphic copy of it is in every *-model of
ZF^2, CH has the same truth-value in all such models of ZF^2. This is fairly
faithful rendering of Kreisel's argument.

Although the proof of almost categoricity will not be reproduced here,[7]
we note some features of it which bear on the conclusions Kreisel draws. In
particular, we notice that considerable set theory is used, including, of
course, the *-ness condition on models. Slightly paraphrased, this condition
says that for each model $M = \langle U, C, R \rangle$, C contains *all subsets of* U. I re-
emphasize that this is a non-trivial condition, since there are models of ZF^2
which do not satisfy it, and if we take them into account, ZF^2 is not even
almost categorical.

Of course, if we are satisfied that we understand the "all subsets" condition then there is no reason not to use it and accept the proof. If, as
Kreisel believes, there is a unique intended interpretation of ZF^2 — a unique
CRS — then we certainly do understand the condition. But suppose that, as
the alarmists maintain, there are two or more interpretations of the axioms
of ZF which are equally natural candidates for "the" intended interpretation, but which are not almost isomorphic. Then the "all subsets" condition
is ambiguous, for what may be "all subsets" in one interpretation need not
be "all subsets" in another. That is, suppose that two mathematicians have
in mind *different* (not almost-isomorphic) natural interpretations of ZF^2.
Then each could use his notion of set to prove the almost categoricity of
ZF^2 according to his own interpretation of that theory, and yet, their interpretations are, by hypothesis, not almost isomorphic. So the almost
categoricity argument for ZF^2 fails unless we know in advance that there is
indeed a unique intended interpretation of it. As it stands, the argument
merely shows that the CH has the same truth-value in the group of
*-structures associated with each natural interpretation.

On the other hand, if there is a unique intended interpretation of ZF,
then this fact *alone* settles the question of whether CH has a truth-value. CH
is true (or false) according to whether it is true (or false) in the rank
$R(\omega + 3)$ of that interpretation. In fact, we can even dress up this point a

bit to use the formal notions of truth and satisfaction, as follows: If we are
assured that there is a unique CRS, then even though there are difficulties in
the model theory of the full CRS, we can quite readily use the rank
$R(\omega + 3)$ from that CRS. Then the CH is true if and only if it is satisfied
($\models$) in the model theoretic structure $M = \langle R(\omega + 3), \in \rangle$. That is, there is
sentence S of the language of set theory which is the translation from Eng-
lish of the CH such that

(7)   The CH is true if and only if $M \models S$

(8)   The CH is false if and only if $M \models \sim S$.

and

(9) $M \models S$ or $M \models \sim S$.

Therefore

(10)   The CH is true or the CH is false.

Premises (7) and (8) merely express the correspondence between
(informal) truth and (formal) satisfaction for CH, given that it concerns the
relevant parts of *the* CRS. Premise (9) is a trivial theorem of model theory.
(7), (8) and (9), therefore (10) is an obviously valid argument and its
premises are a small subset of those of Kreisel's. What is more, all of the
elaborate and shaky trappings of second order logic are absent.

IV

Second-order logic might still be of interest in connection with CH if it
could be proved or refuted in ZF^2. Unfortunately, this isn't the case. To my
knowledge, no one has published an independence proof for ZF^2, but it is
straightforward to adapt L. Tharp's proofs ([21]) for the first-order theory
VBI to ZF^2.

VBI (for von Neuman–Bernays–Impredicative set theory, also called
Kelly–Morse set theory) is an extension of the more familiar (first order)
VBG obtained by strengthening that theory's class comprehension schema.
The relation between VBI and ZF^2 is closely analogous to that between $T^{2'}$
and T^2. Since identity of classes is expressible in VBI, but identity of predi-
cates is not expressible in ZF^2, we introduce the following definition for
ZF^2:

290   THOMAS WESTON

$$P = Q \text{ for } (x)(P(x) \equiv Q(x)).$$

With this definition, it is easy to establish that the axiom of extensionality and the logical axioms of a theory with equality are (abbreviations of) theorems of ZF^2. Then we have the following:

PROPOSITION: If ϕ is any formula of VBI and $\phi*$ is the formula of ZF^2 which results from replacing every atom of ϕ of the form $x \in X$ by the predicate variable $X(x)$, then $\phi*$ is a theorem or the abbreviation of a theorem of ZF^2 if and only if ϕ is a theorem of VBI.[9]

Now we need only note that the CH contains only set variables, and no upper case (class or predicate) variables, so that CH* = CH. Since Tharp has shown that neither CH nor its negation is a theorem of VBI (if VBI is consistent), neither is a theorem of ZF^2 (provided that ZF^2 is consistent).

It is curious that Kreisel never, as far as I have been able to find, mentions the fact of the undecidability of CH in ZF^2, although he chides other authors for failing to discuss his categoricity argument ([10], p. 99).

V

As we have seen, if Kreisel's argument about the truth-value of the CH is needed at all, then it is question-begging — the prior question is that of the existence of a unique intended interpretation. It is interesting that Kreisel actually stresses his belief that there is such an interpretation and that it can be unambiguously described informally along the lines of the explanation of the CRS above ([13], p. 93). In fact, this particular example occupies center stage in Kreisel's defense of the clarity, utility and reliability of "informal rigour" in mathematics. In one place in this defense Kreisel even appears to acknowledge that the interpretation must be settled before the second order argument is made. Thus he introduces a section containing his almost categoricity argument with the statement:

This section takes the precise notion of set (in the sense of the cumulative [rank] structure of Zermelo . . .) as starting point and *uses* it to formulate and refine some intuitive distinctions ([9], p. 14).

Here, Kreisel appears to recognize certain of the points argued above, although he still attaches great significance to second order set theory. It will prove interesting to ask just why he puts such stock in it.

Although he gives no comprehensive defense of SOL, Kreisel claims various advantages for it over FOL. He argues that SOL is more natural for the "ordinary" mathematician and for earlier investigators in foundations ([11], p. 237) that SO axioms provide evidence for FO axiom schemata ([9], p. 148) and that SO formulations are heurestically fruitful in finding FO ones ([9], p. 150). I have argued elsewhere [23] that these advantages are slight when they are not imaginary, and I doubt that they explain Kriesel's attachment to SOL. It is fairly clear that the one *important* property of SOL in Kreisel's view is categoricity. This is indicated not only by his rejection of "first-order foundations" of mathematics because categoricity is lacking, but by his virtual identification of "categorical" with "second order":

> ... the first axiomatic theories, e.g. of Dedekind, [were] intended to be *categorical, and, as we now say, second order* ... ([10], p. 101, emphasis added).

Why is categoricity important? Why does lack of categoricity show the "inadequacy of first order foundations"? Kreisel's answer is that if a theory is not categorical, it will have some non-standard models, i.e., models clearly not equivalent (isomorphic) to the standard, intended one. He says that "non-standard models ... *are* important by showing the inadequacy of first order languages." ([9], p. 166) Thus Kreisel's claim is that if a language *can be* interpreted in an unintended way, there is something wrong with it.

Kreisel is not alone in maintaining this strange position. Paul Bernays, for example, has written that

> from our experience with non-standard models, it appears that a mathematical theory, like number theory, cannot be fully represented by a formal system; and this is the case not only with regard to derivability, as Godel's incompleteness theorem has shown, *but already with regard to the means of expression* ([2], p. 111 emphasis added).

Godel's theorem aside, this cannot be right; it is simply not the case that the possibility of non-standard or unintended interpretations shows that formal languages are defective in their "means of expression."

The extremely simple structure of the artificial languages of logic has made it easy to develop a model theory which can produce down right bizarre interpretations of a single sentence or an entire theory. But I see no philosophically interesting reason why we should not eventually be able to develop a model theory for *English* which would allow us to produce, for any given body of consistent English sentences, unintended interpretations under which those sentences would come out true.

292   THOMAS WESTON

Thus we might systematically reinterpret a body of English sentences so that, in the sentence 'Quine is a philosopher', 'Quine' refers to Ford and 'is a philosopher' has the usual sense of 'is a politician'. But the fact that such reinterpretation is possible does not show that 'Quine is a philosopher' does not, in fact, say that Quine is a philosopher. At most, it shows that these words *might have* been understood in a way in which they are not.

So, if non-categoricity and the resulting non-standard interpretations show that artificial languages are defective in their "means of expression", then they show the same thing about English. Not only is such a conclusion absurd, but it cuts directly against Kreisel's own defense of informal rigour and "intuitive" explanations. Perhaps Kreisel should complain that theories in *English* are not categorical!

Logicians do sometimes study theories with no intended interpretation, but it is more usual to study a theory *because of* its intended interpretation. In fact Kreisel would be the first to argue that we investigate number theory and set theory because we are interested in numbers and sets. For theories of this sort, the intended interpretations of the constants, function symbols, etc. are invariably explained in the same way as the natural language technical terms of any part of mathematics or natural science. One explains the intended sense or "points to" the intended reference in some natural language the speaker and his audience both understand. If these explanations and supplementary examples and audience guesses are inadequate to determine meaning or reference in the formal language case — as, no doubt, they sometimes are — they will certainly fail for natural languages as well.

As I see it, there is only one very peculiar set of circumstances in which categoricity of a theory might be really reassuring. This would be a case in which we knew that a theory had some correct interpretation, but we didn't know what it was. If we knew that such a theory was categorical, we would relieve ourselves of the necessity of selecting a particular model, since all would be isomorphic.

For myself, I find it hard to see how we could be in such circumstances, but some of the "alarmists" quoted at the beginning of this paper apparently think that is just where we are in set theory today. His insistence on categoricity makes me suspect that Kreisel vacillates in their direction on this all-important question.

This inferred suspicion is confirmed by direct statements of Kreisel's.

In one passage, he says that:

Denying the (alleged) *bifircation* or *multifircation of our notion of set of the cumulative hierarchy* is nothing else but asserting the properties of our intuitive conception of the cumulative [rank] structure . . . ([9], p. 144)

and he is evidently willing to make this assertion. Elsewhere, however, he includes a highly significant qualification:

. . . *unless one has theoretical or empirical reasons agains naive judgement,* in particular against [CRS], the precise notion of set [of the CRS] is a foundation for [Zermelo's axioms] . . . ([14], p. 174 emphasis added).

Kriesel doesn't tell us what to do if we have such reasons against "naive judgements."

VI

Various alarmist writers do offer reasons against "naive judgements," although not, in my view, very good ones. Mostowski, who is one of the more careful defenders of such views, claims that:

Models constructed by Godel and Cohen are important not only for the purely formal reasons that they enable us to obtain independence, but also because they show us various possibilities which are open to us when we want to make more precise the intuitions underlying the notion of set. ([15], p. 94)

and elsewhere that

. . . The intuitive notion of a set is too vague to allow us to decide whether that axiom of choice and the continuum hypothesis are true or false. ([15], p. 89)

Mostowski believes that Godel's notion of constructible set is one such candidate clear interpretation for 'set'. But since "we possess as yet no clear intuition of generic sets" as opposed to sets generic w.r.t. some particular model, he believes we don't yet have a clear interpretation based on Cohen's ideas, although he evidently hopes for one.

My view is that Mostowski is wrong on both "possibilities . . . to make precise the intuitions underlying the notion of set." Briefly and dogmatically put, my objections are as follows: the virtual consensus that the constructible universe is not *the* universe (and the various reasons for this consensus) show that the constructible universe is not one among a group of equally plausible candidates for "the" universe of set theory. His case is

294   THOMAS WESTON

even worse for the plausibility of Cohen models as candidates for "The" interpretation of set theory, since Cohen's (and more recent) techniques for producing models of ZF + axiom of choice + not-CH only work when we know that neither the model we start with nor the ones we produce are the real universe. Typically, we need to know that some cardinal of the models in question is "too small" — that is we have to be able to see from outside the model that cardinals of the model are smaller than they appear from inside.[10] But if, for example, the set which is assigned to an uncountable cardinal by an interpretation is not, in fact, uncountable, that interpretation is *not* the intended interpretation of set theory.

Even if all the stated defenses for alarmist views are weak — and Kreisel certainly thinks so (see [12]) — it is certainly worthwhile to do something other than throwing the burden on the other guy and showing he can't hold it up. My plan for the rest of this paper is to focus on Mostowski's "diagnosis" of independence results — that the notion of set is too vague to decide the independent sentences — and argue that it is actually wrong, not just inadequately supported.

In outline, the argument is this. At each rank from $R(\omega)$ on up we find or could find sentences independent of ZF. At the first few ranks, $R(\omega)$ and $R(\omega + 1)$, we can show that there is no confusion about which structures '$R(\omega)$' and '$R(\omega + 1)$' designate. Hence Mostowski's diagnosis — vagueness in the notion of set — is clearly not correct for these cases. But if we must attribute independence to something other than vagueness at lower ranks, then it is at very least a reasonable guess that vagueness isn't the trouble at higher ranks. Now for some details and elaboration of the argument.

$R(\omega)$ is the set of all hereditarily finite sets, that is, sets whose members, members of members, etc., are all finite. It should not be surprising that $R(\omega)$ is isomorphic — in a simple and natural way — to the natural number sequence 0, 1, 2, 3 . . . . Two things follow from the existence of this simple isomorphism. First, we know which structure '$R(\omega)$' designates provided we know which structure the sequence of natural numbers is. But surely if there is *any* mathematical structure we understand, it is this one. Second, the arithmetic sentence that Godel's theorem guarantees independent of ZF (if ZF is consistent) is (via the isomorphism) an independent sentence *about $R(\omega)$*.

Mostowski thinks that this independence result is essentially different from the CH case because we can eliminate independence in the former case but not in the latter by adding an obviously sound proof rule with infinitely many premises ([15], p. 94). I have argued elsewhere ([24]; [23], chap. I) that this is an important difference, but this contention merely underlines the fact that the correct diagnosis of independence in $R(\omega)$ is *not* vagueness.

At the next rank up, we find that $R(\omega + 1)$ is naturally isomorphic to the continuum, the set of real numbers. I offer two sorts of arguments that we know which structure this is. The first, might be called the argument from sociology of mathematics. Here we rely on the fact that the actual practice of mathematicians reveals some alternative *theories of* the real numbers ("classical" versus various "constructivist" accounts), but *not* different real number structures being studied.

To see why this is so, we need only examine a typical constructivist treatment of the reals as convergent sequences of rationals, $\{x_n\}$. Such sequences are required to satisfy some effective convergence condition such as:

$$|x_m - x_n| < \frac{1}{f(m, n)}$$

where f is an effective function of m and n (see [3], [22]). Speaking as a "classical" mathematician, however, it is obvious that every real number (and of course, only real numbers) can be defined by such a sequence. For each real, one need only consider the sequence $\{x_n\}$ of initial segments of its decimal expansion, and let $f = 10^{\min(m, n)-1}$.

For the second argument that we know what the continuum is, we note that this is a physically realized structure, and hence can be referred to in ways which might be problematic for a more abstract structure. Instants of time, locations in space, and a host of physical quantities take a continuum of values (yes, Virginia, even in quantum mechanics). Thus I can say perfectly well what the continuum is without worrying about special problems of reference to mathematical objects; it's the sequence of points traced out by the center of mass of my pen as I write. In putting things just this way, I am taking a strongly realist position about physics. Without apologizing for this, however, I point out that even an instrumentalist may find some comfort in the amazing utility of the real number continuum as a part of a "predictive device" within physical theory.

Neither of these two arguments shows that it is logically impossible that there are essentially different notions of the continuum which may some

296   THOMAS WESTON

day be recognized. What I claim is that together they give quite good
reasons to believe that we know what we are talking about when we say
"all real numbers." Alarmists are still free to invent a Cartesian demon argument that the notion *might be* vague — but that won't be a real objection
here anymore than it would be to a tenet of physical theory.

Since we know what the continuum is, then if Mostowski's diagnosis of
independence results were correct, we should not expect to find any independent statements about $R(\omega + 1)$, or at least not any unremovable by
technical devices such as infinite premise proof rules. Unfortunately, we do
find such independent statements, and they are: (1) interesting, (2) discovered in course of work on CH and (3) of unknown truth value.

As an example, let us take the statement ERC: "Every real number is
constructible." Evidently, this statement is a more modest version of
Godel's axiom of constructibility which says that every set lies in a hierarchy of sets analogous to the CRS. Unlike the CRS, each level of the constructible heirarchy adds not all subsets of sets at lower levels, but only
those which are definable in a certain way.

Addison has shown ([1], p. 356) that ERC has an equivalent formulation
which concerns only elements of $R(\omega + 1)$. ERC is independent of ZF, it is
(obviously) implied by the axiom of constructibility and its negation is
implied by various "large cardinal" axioms ([19]). As Mostowski notes
([15], p. 88), most people find ERC implausible, but the question is far
from definitely settled. Our long argument can now be very simply summar-
ized as follows: Mostowski's diagnosis of the independence of ERC is
wrong, so his diagnosis for the case of CH is highly implausible.

It has probably entered the reader's mind that we ought to continue
analyzing the CRS through $R(\omega + 2)$ to $R(\omega + 3)$ and the CH. I won't
attempt to do this because I doubt that the sort of argument given here for
$R(\omega)$ and $R(\omega + 1)$ could be advanced with equal confidence for $R(\omega + 2)$.
The difficulty is that the earlier arguments have made essential use of the
fact that structures isomorphic to $R(\omega)$ and $R(\omega + 1)$ can be readily identified in disciplines of mathematics — and physics — which are relatively
independent of set theory. In contrast, the interesting objects in $R(\omega + 2)$,
the real valued functions, were first given a reasonably precise characterization by the development of set theory. Of course, other sorts of arguments against vagueness may be possible, but I think the arguments above
are sufficient for the purpose at hand.

KREISEL, THE CONTINUUM HYPOTHESIS   297

NOTES

[1] Thanks to Professor Stan Weissman for criticism of an earlier draft of this paper.

[2] In this paper, we take the following as the formulation of CH: "For every infinite set of integers x, there exists a set of ordered pairs which gives a one-to-one correspondence between x and either ω or $P(\omega)$, the set of integers or the set of sets of integers."

[3] The results cited in this paragraph can be found in a number of places. An elegant presentation is in [7].

[4] This argument is modeled on that in [6].

[5] The proof: Take T^2. Add the negation of the Godel sentence for T^2. The resulting theory is consistent but since the added sentence is false, it has no $*$-models.

[6] The details on this theory are in [17], section 8.5.

[7] The proof requires only small modifications in the arguments of [16] or [25].

[8] The rank calculation (see the statement of CH in note 2): All members of ω are in $R(\omega)$, so ω and its subsets are in $R(\omega + 1)$. Thus $P(\omega) \subseteq R(\omega + 1)$, so $P(\omega) \in R(\omega + 2)$. The one-to-one correspondences are sets or ordered pairs from $R(\omega + 2)$. With a suitable (but unorthodox) definition of ordered pair, the pairs are also in $R(\omega + 2)$, and the correspondences themselves are members of $R(\omega + 3)$.

[9] The proof of this proposition is just quantification theory. Details are in [23], Appendix C.

[10] The relevant techniques are reviewed in [23], pp. 143–150. The weakest practical set of conditions seem to be those of [7], lemma 55 and p. 64.

REFERENCES

[1] J. W. Addison, "Some Consequences of the Axiom of Constructibility," *Fundamenta Mathematicae* **46** (1959).

[2] P. Bernays, "What Do Some Recent Results in Set Theory Suggest?", in Lakatos, *The Philosophy of Mathematics*, North-Holland Publishing Co., Amsterdam, 1967.

[3] E. Bishop, *Foundations of Constructive Analysis*, McGraw-Hill Publishers, New York, 1969.

[4] A. Church, *Introduction to Mathematical Logic*, Princeton University Press, 1956.

[5] K. Godel, "What is Cantor's Continuum Hypothesis?", in Benecerraf and Putnam, *Philosophy of Mathematics*, Prentice-Hall, Englewood Cliffs, 1964.

[6] L. Henkin, "Completeness in the Theory of Types," *Journal of Symbolic Logic* **15** (1950).

[7] T. J. Jech, *Lectures in Set Theory*, Springer-Verlag, New York, 1971.

[8] L. Kalmar, "On the Role of Second Order Theories" in Lakatos, *loc. cit.* in [2].

[9] G. Kreisel, "Informal Rigour and Completeness Proofs," in Lakatos, *loc. cit.* in [2].

[10] G. Kreisel, "Comments on Mostowski's Paper," in Lakatos, *loc. cit.* in [2].

[11] G. Kreisel, Appendix II, in S. MacLane, ed., *Reports of the Midwest Category Seminar* **III**, Springer-Verlag, New York, 1969.

[12] G. Kreisel, "Observations on Popular Discussions of Foundations," in D. Scott, *Axiomatic Set Theory*, American Mathematical Society, Providnece, R.I., 1971.

298   THOMAS WESTON

[13] G. Kreisel, "Two Notes on the Foundations of Set Theory, " *Dialectica* 23
(1969).
[14] G. Kreisel and J. L. Krivine, *Elements of Mathematical Logic (Model Theory)*,
North-Holland Publishing Co., Amsterdam, 1967.
[15] A. Mostowski, "Recent Results in Set Theory," in Lakatos, *loc. cit.* in [2].
[16] J. Shepherdson, "Inner Models for Set Theory," II, *Journal of Symbolic Logic* 17
(1952).
[17] J. R. Shoenfield, *Mathematical Logic*, Addison-Wesley Co., Reading,
Massachusetts, 1967.
[18] R. Smullyan, "Continuum Hypothesis," in *The Encyclopedia of Philosophy*,
P. Edwards, ed., Macmillan Co., New York, 1967.
[19] R. Solovay, "A Δ_3' Nonconstructible Set of Integers," *Transactions of the A.M.S.*
(1967).
[20] P. Suppes, "After Set Theory, What?", in Lakatos, *loc. cit.* in [2].
[21] L. Tharp, *Constructibility in Impredicative Set Theory*, unpublished Ph.D. dissertation, Massachusetts Institute of Technology, 1965.
[22] A. S. Trolestra, *Principles of Intuitionism*, Springer-Verlag, New York, 1969.
[23] T. Weston, *The Continuum Hypothesis: Independence and Truth Value*, unpublished Ph.D. dissertation, Massachusetts Institute of Technology, 1974.
[24] T. Weston, "Theories Whose Quantification Cannot Be Substitutional." *Nous* 8
(1974).
[25] E. Zermelo, "Ueber Grenzzahlen und Mengenbereiche," *Fundamenta
Mathematicae* 16 (1930).

[19]

HISTORY AND PHILOSOPHY OF LOGIC, 6 (1985), 75–89

Skolem and the Löwenheim-Skolem Theorem: A Case Study of the Philosophical Significance of Mathematical Results

ALEXANDER GEORGE

Department of Philosophy, Harvard University, Cambridge, Massachusetts 02138, U.S.A.

Received 21 September 1984

The dream of a community of philosophers engaged in inquiry with shared standards of evidence and justification has long been with us. It has led some thinkers puzzled by our mathematical experience to look to mathematics for adjudication between competing views. I am skeptical of this approach and consider Skolem's philosophical uses of the Löwenheim-Skolem Theorem to exemplify it. I argue that these uses invariably beg the questions at issue. I say 'uses', because I claim further that Skolem shifted his position on the philosophical significance of the theorem as a result of a shift in his background beliefs. The nature of this shift and possible explanations for it are investigated. Ironically, Skolem's own case provides a historical example of the philosophical flexibility of his theorem.

> Our suspicion ought always to be aroused when a proof proves more than its means allow it. Something of this sort might be called 'a puffed-up proof'.
>
> Ludwig Wittgenstein, *Remarks on the foundations of mathematics* (revised edition), vol. 2, 21.

1. Introduction

If theories are not to go the way of science fiction, then they must be subject to certain constraints. I take this to be obvious as well as the view that, however little consensus there is on what the constraints of philosophical theories could be, there is recognition that philosophical reflection is also subject to this requirement. Unbridled philosophical fantasies, like feet on a frictionless floor, get nowhere.

Constraints, however, needed as they are, do not suffice to make philosophical reflection into the kind of inquiry many of its practitioners long for it to be. Our feet, though now secure on a frictionful floor, may take off in any number of opposing directions. What is needed, in addition, is a commonality of the constraints taken to be applicable to philosophical theories. Without a consensus on the considerations germane to a theory's confirmation, adjudication between competing views will prove futile. In fact, even talk of *competing* views in such cases becomes problematic. Philosophers must learn to walk along the same paths.

These requirements have been recognized only tacitly in the philosophy of mathematics. To date, there have been few responses to these demands. The view many thinkers have taken often is something akin to the one once voiced by an enthusiastic Bertrand Russell (*1901*, 75):

> In the whole philosophy of mathematics, which used to be at least as full of doubt as any other part of philosophy, order and certainty have replaced the confusion and hesitation which have formerly reigned. Philosophers, of course, have not yet discovered this fact, and continue to write on such subjects in the old way. But mathematicians, at least in Italy, have now the power of treating the principles of mathematics in an exact and masterly manner by means of which the certainty of mathematics extends also to mathematical philosophy.

On this view, some results in mathematics carry philosophical import of their own and can serve as fixed-points relative to which competing philosophies of mathematics can be judged. This idea was carried to the limit by some Hilbertians who looked forward to the day when, as Hilbert put it, 'Mathematics in a certain sense develops into a tribunal of arbitration, a supreme court that will decide questions of principle' (*1925*, 384). Von Neumann, for example, stressed 'the fact that this question [of the origins of the generally supposed absolute validity of classical mathematics], in and of itself philosophico-epistemological, is turning into a logico-mathematical one' (*1931*, 61).

The following discussion, focussing on the Löwenheim-Skolem Theorem (LST), is part of a critique of the view that mathematics provides a suitable source of constraints on philosophical reflections about it. It is the beginning of an examination of whether the LST yields support for a philosophical position, as many, Thoralf Skolem in particular, have claimed it does, or whether it only appears to because key aspects of the position have been tacitly assumed in the interpretation of the result. If the latter regularly turns out to be the case, then one might be tempted to conclude that the LST lacks independent philosophical significance and, taken alone, is irrelevant to most traditional philosophical disputes.

My inquiry will be limited to an examination of the very first attempt, Skolem's, to foist the LST into the philosophical fray. In the process, I will urge that a significant shift in Skolem's use of the LST took place. Aside from its general historical interest, this shift provides an actual, and therefore all the more striking, example of the LST's philosophical flexibility.

2. The Löwenheim-Skolem Theorem

The LST states that if there exists a model for a countable collection of sentences of some first-order language, that is, an interpretation that makes them true, then there exists an enumerable model for this collection, that is, a model whose universe of discourse contains at most denumerably many elements. A formal system any two models of which are isomorphic is called *categorical*. It follows from the LST that no formal system that has a non-denumerable model is categorical.

Skolem gave two proofs of this theorem. The first proof, published as *1920*, used the Axiom of Choice to *pare down* the universe of discourse of the original model to a countable number of elements. The second proof, published two years later (Skolem

1922) and closer in spirit to Löwenheim *1915*, does not make use of the Axiom of Choice. In fact, the second proof was offered to show the dispensability of this axiom for establishing the LST. Leaning crucially on the Axiom of Choice, the earlier result is stronger than this because the countable model it guarantees is a restriction of the original uncountable one. The second result provides no such guarantee and the countable model that is *built up* need bear no relationship to the original one. Really then, there are two distinct results and it is misleading to speak of *the* LST. Skolem, in particular, was at great pains to distinguish the two versions (see, e.g., *1922*, 293; and *1941*, 457–458) since he felt that inquiries into the foundations of set theory were best pursued agnostic with respect to its more controversial components, e.g. the Axiom of Choice. For this reason, Skolem confined his culling of foundational and philosophical consequences to the later result. In line with this, I will intend the 1922 result by 'LST'.

Skolem, notoriously, drew conclusions concerning the 'relativity' of set-theoretic notions. According to him, the relativity resides in the fact that a set may have a property in one model but lack it in another. For example, the set of all subsets of the natural numbers, the referent of '$\mathfrak{P}(\omega)$', is not enumerable according to the model guaranteed by the LST; this is so because a model renders every theorem of the theory true, and one of these states '$\mathfrak{P}(\omega)$ is not enumerable'. But since this model is countable, every set in its universe, including the referent of '$\mathfrak{P}(\omega)$', is countable as well.

This curious state of affairs is known as 'Skolem's paradox', despite Skolem's never having considered the situation paradoxical. Its 'resolution', presented by Skolem in his 1922 Address (*1922*, 295), consists in noting that to claim that a set s is countable is tacitly to make an existence claim of the form 'there exists a one-to-one function with domain the natural numbers and range s'. The referent of '$\mathfrak{P}(\omega)$' in the countable model is countable because there exists such a mapping between it and the natural numbers which, however, does *not* exist in the countable model. The countable model is blind to the fact that the set it has '$\mathfrak{P}(\omega)$' denote is countable in the intended model. This model's claim that $\mathfrak{P}(\omega)$ is uncountable, that is, that there exists no mapping of the appropriate kind (in the countable model), is true. In so far as claims of denumerability, equinumerosity, and finitude are tacitly existence claims, the corresponding properties, relations, as well as their negations are also said to be relative. Even the relation of equality may be said to be relative in this respect.[1]

Perhaps the following consideration will make the result seem even natural. The only subsets of a set that a model need 'see' are the ones that arise out of repeated (but at most countably many) applications of the set-construction operations permissible

---

1   Given the Axiom of Extensionality, we have

$$\alpha = \beta \longleftrightarrow (z)(z \varepsilon \alpha \longleftrightarrow z \varepsilon \beta).$$

However, in a non-transitive model (the existence of which is guaranteed) not all members of elements of the domain of the model are elements of the domain. Therefore, in such a model '$\alpha = \beta$' could be satisfied by $s1, s2$ whereas it might not be satisfied by this ordered pair in a transitive model if $s1$ and $s2$ do not share all their members.

in the theory. In a countable formalism, there will be at most a countable number of such operations. Therefore, this procedure can yield at most countably many sets. It is in this sense that the LST is an artifact of the countable nature of formal systems.

Before I begin, a few words of some historical interest aimed to prevent confusion when I turn to some of Skolem's remarks below. From our perspective, it is obvious that the possession of set-theoretic properties, if relative to anything, is relative to the model one is using to interpret the formal system. Several writers[2] have noted that this was perhaps not obvious to Skolem who, they claim, often interpreted the LST as demonstrating the relativity of set-theoretic notions to the axiom system itself (see, e.g., his *1929b*, 293; and *1958*, 635–637). This confusion, on first thought historically minor, gains in interest when one realizes that it went hand in hand with Skolem's inability to keep sharp the distinction between a syntactic formal system and its semantic interpretation, between 'is (un-)satisfiable' and 'is consistent' ('is contradictory'). Bernays (reported in Skolem *1970*, 22) has suggested that this was due to Skolem's training in the Boole/Schröder/Löwenheim/Korselt tradition of logic. This school did not consider logic to be a deductive system, with the consequences (i) that the difference between syntax and semantics was blurred and (ii) that the required sensitivity to the distinction between metalogic and logic—perhaps even the very idea of such a distinction—was lacking.

Undoubtedly, this state of affairs was responsible for Skolem's failure to prove the completeness theorem for quantificational logic (or even to consider the issue of completeness), even though he had the mathematical essentials for its proof eight years before Gödel presented his.[3] Viewed in this context, Skolem's (mis)formulations of the LST are not best looked upon as curiosities possessing whatever interest a great logician's confusions or errors may have. Rather, they are of significant historical interest in illuminating the conception of logic that Skolem had, perhaps one that was dominant in the early decades of this century.

Throughout his career, Skolem claimed that the LST had profound implications for the philosophy of mathematics.[4] It is often assumed that he always drew the same consequences from the LST. I will argue, however, that there was an important shift in his use of this result. A study of this shift will be quite suggestive in determining what the philosophical consequences of the LST might be.

---

2   This is suggested by W. Hart (*1970*, 107). The 'eminent logician' who also holds these views is presumably Hao Wang; see his 'A survey of Skolem's work in logic', in Skolem *1970*, 17–52 (p.40).
3   The reader can consult W. Goldfarb's *1979* for additional information.
4   Skolem, to repeat, was not alone in this belief. For example, von Neumann (*1925*, 412) wrote that

> The consequences of all this is that no categorical axiomatization of set theory seems to exist at all [...]. And since there is no axiom system for mathematics, geometry and so forth that does not presuppose set theory, there probably cannot be any categorically axiomatized infinite systems at all. This circumstance seems to me to be an argument for intuitionism.

For a more contemporary example, see Putnam *1980*.

4. Skolem's earlier views

During his early career, Skolem was an intuitionist of sorts. As will be shown shortly, he seemed to be committed to the view that, unless our faculty of mathematical understanding or intuition was capable of apprehending with clarity the existence of given objects or the validity of given inferential operations, these objects and operations could not be countenanced. He was an intuitionist in the sense that he believed that constraints in our capacity for creative mathematical reflection were constraints on which propositions could be asserted intelligibly and truthfully. Mathematical reflection did not merely discover facts but also determined which facts were around to discover.[5]

Unlike traditional intuitionists, however, Skolem accepted the idea of a foundation for classical mathematics. In his 1922 attack on the adequacy of axiomatic set theory (AST) for founding mathematics, he does not seem to question the need for one. Indeed, in 1919 he himself attempted to found arithmetic using the non-formal means of 'the recursive mode of thought' (*1923*, 304). Rather, he faulted AST and other formal systems because they failed to meet his criterion of adequacy for any foundation of mathematics. Roughly, if a system is to provide an adequate foundation for some domain, then (at least) it must be the case that all properties of the founding system can be considered properties of the founded domain.[6] The primary characteristics of the truths we are made aware of through the use of our faculty of intuition were clarity and absoluteness. Once arrived at, a truth was unequivocal. Consequently, the mental objects of mathematics, e.g. integers, and the rules of inference employed, e.g. mathematical induction, were 'immediately clear, natural, and not open to question' (*1922*, 299).

Skolem believed the LST guaranteed that AST failed the criterion of adequacy for foundational systems. AST, dealing as it does with relativized notions (such as 'is uncountable'), could not lay claim to the clarity, naturalness, and absoluteness required of a foundational system of arithmetic, and hence of all mathematics. For Skolem 'it was so clear that axiomatization in terms of sets was not a satisfactory ultimate foundation of mathematics' (*1922*, 300–301).

It is interesting to note that Skolem never considered the possibility of a foundation erected on the basis of some non-axiomatized notion of set and its properties, or the possibility of the development of some conception of set that would evade relativization and satisfy his criterion of adequacy. Why did he seem to rule out the possibility that our intuitive faculty of mathematical reflection might lead to some future theory of sets that would provide a foundation for mathematics, a foundation, the LST would then be taken to indicate, no axiomatized formal system could represent?

5   I mention, only to put aside, the interesting issue of whether Skolem's intuitionism was a product of Brouwer's influence or had some other source, perhaps the ideas of the school within which he was trained (see the text above). Since, however, Skolem remarked in (*1929a*, 217) that the ideas of his 1923 paper were developed 'independently of Brouwer and without knowing his writings' I am inclined to discount the first source.

6   Many qualifications are in order. For example, the properties in question should be non-epistemological, for clearly one wants to permit occasions when the founding system is epistemologically more perspicuous than the founded system or vice versa (e.g., what may be called 'Russell's trickle-down theory of psychological plausibility': see his *1913*). More could be said, but I think that this formulation is sufficient for the purposes to which I wish to put the criterion in this paper.

I conjecture that Skolem saw the set-theoretic paradoxes (e.g., Russell's) as proof positive that our intuitive mathematical faculty leads us astray about sets and that a reconstruction of naive set theory would have to begin, if it were to begin at all, by taking some linguistic structure as the touchstone of truth and existence. Skolem believed (*1950*, 524) that

> the set theoretic antinomies [. . .] scattered [*sic*—shattered?] the conviction that it was possible to find logical principles which were reliable. But, certainly, the mistake that the naive set theory was reliable does not prove that it should not be possible to detect the error in the classical set theoretic thinking and perhaps formulate a really correct reasoning,

relying essentially, according to Skolem, on some formal axiomatized system.[7]

This point reinforces my claim that the early Skolem was an intuitionist, albeit an idiosyncratic one. If mathematical intuition were viewed only as an instrument of discovery, there would have been no reason for him to infer the unintelligibility of a set's *really* being uncountable from the LST. That some truths may be undetectable by our mathematical telescope (and by the linguistic means available) would then be irrelevant to their status as truths. The relativization of all set-theoretic notions is a consequence of the LST *and* the view that, because our intuition leads to paradox when applied to these notions, formal axiomatized systems are the only handle we have on them.

In support of this analysis, I note the following sharp and revealing asymmetry. In *1922*, 295–296, Skolem remarks that the number sequence, defined as the intersection of all sets having the same inductive property, may be different in different models of ZF, and he always believed that 'this definition cannot [. . .] be conceived as having an absolute meaning, because the notion subset in the case of infinite sets can only be asserted to exist in a relative sense' (*1955*, 587). Yet, in contrast to his opposition to, or neglect of, set-theoretic foundations, such considerations did not prevent him from advancing a foundation for mathematics based on the integers, 'inductive inferences and recursive definition' (*1923*, 299–300). The reason is that Skolem did not believe we are forced to rely on formal linguistic characterizations of these notions; our faculty of intuition guides us along securely in our inquiry into their properties. The asymmetry, stems from his judgment, that 'the logical intuitions which gave rise to naive set theory are rather uncertain whereas the arithmetical

----

7   See, for example, (*1922*, 291) or (*1941*, 460) where he writes:

> La découverte des antinomies ayant montré clairement que la théorie simple des ensembles, due à Cantor, ne peut pas être maintenue, on a entrepris la restauration de la théorie des ensembles, soit par la voie axiomatique, soit au moyen de systèmes logico-formels. Les deux essais reviennent au fond au même.

> According to Skolem, Cantor's naive set theory, the product of intuitive reflection on the notion 'set', sinned doubly by being vague and inconsistent (*1941*, 469), in contrast to arithmetic which, as noted, he considered 'clear, natural, and not open to question' (*1922*, 299).

ones known as recursive or inductive reasoning are quite clear and completely safe' (*1953*, 544).[8]

It is easy to see that Skolem and others might naturally have taken these reflections on the LST to reinforce a belief in our possession of a faculty of mathematical intuition whose creative but secure use was what enabled the doing of mathematics. If linguistic structures such as axiomatized formal systems could not supply the absoluteness that seemed so patent in many areas of mathematics, then its source must be sought elsewhere. What with the paucity of plausible or prominent candidates, this would naturally lead to a securing of intuition's position as guarantor of clarity, security and truth in mathematics. Indeed, this line of thought appears to have attracted von Neumann.[9]

Yet, a believer in the absoluteness of set-theoretic notions and the knowledge-independence of set-theoretic truth would find such use of the LST very suspect. Such an individual would not grant that the paradoxes of naive set theory gave our current linguistic structures the last word on which sets exist and which properties they may have. Perhaps our intuition can be 'corrected', or other as yet undreamt of means of inquiry may become available to us, or, finally, we just may never know what the facts about sets really are, these being accessible only to creatures with quite different constitutions. Skolem's move, from the paradoxes to making then current AST the arbiter of set-theoretic truth, *already* assumes the tacit rejection of this picture, a picture which he might have felt the LST undermined. On this reconstruction, it seems as if the LST's support of the naive intuitionist, over the naive realist, is purchased at the cost of tacitly assuming key elements of the former position while rejecting aspects of the latter.

However natural it would have been for Skolem to have believed the LST capable of adjudicating this dispute, it is difficult to be certain, so I will turn to an issue on which the early Skolem certainly felt the LST bore, namely, whether AST is an appropriate foundation for mathematics. On this front, his opponent is one who believes that all thoughts, mathematical or other, must be expressible in language. Although it may be very difficult to place one's thoughts into words, it is incoherent to

8   Skolem's basic position on these issues has been subject to misinterpretations. E.g., Resnik states that 'Skolem's own conclusion [. . . was] that the standard axiomatic set theories contain sets which are uncountable only relative to these set theories *but which are countable from an absolute point of view*' (*1965*, 425, italics inserted). The italicized clause is simply false. Skolem always writes of the relativity of all set-theoretic notions (see, for instance, some of the quotations reprinted below). It would be antithetical to his whole approach to set theory (as opposed, for example, to arithmetic) to suppose the intelligibility of 'an absolute point of view' from which one may determine what the properties of a given set really are. (If Skolem does somewhere write of 'absolute countability', then, instead of taking him to be making reference to the set-theoretic notion of countability, one would have to interpret him as referring to some notion of countability given to one through reflection on the construction of the number sequence by one's faculty of intuition. The issue may not arise since Resnik cites no references, and I know of none.)

9   See footnote 4. That this is in fact what happened is very hard to substantiate directly. It is, however, completely consistent with the available evidence. Skolem, according to his own report, proved the LST in 1915–1916 (*1922*, 300–301), though he only published it in 1920; the beginnings of his attempt to found mathematics on the basis of his non-formal 'recursive mode of thought' were completed by 1919 (though only published in 1923; see *1923*, 332).

countenance thoughts which must forever elude such linguistic cloaking. Such an opponent deems it illusory to imagine that we possess a magical faculty of mathematical intuition that permits us to apprehend thoughts that cannot be carried by any linguistic vehicle. In mathematics, such a characterization would be in terms of some formal axiomatized system like AST. This opponent would insist that if the notion of foundation is to play any role at all, then it cannot consist of truths that must resist characterization by all linguistic means, in particular by all formal axiomatizations of our knowledge of the domain in question. Such an opponent need not question the relativity of all set-theoretic notions; nor need she reject Skolem's criterion of foundational adequacy. Rather, she would point to the fact that Skolem enters the argument with a particular conception of the nature of mathematical truth, namely, that, in some cases, we are made aware of it through the clear and secure operation of a faculty of intuition which permits us to apprehend objects and thoughts that resist complete linguistic characterization. It seems as if this background position is needed to get the LST to render Skolem's verdict on the case of AST's foundational adequacy. But this assumption would be hotly disputed and, indeed, is part of what is at issue. In short, if someone were convinced by his argument, then it would be dispensable since he already must have assumed something like its conclusion in order to make it cogent.

Again, it is not clear that the LST can be brought to decide between these competing positions without begging the question. Skolem, at least, was unsuccessful in doing this. It is so ironic that a prime example of a historical figure who adopted this anti-early-Skolem position is Skolem himself. He shifted his use of the LST from a weapon against the foundational adequacy of AST to one wielded against the intelligibility of the language-independence of mathematical truth. I will now turn to an articulation and a defense of this claim.

4.   Skolem's later position

Over the years, Skolem abandoned his early conclusions about the adequacy of formal systems for foundational purposes. Whereas before, he believed that AST was not a suitable basis for mathematics, later (*1958*, 635), he could

> not understand why most mathematicians and logicians do not seem satisfied with this idea of sets defined by a formal system, but, on the contrary, speak of the insufficiency of the axiomatic method. Naturally, this idea of set has a relative nature since it depends on the chosen formal system. But if this system is suitably chosen, one can nevertheless develop mathematics taking it as a basis.

Skolem not only left open the possibility that formal systems, like AST, are of some use in foundational research, but went further and urged that they *should* be employed in any serious analysis of 'mathematical thought'. He declared that his 'point of view is [. . .] that one *should* use formal systems for the development of mathematical ideas' (*1958*, 634, italics added; see also p.637). In fact, his shift on the foundational adequacy of formal systems was so great that he could claim that 'one of the most important achievements in modern foundational research is the perfection of the axiomatic method known as the notion of formal system or logical language' (*1953*, 545–546) and declare that 'his conception [of 'the fundamental mathematical notions'] is founded above all on the idea of *systems* or *formal language*' (*1958*, 633).[10]

Why did Skolem repudiate his earlier conclusions on this issue? It is tempting to think that his work on, and ultimately his proof of the existence of, non-standard models of arithmetic in *1933* and *1934* shook his faith in the intuitive grasp he felt we had on the numbers and their properties, thus leading him to lay more foundational emphasis on formal systems.

This explanation fails on several counts. For one thing, it is false to the historical facts since Skolem, early and late, held that our faculty of intuition was at home with the integers. (For supporting passages and a discussion of how consistent this position is with some of his other later views, see below.) For another thing, this explanation simply begs the relevant question since he never would have felt our intuitive grasp on the numbers threatened by the existence of non-standard models *unless* he already felt our understanding was given us by some axiomatic system, in this case the Dedekind-Peano axioms. Recall how unperturbed Skolem was in *1922* when noting that different models of ZF may take different sets to be ω.

One highly relevant factor was the increasing importance formal systems took on in foundational studies and mathematical logic. The intensive work on Hilbert's program by many young and gifted mathematicians in the late 1920s is surely significant. Out of this research came not only particular results, but also a greater understanding of what a fully formalized and axiomatized system is. When these investigations culminated with Gödel's epochal paper *1931*, it would have been quite difficult for a researcher in the foundations of mathematics to deny the importance of formal systems. It really was Gödel's work (his *1931* and his work in *1930* on the completeness of quantificational logic) that signalled the general understanding of the modern distinctions between syntax and semantics, and between theory and meta-theory, and, hence, of a formal system.

The waxing of this conception saw the waning of the older tradition of logic, the one in which Skolem began his career. The reversals in his estimation of the foundational value of formal systems in general, and AST in particular, illustrate well the extent to which he was a transitional figure in the changing conception of logic that took place in the 1920s and 1930s.[11]

In sum, Skolem now believed that there were no linguistically disembodied mathematical thoughts. Although he continued to hold that mathematical objects were the creations of human minds, he no longer claimed that this activity and its results were undescribable by linguistic means. In (*1958*, 636), he wrote that 'It is a misunderstanding to speak of the insufficiency of the axiomatic method. Because mathematical objects are nothing but human thoughts and therefore the existence of

---

10 Failure to note this shift seems universal. It is, for example, a drawback of Hart *1970*. Goldfarb, also, repeatedly, but incorrectly, claims that 'Skolem was a constant opponent of all formalist and logicist foundational programs' (*1979*, 358; see also p.364).

11 The reader is advised to consult Goldfarb *1979*.

84   *Alexander George*

these objects naturally is limited as are the possible logical operations'. He now deemed axiomatized theories 'sufficient' to express all truths about mathematical objects and 'the possible logical operations' that could act on these truths to generate yet further truths. Truths and 'operations' not so formulable were considered 'impossible' (*1941*, 470).

Putting aside explanations for this about-face, what implications did it have for Skolem? Firstly, one should expect it to have consequences for the kind of mathematical work that he would consider undertaking. Indeed, in 1950, Skolem, who in 1922 criticized those who thought AST provided a foundation for arithmetic, reports that he had 'many years ago made an attempt to base arithmetic on RTT [ramified type theory] but did not succeed very well at that time. Recently I tried again with better results' (*1950*, 527).

Secondly, and more relevantly, one should expect this change of background to alter the philosophical consequences drawn from the LST. It would seem that if formal systems like AST provide suitable foundations for mathematics, perhaps even necessary ones, then, on the assumption that the LST entails the relativity of set-theoretic notions, we should abandon a language-independent view of mathematical truth. In fact, this is exactly the position that Skolem took. In (*1941*, 468), he wrote that 'The true significance of Löwenheim's theorem [the LST] is precisely this critique of the undemonstrable absolute'. It makes as much sense to ask whether a mathematical entity *really* has some property or not (e.g., whether a set *really* is finite or not) as it does to ask whether the temperature of a liquid *really* is 0° or not. Relatedly, for Skolem, there is no such thing as *the* set of all real numbers *tout court*. Relative to a choice of a formal system and a choice of a model, we can speak of the referent of a syntactic expression, in this case ' $\Re$ ', of the theory; until then, 'one does not know what the author really means' (*1953*, 583).[12]

In general, Skolem later viewed the LST as dealing a death-blow to the plausibility that there are *any* absolute (in particular, language-transcendent) mathematical truths to be captured or given a foundation. On the contrary, 'A consequence of this state of affairs [the LST] is the impossibility of absolute categoricity of *the fundamental mathematical notions*' (*1958*, 635, italics added). According to him, it followed that 'All the notions of set theory, *and consequently of all of mathematics*, find themselves in this way *relativized*. The meaning of these notions is not absolute; it is relativized to the axiomatic model' (*1941*, 468; italics added). He argued that 'if one analyzes mathematical reasoning in such a way as to formulate the fundamental modes of thought as axioms'—something the later Skolem urged (see, e.g., *1941*, 470; *1958*, 634, 635, 636—all quoted in the text)—'then the relativism is inevitable because of the general nature of Löwenheim's theorem' (*1941*, 468). He concluded

12 Resnik *1965* saddles the 'Skolemite' with the view that the set of real numbers is absolutely countable. As far as Skolem, the arch 'Skolemite', is concerned, this is doubly misguided. On his later view, the referents of many expressions in a theory, as well as *all* set-theoretic notions, are relative to choices of formal system and interpretation. Pending these, Skolem just 'does not know what the author really means' (*1955*, 583) by such words as 'is countable' or 'the set of real numbers' (see footnote 8). Independently of such choices, there is no right answer to the questions what the correct notion of countability is or what the real set of real numbers is.

that a relativist conception of the fundamental mathematical notions 'is clearer than the absolutist and platonist conception that dominates classical mathematics' (*1958*, 633). In (*1941*, 470), in an apparent reference to his earlier views, he wrote:

> That axiomatization leads to relativism is sometimes considered to be the weak point of the axiomatic method. But without any reason. Analyzing mathematical thought, and fixing the fundamental hypotheses and modes of reasoning could not but be advantageous to the science. It is not a weakness of a scientific method that it cannot give us the impossible.

That this is the best reconstruction of Skolem's later views is obscured by the fact that the historical Skolem lapses into incoherence. I will take a moment to explain why I think this is so, and why I believe that the above analysis resolves the incoherence in the most satisfactory manner.

The problem arises because Skolem, on a few occasions, falls back into talk of our clear and secure intuition which provides us with *the* foundation for arithmetic. In (*1950*, 527), discussing his disappointment that so few logicians had attempted to develop as a foundation for mathematics the system of primitive recursive arithmetic that he laid out in his *1923*, he wrote: 'When I wrote my article I hoped that the very natural feature of my considerations would convince people that this finitistic treatment of mathematics was not only a possible one but *the* true or correct one—at least for arithmetic'. It is true, he continued, that, adopting such a foundation, much of present-day mathematics would then be lost. 'The question is, however, what we shall lose or gain by such a change. As to clearness and security we certainly only gain much' (*1950*, 527). This, together with his other views, generates an inconsistency. One cannot hold simultaneously (1) that AST can and should provide a foundation for all of mathematics (including arithmetic); (2) that the LST guarantees the relativity of all set-theoretic notions; (3) that the truths of the founding system are of a kind with the truths of the founded domain (i.e., the criterion of foundational adequacy); and (4) that the truths of primitive recursive arithmetic can be seen to be clearer and more secure than any others (in particular, than those of AST). The interpretive principle that enjoins one to do minimum damage to Skolem's views requires that we restrict our discussion of what to reject to (3) and (4). I shall argue that, though either choice involves problems, we should reject (4).

The costs involved in reinterpreting Skolem as rejecting (4) are rather straightforward; we must explain away the odd occasions when Skolem writes as quoted above. This task is facilitated slightly once one realizes that the above quotation mildly misleads in several respects. First, it is not clear whether Skolem merely was reporting what his intentions were in *1923* or whether, in addition, he was endorsing and reaffirming them. In support of the first interpretation, we find Skolem assuring us later that his is the spirit of tolerance. 'I am no fanatic', he wrote (*1950*, 527), 'and it is not my intention to condemn the nonfinitistic ideas and methods'. His point then was not so much one of 'good or bad' but of 'better or worse'. Secondly, when Skolem wrote of 'finitistic mathematics', although it seems that he was referring to his non-formal 'recursive mode of thought' of *1923*, he really had *formalizations* of primitive recursive arithmetic in mind. He approvingly cited Curry's paper on 'A formalization of recursive arithmetic' (*1950*, 526) and himself referred to recursive arithmetic as a formal language (*1958*, 634). Skolem's assertion now reduces (at worst) to the claim that, though many formal systems may do the foundational trick, a foundation based on a formalization of primitive recursive arithmetic would be clearer and more secure, this clarity and security perhaps being guaranteed by some faculty of mathematical intuition. Though this claim still does not sit very well with (1)–(3), the costs involved in rejecting it are not intolerably high; that is, Skolem's position remains an interesting one and some account can be given of why occasionally he seemed to favor primitive recursive arithmetic as a foundation over other (now formal) systems. We can, at least in part, attribute this partiality to the fact that he was the originator of this arithmetic and that it had received, at least in his opinion, insufficient attention (*1950*, 526–527).

The situation is quite different when we consider rejecting (3). In the first place, doing so leaves one without any account of why AST's being a good (even a necessary) foundation for mathematics leads to the relativity of 'fundamental mathematical notions' (*1958*, 635). The criterion of adequacy assured us that if *F* provided an adequate foundation for some domain *D* and the truths of *F* have property *P*, then the truths of *D* have property *P*. We have seen that Skolem, early and late, assumed some version of this.

More deeply, the criterion of adequacy seems to be built into the historical notion of a foundation in that, usually, *F* is offered as a foundation for *D* when the nature of the truths of *D* with respect to some property *P* is not known but there is confidence about whether the truths of *F* possess *P* or not. Thus, Frege believed that a reduction of arithmetic to logic would decide the issue whether the truths of arithmetic were analytic or not since the truths of logic clearly were. And Hilbert attempted to found classical mathematics on finitary mathematics (by using the latter to give a consistency proof of the former) with the expectation of showing that all finitary truths derived classically could be proved by finitary means alone. If one rejects the criterion of adequacy, then one prominent historical motivation for foundations is lost, since the nature of the truths of the founding system may differ utterly from those of the founded system.

In short, the costs of securing the consistency of Skolem's position by jettisoning (3) are very great. Doing so would leave Skolem with a rather uninteresting position philosophically, since a historically important philosophical rationale for seeking mathematical foundations would be undercut, a rationale which he always accepted tacitly. For these reasons, the above reconstruction of his views best preserves their philosophical interest and does least damage to the historical record (though, of course, any reconstruction that renders Skolem consistent will have to do some such damage).

What is of particular interest in this context is the shift in Skolem's background assumptions and the concomitant shift in his philosophical employment of the LST. Whereas the early Skolem began by assuming the language-transcendence of certain mathematical truths, e.g. those of arithmetic, and argued that the LST rendered doubtful the foundational utility of AST, the later Skolem affirmed the foundational indispensability of formalized, axiomatized theories and used the LST to urge that all mathematical truth is relative and language-dependent. Naturally, the later Skolem's argument has no more force against one who starts with a language-transcendent view of mathematical truth, than his earlier arguments have against a committed 'relativist'. The tacit assumption that mathematics can do with (and, indeed, cannot do without) a formalized, axiomatized foundation, where it is understood that an acceptable foundation must meet the criterion of foundational adequacy, would be rejected by the believer in the language-independence of mathematical truth; it seems as if the premises needed to draw the conclusion of relativism from the LST would be as unpalatable to the absolutist as the conclusion itself.

This historical case study exhibits the philosophical leeway of the LST and suggests that its philosophical consequences are parasitical on the philosophical views one conjoins with it. In the writings of both the early and the late Skolem, there are instances of philosophical positions deriving support allegedly from a mathematical result that seems philosophically lifeless without the infusion of precisely those or related positions.

5. Conclusion

Obviously, not all the philosophical possibilities of the LST have been explored, only those Skolem thought it had.[13] Though further exploration is in order, one should expect that other attempts to distill potent philosophical spirits directly from the bare LST (or related results) will fare no better than Skolem's. The more one examines the case of the LST, the more skeptical one becomes of the existence of precious philosophical nodules lying beneath the surface just waiting to be dug up.

As mentioned earlier, some thinkers, following Russell's lead, have hoped that mathematics could provide constraints for the philosophy of mathematics; that is, they have hoped that if philosophers just could formulate their questions precisely enough and attend to the doings of mathematicians carefully enough, then answers would be forthcoming finally to their problems. Yet, one must recognize that mathematics is just another aspect of the total human experience that so perplexes and leads to philosophy. Mathematical experience is no more philosophically self-interpreting than any other. For the most part, we do not understand the full philosophical significance of a mathematical truth (or any other truth for that matter) until we understand the role it plays in an articulated theory that brings together in harmonious fashion various central conceptions.

A mathematical result shines only when illuminated by other views. When these are simple as they are in Skolem's use of the LST, the shine is more like a crude reflection, as sections 3 and 4 suggest. It remains to be seen whether aspects of our mathematical experience, including, perhaps, particular of its products, can be brought together in a sophisticated, satisfying, and self-reinforcing manner.

---

13   For example, Putnam's recent use *1980* of this theorem has not been addressed.

88   Alexander George

Acknowledgements

Many thanks to William Aspray, Burton Dreben, Michael Dummett, Warren Goldfarb, James Higginbotham, Daniel Isaacson, Mary McComb, Thomas Pogge, Hilary Putnam, and Michael Resnik for helpful discussions and comments. A version of this paper was read to Oxford University's Occam Society in Michaelmas Term 1983, and to Robin Gandy's Foundations of Set Theory seminar in Hilary Term 1984; I am grateful for these opportunities and to those who attended for their questions and suggestions. Several drafts of this paper were written while I was a Fulbright Scholar at New College, Oxford in 1983–1984; I would like to thank the American and British peoples for making that wonderful year possible.

This essay is dedicated to the cherished memory of Nelson.

Bibliography

'†' indicates the edition of a work cited by page number in the text.

Benacerraf, P. and Putnam, H. *1984 Philosophy of mathematics: selected readings*, 2nd edition, Cambridge (University Press).

Gödel, K. *1930* 'The completeness of the axiom of the functional calculus of logic', translated in van Heijenoort *1967*, 582–591.

—— *1931* 'On formally undecidable statements of *Principia Mathematica* and related systems', translated in van Heijenoort *1967*, 596–616.

Goldfarb, W. *1979* 'Logic in the twenties: the nature of the quantifier', *Journal of symbolic logic*, **44**, 351–368.

Hart, W. *1970* 'Skolem's promises and paradoxes', *Journal of philosophy*, **67**, 98–108.

van Heijenoort, J. (ed.) *1967 From Frege to Gödel*, Cambridge, Mass.

Hilbert, D. *1925* 'On the infinite', translated in van Heijenoort *1967*, 367–392.

Löwenheim, L. *1915* 'On possibilities of the calculus of relatives', translated in van Heijenoort *1967*, 228–251.

von Neumann, J. *1925* 'An axiomatization of set theory', translated in van Heijenoort *1967*, 393–413.

—— *1931* 'Die formalistische Grundlegung der Mathematik', *Erkenntnis*, **2**, 116–121; translated in Benacerraf and Putnam *1984*, 61–65.

Putnam, H. *1980* 'Models and reality', *Journal of symbolic logic*, **45**, 464–482.

Resnik, M. *1966* 'On Skolem's paradox', *Journal of philosophy*, **63**, 425–437.

Russell, B. *1901* 'Mathematics and the metaphysicians', in his *Mysticism and logic*, 1957, London, ch. 5†.

—— *1913* 'The philosophical importance of mathematical logic', reprinted in his (ed. Lackey, D.) *Essays in analysis*, 1973, London, 284–294 [and mistitled]†.

Skolem, T. *1920* 'Logisch-kombinatorische Untersuchungen über die Erfüllbarkeit und Beweis barkeit mathematischen Sätze nebst einem Theoreme über dichte Menge', *Skrifter, Videnskabsakademiet i Kristiania I*, 4, 1–36; reprinted in Skolem *1970*, 103–136; excerpts translated in van Heijenoort *1967*, 252–263†.

—— *1922* 'Einige Bemerkungen zur axiomatischen Begründung der Mengenlehre', *Proceedings of the 5th Scandinavian Mathematical Congress*, Helsinki, 217–232; reprinted in Skolem *1970*, 137–152; translated in van Heijenoort *1967*, 290–301†.

—— *1923* 'Begründung der elementären Arithmetik durch die rekurriende Denkweise ohne Anwendung scheinbarer Veränderlichen mit unendlichem Ausdehnungsbereich', *Skrifter, Videnskabsakademiet i Kristiania I*, 6, 38 pp.; reprinted in Skolem *1970*, 153–188; translated in van Heijenoort *1967*, 302–333†.

—— *1928* 'Über die mathematische Logik', *Norsk Matematisk Tidsskrift*, **10**, 125–142; translated in van Heijenoort *1967*, 508–524†.

Skolem and the Löwenheim-Skolem Theorem   89

────── *1929a* 'Über die Grundlagendiskussionen in der Mathematik, in *Proceedings of the 7th Scandinavian Mathematics Congress*, Oslo, 3−21; reprinted in Skolem *1970*, 207−225[†].

────── *1929b* 'Über einige Grundlagen fragen der Mathematik', *Skrifter, Vitenskapsakademiet i Oslo I*, **4**, 1−49; reprinted in Skolem *1970*, 227−273[†].

────── *1933* 'Über die Unmöglichkeit einer Charakterisierung der Zahlenreihe mittels eines endlichen Axiomensystems', *Norsk, Matematisk Forening, Skrifter*, (2) No. 1−12, 73−82; reprinted in Skolem *1970*, 345−354[†].

────── *1934* 'Über die Nichtkaracterisierbarkeit der Zahlenreihe mittels endlich oder abzählbar unendlich vieler Aussagen mit ausschliesslich Zahlenvariablen', *Fundamenta mathematica*, **23**, 150−161; reprinted in Skolem *1970*, 355−366[†].

────── *1941* 'Sur la porté du théorème de Löwenheim-Skolem', in *Les Entretiens de Zürich*, 6−9 December 1938, Zürich, 25−47, discussion, 47−52; reprinted in Skolem *1970*, 455−482[†].

────── *1950* 'Some remarks on the foundation of set theory', in *Proceedings of International Congress of Mathematicians*, Cambridge, Massachusetts, 695−704; reprinted in Skolem *1970*, 519−528[†].

────── *1953* 'The logical background of arithmetic', *Bulletin de la Societé Mathématique de Belgique*, 23−34; reprinted in Skolem *1970*, 541−552[†].

────── *1954* 'Peano's axioms and models of arithmetic', in *Studies in logic and the foundation of mathematics*, Amsterdam, 1−14; reprinted in Skolem *1970*, 587−600[†].

────── *1955* 'A critical remark on foundational research', *Kongelige Norske Videnskabsselskabs Forhandlinger*, Trondheim, **20**, 100−105; reprinted in Skolem *1970*, 581−586[†].

────── *1958* 'Une relativisation des notions mathématiques fondamentales', in *Colloques internationaux du Centre National de Recherche Scientifique*, Paris, 13−18; reprinted in Skolem *1970*, 633−638[†].

────── *1970 Selected works in logic*, edited by Fenstad, J.E., Oslo (Universitetsforlaget).

[20]

SKOLEM AND THE SKEPTIC

I—Paul Benacerraf

In 1922 Thoraf Skolem published 'Some Remarks on Axiomatized Set Theory'. This paper contains a powerful skeptical argument which, siren-like, has lured philosopher after philosopher at least to flirt with the disaster toward which it beckons. One of the argument's most recent and worthy *naufragés* is Hilary Putnam, who has deployed Skolem's argument as a major move in the realism/anti-realism wholly wars (Putnam [1977]).

The present paper is about Skolem, an imaginary being (The Skeptic), their respective views on set theory as a foundation for mathematics and on the foundations of set theory itself, and by way of commentary and afterthought, some uses to which Putnam has recently put these views and arguments.

In outline:
  I   Introductory remarks
  II   The Skeptic
  III Skolem [1922]—an analysis of the structure of that paper
  IV   Why the turnabout?
  V   Discussion of the Skeptic's central argument
       (A) As used in [1922]
       (B) More generally
  VI   An application of the Skeptic's reasoning: Hilary Putnam

I

Introductory remarks

I presuppose some familiarity with the basic concepts of set theory, and, to some extent, with Zermelo [1908a]. For the reader without the latter, I include in an Appendix a list of Zermelo's axioms, as well as a very sketchy summary of the philosophical views he advances there, too sketchy to do them justice, but I hope sufficiently ample to make the present paper intelligible. The interested reader should consult Zermelo's

86   I—PAUL BENACERRAF

paper for a very straightforward presentation (of all except perhaps for the disputed aspects of Axiom III—see the discussion below).

My discussion will revolve around my reading of the extremely fecund Skolem [1922]:

It contains, among other things, (1) the 'syntactic' explication of Zermelo's notion of 'definite property' as employed in Axiom III, the *Aussonderungs* axiom; (2) a suggested strengthening of Zermelo's axioms by the addition of the axiom of substitution —whose customary attribution solely to Fraenkel is codified in the customary name 'ZF'; (3) a simplified proof of Löwenheim's theorem (amounting to an almost-proof of the Completeness Theorem for First Order Logic); (4) the renowned application of the Löwenheim-Skolem theorem in the argument that the concepts of axiomatic set theory are inexorably relative—most particularly, that there is no absolute concept of non-denumerability;[1] (5) that ZF (without the axiom of Foundation) has non-isomorphic models, specifically ones with infinite descending $\in$-chains; and more.

All this is well known.

But I read Skolem [1922] as a paper written to make a philosophical point. I will try to bring out what that point is and locate it in the development of Skolem's own thought and in current debates on the foundations of set theory, mathematics, and philosophy.

II

The Skeptic
For ease of reference, I introduce 'The Skeptic'. The Skeptic will turn out to be largely a skeptic, of sorts. But nothing will hang on that, and no effort will be expended in defense of the view that The Skeptic is a skeptic—i.e. that his views are in fact *skeptical* views. (That would require a deeper understanding of philosophical skepticism than I possess. And besides, as I said above, nothing will hang on that issue.)

---

[1] I will use 'countable', 'enumerable', and 'denumerable' as stylistic variants of one another (and similarly for their cognates). M is countable if and only if it is the range of a function (partial or total) whose domain is a subset of the natural numbers. (Thus, harmlessly, 'countable' includes 'finite'. Where it matters, we will specify 'countably infinite'.)

The Skeptic believes several things, most notoriously that the concepts 'denumerable set' and 'non-denumerable set' must be viewed as relative, not absolute, in a sense of 'relative' and 'absolute' that can be culled from an examination of his arguments. Indeed, the relativity in question is claimed for *all* set-theoretic concepts, but attention is usually focused on the cardinality concepts and the ones used to define them (finite, non-denumerable, denumerable, subset, etc.) largely, I think, because of their prominence in the development of Cantorian (and post-Cantorian) set theory. Set theory *is* a theory of cardinality (and ordinality), to put it contentiously.

The Skeptic also believes that set-theoretic concepts should be presented axiomatically, in first-order formalization. As we will see, expressing the view in terms of the notion of *formalization* may be a bit artificial and somewhat controversial. But 'first-order' is a necessary ingredient of the view; it is normally taken to be a *syntactical* concept which may itself require for its explanation at least a flirtation with the notion of formalization, if not a full employment of it. What 'formalizability as a first-order theory' amounts to—how it is to be explained—is a question of critical import, as the very applicability of the Skeptic's principal argument depends on it.[2]

The Skeptic then does his familiar number:

---

[2] It would be interesting, both historically and philosophically, to examine the role of formalization—representation in a logistic system—in the development of logic and its philosophy. *We* think of the Löwenheim-Skolem Theorems as ones that apply exclusively to formal languages, or at least to languages whose syntax is sufficiently well specified to be clearly first order languages—whatever that may amount to. But it's not entirely clear that Skolem did.

It might well be (though this is hardly the place to examine the issue) that the concept of a logistic system and the notion of formalization on which it depends is an outgrowth of the Frege/Russell-Whitehead/Hilbert-Bernays/Carnap tradition in logic, whereas Löwenheim, Skolem, and others were bred from another, more algebraic strain, the Boole/Schröder tradition. There is philosophical substance here bearing directly on how the paper that is before us—Skolem [1922]—is to be interpreted: the intended thrust of Skolem's arguments will vary with the background 'premises' he takes himself to be granted without argument. As will soon emerge, I claim Skolem [1922] to be arguing *against* The Skeptic that he is later to become. His paper will not make sense unless we suppose him to be granting himself premises (e.g. that some intuitive principles of model theory make sense) that The Skeptic would find unintelligible. But *these* principles (used in the proofs and subsequent interpretations of Löwenheim's theorem and of Skolem's generalization of it) are more intelligibly seen as in the Boole/Schröder algebraic tradition than in the strain to which Frege, Carnap, *et al.* belong.

88   I—PAUL BENACERRAF

The Skeptical Argument. Suppose that it has been 'proved' in
some first-order set theory that there exist non-denumerable
sets—perhaps by proving a theorem to the effect that some
particular set, the power set of the integers (henceforth PZ_o), say,
is non-denumerable. If the theory is consistent, it has a
countable model in which sets 'proved' to be *un*countable *within*
the theory must be denumerable, since the entire domain of the
model can be enumerated. Hence 'uncountable' as determined
(defined?) by the axioms applies to an object (a 'collection'),
PZ_o, which is *provably countable* from another standpoint. (We
will, of course, return to this argument.)

Finally (one of his less publicized views): The Skeptic believes
that set theory, thus construed, makes a suitable foundation for
mathematics: mathematical axioms and concepts are to be
reduced to those of set theory. Grafted onto such a reduction, the
Skeptical argument banishes the (absolutely) non-denumerable
from mathematics altogether.

Who ever held this view? Many, I suppose; including Skolem,
of course. Several Skolems, actually. A typical one is the 1941
Skolem. In his paper 'Sur la portée du théorème de Löwenheim-
Skolem' he argues that the Löwenheim-Skolem Theorem
(henceforth 'LS', except where it matters which version is at
issue) implies the relativity of cardinality and concludes:

> Since all reasoning in axiomatic set theory or within a
> formal system is carried out in such a way that the
> absolutely non-denumerable does not exist, the claim that
> absolutely non-denumerable sets exist should be considered
> a mere pun; this absolute non-denumerability is therefore
> only a mere fiction. The true import of Löwenheim's
> theorem is precisely this critique of the absolutely non-
> denumerable. In short: this critique does not reduce the
> higher infinities of the simple theory of sets *ad absurdum*, it
> reduces them to non-objects. ([1941a], p. 468)

Two pages later:

> . . . that axiomatization should lead to relativism is a fact
> sometimes considered to be a weak point of the axiomatic
> method. But without reason. The analysis of mathematical
> thought, the fixing of its fundamental hypotheses and

modes of reasoning can only be an advantage for science. It is not a weakness of a scientific method that it doesn't yield the impossible. But it appears that most mathematicians are terrified that the absolute theory of sets should turn out to be an impossibility. ([1941a], p. 470)

A near cousin, the Skolem of 1958 says (in 'Une relativisation des notions mathématiques fondamentales'):

My point of view is therefore that one should employ formal systems for the development of mathematical ideas. ([1958a], p. 634)

I do not understand why most mathematicians and logicians do not seem to be satisfied with this notion of set as defined by a formal system, but on the contrary speak of the inadequacy of the axiomatic method. Naturally this notion of set has a relative character: for it depends on the formal system chosen. But if this system is suitably chosen, one can, nevertheless, develop mathematics on its basis. ([1958a], p. 635)

To summarize, the Skeptic holds that:

(*i*)   Set-theoretic concepts must be presented in axiomatic form, in first-order formalization, and that
(*ii*) Thus presented, set theory constitutes an adequate foundation for mathematics.

Applying LS, he then concludes that

(*iii*) set-theoretic notions [and the ensuing mathematical notions defined in terms of them] are *relative*, not absolute.

Notably, no 'set' is absolutely 'non-denumerable', because any 'set', 'non-denumerable' in some model of set theory, is 'denumerable' in some other.

So much for The Skeptic and the Skolems of [1941a] and [1958a]. They are a perfect match.

III

But what of the Skolem who first advanced the relativistic argument: Skolem [1922]? His intriguing paper is a blistering attack on Zermelo's 1908 paper 'Investigations in the

90   I—PAUL BENACERRAF

Foundations of Set Theory', in which Zermelo makes three
important claims:

(*a*) It is the task of set theory to serve as a foundational
discipline for mathematics.
(*b*) Axiomatization is the way of salvation for set theory, its
intuitive underpinnings having been cut away by the
paradoxes.
(*c*) His axioms (now familiar to us) should be adopted as
the appropriate foundation for this foundation.

Skolem takes exception to *all three* of these claims, with the
conjunction of (*a*) and (*b*) bearing the brunt of his attack.
Indeed, his principal point can be put:

Set theory, *axiomatically presented*, cannot serve as a
foundation for mathematics.

Once more, by 'axiomatically presented' I take Skolem to mean
something like 'presented as a theory formalized (or formalizable) in the Lower Predicate Calculus (LPC)', where *that*
serves to guarantee that the meaning of the logical constants,
and only the logical constants, is fixed. It is otherwise extremely
unclear how LS (particularly the *numerical* version Skolem
invokes) could be applied in the argument.
    Skolem [1922]'s *adversary*—a near twin of Zermelo—occupies
the following position:

(*a*) Set theory (and its fundamental concepts) must be
presented axiomatically in first-order formalization.
(*b*) Thus presented, set theory (in particular, the system Z
of Zermelo [1908a], suitably amended and dressed up as a
first order theory) is an adequate foundation for mathematics.

Despite its striking resemblance to The Skeptic's position
(and to that adopted by the Skolem of [1941a] and [1958a]), *this*
is the view Skolem [1922] attacks. To be perfectly explicit, what
I am claiming is that, far from adopting The Skeptic's position,
Skolem in [1922] is arguing directly against it. I was long misled
on this very issue because, seeing the relativity argument in
[1922] I read back into that paper the view which that argument
is normally used to support. But it is the burden of the first

portion of the present paper to argue that the relativity argument appears in [1922] as just one more nail in the coffin of Zermelo's view—as a *reductio* of what happens when you maintain that the set-theoretic concepts are implicitly defined by the axioms: they end up relative. Skolem's paper has eight sections and a conclusion. I will comment briefly on each.

§1. Presented axiomatically, Z (Zermelo's Theory) can't serve as a foundation for set theory because to present a theory axiomatically is to presuppose the transparently set-theoretic notion of a *domain*:

> The entire content of this theory is, after all, as follows: for every domain in which the axioms hold, the further theorems of set theory also hold. But clearly it is somehow circular to reduce the notion of set to a general notion of domain. (p. 292)

He adds the further remark that it won't do in reply to replace the general concept of domain with that of the particular universe B of sets; for if that were permissible, then if a question arose 'about some unspecified set' we could, *mutatis mutandis*, waive the general notion of set and say: 'No general notion of set is needed, but only the idea of a single set that we assume to be given.' (p. 292)

§2 contains a critique of Zermelo's notion of 'definite proposition' in Axiom III, the Aussonderungs axiom ('Probably no one will find Zermelo's explanation of it satisfactory' (p. 292)). Skolem proposes its replacement by the well-known syntactic definition, in which the troublesome notion of a 'definite property' is replaced by the *syntactic* concept of an open sentence with a single free variable. (For a little more on this, see fn. 4 below.)

For a brief moment, this will have the appearance of a *constructive* suggestion.

§3. Skolem offers a simplified proof[3] of the following form of the Löwenheim-Skolem theorem:

---

[3] The 'simplified proof' is one which avoids the axiom of choice (which Löwenheim employed in his original proof). Skolem motivates this avoidance by philosophical scruple: '. . . where we are concerned with the foundations of set theory, it will be desirable to avoid the principle of choice as well.' (p. 293) That sounds unexceptionable. Still, it is worth noting in passing that such scruple is often excessive, and may be so in

92   I—PAUL BENACERRAF

> If a denumerable set of First Order propositions is
> consistent (i.e., has a model), it has a model in the integers.

After presenting this proof, he notices that his 'clarification' in
§2 of Zermelo's concept of a definite proposition has insured that
the axioms of Z constitute 'a denumerable set of First Order
propositions' (by reducing what was formerly an indefinite, or
circular, or second order axiom to an infinite list of first order
axioms). He concludes that Zermelo's axiom system, 'when
made precise' (these are his words), must itself have a model in
the integers, if it has any model at all.[4]

This sets the stage for the famous passage, the *locus classicus* of
so-called 'Skolem Paradox' and the origin of the doctrine of the
relativity of set-theoretic concepts:

> So far as I know, no one has called attention to this peculiar
> and apparently paradoxical state of affairs. By virtue of the
> axioms we can prove the existence of higher cardinalities,
> of higher number classes, and so forth. How can it be, then,
> that the entire domain B can already be enumerated by
> means of the finite positive integers? The explanation is not
> difficult to find. In the axiomatization, 'set' does not mean
> an arbitrarily defined collection; the 'sets' are nothing but
> objects that are connected with one another through
> certain relations expressed by the axioms. Hence there is
> no contradiction at all if a set M of the domain B is
> nondenumerable in the sense of the axiomatization; for this
> means merely that *within* B there occurs no one-to-one
> mapping Φ of M onto Z_0 (Zermelo's number sequence).
> Nevertheless there exists the possibility of numbering all
> objects in B, and therefore also the elements of M, by

---

this case. If the conclusion of the investigation is to be a skeptical one—if the form of the
argument is a *reductio* of a set of propositions accepted by his adversary and that include
AC among them, as in the present case, it is perfectly legitimate, *ad hominem*, to use
propositions from the questionable set in the argument. However there do remain
excellent reasons not to use AC in the present case. I mention two. (1) The relativistic
conclusions are meant to apply to versions of set theory that don't include AC, and (2)
the argument is more informative if it can be presented with weaker premises.

Although the use of AC in the proof of LS is largely (although not entirely) tangential
to the matters that concern us here, I mention the passage to highlight the *philosophically*
contentious character of Skolem's paper.

[4] *See Appendix I.*

means of the positive integers; of course, such an enumeration too is a collection of certain parts, but this collection is
not a 'set' (that is, it does not occur in the domain *B*.) (p.
295)

Before returning (in section E below) to discuss this argument
in some detail, I will complete my survey of Skolem's paper. We
are still in Section 3 and Skolem has a few more shots to take. He
suggests (somewhat obscurely) what one might take as the first
hint (1) of the existence of non-standard (non-isomorphic)
models for the Zermelo integers and (2) of a good reason for it.

Likewise, the notion 'simply infinite sequence' or that of
the Dedekind 'chain' has only relative significance. If Z is a
set having the property required by Axiom VII, Zermelo's
number sequence Z_0 is defined as the intersection of all
subsets of Z that have the same (chain) property. But being
a subset of Z is not merely to be in some way definable, and
nothing can prevent a priori the possibility that there exist
two different Zermelo domains B and B' for which different
Z_0 would result. (pp. 295–6)

He concludes §3: 'Thus *axiomatizing set theory leads to a relativity
of set-theoretic notions, and this relativity is inseparably bound up with
every thoroughgoing axiomatization*' (p. 296; his emphasis). Given
Skolem's own work in model-theory and his earlier remarks
about the set-theoretic character of model-theoretic reasoning,
it is difficult to escape the conclusion that his target here is *not* set-
theoretic concepts in general, but only *axiomatically presented (i.e.,
implicitly defined)* set-theoretic concepts. He ends the section with
two remarks, both of which seem to me to support this
interpretation. Again, he does best speaking for himself:

With a suitable axiomatic basis, therefore, the theorems of
set theory can be made to hold in a merely *verbal* sense, on
the assumption, of course, that the axiomatization is
consistent; but this rests merely upon the fact that the use of
the *word* 'set' has been regulated in a suitable way. We shall
always be able to define collections that are not called sets;
if we were to call them sets, however, the theorems of set
theory would cease to hold. (p. 296)

94   I—PAUL BENACERRAF

In other words, implicitly define 'set' and, modulo consistency, you can't go wrong. But the arguments of §3 show that any such fixing of the concept 'set' will leave out collections which, relative to that fixing, we don't happen to be calling 'sets' (but which deserve the name nonetheless).

There is a gap in the argument here, even construed as *ad hominem* against Zermelo. At most what has been shown in §3 is that given any denumerable set of first order axioms A (presumably that are in some sense recognizably *set-theoretic*), and *any model* M for these axioms, there are collections $(B_1, \ldots, B_i, \ldots)$ definable on M, that don't belong to M. It has not been shown that the B_i aren't sets *in the sense defined by the axiomatization* (whatever that may be). The B_i depend not only on A, but on M as well, and no argument has been given that there is no model M′ to which *these very* B_i belong, thus legitimizing their sethood. This opens the question of the very meaning of the claim of relativity: Is it relativity to an axiomatization? To a model, given an axiomatization? Both?

Without straining charity, we can fill the gap and interpret him as saying that any denumerable system A of first-order axioms has denumerable models if it has any at all; now, given A and a denumerable model M for A, there are collections not provided for by M. Hence defining 'set' by fixing A won't adequately fix the meaning of 'set'.

This, remember, is Skolem.

§4 is a constructive attack on Z, pointing out that although

$$Z_o, \; PZ_o, \; PPZ_o, \; \ldots$$

all can be proved to exist, the union of all the members of the sequence cannot. He suggests what we now know as the Axiom of Substitution, in the form: The range of any function whose domain is a set is itself a set. This is the axiom that we think of as Fraenkel's contribution (and which Fraenkel did indeed obtain independently).

§5 is a complaint against the impredicativity of Zermelo's axioms considered as generating principles. And the complaint is a narrowly philosophical one: The impredicative nature of the *Aussonderungs* axiom, in the presence of the axiom of infinity (he notes that without it there is a model in the finite sets), renders proof of consistency by constructive methods unlikely (remember, consistency for Skolem is a semantical concept—it means: having a model). He appears to make two points. (1) It is unlikely that a constructive proof can be found, i.e., that a suitable model can be *constructed* and (2) even if one were found, it wouldn't be available to the likes of Zermelo, Russell, Whitehead, etc., since it would depend on the concept of integer that they all require be defined in set theory—once more Skolem is arguing *ad hominem* against those who would reduce mathematics to some version of set or type theory.

Commenting on what appears to be this passage, Jon Barwise astutely notices Skolem's anti-Skeptic bent and remarks:

> Skolem's suggestion . . . that we use 'first order property' for 'definite property' explicitly presupposes that one is willing to form the universe V of all sets,
>
> $$V = \bigcup_{\alpha} R_{\alpha}$$
>
> (the union taken over all ordinals) and consider it a meaningful mathematical totality. In fact this was one of Skolem's arguments against using set theory as a foundation for mathematics.[5]

[5] K. J. Barwise ([1972], p. 593). It is perhaps worth noting that Barwise's remark appears to be too strong in two ways. (1) Unquestionably, the natural reading of the resulting theory requires one to *quantify* over all sets. Whether this amounts to a willingness to form $\bigcup_{\alpha} R_{\alpha}$ and consider it to be a meaningful mathematical totality depends upon whether the right interpretation of quantifiers involves such a willingness. I hope not. (2) Secondly, the impredicativity appears to infect *numerical models* as well, as Skolem was perhaps aware (hence the 'difficulty' of producing a Löwenheim model which, for Skolem, would have been a numerical one). It appears to depend on the first order structure of Z (ZF) not on its interpretation in terms of sets.

It is unclear what this comes to. Someone might counter, maybe Barwise himself: 'The impredicativity argument is often directed at the ZF axioms (and others) when considered as *generating principles* for the elements of the domain. Viewed as descriptive of the relations that obtain among the elements of the domain, they are unexceptionable (See Gödel [1944]). So the numerical models are immune because we have reasons independent of the ZF axioms for believing in the existence of the numbers.' But I would find such a reply unconvincing, because what's at issue in an alleged numerical model for ZF may well be the truth of the claims of the model *about the integers*. What reason could we have for thinking that the integers of the model are related to one another as the axioms specify? What are the properties attributed by the axioms? Any 'construction' of a numerical model would be a proof that the axioms held in some domain consisting of integers and would thus imply the consistency of ZF. It would therefore be a mathematical theory of considerable complexity.

96   I—PAUL BENACERRAF

§6 contains another pot shot: Your axioms don't 'uniquely
determine' your domain, pointing out the possibility of two
kinds of models; those that are well-founded and those that are
not. If 'uniquely' is to be interpreted as 'up to isomorphism', the
cardinality arguments already establish that. However, the
dialectical position is such that he may not have wanted to avail
himself of notions he had argued were unavailable to his
opponent—such as the cardinality concepts. In any event,
although it is not clear that he had a proof of it, the result is of
interest in its own right and independent of cardinality and
holds within each cardinality.

§7 contains a discussion of the difficulties Zermelo's view
poses for foundational investigations of set theory (consistency
and independence proofs): These can proceed set-theoretically,
by constructing set models and counter-models, in which case
they beg the question; or they can proceed syntactically, in
which case they presuppose the concept of an 'arbitrary finite
number' [of applications of the axioms], and thereby, as we have
seen Zermelo [1930] was later also to argue, concepts destined to
be reduced to set-theoretic terms. This would make them
viciously circular. There follows a beautifully anti-Zermelo,
anti-Skeptic passage:

> Set-theoreticians are usually of the opinion that the notion
> of integer should be defined and that the principle of
> mathematical induction should be proved. But it is clear
> that we cannot define or prove ad infinitum; sooner or later
> we come to something that is not further definable or
> provable. Our only concern, then, should be that the
> initial foundations be something immediately clear,
> natural, and not open to question. This condition is
> satisfied by the notion of integer and by inductive
> inferences, but it is decidedly not satisfied by set-theoretic
> axioms of the type of Zermelo's or anything else of that
> kind; if we were to accept the reduction of the former
> notions to the latter, the set-theoretic notions would have
> to be simpler than mathematical induction, and reasoning
> with them less open to question, but this runs entirely
> counter to the actual state of affairs. (p. 299)

He ends the section with a similar diatribe against Hilbert.

§8 is a comment about the Axiom of Choice, that further reveals his own bias—evident throughout the paper—for an intuitive and probably predicative conception of set. It is brief and can be quoted in full.

> 8. So long as we are on purely axiomatic ground there is, of course, nothing special to be remarked concerning the principle of choice (though, as a matter of fact, new sets are *not* generated *univocally* by applications of this axiom); but if many mathematicians—indeed, I believe, most of them—do not want to accept the principle of choice, it is because they do not have an axiomatic conception of set theory at all. They think of sets as given by specification of arbitrary collections; but then they also demand that every set be definable. We can, after all, ask: What does it mean for a set to exist if it can perhaps never be defined? It seems clear that this existence can be only a manner of speaking, which can lead only to purely formal propositions—perhaps made up of very beautiful *words*—about objects *called* sets. But most mathematicians want mathematics to deal, ultimately, with performable computing operations and not to consist of formal propositions about objects called this or that. (p. 300)

In his Conclusion, which is certainly susceptible to mis-understanding if you read only the first sentence, 'The most important result above is that set-theoretic notions are relative', Skolem makes it plain that he is not The Skeptic: Although he had obtained these results six or seven years back, he hadn't published them because he 'believed that it was so clear that axiomatization in terms of sets was not a satisfactory foundation for mathematics that mathematicians would, for the most part, not be very much concerned with it'. But they were.

IV

Why the turnabout?
Somewhere between 1922 and 1941 Skolem *drops* his reliance on intuitive mathematics, and with it, the anti-reductionist arguments; but he continues to accept all the arguments *for* relativity and non-categoricity. Whereas in 1922 these latter arguments had functioned as *reductio* arguments of *Zermelo's*

view, they are now used to point to the inescapable consequences
for the philosophy of mathematics of the only tenable
foundational position: the very brand of reductionism he *attacked*
in [1922].

I don't know why he changed.[6] I don't even know of any
passage in which he explicitly recants, although it appears that
[1929b] may be a transitional paper in precisely this regard. Its
first two sections are entitled 'Proof of (the) set-theoretical
relativity' and 'Proof of the Relativity Without Appeal to
Löwenheim's Theorems' and contain arguments that are
largely the same as those offered in [1922]. Unlike [1922], in
which 'axiomatic' was present as a *qualifier* to the set theory
whose concepts were being shown to be relative, here it is
dropped: It is simply 'set theory' whose concepts are said to be
relative; not 'axiomatic' set theory (recall: '. . . *on an axiomatic
basis higher infinities exist only in a relative sense* ([1922], p. 296; his
emphasis)). The argument remains the same, but greater
generality is claimed for the conclusion—as if Skolem had, in the
interim, accepted the proposition that the only viable way to do
set theory was axiomatically (in Zermelo's sense).[7] The dropping
of the qualifier appears at this point to have been simply a
slip—almost as if he had been persuaded of the relativist
conclusion and simply detached it whereas before he had
contraposed: As someone very astute once pointed out: One
person's *modus tollens* is another's *modus ponens*.

Although this may not help *explain* the change, it seems clear
that Skolem became wedded to formalizing theories in LPC
(note his work on reducing type-theoretical formalisms in
LPC—in [1961a] he brandishes the result of Gilmore [1957],
that the simple theory of types is 'translatable' into a first order
theory, in an effort to ward off objections to his work based on
second-order versions of ZF). When he noted [1933d, 1934b]
that *arithmetic*, which until that point had played the role of an

---

[6] Perhaps it's not worth speculating about. Addressing this question, G. Kreisel
remarked 'You should not ask this; you see, Skolem didn't have a philosophical bone in
his body.'

[7] In a way, he *argues* for this, but only in [1941], where he surveys only three kinds of set
theory as if these were all there were: 'simple'—i.e. simple-minded (= naive);
axiomatic—i.e. à-la-Zermelo; and 'logico-formal'—i.e. Russell-Whitehead Types. The
intuitive 'ordinary set theory' of [1922] has dropped away, even as an option to be
rejected.

intuitive foundational discipline, was similarly non-categorical, and that the deeper reason for it was what *we* might call the 'unformalizability' of the induction axiom because of its reference to *all* subsets of the integers, he accepted the obvious consequences. There was no longer a foundational discipline left which could rival axiomatic set theory by being free of *its* defects. Inductive inferences, which he had declared in 1922 to be 'clear, natural, and not open to question' were now infected with the same diseases as was set theory—perhaps because of their own 'dependence' on an unformalizable notion of set.[8]

This is a turning point. He could take this new evidence (the unformalizability of the concept of number) as a reason for rejecting *as too demanding* relativistic arguments based on the unformalizability of the key notions. Or he could retain the arguments and extend his 'skepticism' to whatever concepts were ever to be found to be similarly unformalizable, including those he had previously accepted as perfectly clear and unproblematic. He evidently chose the latter. So, we find him in 1958 unable to 'understand why most mathematicians and logicians don't seem to be satisfied by the notion of set as defined by a formal system'.

But that this is in fact how he reasoned is sheer speculation.

V

A. The Skeptic's central argument
This section will contain an analysis of the passage from Skolem [1922] quoted above, and some general remarks concerning ways that Skolem's argument (which I will find unsatisfactory) could be strengthened, *as an argument against Zermelo*, in an effort to discover both its appeal and its Achilles heel.

---

[8] It is of some interest to speculate what should be made of the reductionistic arguments—more specifically those that reduce all mathematical concepts to those of some *formalized* axiomatic set theory—in a setting in which the very notion of *formal system* is drawn into question by the newly discovered instability in the concept of number. It is no longer just the alleged *circularity* that is in question, but the very definiteness of the formal systems to which the reduction is meant to be carried out. It can no longer be argued that the concepts of mathematics (including that of 'finite iteration') are relative to their particular embodiment in a given specified formal system: *the very notion of a definitely specifiable formal system has now been undermined by the very relativistic and reductionistic arguments that led Skolem to use it as a grounding conception.* Suppose, for example, that some provable sentence ('0 = 1', say) had a non-standard integer as the number of lines in its proof. . . . I owe this example to Hilary Putnam.

Let us return then to Skolem's central argument for relativity, since that is perhaps the most distinctive feature of The Skeptic's position. In the classical passage in which it is presented Skolem is explaining how it can be that a set—PZ_0, say—can be provably 'non-denumerable' and yet, because of its embedding in a denumerable model, be denumerable as well. This is the core of the argument that cardinality concepts are relative: M is non-denumerable *in the model B* but denumerable from some external standpoint. As it stands, the argument is inconclusive, because there is an obvious equivocation on 'denumerable' and related notions. *In the model B*, something is denumerable if it is an integer that bears a certain relation—the '$\in$' of the model—to another integer—the designation in B of the term 'Z_0', the term that represents the 'set' of Zermelo integers within the theory. But what is it for a set to be non-denumerable? To see, we must look at the definition of *denumerability*:

Den $(x) =_{df}$ $\exists f(f$ is a 1-1 function with domain Z_0 and range $= x)$

So, putting 'M' for 'x', M is *non*-denumerable iff there is *no* such f. So much for what happens within the theory, and therefore *within the model B*. The argument continues, however, by showing that, since the domain B is itself denumerable, *and since M is a subset of it*, M must also be denumerable, no matter *what* the theory says.

At this point, Zermelo should resist, since the latter part of Skolem's argument employs the outlawed *intuitive* set-theoretic reasoning, treating B as a real set, M as a subset of it, etc . . . M started out as a number, but all of a sudden it has sprouted 'elements', without which the 'therefore' of the last line on p. 92 above would be very hard to justify. As it stands, therefore, the argument is merely a complicated equivocation. To give it some semblance of plausibility Skolem must do one of two things:

(*a*) Run it entirely in terms of some intuitive notion of set—which would render the argument useless against an antagonist like Zermelo, who rejects such notions, or (*b*) move to eliminate the intuitive concept of set altogether and run the argument through entirely in terms of numerical models, continuing to base it on the numerical version of the Löwenheim-Skolem Theorem.

That too risks making it useless against Zermelo (or another reductionist), but for a different reason: it employs a concept of *number* available only through reduction to the theory itself. If, however, it could be run within ZF or a near cousin of it, in terms of that theory's own 'numbers,' the reductionist's objection may be blunted.

I will explore both possibilities, somewhat deviously, by 'trying' to generate a genuine paradox and seeing where the attempt fails (for there is none); I will then explore what went wrong and to what extent the arguments can be patched up, if not to produce paradox, at least to obtain some form of relativity. Perhaps we can help Skolem in his fight against Zermelo.

First, let me state two versions of the Löwenheim-Skolem Theorem which it will prove useful to have before us:

> N (the numerical version); Any consistent countable set of First Order sentences has a model in the integers.

> SMT (a transitive submodel version): Any transitive model for ZF has a transitive countable submodel. (A model is *transitive* if and only if each element of each set in the model belongs to the domain of the model.)

What follows is an (admittedly feeble) attempt at generating a paradox from the elements at our disposal.
A first try:

> (*i*) If ZF is consistent, it has a model M whose elements are 1, 2, 3, . . . (by N, above).

> (*ii*) But we can prove in ZF that non-denumerable sets exist, more particularly, that the power set of the integers is non-denumerable:

> $$\vdash_{ZF} \neg \, \mathrm{Den}(PZ_0) \quad \text{[call this sentence 'T']}$$

> (*a*) T is therefore true in every model of ZF; and
> (*b*) T *says* that PZ_0 is non-denumerable, therefore
>   (1) T is true iff PZ_0 is non-denumerable, and
>   (2) T is true *in a given model M* iff $(PZ_0)_M$ is non-denumerable.

102   I—PAUL BENACERRAF

(*iii*) But no element of a denumerable mo⁻¹el for ZF can be
non-denumerable.

Therefore
(*iv*) If ZF is consistent, it has a model M in which $(PZ_0)_M$
is denumerable (by (*iii*)) and non-denumerable (by (*ii b*
2)).

This is unacceptable; but happily, glaring blunders abound.
To begin with, (*iii*) is clearly false (although Skolem equally
clearly assumes it). There is no result in model theory that
implies that the set of real numbers in the unit interval cannot be
an element of some countable model for ZF—so long as its
members aren't *also* elements of that model, i.e., so long as the
model isn't *transitive*. Furthermore, it is hard to maintain that
there is any recognizable sense in which T *says* that $(PZ_0)_M$ is
non-denumerable (*ii* above), if M is a *numerical model*. So if we are
to generate even an air of paradox, we should scrap (*iii*), moving
to transitive models, and also leave the realm of *numerical* models
for the more rarified one of set models. We will return to
numerical models below.

A second try:
(*i′*) Any transitive set model M for ZF has a transitive
denumerable submodel M⁻ in which $\in_M\text{-}\subseteq\in_M$.
(*ii′*) T is a theorem of ZF and says that PZ_0 is non-
denumerable in any model whose domain consists of sets
and in which '∈' denotes the membership relation. T is
therefore true in such a model M⁻ iff $(PZ_0)_M$- is non-
denumerable.
(*iii′*) No element of a transitive countable model for ZF is
non-denumerable.

Therefore
(*iv′*) $(PZ_0)_M$- is non-denumerable (by (*ii′*)), and denumerable (by (*iii′*)).

This looks more serious. (*i′*) is essentially what we have called
SMT, the transitive sub-model theorem. Further, since in a
transitive set model every element of the domain is a subset of

the domain (that is the precise import of transitivity), (iii') is true as well. This leaves (ii') as the obvious culprit.

Now it pays to look at T.

T) $\neg \exists f[f$ is 1-1 & $\text{Dom}(f) = Z_0$ & $\text{Range}(f) = PZ_0]$

So, the interpretation of T in a denumerable model—even in a denumerable transitive submodel of the standard model—is not sufficient to guarantee that T, on that interpretation, *says that* PZ_0 *is non-denumerable*—i.e. not *every* such interpretation of T is one in which T says *of* the referent of 'PZ_0' in that interpretation that it is non-denumerable. That's a proposition that is not invariantly expressed by T over all transitive submodels of a given model, even if that model starts out as somehow 'standard'—i.e. as the (an) 'intended' model. The conclusion is obvious; whether T *says* that a set is non-denumerable depends on *more* than whether the interpretation is over a domain of sets, '$\in$' of the interpretation coincides with membership among *those* sets, and every element of any set in the model is also in the model. The universal quantifier has to mean *all*, or at least *all sets*—or at least it must range over a domain wide enough to include 'enough' of the subsets of Z_0.

But as we have already seen, this first reply is useless to Zermelo against Skolem since Zermelo must do without the intuitive concept of set in terms of which it is framed—just as the argument exhibited there is useless to Skolem against Zermelo. Does Skolem have any cards left to play, or must he rely upon an intuitive concept of set in forcing a relativistic conclusion on Zermelo?

The following might help. (Call this the *Super*model Theorem, as we start with a model N and build it up to a supermodel N'):

If ZF is consistent it has
   (a) *numerical* models N and N' such that N⊆N',
   i.e. [Dom(N)⊆Dom(N') and $\in_N \subseteq \in_{N'}$], and
   (b) an element P of N (and N') which is $\text{Den}_{N'}$ and $\text{Den}_N.$[9]

---

[9] For a proof of the Supermodel theorem, see Clifton McIntosh's fine treatment of some of these problems in *Nous*, 1979; although we do not have exactly the same view of Skolem, I am indebted to his discussion in a number of places.

104   I—PAUL BENACERRAF

So, 'denumerability' is relative to the model, since the very same object is 'non-denumerable' from the standpoint of one model (N) and yet 'denumerable' from the standpoint of a supermodel (N'). This avoids the submodel version of the Löwenheim-Skolem theorem, and is therefore usable against Zermelo. In both cases 'denumerable' and 'non-denumerable' are *numerical* concepts, hence Skolem's equivocation between a 'numerical' and a set-theoretic sense of the set-theoretic concepts doesn't obstruct the argument (as an *ad hominem* argument against Zermelo).

One final note: all of these arguments, including Skolem's, can be formalized in ZF+ (ZF augmented by a truth-predicate), which Zermelo would (presumably) accept. So there is a sense in which they are *all* acceptable to Zermelo. The problem then becomes, of course: How are we to understand the formulas of ZF+ which express the premises and conclusions? Set-theoretically? Or numerically? If they are consistent, they have a numerical model. . .

B. *General Remarks*

In this subsection I will offer a few remarks by way of commentary. They are meant to indicate the direction in which I feel we should grope for a resolution, not as that resolution itself.

One important reason for constructing a theory of sets is to represent the intuitive content of Cantor's Theorem. That content is not preserved in a restricted model of some First Order Theory whose existence is guaranteed by one of the Löwenheim-Skolem Theorems. Any interpretation I of ZF on which one can define a function from $(PZ_o)_I$ onto Z_{oI} is *eo ipso* an inadequate interpretation. There *should be* no such functions.

Therefore, not every model of the axioms is an admissible interpretation—if one is doing set theory (although they might be for other purposes).

Some pictures will make this *seem* unduly restrictive. Once we have written down some axioms, they should be able somehow to stand on their own. Yet they don't wear their interpretation upon their sleeves. So, perhaps we should supply an interpretation, to forestall misunderstandings.

But an interpretation is just a further sentence correlating predicates with extensions in some fixed domain. A kind of super axiom.

Generalized, this picture can prove fatal. But the *apparent need* for such a further sentence stems from confusing a language without an interpretation explicitly and visibly attached to it with an *un*interpreted language—with one that is open to all models (and non-models) as admissible *interpretations*. That is a confusion—an understandable confusion but a confusion nevertheless. To point it out should suffice to block the arguments for relativity, because it takes the language out of the range of applicability of the Löwenheim argument. Only the axiomatic presentation of a theory *meant to be understood model-theoretically* is subject to Skolem's relativistic claims. And even then.

Skolem [1922] argued against a formalist or conventionalist position in mathematics, defending [an unspecified portion of] intuitive mathematics as the core on which all of it is based. He saw the axiomatization and formalization of set theory as attempts to escape from intuitive mathematics and found mathematics on an empty shell. It makes no sense, he argues, because (1) it presupposes for its very formulation some intuitive model-theoretic notions that look suspiciously set-theoretical: A domain and relations on that domain—satisfaction of the theorems in every model of the axioms; (2) it presupposes for its *investigation* (the consistency problem) the acceptance at least of independently established principles, either of set theory, of arithmetic, or of (what we now call) syntax [indeed Skolem argues that it presupposes these for its *formulation* (see (4) below)]; (3) as a foundational enterprise, it is clearly less secure than the mathematics it attempts to found; (4) the particular axiomatization chosen suffers from a number of unclarities and seemed insufficient to do justice to 'the usual theory of sets'. These unclarities concerned Zermelo's notion of 'definite property' and one of the insufficiencies was the weakness to be remedied by the axiom of substitution. Both depend, in the formulation Skolem gives, on the concept of an arbitrary finite number (which is used to define the concept of formula needed to replace the notion of definite property or function). At the end (e.g. [1958a]) we see Skolem holding precisely the position he had criticized in 1922, hailing as virtues what he had condemned as some of its most glaring faults; he nevertheless recognizes what he had not explicitly recognized in 1922—that the complete characterization of the number sequence is parasitic on the set-theoretic notions he claims are relative, thereby dragging the numbers down into the mire as well (if to lack a complete first-order characterization be to be bemired). It is hard to say how seriously he takes this, as he seldom speaks of the *integers* as 'shadowy non-entities' which, chameleon-like, change their plumage with each new interpretation of set theory. He says as late as 1955 in 'Peano's Axiom and Models of Arithmetic':

> It was then to be expected that if we try to characterize the number series by axioms, for example by Peano's, using the reasoning with sets given axiomatically . . . we would not obtain a complete characterization . . . This fact can be expressed by saying that *besides the usual number series other models exist of the number theory given by Peano's axioms or any similar axiom system."* ([1955d] p. 587, my italics).

Given the philosophical view he held at that time, he has no right to put the matter in this way, because he can have no concept of 'the usual number series', any more than he has a concept of *all* subsets of the integers. Yet he is clearly right—that *is* the right way to put the point. Just as the correct reply to Zermelo was that allowing just *any* model of his axioms to count as a domain of sets won't get you much set theory back (though 'thin' models are a wonderful tool for the foundational examination of those axioms). The meaning of '∈' and the range of the quantifiers must constrain the class of permissible interpretations if the formalized version is to retain the connection with intuitive mathematics with which set theory began—if it is to be a formalization *of set theory*.

In 1958, Alfred Tarski[10] attempted a reply to Skolem along similar lines, but one that, in my opinion, ultimately fails. Let me put Tarski's point and the reply I feel Skolem should make (no reply was recorded) somewhat fancifully, in the form of a brief dialogue:

[10] In the discussion following Skolem [1958a], p. 638. He also makes the 'upwards' LS reply discussed below. Skolem had complained about that reply (in [1955c], p. 583) that, given relativity, what could the replier *mean* by non-denumerable?

Tarski: It is hardly surprising that treating '∈' as an uninterpreted symbol of the theory leaves you with the unintended denumerable models. If, however, you treated '∈' as one of the *logical constants*, on a par with the quantifiers, negation and material implication, the denumerable models would no longer be possible. The meaning of '∈' would have to be preserved in each interpretation.

Skolem: As you know, Löwenheim's theorem (I have always resisted grafting my own name onto it) comes in a *submodel* version: To any infinite model M of a theory T there corresponds a denumerable submodel M′ of T, in which the domain of M′ ⊆ domain of M and the properties and relations of M′ are just those of M, cut down to the narrower domain. So, even models in which '∈' means *membership*[11] can be denumerable. Treating '∈' as a logical constant isn't enough—unless part of fixing *its meaning* involves stocking the quantifier domain with the right materials, in this case, enough sets. But this would set '∈' sufficiently apart from the 'other' logical constants to make it questionable that it should be called one at all. Besides, you yourself were the first to point out in a very illuminating way the *arbitrariness* of the choice of logical constants, in the absence of a theory that would pick them out. We still lack such a theory, so your suggestion strikes me as very much *ad hoc*.

There are, of course, other replies to The Skeptic. Perhaps the most interesting come from the radical wing and invoke the 'upwards' Löwenheim-Skolem Theorem: If a theory has any infinite models at all, it has models of *every* infinite cardinality. Although there isn't space to discuss them fully, it might be helpful to make a remark or two.

First, the use of this theorem serves well to bring out of the closet The Skeptic's bias for the finite, or at least the at-most-denumerable: Less is better, or at least more intelligible. To heighten the contrast, the Upwards Skeptic chooses his

---

[11] Skolem explicitly takes the position that the meaning of the predicates is preserved in the submodel version—cf. [1941a], p. 455.

108   I—PAUL BENACERRAF

favorite infinite cardinal, say aleph$_{17}$, and argues, *mutatis mutandis*, that there is no absolute concept of non-aleph$_{17}$-infinite, since every attempt to specify such a cardinality by axioms must fail: If the theory in which the set in question is embedded is not one all of whose models are finite, then it has a model in which the set has aleph$_{17}$ members. Who is to say otherwise? Failing a convincing *foundational* argument, we seem to be at a standstill. This, in itself, is a fairly good *indirect* argument against the relativity that The Skeptic is urging on us. His argument is *too* good: By keeping exactly the same things fixed (the meanings of the logical constants) we seem to be able to show that any infinite cardinality other than aleph$_{17}$ is relative.

But, of course, this is grist for the *clever* Skeptic's mill: If the Upwards Skeptic has a point at all, things are even worse than even The Skeptic had suspected; if he does not, since the entire argument is interpretable with the denumerable models The Skeptic prefers, it shows nothing at all. It is for this reason that I don't find the Upwards Skeptic a fully convincing tool to wield against the convinced (downwards) Skeptic.

Interestingly enough, nowhere in Skolem have I found an argument that supports or even argues for the conclusion that only first-order structure matters, or makes mathematical sense. Nor, by the way, is there any argument to support the related underlying claim that first-order structure itself is *sacrosanct*: Why isn't there a *further* relativity due, say, to the possibility of interpreting quantifiers and negation intuitionistically? But to explore the implications of this line of reasoning would take us too far afield.

The reason we don't have a Skolem Paradox, and the reason set-theoretic notions *aren't* relative in Skolem's sense, is that there is no reason to treat set theory model-theoretically—which is *not* to say that there is no reason to study its model theory. What a set-theoretical statement *says* can simply not be identified with what is invariant under all classical models of the theory (i.e., models that allow '∈' to take on whatever binary relation will satisfy the axioms). Tarski's point, when divorced from the suggestion that '∈' is a logical constant, is precisely the right one. Axiomatized set theory took its roots in informal mathematics and in order to retain its sense through axiomatization and formalization it must retain these points of contact.

After all, Zermelo (and Skolem, as we saw in connection with the axiom of replacement) chose *these* particular axioms because they give us *these* theorems, which have an independent intuitive meaning of their own. Why turn our backs on this ancestry and pretend that the sole determinant of meaning of the axioms is their first order structure? We saw above that even when he is trying very hard to think about the problem in this formalistic way (in connection now with number theory) Skolem lapses into sense and describes the situation quite accurately—but in a way that would make no sense if his general position were correct.

The Skolem I presented to you is a man divided—seduced by a misunderstanding of the import of his own arguments. What his view needs to substantiate it is no less than the principle that the *only* mathematical concepts we have are those that are embodied by all models of our mathematical theories under first order formalization. But *that* is a principle whose very expression requires mathematical concepts (models, first order theories, etc.) which themselves have—and must be taken to have— intuitive content if it is itself to make sense. Clearly the Skolem of [1922]—the *locus classicus* of The Skeptic's most seductive argument—did not hold that view.

Equally clearly, the Skolem of [1958a] did. Paradoxically, (and this is the real Skolem paradox) he adopted it after brilliantly unfolding the disastrous consequences of that position in its earlier Zermeloite incarnation.

VI

Putnam: Logocentricity
Putnam (or The Skeptic) will argue that the shoe belongs on the other foot—that in urging The Skeptic's 'paradoxes' on us they are not claiming, at least not *directly*—that the only concepts that make sense are those imbedded in theories whose first order axioms determine their models up to isomorphism, or that what sense others make is limited to what structure is preserved by all models. The more moderate (and charitable) view of their position is one that sees their flaunting of this 'paradox in the philosophy of language', as Putnam calls it, as a means of posing a challenge: it is up to us 'platonists' or 'realists' to *make*

clear what we mean when we present theories whose model theories don't have that form of pristine simplicity. Witness:

> The philosophical problem appears at just this point. If we are told, 'axiomatic set theory does not capture the intuitive notion of a set', then it is natural to think that *something else*—our 'understanding'—does capture it. But what can our 'understanding' come to, at least for a naturalistically minded philosopher, which is more than *the way we use our language*? And the Skolem argument can be extended . . . to show that total use of the language (operational plus theoretical constraints) does not 'fix' a unique 'intended interpretation' any more than axiomatic set theory by itself does. (Putnam [1977], p. 424)

But this gives the show away. Of course no explanation can be satisfactory. It is obvious why. The reason resides in our logocentric predicament: Any explanation must consist of additional words. Words which themselves are going to be said to need interpretation. His strategy has a wondrous simplicity and directness. He will construe any account we offer as an *un*interpreted extension of our already *de*interpreted theory—by explaining we merely produce a new theory which, if consistent, will be as subject to a plethora of (true) interpretations as was the old. ('Any *interpretation* is itself susceptible to further interpretation.')

To a point, the challenge is well taken. We *do* need a metaphysically and epistemologically satisfactory account of the way mathematical practice determines or embodies the meaning of mathematical language. (We could also use a satisfactory account for areas other than mathematics.) We may even need to devise new concepts of meaning to forge such an account.

No one, to my knowledge, has made satisfactory, comprehensive, sense of both the constructivist and classical traditions in the philosophy of mathematics. They pull in opposite directions. Each picture has its attractions, but neither seem fully to meet the challenge of the other. I see Putnam's Skeptical arguments as a challenge laid down to the Platonist. That is their strength. But they fall short of establishing a contrary position. They aren't even strong enough to refute the Platonist, since they depend so completely on this question-begging Skolemization of our 'theory of the world'. Even if we agree with Putnam that the determinants of mathematical meaning must lie somewhere in our *use* of mathematical language, why think use to be captured (or capturable) by axioms (we *do* argue about new axioms)? Or that the proper way to treat the *explanation* is to de-interpret it and add it to our axioms to form a new first order theory of the world, an extension of our old theory, itself de-interpreted and a candidate for Skolemization? His view *builds in* the systematic undercutting of any ground there could be for explaining not only *what* we mean, but *what it is to mean anything at all*. That is the effect of adding the explanation to the theory and ignoring the meaning of the terms in the extended theory in order to treat it model-theoretically.

I wish I had a positive account to offer that had any hope of being objectively satisfactory (clearly none can satisfy Putnam). For now, I fear we must be content with the insight Sweeney expressed when, struck by his own logocentric predicament, he said,

> . . . I gotta use words when I talk to you
> But if you understand or if you don't
> That's nothing to me and nothing to you
> We gotta do what we gotta do . . .

Despite the imagined possible misunderstandings, mathematical practice reflects our intentions and controls our use of mathematical language in ways of which we may not be aware at any given moment, but which transcend what we have explicitly set down in any given *account*—or may ever be able to set down.

With Gödel, I incline toward this view. But I am sufficiently aware of its vagueness and inadequacy not to be tempted into thinking it constitutes a *view*. It is merely a direction.[12]

[12] Ancestors of this paper were delivered at New York University, Stanford University, the University of Colorado (Boulder), and a meeting of the Association for Symbolic Logic. It was written in part as the author was (simultaneously) supported by leave from Princeton University and as Fellow of the Center for Advanced Study in the Behavioral Sciences, Sloan Foundation Fellow, Fellow of the National Endowment of the Humanities. This generous support is gratefully acknowledged. Although a more specific acknowledgement would be difficult to make, I have also profited immensely from Hao Wang's lucid and sensitive Introduction [1970] to a partial edition of the works of Skolem.

112   I—PAUL BENACERRAF

BIBLIOGRAPHY

Note: References to Skolem [1922] are to the translation appearing in van Heijenoort [1967]; references to his other works are to their appearance in Skolem [1970]. Where the articles do not appear in English, quotations are my own translations. With the exception of [1922] and [1970], a date followed by a letter in Skolem's works is as they appear in the Bibliography in Skolem [1970].

Barwise, K. J., [1972] 'The Hanf Number of Second Order Logic', *JSL* 37, No. 3, September 1972.

Benacerraf, P., and Putnam, H., Eds., *Philosophy of Mathematics*, 2nd edition, Cambridge University Press, New York 1983 ('B&P' below).

Fraenkel, A. A. [1925] 'Unterschungen über die Grundlagen der Mengenlehre', *Mathematische Zeitschrift 22*, 250–273.

Gilmore, P. C. [1957], 'The Monadic Theory of Types in the Lower Predicate Calculus', *Summaries of Talks Presented at the Summer Institute of Symbolic Logic in 1957 at Cornell University*, pp. 309–312.

Gödel, K, [1944], 'Russell's Mathematical Logic', in Schilpp, P. A., ed., *The Philosophy of Bertrand Russell*, The Library of Living Philosophers, Evanston, Ill. (Evanston & Chicago: Northwestern University, 1944), reprinted in B&P, pp. 447–469.

McIntosh, C., 'Skolem's Criticisms of Set Theory', *Nous 13*, 1979, pp. 313–334.

Putnam, H., [1977] 'Models and Reality', Presidential Address to the Association for Symbolic Logic, December 1977, reprinted in B&P, pp. 421–444.

Skolem, T., [1922] 'Some Remarks on Axiomatized Set Theory', in van Heijenoort [1967], pp. 290–301.

—[1929b] 'Über einige Grundlagenfragen der Mathematik', in Skolem [1970], pp. 227–273.

—[1930b] 'Einige Bemerkungen zu der Abhandlung von E. Zermelo: 'Über [der Begriff] die Definitheit in Axiomatik'' in Skolem [1970], pp. 276–79 ('der Begriff' does not appear in Skolem's version of Zermelo's title).

—[1933d] 'Über die Unmöglichkeit einer Characterisierung der Zalenreihe mittels eine endlichen Axiomensystems', in Skolem [1970], pp. 345–354.

—[1934b] 'Über die Nichtcharacterisiertbarkeit der Zahlenreihe mittels endlich oder abzählbar unendlich vieler Aussagen mit ausschliesslich Zahlenvariablen', in Skolem [1970], pp. 355–366.

—[1941a] 'Sur la Portée du Théorème de Löwenheim-Skolem', in Skolem [1970], pp. 455–482.

—[1955c] 'A critical remark on foundational research', in Skolem [1970], pp. 581–586.

—[1955d], 'Peano's axioms and models of arithmetic', in Skolem [1970], pp. 587–600.

—[1958a] 'Une relativisation des notions mathématiques fondamentales', in Skolem [1970], pp. 633–638.

—[1958d] 'Reduction of axiom systems with schemes to systems with only simple axioms', in Skolem [1970], pp. 645–652.

SKOLEM AND THE SKEPTIC 113

—[1961a] 'Interpretation of mathematical theories in the first order predicate calculus', in Skolem [1970], pp. 673–680.

—[1970] *Selected Works in Logic*, Fenstad, E. J., ed., Universitetsforlaget, Oslo 1970.

Tarski, A. [1958] Remarks on Skolem [1958a], in Skolem [1970], pp. 637–638.

van Heijenoort, J. [1967], ed. *From Frege to Gödel*, Harvard University Press, Cambridge 1967.

Wang, H. [1970] 'A Survey of Skolem's Work in Logic', in Skolem [1970], pp. 17–52.

Zermelo, E., [1908a] 'Investigations in the Foundations of Set Theory', in van Heijenoort [1967].

—[1929] 'Über der Begriff der Definitheit in der Axiomatik', *Fundamentae mathematicae 14*, 339–344.

—[1930] 'Über Grenzzahlen und Mengenbereiche', *ibid. 16*, 29–47.

APPENDIX I

It is uncertain whether Zermelo actually held a view to which Skolem's arguments will apply. In particular, although Zermelo unquestionably held that set theory should serve as a foundation for mathematics, it is *prima facie* questionable whether a *first-order* axiomatization was the way he had in mind: Either (a) Aussonderungs is meant as a *second-order* sentence; or, what is more likely, (b) no first-order/second-order distinction was intended, or (c) even if a first-order *axiomatization* is acceptable Zermelo would not have agreed that *formalization* within LPC is the proper means of presenting those axioms.

Interestingly enough, Zermelo [1929] helped settle the first issue when he rejected Fraenkel's [1925] syntactic first-order formulation of replacement (apparently he still hadn't seen Skolem [1922], which first appeared in Norwegian) in favor of an attempted *axiomatization* of the notion of 'definite property', a kind of extension of the axioms of Z. This proposal is mute on the matter of the first-order/second-order distinction, but addresses itself instead to the rejection of a *syntactic*, i.e. *formal*, rendering of the notion of 'definite property'. Helpful to a fault, Skolem seized upon Zermelo's account, pointed out its obscurity and suggested a 'natural' clarification which, as luck would have it, rendered the patched up proposal equivalent with the Skolem-Fraenkel version and therefore equally subject to the deployment of the Löwenheim-Skolem argument. If it applied before (to Z), it still applies (to the newly-created ZF). Zermelo objected to *formalization*, arguing that formalization requires recursive definitions (of the syntactic concepts), and thus presupposes syntax, which itself presupposes the concept of 'finite iteration'. Since in his view this concept was destined in turn to be reduced to set theory, he felt that the presentation of set theory could not presuppose it. See Hao Wang's Introduction to Skolem [1970], esp. pp. 35–37.

Of course, this bears importantly on whether Skolem's arguments reach *the*

114   I—PAUL BENACERRAF

real Zermelo, but somewhat more circuituously, since Skolem [1922] does not himself speak in terms of formalization. His early views are innocent of most of these concepts of logistic systems. (Skolem [1929b] and, esp. [1941a] and [1958a] are more explicit on the matter.) My reconstruction of his 1922 paper has Skolem attributing to Zermelo a formulation of Zermelo's view that he (Skolem) will find easy to attack with the model-theoretic tools near at hand. As we just saw, Zermelo can (and does) defend himself against this attack by repudiating such a view. But at what price? Set theory is a foundation for mathematics; all mathematical concepts must be defined in set-theoretic terms. Set-theoretic concepts are characterized (implicitly defined) by the axioms—axioms whose very consistency cannot be discussed except in terms of the concept of finite iteration, a concept which he requires us to reduce to the membership relation being 'defined' by those very axioms. With such a defense, Zermelo might fare better taking his chances against Skolem.

If my rendering of Skolem's construal of Zermelo doesn't at least come close to the way Skolem actually thought of it in [1922], the actual situation had to be far more complicated still. My rendering of Skolem's construal of Zermelo may be the only hope we have of lending Skolem's case any initial degree of plausibility.

APPENDIX II

ZERMELO 1908a:

The Philosophical View

A. Set theory must serve as a foundational discipline for mathematics, in the strongest sense: All mathematical concepts and proofs must be reduced to set-theoretical ones.
B. Intuition having proved bankrupt, set theory must itself be presented axiomatically.[13]
C. The following axioms are the appropriate foundation for this foundation:

The system Z

I.   *Extensionality*. A set is determined by its members.
II.   *Axiom of elementary sets*: Null set, unit sets, pairs.
III. *Aussonderungs*: Given a set M and a propositional function P that is definite* for all members of M, there is a set M_P of all members of M satisfying P.
       *'An assertion is *definite* if the fundamental relations of the domain, by means of the axioms and the universally valid laws of logic, determine without arbitrariness whether it holds or not.'
IV.   *Power set*: The power set (PM) of M is the set of all subsets of M.

SKOLEM AND THE SKEPTIC   115

V.   *Union, or Sum*: the set of all members of members of a set M.

VI. *Choice*: Given any set M of disjoint non-empty sets, there exists a set with
     exactly one member from each member of M.

VII. *Infinity*: There exists a set A which contains the null set and the unit set of
     each member of A.

[13] I am grateful to Zlatan Dannjanović for pointing out to me that some of Zermelo's
other writings in this period suggest that he clearly had an intuitive conception of set that
he took himself to be recording in the axioms listed below: He did not see the
specification of these axioms as something like giving an implicit definition of '∈'. But
Skolem's arguments make it equally clear that this is how *he* interpreted Zermelo.
Rather than try to unravel these further intricacies here, I will simply acknowledge that
B, as I have interpreted it in the text, appears to have been a figment of Skolem's
imagination. A and C remain uncontroversially Zermelo's.

[21]

SKOLEM AND THE SKEPTIC

II—Crispin Wright

I shall not try to add to Professor Benacerraf's illuminating presentation and diagnosis of the evolution in the views on these difficult matters of the actual historical Skolem. Such (largely inexpert and simple-minded) remarks as I have to make will mainly concern the 'direction' with whose commendation Benacerraf concludes his paper.

I

Like Benacerraf, I have found it difficult to see how a cogent version of the argument for 'set-theoretic relativity' might seem to run. But it is worth separating what I think are two quite different lines of thought.

Moral relativism, for instance, is (or could be) the view that there are no absolute moral standards, and (hence?) that moral judgements take on a content which is conditioned, in part, by the moral standards of the culture(s) to which their makers belong. Someone who took this view might hold, e.g. that there is no genuine conflict between enlightened English-speaking opinion on capital punishment and the views of the followers of the Ayatollah Khomeini, since there is sufficient cultural diversity between the respective sources of these judgements to ensure that the moral concepts involved are not the same. Deeply confused as this thinking may be, it at least serves to remind us that relativism is a classic form of escape from *contradiction*: the relativist postulates an ambiguity in order to avoid a collision.

The first line of thought belongs in this sort of relativistic tradition. The Löwenheim-Skolem theorem (LST) *about*, say, ZF set theory is presented as contradicting Cantor's theorem *in* ZF. Cantor's theorem entails that the power set of the integers cannot be put into 1-1 correspondence with the integers; whereas LST entails—(at this point we begin to struggle)—that ZF deals in at most countably many objects, and hence that the power set of the integers, as an object dealt with in the theory, is

118   II—CRISPIN WRIGHT

itself at most countably infinite. As Benacerraf says, this is, so far,
a tissue of confusion. Bluntly: LST entails neither that ZF deals
only in a countable infinity of objects nor that it deals only in
objects which are at most countably infinite. It says that ZF, if it
has a model at all, has a countable model. There is not even the
appearance of a contradiction. All we have—or so the
Cantorian is so far at liberty to contend—is a proof that ZF, if
consistent, has non-standard models.

In order to get a line worth considering we must I think, go
for something like the more sophisticated reconstruction of the
argument which Benacerraf builds on the transitive countable
sub-model version of LST (=SMT). The intended interpretation
for ZF involves, I take it, a transitive model: that is, every
member of every set which the intended interpretation would
include in the subject matter of set theory is likewise part of that
subject matter. According to SMT, then, if ZF can sustain its
intended interpretation at all, it may be interpreted in a
countable sub-domain of the sets involved in the intended
interpretation, in such a way that '$\in$' continues to mean set
membership and every set in the domain of the new interpretation is itself at most countably infinite. Even now it is apt to seem
pretty unclear how exactly there is supposed to be an
appearance of contradiction. The thought, presumably, will be
something like the following. Consider the Power Set Axiom:

$$(x) \ (\exists u) \ (t) \ [t \in u \equiv (y) \ (y \in t \supset y \in x)]$$

And let x be Z_0, the set of positive integers. (It could, in fact, be
any denumerably infinite set which features both in the domain
of the intended interpretation and in that of the countable submodel—there has, presumably, to be some such set.) Let D be
the domain of the intended interpretation and D′ that of the
sub-model. Since—the argument is supposing—Z_0 is an element
both of D and D′, and since the Power Set Axiom holds *under its
intended interpretation* with respect both to D and D′—(for,
remember, '$\in$' is not re-interpreted in the sub-model and is the
sole non-logical constant in the Power Set Axiom)—the power
set of Z_0, PZ_0,—the u whose existence the Power Set Axiom
stipulates—must be an element both of D and D′. And now we
do appear to have shown contradictory results about it. Cantor's
theorem, proved within ZF, entails that there is no 1-1

correspondence between Z_0 and PZ_0; but if PZ_0 is an element of a transitive, countable sub-model of the ZF-axioms, it must itself be countable, i.e. there must indeed be such a 1-1 correspondence.

One response would be to regard the situation as demonstrating that ZF, like any set theory adequate for the demonstration of Cantor's theorem, is, if not actually inconsistent, at least *unsound*: a misdescription of what is true of sets. The relativist response, in contrast, would be to introduce an ambiguity, so that the two results are not really in conflict. Cantor's theorem, proved within the system, should—the relativist would contend—be seen as a result about the mapping-defining power of the system: no 1-1 correspondence between Z_0 and PZ_0 can be defined *within* the system. And this claim is quite consistent with the reflection that such a 1-1 correspondence can be defined *outside* the system, using methods, presumably, for whose formalization ZF is inadequate.

Well, that this could not be a satisfactory response is evident, it seems to me, from the mundane reflection that it is so far utterly unclear why the purported $Z_0 : PZ_0$ mapping, mysteriously unformalizable in ZF set theory, is not confounded by Cantor's reasoning when presented *informally*. For Cantor's reasoning *seems* entirely general: it invites us, presented with any purported 1-1 correspondence, R, between the integers and their subsets, to form a set of exactly those which are not members of their correlates under R. Intuitively, there *ought* to be such a set when the relation in question is the one which establishes—allegedly—the extra-systematic countability of PZ_0; and, on pain of contradiction, this set cannot itself be correlated with an integer under R. So something is wrong here: if the relativist argument were correct up to the point where there is an apparent contradiction and a relativistic response might seem to be called for, that response could not possibly be intuitively adequate.

Benacerraf seems to me to bring out perfectly what has gone wrong. The trouble is in the supposition that if Z_0 is an element both of D and D′, then the fact that the Power Set Axiom holds for both domains, without reinterpretation of its sole non-logical constant, is a guarantee that PZ_0 is an element of both domains as well. This is simply incorrect. What the Power Set Axiom

120   II—CRISPIN WRIGHT

guarantees is that each domain will contain, for Z_o, a set u which
contains every t in that domain which satisfies the condition on
the right-hand-side of the biconditional *vis à vis* Z_o. But whether
these should intuitively be regarded as *the same* set depends, of
course, on the respective ranges of t in the two domains. If those
ranges are not the same, it cannot be assumed that the 'PZ_o'
which is an element of D′ is indeed PZ_o, the full-blown power set
of the integers. And without that assumption, the transitive
countability of D′ is in no sort of tension with the uncountability
of PZ_o. So we simply do not get to the point when we appear to
confront contradictory claims about one and the same object—
the point where relativisation of the content of those claims
might seem to be an appropriate strategy for dissolving the
problem. We don't get to that point because the supposition is so
far totally unjustified that it is one and the same object with
which the claims are concerned.

   This seems to me a decisive and helpful point, and I think that
Benacerraf's conclusion, in effect that there simply is no
coherent relativistic argument, of the classic sort, to be gleaned
from SMT (or LST), is justified. However, I think the situation
with the second train of thought is different.

II

Here is a passage from a justly influential article:

> . . . numbers are not objects at all, because in giving the
> properties . . . of numbers you merely characterise an
> *abstract structure*—and the distinction lies in the fact that the
> 'elements' of the structure have no properties other than
> those relating them to other 'elements' of the same
> structure. . . .
> . . . That a system of objects exhibits the structure of the
> integers implies that the elements of that system have some
> properties not dependent on structure. It must be possible
> to individuate those objects independently of the role they
> play in that structure. But this is precisely what cannot be
> done with the numbers. To *be* the number 3 is no more
> and no less than to be preceded by 2, 1, and possibly
> 0, and to be followed by 4, 5, and so forth. And to
> *be* the number 4 is no more and no less than to be

preceded by 3, 2, 1, and possibly 0, and to be followed by. . . .

. . . *Any* object *can play the role of* 3; that is, any object can be the third element in some progression. What is peculiar to 3 is that it defines that role—not by being a paradigm of any object which plays it, but by representing the relation that any third member of a progression bears to the rest of the progression.

Arithmetic is therefore the science that elaborates the abstract structure that all progressions have in common in virtue of being progressions. It is not a science concerned with particular objects—the numbers. The search for which independently identifiable particular objects the numbers really are (sets? Julius Caesars?) is a misguided one.[1]

The argument which precedes these conclusions is well known. Suppose that someone has received satisfactory definitions of '1' (or '0'), 'number', 'successor', '+', and '×' on the basis of which the laws of arithmetic can be derived. Let him also have confronted, if you like, an explicit statement of Peano's axioms. Suppose further that he has been given a full explanation of the 'extra-mathematical' uses of numbers[2]—principally, counting —and has thereby been introduced to the concepts of cardinality and of cardinal number. Then it is possible to claim that, *conceptually* at least, the subject's arithmetical education is complete: he may be ignorant of all sorts of aspects of advanced (and less advanced) number theory, but his deficiencies, if any, are not in his *understanding*—at least not if he has followed his training properly. Yet the striking fact is—the argument runs—that someone who in this way perfectly understands the concept of (finite cardinal) number has no basis for (non-arbitrary) identification of the numbers which any objects given in some other way. Benacerraf makes the point vivid by comparing two hypothetical logicist-educated children, each of whom takes zero to be Λ but one of whom identifies successor

[1] P. Benacerraf, 'What Numbers Could Not Be', *Philosophical Review* 74 (1965). The above quotation is taken from the reprint in P. Benacerraf and H. Putnam, eds., *Philosophy of Mathematics*, 2nd edition, Cambridge University Press 1983, p. 291.

[2] Cf. Benacerraf, *loc. cit.*, p. 277.

122   II—CRISPIN WRIGHT

(following Zermelo) with the unit set operation while the other (following Von Neumann–Berneys–Gödel) identifies the successor of a number with the set consisting of that number and all its elements. Each of the set-theoretic frameworks is perfectly adequate for the explanation of the arithmetical primitives, and the derivation of the Peano axioms, and supplies the background against which the applications of arithmetic can be satisfactorily explained. Yet a dispute between the two children as to the *true* identity of the numbers is intractable.

The moral is that the concept of number has no content sufficing to resolve such disputes, has indeed no content sufficing genuinely to individuate the numbers at all. When the explanations, formal and informal, are in, a good deal will have been said to characterise the *structure* which the numbers collectively exemplify, and which, in Benacerraf's view, is the real object of pure number-theoretic investigation; but nothing will have been said to enable a subject to know which, if any, sets the numbers are—or which, if any, objects of any sort they are. However if the numbers *really were* objects of some kind, surely someone who perfectly understood the concept of number should be able, at least in principle, to identify them. Since we do not have the slightest idea how such an identification might be defended, we ought to contrapose. Whence Benacerraf's anti-platonist conclusion.

I have tried to give reason elsewhere[3] for thinking that the force of this argument is qualified both by certain internal weaknesses—the concept of finite cardinal number is determinate in ways the argument overlooks—and by the company it keeps—for instance Frege's 'permutation' argument and the various Quinean arguments for inscrutability of reference. But what is the point of reminding you of this argument here? Simply that it may be contended to furnish somewhat strange company for the direction Benacerraf would have us take in response to his 'Skeptic'. He writes

The meaning of '∈' and the range of the quantifiers must constrain the class of permissible interpretations if the

[3] C. Wright, *Frege's Conception of Numbers as Objects*, Aberdeen University Press 1983, especially Section xv, pp. 117–29.

formalised version is to retain the connection with intuitive mathematics with which set theory began—if it is to be a formalization *of set theory*. (this volume, p. 106)

And later

> The reason we don't have a Skolem paradox, and the reason set-theoretic notions *aren't* relative in Skolem's sense, is that there is no reason to treat set theory model-theoretically—which is *not* to say that there is no reason to study its model theory. What a set-theoretical statement *says* can simply not be identified with what is invariant under all classical models of the theory (i.e. models that allow '∈' to take on whatever binary relation will satisfy the axioms). . . . Axiomatized set theory took its roots in informal mathematics and in order to retain its sense through axiomatization and formalisation it must retain these points of contact.
>
> After all, Zermelo . . . chose *these* particular axioms because they give us *these* theorems, which have an independent intuitive meaning of their own. Why turn our backs on this ancestry and pretend that the sole determinant of meaning of the axioms is their first order structure? (ibid, p. 108–9)

In common, I imagine, with many, I find the general train of thought here attractive; at least, if there is a coherent line of resistance to 'Skolemism' in general, it must surely involve taking issue with the assumption that the meaning of a sentence, or class of sentences, can aspire to no greater determinacy than is somehow reflected in the common ground between its, or their, defensible (by some criterion or other) interpretations. But I am not clear why, on the *specific* issue concerning uncountability, the Skolemite—Benacerraf's Skeptic—needs to take this line on determinacy of sense. His argument is better put, it seems to me, in a form which, superficially at least, closely resembles Benacerraf's own argument against arithmetical platonism. Let somebody have as rich an informal set-theoretic education as you like—which, however, is to stop short of a demonstration of Cantor's theorem, or any comparable result, since these findings

124   II—CRISPIN WRIGHT

are, after all, supposed to be available by way of *discovery* to
someone who has mastered the intuitive concept of set. And let
him recognize the axioms of ZF as a correct formal digest of his
informal notion. Then SMT entails that a commitment to these
axioms, plus possession of the intended grasp of '∈'—the sole
non-logical primitive—need involve no commitment to the
uncountable; someone who rejects it need not say or do anything
at variance with acceptance of all the standard set-theoretic
axioms interpreted *as* set-theoretic axioms. In other words: just
as—according to Benacerraf's argument—a full and complete
explanation of arithmetical concepts is neutral with respect to
the identification of the integers with any particular objects,
so—the Skeptic will urge—a full and complete explanation of
the concept of set is neutral with respect to the existence of
uncountable sets. But if there really were uncountable sets, their
existence would surely have to flow from the concept of set, as
intuitively satisfactorily explained.

Here, there is, as it seems to me, no assumption that the
content of the ZF-axioms cannot exceed what is invariant
under all their classical models. It is granted that they are to
have their 'intended interpretation': '∈' is to mean set-
membership. Even so, and conceived as encoding the intuitive
concept of set, they fail to entail the existence of uncountable
sets. So how can it be *true* that there are such sets?

Benacerraf's reply is that the ZF-axioms are indeed faithful to
the relevant informal notions only if, in addition to ensuring that
'∈' means set-membership, we interpret them so as to observe
the constraint that 'the universal quantifier has to mean *all* or
at least *all sets*' (p. 103). It follows, of course, that if the
concept of set does determine a background against which
Cantor's theorem, under its intended interpretation, is sound,
there is *more* to the concept of set that can be explained by
communication of the intended sense of '∈' and the stipulation
that the ZF-axioms are to hold. And the residue is contained,
presumably, in the *informal* explanations to which, Benacerraf
reminds us, Zermelo intended his formalization to answer. At
least, this must be so if the 'intuitive concept of set' is capable of
being explained at all. Yet it is notable that Benacerraf nowhere
ventures to supply the missing informal explanation—the story
which will pack enough into the extension of 'all sets' to yield

Cantor's theorem, under its intended interpretation, as a highly non-trivial corollary.

The dialectical position then—according to this second 'Skeptical' line of thought—is not that we have been given what is, by ordinary criteria, a perfectly satisfactory explanation of the intuitive concept of set—but an explanation whose very expression some uncooperative individual now persists in 'Skolemising'. Rather, it is unclear whether the 'intended interpretation' has been satisfactorily explained, formally or informally, at all. At any rate, if the 'intuitive conception of set' *is* satisfactorily explained by informal characterization of the meaning of '∈' and stipulation of the ZF-axioms, then the Cantorian—Benacerraf?—owes an explanation, I believe, of why the Skeptic about uncountability does not have an argument strategically identical to that which Benacerraf brought against arithmetical platonism. And if, on the other hand, such is not a satisfactory explanation of the intuitive concept, and more, in particular, needs to be packed into the interpretation of the quantifiers, then an appropriate informal explanation is owing of the residue. (I shall return to this.)

A successful skeptical argument has somehow to provide a bridge from the non-categoricity of the ZF-axioms to the realisation that all is not well with the classical concept of set in general or indenumerability in particular. Benacerraf represents the Skeptic as seeking to effect the bridge via some such assumption as that 'the *only* mathematical concepts we have are those that are embodied by all models of our mathematical theories under first order formalisation' (p. 109) and hence that the content of the ZF-axioms cannot have a greater determinacy than that reflected by whatever properties are invariant through all their models. He complains, rightly as it seems to me, that such a contention not merely seems quite arbitrary but has in addition a self-defeating character: the very formulation of the contention requires recourse to certain concepts—'model', 'first-order theory', etc.—which it has to take to have *intuitive content*. The last thing which the Skeptic intends, presumably, by such a contention is something dilute enough to be shared by all admissible interpretations of it! If the Skeptic is happy to rely on specific intuitive concepts in the formulation of his claim, why does he refuse his Cantorian opponent the same right?

126   II—CRISPIN WRIGHT

This is surely a fair point *ad hominem*. But, whatever Skolem or any other actual relativist may have said, the second line of thought I have described would not attempt to give the bridge this shape. The non-categoricity of the ZF-axioms must concern anyone who, while granting Benacerraf that there is a genuinely intuitive, informal concept of set to be had, inclines to view the standard axiomatization as something akin to an *explication* in Carnap's sense. The essence of explication, of course, is that the *explicans* preserves everything essential to the *explicandum*; unlike strict analysis, or definition, however, it is permissible that elements of determinacy, or precision, be introduced. At any rate, a good explication cannot be *weaker* than the intuitive concept it supplants, cannot be neutral on points on which the latter is committed. Accordingly if the ZF-axioms, with '$\in$' interpreted as set membership, did constitute a satisfactory explication of the extension of the intuitive concept of set, the fact that they do not, so interpreted, entail the existence of uncountable sets would force the conclusion that there is no such entailment from the intuitive concept of set either. So I do not think it is enough for Benacerraf to urge, on behalf of the Cantorian, that we recognize that set theory is answerable to intuitive concepts. So much *is* recognised by one who looks to the ZF-axioms for an explication, and the difficulty still arises. The Cantorian claim has to be something stronger: that the intended range of the quantifiers in the ZF-axioms is precisely *not* susceptible to explication by axiomatic stipulation at all. It is not so much a matter of paying due heed to the informal roots of the discipline as recognising, in the Cantorian view, that no satisfactory formal explication of the concept of set can be given, even when we hold fixed the interpretation of the only set-theoretic primitive—'$\in$'—which need feature in such an explication. It does not seem to me that only a theoretical position which drew on the assumption which Benacerraf complains about could inspire misgivings about this.

III

One reason why it can be hard to hear the modest criticisms which, in my view, SMT can properly encourage, is because of the clamour raised by the much less modest generalisation of

'Skolemism' developed by writers such as Putnam. The last third of a century has, indeed, been a bad time for the notion of determinacy of meaning in general. Quine, Putnam, and Kripke (on behalf of Wittgenstein) have all developed arguments which, notwithstanding important differences, would each, if sustained, have the effect that the traditional notion of meaning simply could not survive. Quine, Kripke's Wittgenstein, and Putnam all seem to believe that we can, and should, pay this price: whether by a more resolute adherence to the models of explanation displayed in theoretical physics, or by the adoption of some form of 'skeptical solution', or by a shift to a more constructivistic conception of meaning, intellectual life can flourish without the traditional notion of meaning. Like many, I am skeptical whether this is so. To mention but one cause of anxiety: the notion of meaning connects, in platitudinous ways, with that of truth—whether a particular utterance expresses a truth is a function of what, in context, it succeeds in saying, which is in turn a function of the meaning of the type-utterance of which it is an instance. If we really decided that there is no such thing as determinate meaning, *what* would be the determinants of the truth of an utterance? Nothing is more likely to make us sympathise with the kind of response which Benacerraf wants to make to 'Skolemism' about the classical notion of set than the belief that the argument involved is simply a restricted precursor of the grand-skeptical arguments which have commanded so much recent philosophical attention. We are not, most of us, in the market for purchase of the conclusion of these arguments; they can at best have a status of paradoxes, rather than findings, even if there is no general agreement about how to disinfect them. So, for as long as it seems that Skolemistic doubts about the determinacy of the classical concept of set can be sustained only if we are prepared to admit their wholesale generalisation, we are likely to want to sympathise with Benacerraf's point of view: Cantorianism may carry its own intellectual discomforts but it's a good deal more comfortable than what appears to be the alternative.

It seems to me that the dichotomy should be false. I have great sympathy for Benacerraf's 'direction' as a response to *generalised* Skolemism, and I shall try in a moment to indicate why. But this sympathy is quite consistent with the sort of doubt about the

128   II—CRISPIN WRIGHT

good standing of the classical concept of set which, I believe,
SMT may contribute towards motivating.

I have no space here to attempt to review the detail of
Quine's, Kripke's, and Putnam's arguments. But someone who
is familiar with them will recognise, I think, that they involve a
common assumption. The assumption is a version of the later
Wittgenstein's idea that meaning cannot transcend use. One of
Quine's claims is exactly that whatever the extent of his data
about our use of an expression, a radical interpreter would still
be confronted with indefinitely many mutually incompatible
hypotheses about the meaning of that expression each of which
would serve adequately to rationalise the data. Kripke's skeptic
persuades his victim to accept that his previous use of an
expression cannot rationally constrain its interpretation to
within uniqueness, and hence that the fact—assuming, at this
stage of the dialectic, that there is such a fact—in which the
determinacy of what he meant by the expression consists must
be sought elsewhere. And Putnam writes—if I may repeat the
quotation cited by Professor Benacerraf—that

> If we are told, 'axiomatic set theory does not capture the
> intuitive notion of a set', then it is natural to think that
> *something else*—our 'understanding'—does capture it. But
> what can our 'understanding' come to, at least for a
> naturalistically minded philosopher, which is more than
> *the way we use our language*? And the Skolem argument can
> be extended, as we have just seen, to show that the *total use
> of the language* (operational plus theoretical constraints)
> does not 'fix' a unique 'intended interpretation' any more
> than axiomatic set theory by itself does.[4]

The dilemma is this. If we hold that meaning cannot transcend
use, we seem to be committed to the contention that there
cannot be more, or more specificity, to the meaning of an
expression than would be apparent to someone who engaged in
purely rational reflection upon a sufficient sample, or an
otherwise adequate characterisation, of its use. The various
skeptical arguments then all take the form of contending—with
a good degree of plausibility, which is what gives them their

---

[4] See above, p. 110.

interest—that no amount of data about the use of an expression, and no (axiomatic) characterisation of its use, can rationally constrain its interpretation to within uniqueness. But if we lurch to the opposite wing and allow that meaning can somehow transcend use, the price we pay for securing determinacy of meaning that way is that we are beggared for a satisfactory epistemology of understanding (=knowledge of meaning). Meaning, it seems, will have to be the object of some sort of direct intellection, since inaccessible to mere rationalisation of perceived use. So, if the sceptical arguments are sound, the choice would appear to be between dropping the idea that there is such a thing as determinate meaning or retaining it at the cost of an awkward silence when charged to explain how meanings can be public, how they can be known, and so on.[5] In these circumstances, since the response of allowing meaning to transcend use seems so utterly futile, one feels there *has* to be something amiss with the skeptical arguments. But what?

Consider Benacerraf's response to Putnam's generalisation of Skolemism:

> Even if we agree with Putnam that the determinants of mathematical meaning must lie somewhere in our *use* of mathematical language, why think use to be captured (or capturable) by axioms. . ? Or that the proper way to treat the *explanation* is to de-interpret it and add it to our axioms to form a new first order theory of the world, an extension of our old theory, itself de-interpreted and a candidate for Skolemisation? [Putnam's] view *builds in* the systematic undercutting of any ground there could be for explaining not only *what* we mean, but *what it is to mean anything at all*. That is the effect of adding the explanation to the theory and ignoring the meaning of the terms in the extended theory in order to treat it model-theoretically. (p. 111)

Now, Wittgenstein wrote, in response to the skeptical paradox which Kripke interprets, that the solution consists in seeing that 'there is a way of grasping a rule which is *not* an *interpretation*'

---

[5] I have not been able to understand how Putnam's espousal of verificationism gets to grips with the dilemma.

130   II—CRISPIN WRIGHT

(*Investigations* 201). What Benacerraf is urging, comparably, is that there is a way of receiving an explanation which is not an interpretation. So if someone e.g. lays down the ZF-axioms and then adds some informal remarks by way of explanation of the intended concept of set, these remarks, Benacerraf is urging, can have a determinacy of content, and so an explanatory force, which far exceeds the constraints which they impose on any model of their first order union with the original axioms.

But *what* is the way of receiving an explanation which is not an interpretation? Has not Benacerraf in effect simply begged the question against the skeptic, assuming that there is such a thing as determinacy of meaning when the least the skeptical arguments teach us is that the right to suppose so is bought only at extortionate epistemological cost? I do not think so, since I believe we can glimpse the possibility of an alternative to the platonism which, pending disclosure of some internal flaw in the skeptical arguments, was all that seemed available as an alternative to the skeptical conclusion. Wittgenstein is very inexplicit about his 'way of grasping a rule which is not an interpretation' but he does at least say that it is something 'which is exhibited in what we call "obeying the rule" and "going against it" in actual cases' (*ibid*). What we need to win through to, I suggest, is a perspective from which we may both repudiate any suggestion of the platonic transcendence of meaning over use *and* recognise that meaning cannot be determined to within uniqueness if the sole determinants are rational methodology and an as-large-as-you-like pool of data about use. Wittgenstein wanted to suggest that the missing parameter, the source of determinacy, is human nature. Coming to understand an expression is not and cannot be a matter of arriving at a uniquely rational solution to the problem of interpreting witnessed use of it—a 'best explanation' of the data. Still less is it a matter of getting into some form of direct intellectual contact with a platonic concept, or whatever. It is a matter of acquiring the capacity to participate in a practice, or set of practices, in which the use of that expression is a component. And the capacity to acquire this capacity is something with which we are endowed not just by our rational faculties but to which elements in our sub-rational natures also contribute: certain natural propensities we have to uphold particular patterns of judgement and response. Crudely, then, what makes it possible to derive something more specific from an explanation than whatever is invariant through the totality of 'models' of the 'theory' containing that explanation is that one's response to the explanation will be conditioned by other factors than the attempt rationally to interpret that 'theory'.

These remarks too constitute 'merely a direction'. But it is a hopeful direction: it holds out the promise of an explanation of how, without platonising meaning, we can uphold the right to maintain that some particular interpretation—of set theory or of anything you like—is what is intended, and what is communally understood to be intended, without suffering the probably hopeless commitment to defending the claim that *only* that interpretation can rationalise our use of the relevant concepts.[6]

IV

To come back to earth, however—if to resume consideration of the uncountable can be to do that—let me conclude by trying to explain why I still think that the classical concept of set, that concept of set which requires the existence of uncountable sets, is indeed something to stumble over. The Cantorian may, of course, disclaim the need for explanation altogether, though if he does he can scarcely hope for much respect for his views. What we have seen is that if the intuitive concept of set is indeed satisfactorily explicable—and how else could it be communicable?—the explanation has to be, at least in large part, informal; and it will not suffice informally to explain the set membership relation and then to stipulate e.g. that the ZF-axioms are a correct digest of the principles of set existence. If the Cantorian wishes it to follow from his explanation that there are

---

[6] There is, admittedly, little cause to be sanguine that a development of this response could be effective against the strong form of Quinean indeterminacy: the version which holds that there are, *ab initio* as it were, rival interpretations which will each serve to explain any amount of a subject's linguistic behaviour (rather than the weaker claim that for any amount of data about linguistic behaviour, there are rival interpretations . . . etc.). Wittgenstein's idea is addressed to the question how the patterns of linguistic practice constitutive of a particular concept can be apparent in a finite sample. But if the strong claim is right, shared patterns of linguistic practice cannot justifiably be taken to constitute shared concepts. The crucial question, of course, is whether Quine succeeds in presenting cogent reason for this claim.

all the sets which, intuitively, he believes that there are, he has to do something more. What?

He has to say something which entails that there are uncountably many sets. And that is not the same as stipulating an axiomatic framework in which Cantor's theorem may be proved, since the difficulty is exactly that if his preferred set theory can take its intended interpretation at all, it can take a *set-theoretic* interpretation under which Cantor's theorem cannot be interpreted as a result about uncountability. So how does the Cantorian get across to a trainee that interpretation (of the range of the individual variables of the theory) which will constrain conception of Cantor's theorem as a cardinality result? Now, the reason why it can seem as if there is no very great difficulty here is the point noted earlier, that Cantor's theorem, and indeed his Diagonal Argument, is apt to impress us as having unrestricted applicability as a piece of *informal* mathematics. That is why the original relativistic position on the uncountability of e.g. the reals seems so unconvincing: the extra-systematic enumerability of the 'reals' in a countable model for ZF ought, intuitively, to be disruptible by diagonalisation—there ought to be something with, intuitively, perfect credentials to be a real number which is nevertheless not, on pain of contradiction, in the range of the enumerating function. Is it not this that makes us feel entitled to regard any countable set model for the ZF-axioms as an *unintended* model, something that only imperfectly embodies our intuitive set-theoretic concepts? For the fact is that there is an informal set-theoretic result which we can achieve—the beautifully simple result which Cantor taught us—which we can prove *about* this model, which is not to be identified with the corresponding result *within* the system when the latter is interpreted in terms of this model, and which *shows* that the model does not include everything which the *intended* interpretation of the system embraces.

These remarks are intended to be diagnostic. I am suggesting that Cantor's reasoning—it doesn't matter at present whether we concentrate on the power set theorem or the Diagonal Argument—plays a role in the *formation* of our conception of what the intended interpretation of set theory is. Its role is less to articulate a surprising consequence of concepts fixed in some other way than to lead the determination of an inchoate concept of set in a particular direction. At any rate, we have absolutely no right to the idea that a countable set model of the ZF-axioms has to be 'non-standard', a misrepresentation of our intuitive intentions, *unless* we buy Cantor's reasoning as a piece of informal mathematics, something that defeats *any* candidate for a 1-1 correlation of the integers and reals, however characterised in whatever sort of system.

The question, accordingly, is whether Cantor's reasoning is cogent as (contributing to) an *introduction* to the intended conception of the range of the individual variables in set theory; whether it does indeed lead us to a concept of set of which no countable model can be an adequate realisation. Well, the answer, I believe, for reasons which you can probably all too easily anticipate, is that it does not. Let us informally rehearse the Diagonal Argument that the reals are uncountable. We know that the rationals are countable, and that the union of any pair of countable disjoint sets is itself countable. So the reals are uncountable if and only if the irrationals are. Each irrational can be represented as an infinite non-recurring decimal; to suppose that they are countable is therefore to suppose that there is some array

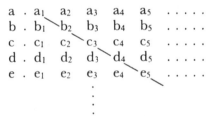

$$a \,.\, a_1 \quad a_2 \quad a_3 \quad a_4 \quad a_5 \quad \cdots\cdots$$
$$b \,.\, b_1 \quad b_2 \quad b_3 \quad b_4 \quad b_5 \quad \cdots\cdots$$
$$c \,.\, c_1 \quad c_2 \quad c_3 \quad c_4 \quad c_5 \quad \cdots\cdots$$
$$d \,.\, d_1 \quad d_2 \quad d_3 \quad d_4 \quad d_5 \quad \cdots\cdots$$
$$e \,.\, e_1 \quad e_2 \quad e_3 \quad e_4 \quad e_5 \quad \cdots\cdots$$
$$\vdots$$

(where each letter to the left of the decimal point represents some numeral, while those to the right represent any of 0-9) in which every non-recurring decimal occurs somewhere. But then, Cantor observes, it is easy to see that no such array can be complete; we have only to reflect that there is a decimal which differs at the *i*th place from each *i*th decimal in the array. However, in order to imbue Cantor's observation with its intended significance, we need to make a substantial assumption. Suppose we restricted our attention to *effectively computable* infinite decimals—those whose every *i*th place can (at least in principle) be calculated; and suppose we stipulated, in addition, that the whole array must itself be effective, i.e. that for each i, it should be effectively determinable what the ith decimal in the array should be. Under these restrictions, the Diagonal Argument takes on a wholly constructive character; we can now effectively compute a decimal which, on pain of contradiction, is no member of the array. But just for that reason we no longer have a result about uncountability. At least, we have no such result if we suppose that Church's Thesis holds. For in that case all the decimals in the array, and each successive diagonal decimal which we care to define, will correspond to recursively enumerable sequences of numerals. And we know independently that the totality of recursive functions is only countably infinite, since they are all definable in a finitely based vocabulary. Under these restrictions, then, Cantor's argument takes on a quite different significance: rather than showing something about uncountability, it shows that there is no recursive enumeration of all recursively enumerable infinite decimals. (The reader will easily see how similar considerations may be applied to the informal proof of the power set theorem.)

What goes for real numbers goes for subsets of natural numbers, since every infinite decimal has a unique binary equivalent, and an infinite binary expansion may be regarded as determining a unique subset of natural numbers in accordance with the principle that the ith natural number will be a member of the subset in question just in case the ith element in the binary expansion is 1. So, under the restrictions, the result is, again, that there is no effective/recursive enumeration of all effectively/ recursively enumerable subsets of natural numbers. And, again, we know that the finite and recursively enumerable infinite subsets of natural numbers form a countable set. So the moral is simple: before the Diagonal Argument, or the informal proof of the power set theorem, can lead us to a conception of the intended range of the individual variables in set theory which will allow us to regard any countable set model as a non-standard truncation, we need to waive the restrictions. And in order to understand the waiver, we need to grasp the notion of a *non-effectively enumerable denumerably infinite subset* of natural numbers.[7]

[7] We could hardly attain the conception of a non-effectively enumerable denumerably infinite array of infinite decimals without grasping this notion first.

This is what, if he is in the business of giving explanations, the Cantorian needs to explain. Now there are, I suppose, two routes into the informal notion of a subset of a given set. One is formally reflected in the *Aussonderungsaxiom*: a subset of a given set is determined by any bona fide *property*. The other route proceeds via the notion of a *selection*: a subset of a given set corresponds to every way, rule-governed or arbitrary, of selecting some of its members. Neither of these routes, it seems to me, holds out any very plausible promise of meeting the Cantorian's needs. Certainly, whichever he chooses, there are extremely awkward challenges to face. For instance, it is evident that the requisite properties cannot all be identified with the content of a possible predication (open sentence) in something we could recognise as a single language. Intuitively, there are indeed non-effectively enumerable infinite subsets of natural numbers which can be characterised in a finite and perfectly definite way; an example would be the set of Gödel-numbers of non-theorems of first order logic. But the Cantorian cannot give us the slightest reason to think that *everything* which he wishes to regard as a subset of natural numbers would allow of an intelligible, finite characterisation in this manner. Let him be granted that such a characterisation need not be constructive, i.e. that there need be no effective way of determining whether a particular natural number qualifies for membership. Still, a property is something which it should be possible to *claim* to be exemplified by a particular object, even if the claim cannot be assessed. If the Cantorian believes that a notion of a subset (of natural numbers) can be gleaned by this route which is sufficiently fertile for his purposes, he commits himself to the belief—if he respects the connection between *property* and *possible predication*—that there are *uncountably many finitely expressible properties*. This is an enormously problematic idea. Naturally, no single language—at least no finitely based language—can be adequate to express them all. Since any intelligible language is, presumably, finitely based, and hence can express at most countably many distinct properties, the Cantorian who takes this line is committed to a potentially uncountable infinity of languages no two of which are completely intertranslatable. Perhaps there is no contradiction in such a conception but it is, at best, a highly unfortunate consequence of something which

was supposed to be an intuitive explanation. And it is, besides, a
very nice question—where inter-translatability is missing—what
would make a pair of the relevant properties, each expressed in
one of these languages in a manner not translatable into the
other, *distinct*. (It would of course be circular in the present
context to attribute their distinction to that of the subsets which
they respectively define.) If, on the other hand, the Cantorian
severs the connection between *property* and *possible predication* he
is open to the charge that he is obscurantising the notion of a
property and thus can use it to explain nothing.

Matters are no better if we focus instead on the notion of a
selection. The selections in question—if they are to generate
enough subsets for the Cantorian's purposes—have to be thought
of as uninformed by any sort of recursive procedure. But they
also have to be thought of as *complete*, since it is only when
infinitely many selections have actually been made—in a non-
recursive way—that we actually have a non-recursively
enumerable infinite subset of natural numbers. So long as only
finitely many selections have been made, the constitution of the
subset so far determined will be consistent with supposing it to
be recursively enumerable. So in order to arrive at the intended
notion of subset, we have to understand what it would be *actually
to complete* an infinite but arbitrary selection from the set of
natural numbers. Of course some philosophers, notably Russell,
have found no difficulty in crediting us with such a conception.
But I do not think that anyone who follows up the literature on
'Supertasks' is likely to feel at ease with the idea. Someone who is
inclined to think there is no problem should ask himself what it
would be like to have reason to think that he had, magically as it
were, acquired the capacity to complete such a selection.
Benacerraf himself writes

> . . . there is probably no set of conditions that we can
> (nontrivially) state . . . whose satisfaction would lead us to
> conclude that a supertask had been performed . . . no
> circumstance that we could imagine and describe in which
> we would be justified in saying that an infinite sequence of
> tasks had been completed . . . there is nothing we can
> describe that it would be reasonable to call a completed
> infinite sequence of tasks.[8]

Can we claim to understand what a certain capacity would be if we do not know what it would be like to have exercised it? And what can the idea of an arbitrary actually infinite selection come to if we do *not* know what the capacity to perform one would be?

I am, of course, aware that what counts as adequate clarity in an explanation is, inevitably, to some extent a subjective business. What I claim is that the analogies which underpin the *further development* of the notion of subset—for that is what it is—with which the Cantorian wishes to work are very, very stretched. If someone is pleased to think that they are nevertheless good enough, I have no decisive counter-argument. But I hope to have reminded you that there are certain not very heavily theoretical—'anti-realist' or whatever—reasons for dissatisfaction with 'the intuitive concept of set', and 'the intended interpretation' of classical set theory. And, more importantly, to have made it plausible that we need not, in sympathising with these reasons, be receiving the thin end of a meaning-skeptical wedge.

---

[8] P. Benacerraf, 'Tasks, Super-Tasks, and the Modern Eleatics', *Journal of Philosophy* Lix (1962), pp. 765–784. The above remarks are taken from the reprint in Wesley C. Salmon, ed., *Zeno's Paradoxes*, Bobbs-Merrill 1970, pp. 125–6.

[22]

NINO B. COCCHIARELLA

PREDICATION VERSUS MEMBERSHIP IN THE DISTINCTION BETWEEN LOGIC AS LANGUAGE AND LOGIC AS CALCULUS

There are two major doctrines regarding the nature of logic today. The first is the view of logic as the laws of valid inference, or logic as calculus. This view began with Aristotle's theory of the syllogism, or syllogistic logic, and in time evolved first into Boole's algebra of logic and then into quantificational logic. On this view, logic is an abstract calculus capable of various interpretations over domains of varying cardinality. Because these interpretations are given in terms of a set-theoretic semantics where one can vary the universe at will and consider the effect this has on the validity of formulas, this view is sometimes described as the set-theoretic approach to logic (see van Heijenoort 1967, p. 327).

The second view of logic does not eschew set-theoretic semantics, it should be noted, and it may in fact utilize such a semantics as a guide in the determination of validity. But to use such a semantics as a guide, on this view, is not the same as to take that semantics as an essential characterization of validity. Indeed, unlike the view of logic as calculus, this view of logic rejects the claim that a set-theoretic definition of validity has anything other than an extrinsic significance that may be exploited for certain purposes (such as proving a completeness theorem). Instead, on this view, logic has content in its own right and validity is determined by what are called the laws of logic, which may be stated either as principles or as rules. Because one of the goals of this view is a specification of the basic laws of logic from which the others may be derived, this view is sometimes called the axiomatic approach to logic.

There is no uniformity among the advocates of either view of logic, incidentally, as to what theory of logical form should be taken as a definitive system of logic. First order logic, for example, might be the favored system for an advocate of either view; but then so might a form of higher order logic as well. Indeed, even the distinction between intensional and extensional logic, as we shall see, fails to mark a clear line between the two views. A set-theoretic semantics for intensional logic is also called possible worlds semantics. We shall in

Synthese **77** (1988) 37–72.

38   NINO B. COCCHIARELLA

this paper only be concerned with one or another version of intensional logic.

Now the question naturally arises regarding the axiomatic approach as to whether or not there can be a complete, recursive axiomatization of the laws of logic of a given theory of logical form. When the answer is affirmative, a set-theoretic semantics may serve in establishing that fact; but in that case, on the view of logic in question, the completeness theorem we establish in terms of such a set-theoretic semantics "is deprived of sense . . . by the very use we make of it. It is a case, in Wittgenstein's figure, of kicking away the ladder by which we have climbed" (see Quine 1968, p. 297). The possibility of a negative answer, however, is another matter altogether, and it is one of the problems we shall consider in this paper.

The historical antecedent of the second view of logic, incidentally, was Leibniz's somewhat visionary goal of a *lingua philosophica* or *characteristica universalis*. For Frege this goal meant the construction of "a logically perfect language" that was not only complete with respect to the laws of valid inference but which also would enable us to give a conceptual or logical analysis of the different possible statements we might make and to which such laws might apply. Instead of logic being an abstract calculus subject to varying interpretations, logic on this view has content of its own and is a language in its own right. That is why this approach is sometimes described as the view of logic as language (see van Heijenoort 1967, pp. 324–30).

This is not to say that all content on this view of logic is logical content, i.e., content expressible in purely logical terms. That is, strictly speaking, the nonlogical content that can be expressed in a logically perfect language, on this view, is really content that can be expressed in an applied form of the theory of logical form characterizing the logico-grammatical structure of that language. In this regard, the language in the idea of logic as language is not intended as a substitute for natural language so much as a way of providing a semantics for natural language. Applied logical forms, in other words, are really semantic structures in their own right, and it is by assigning such logical forms to the expressions of a natural language that one provides an interpretation or semantics for those expressions. Such an assignment, or really translation, amounts, on this view, to a conceptual or logical analysis of those expressions. That is why an advocate of this view might well consider an applied theory of logical form as a *lingua philosophica*.

On the logic as calculus view, needless to say, applied logical forms have no more semantic content of their own than do unapplied logical forms. Both, in other words, are merely expressions of an abstract calculus, and any content they might otherwise be taken as having is always a matter of an external set-theoretical semantics. In this regard, the real contrast between logic as calculus and logic as language is more appropriately seen as a contrast between the external set-theoretical semantics underlying an applied system of logic as an abstract calculus and the internal semantics expressed in the logical forms of that system as a language in its own right. This contrast, I believe, can ultimately be best explained in terms of the difference between a theory of membership in a set on the one hand and a theory of predication on the other.

On the view of logic as language, the most fundamental of all logical forms are the forms of predication, since these are the forms that underlie any possible assertion that we might make in language. Indeed, so fundamental are these forms on the view in question that different theories of logical form as alternative accounts of logic as language are for the most part really based on alternative theories of predication. In the history of philosophy, these alternative theories have generally been presented as theories of universals, the three most prominant being nominalism, conceptualism and realism. This is because on each of these theories a universal is that which can be predicated of things, and in fact it is the predicable nature of a universal that constitutes its universality. Sets, it should be noted, are not universals in this sense, though of course they are abstract entities. Being an abstract entity, in other words, is not always the same as being a universal.

Now we shall not go into the formal differences between different theories of universals here, having already covered that ground in my 1986 and 1988. Nevertheless, we do note that insofar as a theory of logical form is based on a formal theory of predication, which in turn is associated with a theory of universals, then, on the logic as language view of logical forms as semantic structures in their own right, we may take that theory of logical form as being itself the explanation that its associated theory of universals purports to give of the predicable nature of universals, i.e., of that in which the universality of universals consists. In this regard, we shall maintain in what follows that what distinguishes the view of logic as language from that of logic as calculus is that on the former logic is based on a formal theory of predication, whereas on the latter it is based, at least semantically, on membership in a set. This is because on the latter view validity is characterized essentially in terms of a set-theoretic semantics, whereas on the view of logic as language validity is a matter of the laws of logic that are expressed in terms of the theory of logical form in question. This difference, as we shall see, is not unrelated to the problem of the possible incompleteness of a given theory of logical form.

On the set-theoretic approach, it should be noted, predication is interpreted in terms of membership, and not membership in terms of predication. That is, predication has no real significance on this approach over and above its analysis in terms of membership in a set. This is true, moreover, not just for the forms of predication of a logical calculus but for those of natural language as well. That is, one can give a set-theoretic semantics for a recursively specified natural language just as one can for a logical calculus. On the set-theoretic approach, accordingly, there is no need for a theory of logical form as an explanation of the predicable nature of universals, since predication on this approach is ultimately to be analyzed in terms of membership. This suggests that set theory can itself be taken as a *lingua philosophica* or framework within which philosophical analyses can be given; and indeed that is the point of our claim that the real contrast between logic as calculus and logic as language is between a theory of membership and a theory of predication. The problem here, however, is whether any analysis of predication in terms of membership is really adequate, or whether there really is something to the idea of logical forms, especially those regarding predication, as semantic structures underlying the expressions of natural language.

1. THE PROBLEM WITH A SET-THEORETIC SEMANTICS OF NATURAL LANGUAGE

A good example of the supposed superfluous role of logical forms as semantic structures of the sentences of natural language is Montague's 'English as a Formal Language' [EFL]. Here Montague uses set theory to construct the syntax and semantics of a fragment of English in a way that resembles the construction in set theory of the syntax and semantics of a first order modal predicate calculus. For example, syntactic categories are recursively defined for this fragment of English in terms of certain basic expressions of these categories and a system of grammatical rules that generates complex expressions from simpler ones. An expression that belongs to any one of these categories, it turns out, is a meaningful expression of English, and as such it is called a "denoting" expression of its respective category. The meaning of such an expression in any model for the fragment, in other words, will be the denotation it is assigned in such a model (relative to an assignment of values to variables).

A model in Montague's semantics for his fragment of English is a set-theoretic structure consisting of what we are to think of as a set of possible individuals, a set of possible worlds, and certain functions defined on these sets assigning appropriate types of denotations (relative to an assignment of values to the variables) to the basic expressions of the syntactic categories of that fragment. Of course, the goal of the semantics is to assign denotations to all of the meaningful expressions, and not just to the basic ones. But unlike the situation in the predicate calculus, some denoting expressions are ambiguous in that they can be built up from the basic ones in several different ways on the basis of the grammatical rules. This is as it should be, of course, since the point is to formalize a fragment of a natural language where such ambiguity is common. Montague overcomes this problem by first associating with a denoting expression a tree-structure that provides an analysis of how that expression is generated. Ambiguous expressions will then be assigned more than one tree-structure, and the denotation of a denoting expression will be determined relative not just to a model and an assignment of values to variables, but to a tree-structure analysis of that expression as well. The result is a semantical analysis within set theory of both ambiguity and the relation of logical consequence for the fragment of English in question.

Now there are certain important assumptions in Montague's approach in 'English as a Formal Language' that may be challenged. One such assumption, for example, is that "the construction of syntax and semantics must proceed hand in hand" ([EFL], p. 210) and that the basic goal of serious syntax and semantics is the construction of a theory of truth (see Moravcsik 1979, pp. 3–15) for a criticism of this assumption). Another is Montague's use of Frege's principle of compositionality, e.g., that truth-values are to be assigned to sentences by assigning extra-linguistic entities to all of the expressions involved in the generation of those sentences, and in such a way that "the assignment to a compound will be a function of the entities assigned to its components" ([EFL], p. 217; see Hintikka 1980 and 1981, section 6, for criticism of this assumption). We shall not take up these assumptions here ourselves, however, but shall instead consider Montague's use of set theory as a theoretical framework in which to analyze such intensional entities as propositions, properties and relations in intension as these are normally expressed in natural language. For the idea that such analyses can be given amounts in effect to the claim that set theory can be used as a *lingua philosophica*; and indeed Montague was at one time quite clear that that was in fact his goal.

Philosophy, on this view, has "as its proper theoretical framework set theory with individuals and the possible addition of empirical predicates" ('On the Nature of Certain Philosophical Entities' [PE], p. 154). Philosophical analyses, in other words, are to be carried out as definitional extensions of set theory, supplemented when needed with empirical predicates and individuals as concrete *urelements*. In particular, Montague's formalization of the syntax and semantics of a fragment of English is really an analysis carried out by means of a definitional extension of set theory supplemented with the notions of a possible individual and a possible world. On this analysis, where A is the set of possible individuals and I is the set of possible worlds of a model, the set of propositions or possible denotations of formulas is defined as the set $U_1 = \{0, 1\}^I$, i.e., as the set of functions from possible worlds to truth-values (where 0 represents falsehood and 1 represents truth). Similarly, the set of properties or possible denotations of one-place predicates and the set of two-place relations in intension are defined as the sets $U_2 = U_1^A$ and $U_3 = U_1^{A \times A}$, respectively.

Now the question that arises here is how seriously are we to take these definitions? In particular, do propositions, properties and relations in intension really have the kind of dependence on the sets of possible individuals and possible worlds they are assumed to have on Montague's analysis? One consequence of this analysis, for example, is that properties and relations in intension are identical if they have the same extension in every possible world (of a given model), and similarly, that propositions are identical if they have the same truth-value in every possible world. This means, as Montague himself noted, "that if ϕ and χ are logically equivalent sentences (with respect to given analyses f and g), then 'John believes that ϕ' and 'John believes that χ' will turn out also to be logically equivalent (with respect to

LOGIC AS LANGUAGE AND LOGIC AS CALCULUS   43

analyses that contain *f* and *g* as parts)" ([EFL], p. 218). This leads to what Hintikka has called the problems of logical omniscience, i.e., the problem of believing or knowing all of the logical consequences of what one believes or knows, respectively.

We can avoid this consequence, as Montague does in 'Universal Grammar' [UG], by distinguishing between designated and undesignated possible worlds (of a given model), where in the latter logical constants can be given nonstandard interpretations. The logical equivalence of two expressions will then depend only on the extensions they have in designated worlds, while synonymy will depend on the extensions they have in the nondesignated worlds as well (cf. [UG], p. 231). Of course this assumes that some account can be given in purely set-theoretical terms of the distinction between designated and undesignated possible worlds (or of designated and undesignated contexts of use within possible worlds).

But even aside from assuming a new and unexplained distinction between designated and undesignated possible worlds, however, there still remains the problem of the individuation of propositions, properties and relations in intension in terms of the sets of possible individuals and possible worlds of a given model. For this means that within set theory as a purported *lingua philosophica* it is meaningless to talk about the same proposition, property or relation in intension across models that differ, however trivially, in their sets of possible individuals or possible worlds. Certainly, this sort of dependence is contrary to the way intensional entities are represented in natural language. In particular, just as properties, for example, are not individuated in terms of their instances, they are also not ordinarily thought of as being individuated in terms of the instances they have in different possible worlds, designated or undesignated; and the idea that we cannot meaningfully speak of the same property as being represented in different models indicates that something is radically wrong with a set-theoretical analysis of properties.

Of course, much depends here on how we are to understand the notions of a possible world and a possible individual. Montague himself only asked that we *think of* two of the constituents of a model as the sets of possible individuals and possible worlds of that model (cf. [EFL], p. 192); but in doing so he implicitly assumed that we understand the notions of a possible world and possible individual. In order to be articulated, however, such an understanding **would seem to** presuppose a modal or intensional logic either as an adjunct to set theory itself or as a superseding theory of predication in which membership is no longer a fundamental or primitive concept. In either case, set theory, or a theory of membership alone, will not suffice as a theoretical framework for philosophy, or at least not for a philosophy that purports to provide a semantics for natural language. That is, whatever its merits as a *mathesis universalis*, set theory, or a theory of membership alone, it does not suffice as a *lingua philosophica*.

2. INTENSIONAL LOGIC AS A NEW THEORETICAL FRAMEWORK FOR PHILOSOPHY

Montague did not himself remain satisfied with set theory as a *lingua philosophica*, and in the end he proposed instead the construction of an intensional logic as a new theoretical framework within which to carry out philosophical analyses. Thus, according to Montague, "philosophy is always capable of enlarging itself; that is, by metama-thematical or model-theoretic means – means available within set theory – one can 'justify' a language or theory that transcends set theory, and then proceed to transact a new branch of philosophy within the new language. It is now time to take such a step and to lay the foundations of intensional languages" ([PE], p. 155).

Actually Montague went on to construct not one but two intensional logics, but only the first is a theory of logical form in the traditional sense of being based on predication. This was Montague's higher order modal predicate logic, which he originally described as a second order predicate logic with third order predicate constants, but which can be easily extended to include predicate expressions of all finite orders (as is done in Gallin (1975, chap. 3), and as we shall assume it to be here). Montague's second intensional logic, on the other hand, is really a type-theoretical set theory (where sets are represented by their characteristic functions), but extended so as to include a theory of senses as well. Instead of being based on predication, in other words, Montague's second intensional logic is based on membership (as represented by the characteristic function of a set). The distinction may seem unimportant (and we shall not be concerned with it in this section), but, in fact, as we shall argue later, it goes to the very heart of the matter of the difference between logic as calculus and logic as language. For now, we want only to claim that the difference between logic as calculus and logic as language applies as much to intensional logic as it does to extensional logic.

Montague's own reason for dropping his first intensional logic in favor of the second had nothing to do with the difference between logic as calculus and logic as language. Rather, it was a result of the apparent need in the application of his first intensional logic to resort to circumlocution and paraphrase in the analysis of the intensional verbs of natural language. In the application of his second intensional logic, however, Montague was able to describe a precise translation function that provides a direct analysis of intensional verbs, i.e., an analysis that does not resort to circumlocution and paraphrase. (This may be seen as an argument in favor of the set-theoretical approach, at least when the latter is extended to include a theory of senses; for that, as we have said, is what Montague's second intensional logic really amounts to.)

The problem of translation into a theory of logical form seems to be better appreciated by linguists than by philosophers, except perhaps for those who advocate the set-theoretical approach. For unlike a set-theoretical semantics where recursive or effective truth-conditions are given (as translation rules into set theory) corresponding to the recursive construction of the syntactic categories of a natural language, the assignment of logical forms is generally presented as an intuitive process (or as an art that we learn in elementary logic courses) that depends essentially on circumlocution and paraphrase. Montague, for example, relied heavily on such circumlocution in the philosophical analyses he gave in his first intensional logic. For example, in giving an analysis of

> Jones sees a unicorn having the same height as a table actually before him.

that does not involve assuming the existence of sense-data (as some philosophers claim it must), Montague distinguished between 'sees' in its veridical sense and 'sees' in its nonveridical sense. The latter Montague paraphrased as 'seems to see', where 'seems' was analyzed as a 3-place predicate, 'x seems to y to F'. The phrase 'to F' is taken in this analysis as an infinitive or nominalized predicate occurring as an abstract singular term. Using λ-abstracts to represent properties and relations, and as translations of infinitives when occurring in

46   NINO B. COCCHIARELLA

subject positions (where '$[\lambda z\phi]$' is read 'to be a z such that ϕ'), we can describe Montague's analysis of the above sentence as follows:

$(\exists x)[$Table(x) & Before$(x,$ Jones$)$ & Seems(Jones, Jones, $[\lambda z(\exists y)$ (Unicorn(y) & Sees(z, y) & Has-the-same-height-as$(y, x)])])]$.

No sense-data, needless to say, are involved in this analysis, and, according to Montague, an argument for the existence of sense-data based on the content of the sentence in question must therefore fail (cf. [PE], p. 171).

As another example of the apparent need for circumlocution, consider the following two arguments, one of which is clearly valid while the other is clearly invalid.

Jones finds a unicorn; therefore, there is a unicorn.
Jones seeks a unicorn; therefore, there is a unicorn.

The puzzle here is to explain how of two arguments of apparently the same logical form one can be valid and the other invalid.

Montague's approach in this example was to regard 'tries to find' as a circumlocution for 'seeks', where the verb 'tries' is analyzed as a 2-place predicate, 'x tries to F'. Again, the phrase 'to F' in this analysis represents an infinitive or nominalized predicate occurring as an abstract singular term. Thus, whereas the first argument is assigned the following logical form,

$(\exists y)($Unicorn(y) & Finds(Jones, y));
therefore, $(\exists y)$ Unicorn(y),

which is clearly valid, the second argument is assigned the different logical form,

Tries(Jones, $[\lambda z(\exists y)($Unicorn(y) & Finds$(z, y))])$;
therefore, $(\exists y)$Unicorn(y),

which is not valid. In other words, through circumlocution and paraphrase, the two arguments have different logical forms after all; and therefore the validity of the one argument need not carry over to the other.

These analyses, it should be emphasized, occur in Montague's application of his first intensional logic, which is really a theory of logical form that is based on predication. Despite the philosophical usefulness of these analyses, however, Montague subsequently came not to depend on them. In particular, Montague became dissatisfied with the use of circumlocution and paraphrase, and he came to maintain that nothing short of a rigorous theory of translation between a natural language and an intensional logic will do before we can consider the latter as providing the logical forms underlying the expressions of that natural language. It was for this reason that he constructed his second intensional logic, which, as we have said, is really a type-theoretical form of set theory with an adjoined theory of senses. The latter is described in terms of two new operators ^ and ˅, called the sense and denotation (or intension-forming and extension-forming) operators, respectively.

We should keep in mind in this context that natural languages are not to be distinguished from artificial or constructed languages on the grounds that only the latter have recursively constructed syntactic categories. After all, a set-theoretic semantics that bypasses the logical forms of intensional logic is possible for a natural language only by providing grammatical rules that allow us to recursively construct the syntactic categories of that language. Indeed, in [UG], Montague formulated a universal grammar whose purpose was precisely "to comprehend the syntax and semantics of both kinds of language within a single natural mathematically precise theory" (p. 222). As part of this universal grammar, Montague also formulated a mathematically precise theory of translation, and as an example of an application of this theory, he constructed a specific translation function that assigned logical forms from his second or sense-denotation intensional logic to the expressions of the fragment of English he formulated within his universal grammar. That is, Montague showed how one can translate a recursively specified natural language into his sense-denotation intensional logic in no less precise a manner than one can translate that language into set theory by means of recursive truth-conditions. In this way, Montague overcame the objection that one must rely on circumlocution and linguistic intuition in order to associate logical forms with the expressions of a natural language. Of course, given a theory of logical form that is proposed as an alternative to Montague's sense-denotation intensional logic, the objection remains in force – unless a rigorously defined translation function can be defined for that theory as well.

NINO B. COCCHIARELLA

3. THE INCOMPLETENESS OF INTENSIONAL LOGIC
WHEN BASED ON MEMBERSHIP

According to Montague, the point to a rigorously characterized translation into intensional logic is that it induces a more perspicuous
interpretation of the expressions of natural language than is possible
with a purely set-theoretic semantics (see [UG], p. 241). Apparently,
this is because the logical forms of intensional logic give a representation of intensional entities that is both direct and independent of any
antecedently given reference to possible individuals or possible
worlds. That, of course, is a claim that is fully consonant with the view
of logic as language, and at this point one might even maintain that
Montague's sense-denotation intensional logic is an embodiment of
the idea of logic as language. The problem with such a conclusion,
however, is that the logic is essentially incomplete, and this is not
unrelated to the fact that it is based on membership and not on
predication.

We must be cautious here in how we are to understand Montague's
set-theoretic semantics for his intensional logic. For Montague's real
commitment to possible individuals and possible worlds is not in his
set-theoretic semantics but in the intensional logic that he took to
transcend set theory. This intensional logic is really a type-theoretical
form of set theory combined with a theory of senses. Every entity of
whatever type, for example, is taken in this theory as the denotation
(or extension) of a sense (or intension), and properties and relations in
intension are identified as the senses of sets and relations in extension
(as represented by their characteristic functions). Propositions are
similarly identified as the senses of truth-values. Of course, as objects
in their own right (or of their own type), senses are themselves the
denotations of other senses, and consequently there is a whole hierarchy of senses in this framework as well.

In this logic possible individuals are simply the objects indicated by
bound occurrences of the individual variables; and that some of these
individuals might not actually exist is seen in the fact that $(\forall x)^\vee E!(x)$
is not taken by Montague as valid, where $E!$ stands for the property
(or sense) whose denotation is the set of individuals that exists in the
actual world. (The set that is the denotation of this property is
represented by $^\vee E!$, i.e., by applying the extension-forming operator to
$E!$.) Montague's commitment to possible worlds, on the other hand, is

LOGIC AS LANGUAGE AND LOGIC AS CALCULUS   49

really a commitment to what might be called "world-propositions". For example, where P, Q are propositional variables, the property of being a possible world can be defined as the sense of the set of those propositions P that can denote truth and such that for any proposition Q, either P entails Q or P entails the complement of Q:

$$\text{Poss-Wld} =_{df} {}^{\vee}[\lambda P(\diamondsuit {}^{\vee}P \ \& \ (\forall Q)$$
$$(\square[{}^{\vee}P \rightarrow {}^{\vee}Q] \vee \square[{}^{\vee}P \rightarrow {}^{\vee} \sim Q])).$$

Using this notion, Montague's commitment to possible worlds can now be seen in the fact that

$$\square(\exists P)({}^{\vee}\text{Poss-Wld} \ (P) \ \& \ {}^{\vee}P)$$

is a valid thesis of his intensional logic (see Prior and Fine 1977, for an analysis of possible worlds as world-propositions).

Now the validity of the above formula is easily seen by returning to Montague's set-theoretic semantics where relative to certain sets that we are to "think of" as the sets of possible individuals and possible worlds, respectively, senses of entities of a given type are represented by arbitrary functions from the set of possible worlds to entities of that type. Propositions are then represented by functions from possible worlds to truth-values (represented in turn by the sets 0 and 1), or equivalently by arbitrary sets of possible worlds, including all unit or singleton sets of possible worlds (or functions that are true at one and only one possible world). A proposition that is represented by such a unit set is then a "world-proposition" in the above sense.

Note that we speak here of propositions being represented by functions from possible worlds (designated or otherwise) to truth-values, and not as literally being such functions. This is because for an intensionalist possible worlds are to be analyzed in terms of propositions and not propositions in terms of possible worlds; and for an advocate of Montague's sense-denotation intensional logic in particular, possible worlds are world-propositions in the above sense. For such an intensionalist, in other words, functions from possible worlds to truth-values can at best be correlated with propositions, not identified with them.

Of course, such a correlation would still imply that propositions are identical when they necessarily denote the same truth-values, and that result is problematic. This need not affect the argument for the validity of the above formula, however, so long as an intensionalist can assume that there is always at least one (even if not exactly one) proposition corresponding to each function from possible worlds to truth-values. Montague, of course, made the stronger, more problematic assumption of a one-to-one correlation, and, indeed, it appears to be this assumption that is the basis of his claim that we can use set theory to "justify" intensional logic as a framework that transcends set theory itself.

Although Montague assumed that both of his intensional logics transcended set theory, it is really only his sense-denotation intensional logic that contains a type-theoretical form of set theory in its own right. Montague's first intensional logic, again, is really based on predication, and it is only by interpreting predication in terms of membership that Montague was able to think of this logic as containing a type-theoretical form of set theory as well. But an alternative interpretation in which predication is viewed as more fundamental than membership is also possible, and though the logic will contain a theory of classes under that interpretation, these classes will not be sets in the sense of the iterative concept. Indeed, they will instead be classes in the logical sense of the view of logic as language.

Our present concern, however, is with Montague's second or sense-denotation intensional logic, one containing a type-theoretical form of set theory in its own right. This means that a certain necessary condition is imposed on the set-theoretical semantics for this logic; in particular, that the hierarchy of sets (or of their characteristic functions) that is part of this semantics is determined by all finite stages of the operation: $\mathscr{X} \rightarrow \mathscr{P}(\mathscr{X})$, where $\mathscr{P}(\mathscr{X})$ is the power set of $\mathscr{X}$. Or, in terms of functions, and where a and b are arbitrary types of the logic, it is the hierarchy determined by the operation that goes from the universes $\mathscr{U}_a$, $\mathscr{U}_b$ of entities of types a and b, respectively (and as based on given sets of possible individuals and possible worlds), to the set $\mathscr{U}_b^{\mathscr{U}_a}$ consisting of all functions from $\mathscr{U}_a$ into $\mathscr{U}_b$. This is the noncumulative hierarchy of all finite stages generated by the iterative concept of set, and as such it is the basis of what are called the *standard models* or interpretations of higher order logic; but standard, it should be emphasized, only insofar as such a logic is understood as containing a type-theoretical form of set theory. It is well known that when validity is defined in terms of these standard models, i.e., as truth in all standard models, the set of formulas that are thereby determined to be valid is not recursively enumerable. Any higher order logic, in other words, that is to be interpreted as containing a type-theoretical form of the iterative concept of set, such as Montague's sense-denotation intensional logic in particular, is essentially incomplete.

A logic that is essentially incomplete cannot be taken as an embodiment of the view of logic as language, or what we have also called the axiomatic approach; or at least it cannot if the set-theoretic semantics with respect to which it is incomplete provides an adequate external criterion of validity for that logic. For in that case, the set-theoretic characterization of validity cannot be eliminated in favor of an axiomatic characterization; and therefore the set-theoretic characterization must be seen as providing more than a guide to validity. This is true, moreover, even when the logic is intended to be interpreted as transcending set theory, as is the case with Montague's sense-denotation intensional logic. In other words, the set-theoretic semantics used to "justify" this intensional logic in fact provides an essential characterization of validity for it; for it is a characterization that cannot be eliminated by any internal criterion in terms of the so-called laws of that logic. Montague's set-theoretic "justification" of his intensional logic does depend, to be sure, on the problematic assumption that propositions can be adequately represented by functions from possible worlds to truth-values, but even if this assumption were weakened as indicated above, the logic would still be incomplete, since it still must be interpreted as containing a type-theoretical form of set theory.

A completeness theorem would be forthcoming, it might be noted, if we were to reject the restriction to standard models in the characterization of validity and allow what are called general models as well (see Gallin 1975, chap. 1, section 3). In that case, the hierarchy of functions need not go from the universes $\mathcal{U}_a$, $\mathcal{U}_b$ of entities of types a and b, respectively, to the set $\mathcal{U}_b^{\mathcal{U}_a}$ of all functions from $\mathcal{U}_a$ into $\mathcal{U}_b$, but only to some nonempty subset of this set that fulfills certain closure conditions. Such an allowance, however, would amount to rejecting the idea that the logic is to contain a type-theoretical form of set theory; for it would amount to replacing the type-theoretical hierarchy of sets by a hierarchy of classes (or of functions) that is based on a concept other than the iterative concept of set. This latter hierarchy is not an implausible framework for an intensional logic based on predication, or at least so we shall subsequently argue, but it is inappropriate for a logic based on membership. Redefining validity in terms of general models, in other words, is not a viable alternative for Montague's sense-denotation intensional logic.

4. PREDICATION VERSUS MEMBERSHIP IN TYPE THEORY

Both of Montague's intensional logics are theories of logical forms based on one or another version of type theory. This is understandable for Montague's first intensional logic, we maintain, since this logic is based on predication, but it is not really clear why his second, or sense-denotation, intensional logic, which is based on membership, should also be restricted to such a theory of logical form. The historical answer as to why one would adopt a type-theoretical theory of logical form is of course avoidance of Russell's paradox. But in the case of Montague's sense-denotation intensional logic, this answer is somewhat problematic.

Russell's paradox, it will be remembered, really has two forms, one in regard to the class of all classes that are not members of themselves, the other in regard to the property of being a property that is not a property of itself. Russell himself avoided the first form of his paradox by adopting his famous "no classes" theory, according to which all talk of classes is reducible to talk about properties that have those classes as their extensions. He then avoided the second form by imposing type-theoretical restrictions that made it meaningless on grammatical grounds alone to talk about properties being, or not being, properties of themselves. Montague follows Russell in adopting this sort of solution, though the type-theoretical restrictions for his sense-denotation intensional logic are based on Alonzo Church's theory of simple types (supplemented with a hierarchy of types for senses), rather than on Russell's theory of ramified types. Where Montague does not follow Russell in his development of this logic, however, is in not adopting the "no classes" theory. Instead, Montague applies type-theoretical restrictions to talk about classes (or really sets as represented by their characteristic functions) as entities in their own right, as well as to talk about properties and relations in intension as senses of classes and relations in extension.

Now what is problematic about this is that the classes in Montague's intensional logic, as already noted, are really sets or classes in the mathematical sense; i.e., they are classes that are formed in accordance with the iterative concept of set. This means that they are formed in accordance with the limitation of size doctrine according to which sets are not to get too big too fast. In this regard, Russell's paradox is really inapplicable to sets since it assumes a pattern of set-formation that violates the limitation of size doctrine and therefore is not in accord with the iterative concept. Russell's paradox in its first form, in other words, is really a paradox about classes in the logical sense, and not about classes in the mathematical sense; that is, it is a paradox about classes as the extension of properties or concepts, where the latter are based on predication and not on membership. This is why Russell found it natural to avoid this form of his paradox by adopting his "no classes" theory.

Note that unlike sets or classes in the mathematical sense, which have their being in their members, classes in the logical sense have their being in the properties or concepts whose extensions they are. This difference in ontological grounding is not vacuous, moreover, but is based on the difference between membership and predication, or rather on which of these two notions is taken as fundamental. In Montague's intensional logic, for example, membership in a set (as represented by its characteristic function) is fundamental, and the possession of a property, or monadic predication, is analyzed as membership in the set denoted by that property. For example, where δ is a sense whose denotation is a set of entities of a given type, and α is an entity of that type, then the possession by α of the "property" δ is defined by Montague as follows (see [UG], p. 236 and [PTQ], p. 259):

$$\delta\{\alpha\} =df [^{\vee}\delta](\alpha).$$

Since a similar analysis is given for relational predication as well, it follows that predication, whether monadic or relational, is not a fundamental logical form in Montague's sense-denotation intensional logic.

Since membership in a set (as represented by the characteristic function of that set), and not predication, is what is really fundamental in Montague's intensional logic, then his way of avoiding Russell's paradox of predication is really a variant of his way of avoiding Russell's paradox of membership. But the latter, we have said, applies only to membership in a class in the logical sense and is inapplicable to sets. Russell's paradox in either form, in other words, is irrelevant to the kind of framework Montague has in mind in his sense-denotation intensional logic. But then this leaves us without any real motivation

54   NINO B. COCCHIARELLA

for adopting the type-theoretical restrictions Montague imposed on this logic.[1]

Montague's first intensional logic, on the other hand, is a logic in which predication, and not membership, is fundamental; and in this case, once nominalized predicates are allowed to occur as abstract singular terms, Russell's paradox is not irrelevant. Imposing type-theoretical restrictions on the forms of predication in this logic, accordingly, is an understandable ploy as a way of avoiding Russell's paradox. It is noteworthy, moreover, that instead of rejecting Russell's "no classes" theory in his development of this logic, Montague actually formulated an alternative version of such a theory. A class, for example, can be identified in this logic with a property that has the same extension in every possible world, or what we might call a "rigid" property; and an n-ary relation in extension can be similarly identified with an n-ary rigid relation in intension. Rigidity in general (i.e., for each natural number n) can be defined as follows (see Montague 1974, p. 132):

$$\text{Rigid}_n =\text{df } [\lambda F(\forall y_1)\ldots(\forall y_n)(\Box F(y_1,\ldots,y_n) \\ \vee \Box \sim F(y_1,\ldots,y_n))].$$

On this analysis, accordingly, an n-ary relation in extension is simply a rigid_n relation in intension, and a class in the logical sense is simply a rigid_1 property. This notion of a class, needless to say, should not be confused with the mathematical notion of a set.

The law of logic that we need on this analysis in order to account for our talk of the extension of an arbitrary property or relation in intension is the following *principle of rigidity* (or what Gallin (1975, p. 77) calls the principle of extensional comprehension):

(PR) $(\forall F^n)(\exists G^n)(\text{Rigid}_n(G) \& (\forall x_1)\ldots(\forall x_n)$
          $[F(x_1,\ldots,x_n) \leftrightarrow G(x_1,\ldots,x_n)])$.

Every property or relation in intension, in other words, is co-extensive with a rigid property or relation in intension, and all talk of the extension of the former can be analyzed as talk about the latter. Membership in a class that is the extension of a given property, for example, is now analyzed in terms of predication as the possession of a rigid property that is co-extensive with that property. Thus, instead of predication being defined in terms of membership in a set (as represented by the characteristic function of that set), the way it is in

Montague's sense-denotation intensional logic, membership in a class is defined in Montague's higher order modal predicate logic in terms of predication.

One of the noteworthy consequences of (PR) is the commitment to possible worlds in the sense of world-propositions. That is, with (PR) as a basic law of logic,

$$\Box(\exists P)(P \,\&\, (\forall Q)[\Box(P \to Q) \lor \Box(P \to \sim Q)])$$

is provable in higher order modal predicate logic (see my 1986b, Section 11). Given the above analysis of classes in the logical sense, we do not need to rely on a set-theoretic semantics of so-called standard models as a guide to the validity of the above thesis regarding the intensional existence of possible worlds, i.e., the existence of possible worlds as world-propositions.

We should note here, however, that even though predication is fundamental in this logic, and membership is not, Montague himself follows the logic as calculus view and interprets predication in terms of membership in his set-theoretic semantics for this logic. In this way, standard models enter the picture once again; indeed, with respect to such models, the above principle of rigidity, (PR), is easily seen to be valid. But then, of course, with validity defined in terms of standard models we obtain an incompleteness theorem for this logic as well, in which case it too must fail as a candidate for the view of logic as language. That is, insofar as the set-theoretic semantics of standard models provides an adequate external critierion of validity for this logic, then the fact that the logic is incomplete with respect to this semantics shows that the latter provides more than merely a guide to validity – it shows that the characterization of validity that it does provide cannot be eliminated in favor of an internal criterion in terms of the so-called laws of logic.

The weak point in this argument for the view of logic as calculus is the assumption that the standard model set-theoretic interpretation of predication in terms of membership provides an adequate external criterion of validity for this logic. For it is this interpretation that begs the question as to which is fundamental in this logic, predication or membership. In particular, by imposing an interpretation based on standard models, the classes in the logical sense that are definable within the logic in terms of predication are reinterpreted in effect as sets (or rather as constant functions on possible worlds having these

56   NINO B. COCCHIARELLA

sets as their constant values). That is, the notion of a class in the logical sense is simply discarded on this interpretation in favor of the iterative concept of set.

An alternative, of course, is the replacement of standard models in the set-theoretic characterization of validity by general models, and in particular by general models in which the cardinality of the values of the higher order predicate variables is no greater than that of the values of the first order predicate variables. The latter restriction, needless to say, amounts to the assumption that there are no more properties and relations in intension of order $n + 1$ than there are of order n, for each positive integer n. The contrary assumption that there must always be more properties and relations in intension of order $n + 1$ than there are of order n is of course based on Cantor's theorem, which is the central feature of the iterative concept of set; therefore this assumes that the classes of properties and relations in intension of order $n + 1$ must all be sets or classes in the mathematical sense of the iterative concept. But this simply begs the question at issue. For when the classes in question are classes in the logical sense, and in particular when they are merely rigid properties and relations in intension, then there is no reason why their cardinalities must be in accord with the iterative concept of set. Thus, once we reject the idea that classes in the logical sense are actually sets or classes in the mathematical sense, then there is no reason why there should be more second order properties than there are first order properties, or more third order properties than second order properties, etc. That is, the type-theoretical division of properties and relations in intension, as a division designed to avoid Russell's paradox, only divides properties and relations in intension into different types – not into types of different cardinalities.[2]

A completeness theorem is forthcoming with respect to this alternative characterization of validity, needless to say, and in this regard there is no need to think of the set-theoretic semantics of the general models in question as anything more than a guide to validity. Strictly speaking, at least from the standpoint of the view of logic as language, the guidance in this case is really in the opposite direction. For, in order to give the set-theoretic characterization of validity in question, we actually need to rely on certain internal criteria (such as the status of (PR) and the comprehension principle as schematically described basic laws of logic) as constraints that must be imposed on the

LOGIC AS LANGUAGE AND LOGIC AS CALCULUS   57

so-called nonstandard models in question. In other words, what makes the set-theoretic characterization of validity adequate as an external criterion in this case is its dependence on certain basic laws of logic from the internal point of view of logic as language. Of course, given the completeness theorem, the set-theoretic characterization of validity can be by-passed altogether in favor of an axiomatic characterization.

Montague's first intensional logic can serve, accordingly, as a candidate for the view of logic as language, at least as far as the problem of a complete axiomatization is concerned. There are other problems, however, including in particular the objection that this sort of logic requires the use of circumlocution or paraphrase in its analysis of the expressions of natural language. But this objection can be overcome once we are given a precise translation function that can be used as the basis of such an analysis. A strategy that suggests itself here for obtaining such a translation for Montague's recursively specified fragment of English is to use the product of two translation functions, one translating Montague's sense-denotation intensional logic into higher order modal predicate logic, and the other translating the fragment of English in question into the sense-denotation intensional logic. That is, given Montague's precise specification of the latter translation function, all we need is a formal translation of Montague's sense-denotation intensional logic into his higher order modal predicate logic. Since such a formal translation function has been described by Gallin (1975, Section 13) we can put aside this objection to Montague's first intensional logic as a version of the view of logic as language.

There is another objection, however, that applies not only to Montague's higher order modal predicate logic, but to any theory of logical form that is both based on type theory (of third and higher order) and proposed as a version of the view of logic as language. In particular, as a framework for conceptual analyses, such a theory imposes inappropriate restrictions on the meaningful use of predicates in natural language. In general, for example, it is meaningless in a theory of types for a nominalized predicate expression to occur in the subject or argument position of another predicate unless the latter can be assigned a higher type than the former; therefore it is meaningless for any predicate to occur in a nominalized form in its own subject position. Thus, the otherwise unproblematic sentence of English, 'The

property of being a property is a property of itself', is by fiat ruled out
as meaningless, as is the sentence, 'Smith does not think that the
property of being philosophically interesting is itself philosophically
interesting, even if Jones does'. Also ruled out in this way are
otherwise unproblematic sentences whose predicates apply to nominal
expressions of different types, such as, 'Jones thinks that some people
are philosophically interesting, as well as that some propositions,
properties and relations in intension are too'.

The rejection of these and many other meaningful sentences of
English is clearly a defect of any theory of logical form proposed as
representing the view of logic as language. It is one thing to divide up
properties and relations in intension as a way of avoiding Russell's
paradox of predication, and quite another to make the meaningful use
of predicates in natural language actually depend on such a division.
Fortunately, there is a way of achieving the one result without also
imposing the other, and it. is noteworthy that this way involves
returning to the sort of framework that Frege and Russell advocated
at the turn of the century.

5. SECOND ORDER PREDICATE LOGIC WITH NOMINALIZED PREDICATES

At the turn of the century, both Frege and Russell advocated the idea
of logic as language in a form that was in many respects very similar.
Both maintained that logic consisted of what today is called standard
second order predicate logic, but supplemented to include a formal
account of nominalized predicates as well (see my 1986b paper for
a defense of this claim). Of course, at that time only Frege had
actually constructed a theory of logical form, and his formal account
of nominalized predicates was given in his theory of value-ranges
(*Wertverläufe*), or classes in the case of monadic predicates. This was
because, unlike Russell, Frege was an extensionalist, not an intensionalist. That is, Frege's universals "differ only so far as their extensions are different" (*Posthumous Writings* [PW], p. 118), whereas for
Russell two universals could have the same extension (see my 1986b,
Sections 11 and 13).

Frege's theory of logical form is sometimes described as a second
order set theory; but such a view is quite erroneous. For that view
confuses sets or classes in the mathematical sense of the iterative

concept with classes in the logical sense. That is whereas a set, on the iterative concept, has its being in its members, a class as the extension of a concept, according to Frege, "simply has its being in the concept, not in the objects which belong to it" ([PW], p. 183). For Frege the logical forms of predication are more fundamental than that of membership, and this is reflected in the fact that the latter is to be analyzed in terms of the former.

A more appropriate description of Frege's theory is that it is a second order predicate logic with nominalized predicates. Originally, in his *Begriffsschrift*, Frege formulated as an axiom system essentially what today is called standard second order predicate logic (with identity). It is clear that he took this logic as providing a logical analysis of the forms of predication that occur in natural language. Later, in his *Grundgesetze*, Frege added to this logic his theory of value-ranges. This addition was not given as an application of his earlier theory of logical form, it should be emphasized, but as a further development of that theory. In particular, as I have argued in my 1986b article, Sections 4–5, the singular terms generated from formulas by application of the smooth breathing abstraction operator were interpreted by Frege as symbolic counterparts of the abstract singular terms or noun phrases generated in natural language by predicate nominalizations. These nominalizations include not only such familar patterns as 'F-ness', 'F-ity', 'F-hood' and 'being an F', but also infinitives, 'to F', and gerunds, 'F-ing', as well as Frege's own favorite, 'the concept F'. In other words, Frege viewed the theory of logical forms he developed in the *Grundgesetze* as providing a logical analysis not only of the predicate expressions that occur in natural language, but also of the predicate nominalizations that occur therein as well.

Russell's conception of logic at the turn of the century was also essentially that of a second order predicate logic with nominalized predicates, though unlike Frege his account of this logic was presented very informally. (See my 1980 and 1986b articles for a description of Russell's early views.) Russell differed from Frege, however, not only in assuming an intensional theory of universals, but also in maintaining that nominalized predicates denote as singular terms the same entities that predicates otherwise stand for in their role as predicates. Russell rejected, in other words, the Fregean view that universals have an unsaturated nature of their own corresponding to the unsaturated nature of predicates in their role as predicates 1986b, Section 1).

Now we shall not be concerned here with the details of either Frege's or Russell's early views on the nature of logic, having already covered that ground in my 1980 and 1986b articles. But, we do want to note and emphasize that what is common to both Frege and early Russell is the idea of second order predicate logic with nominalized predicates as a paradigm of the view of logic as language. Here we have all of the essentials of what constitutes a theory of logical form as an expression of that view; namely, basic forms of predication, propositional connectives, quantifiers that reach into predicate as well as subject positions, and nominalized predicates as abstract singular terms. These all correspond to fundamental aspects of natural language, and to attempt to do without any of them in a theory of logical form would leave those aspects of natural language unexplained. Of course, these are precisely the features that constitute higher order predicate logic, except that in the latter predicates are grammatically divided into different types, with the constraint on well-formedness that nominalized predicates can occur as subject expressions only of predicates of higher types. This suffices as a way of avoiding Russell's paradox, since the grammatical division imposes a logical division as well. But it turns out that the logical point of the theory of types can be made without the grammatical restrictions, and since the latter impose inappropriate constraints on the meaningful use of predicates in natural language, it is much to be desired that we can make the logical point, and avoid Russell's paradox, without also imposing the grammatical restrictions. What is significant is that this can be done by returning to the original context out of which type theory arose, namely, second order predicate logic with nominalized predicates. We can return, in other words, to the paradigm of the view of logic as language.

We can briefly describe this paradigm as follows, where, for convenience, we take $\rightarrow$, $\sim$, $=$, $\forall$, $\Box$, and λ as primitive *logical constants* and assume the others to be defined in the usual way. We assume the availability of denumerably many individual variables, and, for each natural number n, denumerably many n-place predicate variables. (Propositional variables are taken as 0-place predicate variables.) We shall use 'x', 'y', 'z', with or without numerical subscripts, to refer to individual variables, and 'F^n', 'G^n', 'R^n' to refer to n-place predicate variables. Usually we drop the superscript when the context makes clear the number of subject positions that go with a predicate variable.) Complex predicates will all be generated from formulas by means of the λ-operator. Note that although predicates are not themselves singular terms, they can be transformed into such by deletion of the subject positions that come with them in their role as predicates. Traditionally, this transformation is marked by a deletion of the parentheses (and commas in the case of a relational predicate) that precede and succeed (or separate in the case of commas) the singular terms to which the predicate can be applied. We shall retain this traditional practice here. For example, on the definition that is to follow, $F(x)$ and $R(x, y)$ are formulas in which F and R occur as predicates, but $G(F)$ and $G(R)$ are formulas in which F and R occur as singular terms. In $F(F)$ and $R(F, R)$, of course, F and R occur both as predicates and as singular terms (though no single occurrence can be both as a predicate and as a singular term). As indicated, we shall require predicates to be accompanied by parentheses (and commas) only when they actually occur in a formula as a predicate.

In the definition that is to follow we shall use 0 to represent the type of a singular term, 1 the type of a formula, and $n + 1$, for $n > 0$, the type of an n-place predicate expression. For each natural number n, accordingly, we recursively define the *meaningful expressions* of type n, in symbols ME_n, as follows:

(1)   every individual variable or constant is in ME_0, and every n-place predicate variable or constant is in both ME_{n+1} and ME_0;

(2)   if $a, b \in \mathrm{ME}_0$, then $(a = b) \in \mathrm{ME}_1$;

(3)   if $\pi \in \mathrm{ME}_{n+1}$, and $a_1, \ldots, a_n \in \mathrm{ME}_0$, then $\pi(a_1, \ldots, a_n) \in \mathrm{ME}_1$;

(4)   if $\phi \in \mathrm{ME}_1$, and $x_1, \ldots, x_n$ are pairwise distinct individual variables, then $[\lambda x_1 \ldots x_n \phi] \in \mathrm{ME}_{n+1}$;

(5)   if $\phi \in \mathrm{ME}_1$, then $\sim\phi, \Box\phi \in \mathrm{ME}_1$;

(6)   if $\phi, \psi \mathrm{ME}_1$, then $(\phi \to \psi) \in \mathrm{ME}_1$;

(7)   if $\phi \in \mathrm{ME}_1$, and a is an individual or predicate variable, then $(\forall a)\phi \in \mathrm{ME}_1$;

(8)   if $\phi \in \mathrm{ME}_1$, then $[\lambda\phi] \in \mathrm{ME}_0$; and

(9)   if $n > 1$, then $\mathrm{ME}_n \subseteq \mathrm{ME}_0$.

Note that by clause (9) every predicate expression (without its ac-

62   NINO B. COCCHIARELLA

companying parentheses and commas) is a singular term. This includes
0-place predicate expressions, but not formulas in general. To
nominalize a formula, however, we need only apply clause (8). Thus,
we may read '$[\lambda\phi]$' when it occurs as a singular term as 'that ϕ'. Note
also that although only individual variables are bound by the λ-
operator, we can define its application to arbitrary variables $a_1, \ldots, a_n$
as follows:

$$[\lambda a_1 \ldots a_n \phi] = \mathrm{df}\ [\lambda x_1 \ldots x_n (\exists a_1) \cdots (\exists a_n)$$
$$\times (x_1 = a_1 \ \& \ \cdots \ \& \ x_n = a_n \ \& \ \phi)],$$

where $x_1, \ldots, x_n$ do not occur free in ϕ.

In regard to the laws of logic as understood in the original context
of this paradigm, we need only take the axioms and rules of standard
second order predicate logic, but applied now to all formulas, including those with as well as those without nominalized predicates among
their singular terms (1986b, Section 3). Since we include $\Box$ as a logical primitive, we can add to these axioms and rules those of the S5
modal propositional logic as well. Of course, Russell's paradox is now
derivable as a consequence of the following instance of the comprehension principle:

$$(\exists F)([\lambda x (\exists G)(x = G \ \& \ \sim G(x))] = F).$$

But instead of following Russell and avoiding the paradox by imposing
type-theoretical restrictions on the meaningful use of predicates, we
can avoid the paradox by simply imposing the lesser grammatical
constraint of excluding all λ-abstracts that are not homogeneously
stratified.[3] (See my 1986b article, Section 9, for an explanation of
why heterogeneous stratification is not restrictive enough.) This still
allows for the meaningfulness of $F(F)$ and $\sim F(F)$, or even of $[\lambda x\phi]$
$\times ([\lambda x\phi])$, so long as $[\lambda x\phi]$ is homogeneously stratified. Note however
that since the λ-abstract involved in Russell's paradox is not homogeneously stratified, it is excluded by this weaker grammatical constraint as not being well-formed. This means that the original comprehension principle,

$$(\mathrm{CP}^*_\lambda) \quad (\exists F^n)([\lambda x_1 \ldots x_n \phi] = F),$$

where F^n does not occur free in ϕ, is now restricted to what we shall
call the *homogeneously stratified comprehension principle* $(\mathrm{HSCP}^*_\lambda)$.

Except for the restriction to homogeneously stratified λ-abstracts, it should be emphasized, everything in the original context of our paradigmatic second order predicate logic with nominalized predicates remains as it was. We call this system $\Box\lambda HST^*$ (or just λHST^* if $\Box$ is dropped as a logical constant). In my 1986, it is shown that $\Box\lambda HST^*$ is consistent if weak Zermelo set theory is consistent, and also that it is equiconsistent with the theory of simple types.

It is noteworthy that even the restriction to homogeneously stratified λ-abstracts can be dropped in favor of the original grammar of our paradigm if we modify instead the standard first order logic that was also part of that paradigm and allow for denotationless singular terms; that is, if we switch to a first order logic that is free of existential presuppositions. In this way we can retain the original comprehension principle (CP_λ^*), including the instance involved in Russell's paradox, but note that all that follows by Russell's argument is that

$$\sim(\exists y)([\lambda x(\exists G)(x = G \,\&\, \sim G(x))] = y)$$

is now provable. This does not contradict

$$(\exists F)([\lambda x(\exists G)(x = G \,\&\, \sim G(x))] = F)$$

as an instance of (CP_λ^*), it should be noted, but only requires that we distinguish the role of $[\lambda x(\exists G)(x = G \,\&\, \sim G(x))]$ *as a predicate* from its role *as a singular term*. Russell's contention that nominalized predicates denote as singular terms the same entities that predicates otherwise stand for in their role as predicates must be rejected, and something like Frege's original contention of a distinction between unsaturated concepts and saturated objects retained (see my 1986b, Section 17, for an interpretation of this distinction different from Frege's). All of the properties and relations denoted by singular terms in $\Box\lambda HST^*$ can be retained, however, by adding to our modified context a special axiom $(\exists/HSCP_\lambda^*)$ to that effect 1986b, Section 15 and 1985, Section 7). We call the resulting system $HST_{\lambda\Box}^*$ (or just HST_λ^* if $\Box$ is dropped as a primitive logical constant). In my 1986 it is shown that $HST_{\lambda\Box}^*$ is equiconsistent with $\Box\lambda HST^*$, and therefore with the theory of simple types as well.

Both $\Box\lambda HST^*$ and $HST_{\lambda\Box}^*$, it should be emphasized, are reconstructions of the original context of our paradigm. Both, in their own way, make the logical point of the theory of types, but without also imposing the grammatical restrictions of the latter. This logical point

64   NINO B. COCCHIARELLA

was actually already present in Frege in his hierarchy of unsaturated
concepts and was taken over by Russell in his attempt to avoid his
paradox (1986b, Section 8). In this regard we are not really changing the original context of our paradigm so much as correcting the
way that properties and relations are to be posited in that paradigm.

We should also note here that $\Box \lambda HST^*$ has been used in Montague
grammar by Gennaro Chierchia in place of Montague's own sense-
denotation intensional logic, and that in fact Chierchia has shown
$\Box \lambda HST^*$ to be a more superior semantical framework over all known
alternatives as a way of explaining a variety of issues in linguistics (see
Chierchia 1984, 1985). This application of $\Box \lambda HST^*$ has involved the
construction of a precise translation function from English into
$\Box \lambda HST^*$, and, as a result, the objection that a conceptual analysis of
natural language in terms of $\Box \lambda HST^*$ requires the use of circumlocution and paraphrase is without force. Since such a translation will
apply to $HST^*_{\lambda \Box}$ as well, these results indicate how either system can
be taken as a reconstruction of our original paradigmatic view of logic
as language.

Finally, let us note that the principle of rigidity, (PR), formulated
earlier for higher order modal predicate logic is also well formed in
$\Box \lambda HST^*$ and $HST^*_{\lambda \Box}$. Adding (PR) to either of these systems,
moreover, implies the existence of possible worlds in the sense of
world-propositions, just as it does in higher order modal predicate
logic. But there is another sense of possible world now available as
well – the notion of a possible world as a maximal "class" of compossible facts. Thus, where facts are defined as true propositions (or as
what true propositional forms denote when nominalized), i.e., where

$$\text{Fact} =_{df} [\lambda x (\exists P)(x = P \ \& \ P)],$$

and a class in the logical sense is just a rigid$_1$ property, this notion of a
possible world can be defined as follows:

$$\text{Poss-Wld}_2 =_{df} [\lambda x (\exists F)(x = F \ \& \ \text{Rigid}_1(F) \ \& \\ \Diamond (\forall y)[F(y) \leftrightarrow \text{Fact}(y)])].$$

Now just as (PR) implies the existence of possible worlds in the sense of world-propositions, so too (PR) in either of our reconstructed systems implies the existence of possible worlds in the sense of maximal "classes" of compossible facts. That is,

$$\square(\exists F)(\text{Poss-Wld}_2(F) \ \& \ (\forall y)[F(y) \leftrightarrow \text{Fact}(y)])$$

is provable on the basis of (PR) in either of these systems. We should note here that none of these results depends on the problematic view of properties and relations in intension as being identical when necessarily co-extensive. That is, the principle of intensionality,

$$\square(\forall x_1) \cdots (\forall x_n)(\phi \leftrightarrow \psi) \rightarrow [\lambda x_1 \ldots x_n \phi] = [\lambda x_1 \ldots x_n \psi],$$

need not be taken as a basic law of logic in either of these reconstructions of our original paradigm of the view of logic as language.

6. A SET-THEORETIC SEMANTICS WITH PREDICATION AS FUNDAMENTAL

As reconstructions of the original paradigm of the view of logic as language, both $\square\lambda\text{HST}^*$ and $\text{HST}^*_{\lambda\square}$ take predication, and not membership, as fundamental. This is a feature we should retain, accordingly, even in the set-theoretic semantics we shall construct for these systems. Indeed, as we shall see, with predication given its own special representation within a semantic structure, the distinction between standard and nonstandard models becomes void, which is as it should be since the classes we want to represent in these structures are classes in the logical sense.

One of the notions that we shall use in this semantics is the idea of a Fregean correlation between the entities that predicates stand for in their role as predicates and the objects or individuals that are denoted by their nominalizations when they occur as abstract singular terms. We do not preclude the possibility that this correlation is the identity function on the entities in question, however; that is the semantics is neutral between a Fregean and a Russellian interpretation of nominalized predicates. The semantics is neutral in other respects as well, as we shall see, but these other features will not concern us here. (See my 1978 for a fuller discussion of the issue of a semantics that is neutral between competing versions of logic as language.)

Since the semantics we describe here is a modalized and Fregean modification of John Simms's (1980) semantics, we shall refer to the

66   NINO B. COCCHIARELLA

semantic structures in question as modal S^*-structures. (The present
version of this semantics was first given in my (1986, Chap. 6) and
another, slightly different, version was given in my (1978, Section
11).) Accordingly, where $\mathcal{A} = \langle D_n, E_n, H_i, f_i \rangle_{n \in w,\, i \in W}$, we shall say that
$\mathcal{A}$ is a modal S^*-*structure* if and only if

   (1) $E_n \subseteq D_n$, for all $n \in \omega$;
   (2) $D_{m+1} \cap D_{n+1} = 0$, for all $m, n \in \omega$ such that $m \neq n$;

   (3)   for $i \in W$, $H_i \subseteq \bigcup_{n \in \omega} (D_{n+1} \times (D_0)^n)$;

   (4)   for $i \in W$, f_i is a function from $\bigcup_{n \in \omega} D_n$ into D_0 such that
              for all $d \in E_0$, $f_i(d) = d$;

and

   (5)   for $n \in \omega$, W and D_n are not empty.

By way of explanation, "think of" the members of W as possible
worlds and the sets D_n and E_n as the ranges, respectively, of the free
and the bound variables of type n. Then, what clause (1) requires,
accordingly, is that all values of the bound variables of type n are also
values of the free variables of type n, for all $n \in \omega$. Structures in which
E_n is a proper subset of D_n are free of existential presuppositions
regarding expressions of type n. However, since in our present case
we shall want the comprehension principle of either $\Box\lambda HST^*$ or
$HST^*_{\lambda\Box}$ to be externally valid, we shall exclude those structures in
which $D_n \neq E_n$, for $n > 0$. Note, however, that in $HST^*_{\lambda\Box}$ singular
terms are free of existential presuppositions, and therefore in the
structures characterizing validity in $HST^*_{\lambda\Box}$, E_0 is a proper subset of
D_0. In $\Box\lambda HST^*$, on the other hand, singular terms, including
nominalized predicates, are posited as always denoting, and therefore
in structures characterizing validity in $\Box\lambda HST^*$, $E_0 = D_0$. These
differences between the structures characterizing set-theoretic validity
for $\Box\lambda HST^*$, as opposed to those characterizing set-theoretic validity
for $HST^*_{\lambda\Box}$, it should be emphasized, are determined by criteria
internal to these systems as separate versions of the view of logic as
language.
   What clause (2) in our definition of a modal S^*-structure requires is
that no m-ary universal is an n-ary universal if $m \neq n$, and this of
course is as it should be. Where $i \in W$, clause (3) describes the
set-theoretic relation that we are to "think of" as predication with

respect to I. Thus, in particular, we are to "think of" the first component of the ordered pairs in this relation as an n-ary universal (a member of D_{n+1}), for some $n \in \omega$, and the second component as an n-tuple drawn from the domain of individuals. There is nothing in this description, it should be emphasized, that involves interpreting predication at i as membership; that is, there is no presumption that the relation in question has any similarity whatsoever to membership (of an n-tuple) in a set. Of course, in general, relative to the internal criteria of a given theory of logical form as a version of the view of logic as language, including especially the comprehension principle that is internally valid in such a theory, the relation representing predication at a given possible world will have a more determinate structure than is described in clause (3); and because it is determined by the laws of logic of the theory in question, this more determinate structure will remain invariant across all possible worlds, even though the particular relation assigned to predication at one possible world will in general be different from that assigned at another. This latter feature, i.e., that different relations may be assigned to predication at different possible worlds, is a consequence of the essential indexical nature of predication. That predication is essentially indexical, and not "timeless" or "worldless", as is sometimes claimed, explains why the same property can be predicated of different objects at different possible worlds (or at different times in the same possible world), whereas a set, having its being in its members, cannot have different members at different possible worlds (or at different times in the same world).

Finally, where $i \in W$, the function f_i is to be our Fregean correlation at i of universals with individuals. We allow this correlation to be indexical, i.e., to vary at different possible worlds, only in the semantics that is neutral between alternative theories. For example, in a modal counterpart of Frege's theory, a nominalized predicate is interpreted as denoting at a given possible world the class that is the extension at that world of the universal otherwise assigned to that predicate. But since the same universal will in general have different extensions at different possible worlds, it follows that a nominalized predicate may denote different objects at different possible worlds. In other words,

$$(\exists F)(\exists G)(F = G \ \& \ \Diamond F \neq G)$$

is consistent in such a modal Fregean framework (1986b, Section

16). In $\Box \lambda \text{HST}^*$ and $\text{HST}^*_{\lambda \Box}$, however, Leibniz's law is valid in its unrestricted form, and as a consequence, for all $i, j \in W$, $f_i = f_j$. A modal S^*-structure where this holds is said to be *rigid*. As indicated, set-theoretic validity for $\Box \lambda \text{HST}^*$ or $\text{HST}^*_{\lambda \Box}$ will be defined only with respect to rigid modal S^*-structures that satisfy the basic laws of logic of such a system. (Some principles will be valid in all rigid modal S^*-structures. These are the principles of the minimal system $\Box \lambda M^*$ described in my 1986 (Chap. 5 and 1978, Section 10.)

We shall forego the details here of the definition of satisfaction in a modal S^*-structure at a given possible world (1986, Section 6.5). We should note, however, that unlike the situation in the usual set-theoretic semantics for second and higher order predicate logic (where predication is interpreted as membership), this definition in no way depends on either the notion of a standard model that is "full" (in the sense of the iterative concept of set), or of a nonstandard model that, if not "full", is at least closed under conditions that enable us to assign semantic values to complex expressions. Indeed, with predication taken as fundamental in our semantics as well as our syntax, there is no basis at all for the kind of distinction that obtains between standard and nonstandard models when predication is interpreted as membership. For with predication as fundamental, n-ary universals are no longer analyzed or represented by functions from possible worlds to sets of n-tuples, but are rather directly represented in the same way that possible worlds and individuals are. This cuts the ground completely from the kind of incompleteness theorem that is generated with respect to the so-called standard models of higher order logic; but this is as it should be, for reasons already explained, when it is intended that predication, and not membership, be taken as fundamental in the higher order logic in question. It also cuts the ground from the idea that properties and relations in intension are identical when they have the same extension in every possible world.

Finally, it should be emphasized that except for a certain minimal system of principles that are valid in all (rigid) modal S^*-structures, a completeness theorem is not forthcoming for a system such as $\Box \lambda \text{HST}^*$ or $\text{HST}^*_{\lambda \Box}$ except by restricting the S^*-structures to those in which the basic laws of the system are valid. (A formula is valid in an S^*-structure if it is true at all possible worlds of that structure.) This is not because these systems are incomplete otherwise, but rather because it is only by excluding certain S^*-structures that we can give a set-theoretical (or really mathematical) representation of the additional content of the basic laws of logic that are valid in these systems over and above those of the minimal system. In other words, on the view of logic as language, where predication and not membership is fundamental, the set-theoretical definition of validity for a system such as $\Box\lambda\mathrm{HST}^*$ or $\mathrm{HST}^*_{\lambda\Box}$ provides a strictly external and extrinsic criterion that in no way is essential to the notion of validity that is internal to this system. The real content of our logic, on this view, is not in our set theory, but in the logic itself as a formal theory of predication.

7. CONCLUDING REMARKS

The account we have given here of the view of logic as language should not be taken as a rejection of the set-theoretical approach or as defense of the metaphysics of possibilist logical realism. Rather, our view is that there are really two types of conceptual framework corresponding to our two doctrines of the nature of logic. The first type of framework is based on membership in the sense of the iterative concept of set; although extensionality is its most natural context (since sets have their being in their members), it may nevertheless be extended to include intensional contexts by way of a theory of senses (as in Montague's sense-denotation intensional logic). The second type of framework is based on predication, and in particular developments it is associated with one or another theory of universals. Extensionality is not the most natural context in this theory, but where it does hold and extensions are posited, the extensions are classes in the logical and not in the mathematical sense.

Russell's paradox, as we have explained, has no real bearing on set-formation in a theory of membership based on the iterative concept of set, but it does bear directly on concept-formation or the positing or universals in a theory based on predication. As a result, our second type of framework has usually been thought to be incoherent or philosophically bankrupt, leaving us with the set-theoretical approach as the only viable alternative. This is why so much of analytic philosophy in the 20th Century has been dominated by the set-theoretical approach. Set theory, after all, does seem to serve the purposes of a *mathesis universalis*.

What is adequate as a *mathesis universalis*, however, need not also

NINO B. COCCHIARELLA

therefore be adequate as a *lingua philosophica* or *characteristica universalis*. In particular, the set-theoretic approach does not seem to provide a philosophically satisfying semantics for natural language; this is because it is predication and not membership that is fundamental to natural language. An adequate semantics for natural language, in other words, seems to demand a conceptual framework based on predication and not on membership.

It has been our contention here that Russell's paradox has not really nullified the second type of conceptual framework, i.e., one based on predication. And in fact we have shown that one can return to Frege's and Russell's original paradigm of second order predicate logic with nominalized predicates as a coherent and philosophically useful theory of predication. Being based on predication, this framework is not subject to the incompleteness problem the way a theory of membership is. In addition, we do not need to resort to circumlocution or paraphrase in the application of this framework to natural language. We have described this framework, and motivated our discussion throughout, primarily in terms of logical realism as its associated theory of universals, i.e., with propositions, properties and relations in intension as the basic entities involved in predication, and with possible worlds as either certain kinds of propositions or certain kinds of properties. We have done so because this ontology, together with a commitment to possible individuals, seems to be implicit in natural language. But we do not mean to claim that this is the only theory of universals, or that $\Box \lambda HST^*$ and $HST^*_\lambda \Box$ with or without (PR), are the only formal theories of predication that can provide an adequate semantics for natural language. In other words, there are other theories of universals, as well as other formal theories of predication, and only a comparison of their strengths and weaknesses will help us to decide which we should adopt as a framework for natural language. (It is this sort of study that I have initiated in my (1986) book and my 1988 article.)

We do not maintain, accordingly, that we should give up the set-theoretic approach, especially when dealing with the philosophy and foundations of mathematics, or that only a theory of predication associated with possibilist logical realism will provide an adequate semantics for natural language. In both cases we may find a principle of tolerance, if not outright pluralism, the more appropriate attitude to take.

NOTES

[1] Without type-restrictions, Montague's sense-denotation intensional logic would amount in effect to a first order set theory supplemented with a theory of senses in the form of the sense- and denotation-forming operators $^\wedge$ and $^\vee$, respectively. That is, it would then amount to an *applied* first order theory with ϵ and $^\wedge$ and $^\vee$ as its primitive *nonlogical constants*. That might actually be preferable to Montague's own description of $^\wedge$ and $^\vee$ as logical constants; for unlike quantifiers and sentential connectives, these operators have no counterparts in natural language, and the distinctions they are used to articulate are more in the order of an applied theory of logical form than as fundamental constituents of logical forms themselves. This is especially true of the hierarchy of senses that is represented by iterated applications of $^\wedge$. For what is represented in this hierarchy is really not in the logical forms of the expressions of natural language, but is rather an interpretation of the occurrence of these expressions in indirect discourse. On Montague's theory, it is, after all, the same word or expression of natural language that is assigned not only a direct sense but an indirect sense as well, and an indirect indirect sense, and so on *ad infinitum*.

[2] It is noteworthy that Russell was led to the theory of types by considering Frege's hierarchy of unsaturated concepts (see 1986b, Section 8). In this hierarchy, however, as I have explained in 1985, Section 2 and 1986b, Sections 4–5, the concepts of any given higher level can be mapped one-to-one into the concepts of the preceding level, just as first level concepts can be mapped one-to-one with certain saturated objects called concept-correlates (or Frege's *Wertverläufe*, given his commitment to extensionality). There is no difference in cardinality, between the concepts of any one level and those of any other level of Frege's hierarchy.

[3] A formula or λ-abstract is homogeneously stratified iff there is an assignment t of natural numbers to the set of terms and predicate expressions occurring in ϕ such that (1) for all terms a, b, if $(a = b)$ occurs in ϕ, then $t(a) = t(b)$; (2) for all $n \geqslant 1$, all n-place predicate expressions π and all terms $a_1, \ldots, a_n$, if $\pi(a_1, \ldots, a_n)$ occurs in ϕ, then (i) $t(a_j) = t(a_k)$, for $1 \leqslant j$, $k \leqslant n$, and (ii) $t(\pi) = t(a_1) + 1$; and (3) for all natural numbers m, all individual variables $x_1, \ldots, x_m$, and all formulas ψ, if $[\lambda x_1 \ldots x_m \psi]$ occurs in ϕ, then (iii) $t(x_j) = t(x_k)$, for $1 \leqslant j$, $k \leqslant m$, and (iv) $t([\lambda x_1 \ldots x_m \psi]) = t(x_1) + 1$.

REFERENCES

Chierchia, Gennaro: 1984, *Topics in the Syntax and Semantics of Infinitives and Gerunds*, Ph.D. dissertation, University of Massachusetts, Amherst.

Chierchia, Gennaro: 1985, 'Formal Semantics and the Grammar of Predication', *Linguistic Inquiry* **16**, 417–43.

Cocchiarella, Nino B.: 1986, *Logical Investigations of Predication Theory and the Problem of Universals*, vol. 2 of Indices, Bibliopolis Press, Naples.

Cocchiarella, Nino B.: 1978, 'On the Logic of Nominalized Predicates and its Philosophical Interpretations', *Erkenntnis* **13**, 339–69; corrigendum in vol. 14, pp. 103–04.

Cocchiarella, Nino B.: 1980, 'The Development of the Theory of Logical Types and the Notion of a Logical Subject in Russell's Early Philosophy', *Synthese* **45**, 71–115.

Cocchiarella, Nino B.: 1985, 'Frege's Double Correlation Thesis and Quine's Set Theories NF and ML', *Journal of Philosophical Logic* **14**, 1–39.
Cocchiarella, Nino B.: 1986b, 'Frege, Russell and Logicism: A Logical Reconstruction', in Haaparanta Leila and Jaakko Hintikka (eds.), *Frege Synthesized*, D. Reidel, Dordrecht, 197–252.
Cocchiarella, Nino B.: 1988, 'Philosophical Perspectives on Formal Theories of Predication', in *Handbook of Philosophical Logic*, vol. IV, Gabbay, Dov and Franz Guenthner (eds.), *Handbook of Philosophical Logic*, vol. IV, D. Reidel, Dordrecht.
Gallin, Daniel: 1975, *Intensional and Higher-Order Modal Logic*, North-Holland, Amsterdam.
van Heijenoort, Jean: 1967, 'Logic as Language and Logic as Calculus', *Synthese* **17**, 324–30.
Hintikka, Jaakko: 1980, 'Theories of Truth and Learnable Languages', in S. Kanger and S. Ohman (eds.), *Philosophy and Grammar*, D. Reidel, Dordrecht. pp. 37–57.
Hintikka, Jaakko: 'Semantics: A Revolt Against Frege', in G. Fløistad (ed.), *Contemporary Philosophy: A New Survey*, Martinus Nijhoff, The Hague, pp. 57–82.
Montague, Richard: 1974, *Formal Philosophy, Selected Papers of Richard Montague*, edited and introduction by R. H. Thomason, Yale University Press, New Haven. The papers referred to in this volume are the following:
[EFL]:   'English as a Formal Language'.
[PE]:   'On the Nature of Certain Philosophical Entities'.
[PTQ]:   'The Proper Treatment of Quantification in Ordinary English'.
[UG]:   'Universal Grammar'.
Moravcsik, J.: 1979, 'Grammar and Meaning', in I. Niiniluoto and M. B. Provence Hintikka (eds.), *Essays in Honour of Jaakko Hintikka*, D. Reidel, Dordrecht, pp. 3–15.
Prior, Arthur and Kit Fine: 1977, *Worlds, Selves, and Times*, Duckworth Press, London.
Quine, W. V. O.: 1968, 'Replies', in *Essays in Honor of W. V. Quine*, *Synthese* **19**.
Simms, John: 1980, 'A Realist Semantics for Cocchiarella's T*', *Notre Dame Journal of Formal Logic* **21**, 1–32.

Department of Philosophy
Indiana University
Bloomington, IN 47405
U.S.A.

ANALYSIS 53.3   JULY 1993

Logicism, The Continuum and Anti-Realism

PETER CLARK

1. Introduction

Within the foundations of mathematics it is traditional to see the irreconcilability of realism and constructivism as emerging, at least in its starkest form, at the level of the classical continuum or the real line. However I argue here that there is a sense in which the dichotomy between classical mathematics and anti-realism cannot really be confined to the level of the continuum, for there is a strong argument that it must also penetrate to the level of the theory of the natural numbers.

This can be brought out most effectively by looking at some recent work on the consistency of the formal system implicit in Frege's *Grundlagen* (Boolos [1], [2], [3] and Wright [23]) and, in particular, a defence of a version of number theoretic logicism by Wright. This will be contrasted with a recent attack – also by Wright ([24], pp. 131–37), but in the spirit of Dummett's anti-realism – on the coherence of Cantor's famous diagonal argument ([6], [7]) for the existence of an uncountable set, viz. the continuum, which is surely the primary distinctive result of set theory. What Wright attacks here is the notion of an arbitrary subset of natural numbers, a notion that seems to be needed for us to be able to claim that there is in fact an uncountable set. Yet it is precisely this notion which is crucial to the non-contradictory reconstruction of Fregean logicism which Wright wants to defend as being plausible and thus at least coherent. Thus it seems that, *either* the argument against Cantor shows that the modern logicist account of arithmetic is incoherent along with the Cantorian conception of the classical continuum, incomplete as it may be; *or* the attack on the latter fails. The point here is of general interest and is not just an *ad hominem* argument against Wright, for his position is unusual and challenging in that it tries to combine logicism (admittedly of a sophisticated form) about the natural numbers with constructivism of a very radical kind about the real numbers. This is very much against the general tradition in the philosophy of mathematics which sees as the only viable position constructivism for both or neither. I shall argue that this middle way is incoherent and this for rather deep and general reasons.

ANALYSIS 53.3, July 1993, pp. 129–41. © Peter Clark

2. *Arbitrary Subsets and the Diagonal Argument*

The informal mathematical argument employed by Cantor to demonstrate that the real line, the continuum, is uncountable has two quite separate components. The first component is constructively interpretable and shows that no list can exhaust the real numbers in the sense that for any purported listing of the reals, effective or otherwise, a real number exists which is not on the list. The second component is decidedly not constructively interpretable and relies on the collection of *all* subsets of the natural numbers to form a definite collection $P(\omega)$, the power set of the set of natural numbers. Only by accepting both components of the informal reasoning does one obtain the result that there is a set $P(\omega)$ which is uncountable. Since the two components are distinct, it is possible to accept the constructively interpretable part while rejecting the second, non-constructive component. Indeed this is precisely what Poincaré urged as the correct analysis of Cantor's reasoning.

But what is it in the second, non–constructive component of the proof which is unacceptable? Clearly it is not the basic idea of set formation itself, for there seems to have been no objection *tout court* to the notion of a countable collection. The core of Poincaré's objection was the idea of an *arbitrary subset* of the natural numbers, one specified by an arbitrary property. He objected to the idea that any property whatsoever will in fact determine or specify a subset of the natural numbers. Poincaré was however notoriously unclear as to exactly what he would have regarded as a constructively acceptable restriction on the notion of property of the natural numbers.

One straightforward way of interpreting him (there are others, like those pursued by predicative analysis) is that the only acceptable properties are those generated by what we would now term a computable rule. Thus if we restrict attention to those irrational numbers whose decimal expansion is effectively computable, what the diagonal argument shows is that this collection although denumerable, cannot itself be effectively enumerated (this in effect is the result given in Turing's original paper [22]). For any effective enumeration proceeding according to some law or rule which purportedly lists all effectively enumerable irrationals, it is possible to diagonalize in a completely effective way to produce an irrational number not already on the list, thus defeating the claim that the list contains *all* such irrationals.

Now Poincaré was not possessed of the modern concept of effective enumerability, and cast his objection by focusing attention on those irrationals which can be specified by a finite number of words, which is close to saying that they should be specifiable by a predicate in a countable language.[1] He called the position that insists that all the objects cited by mathematics be definable 'in a finite number of words' *pragmatism* and gives his account of how the pragmatist should understand Cantor's reasoning as a demonstration of the 'disruptibility' of any purported law of correspondence between the points of a line specifiable by a finite number of words (predicable) and the points on a line. It's worth quoting him in full, for what he gives is the paradigm 'constructive' interpretation of Cantor's result. He writes

> For example, the pragmatists admit only objects which can be defined in a finite number of words; the possible definitions, which can be expressed in sentences, can always be numbered with ordinary numbers from one to infinity. According to this reckoning, there would be only a single infinite cardinal number possible, the number Aleph–zero. Why, then, do we say that the power of the continuum is not the power of the integers? Yes, being given all the points in space which we can define with a finite number of words, we can imagine a law which is itself capable of being expressed in a finite number of words and which establishes a correspondence between them and the set of integers. But let us now consider sentences in which the notion of this law of correspondence is involved. A moment ago these sentences had no meaning since this law had not yet been invented, and they could not serve to define points in space. Now they have acquired a meaning: they will permit us to define new points in space. But these new points will not find any room in the classification already adopted, and this will compel us to upset it. And this is what we mean, according to the pragmatists, when we say that the power of the continuum is not the power of the integers. We mean that it is impossible to establish between these two sets a law of correspondence which will be free from this sort of disruption; whereas it is possible to do it, for example, when a straight line and a plane are involved. ([21], p. 68)

Clearly then Poincaré accepts the constructive part of Cantor's proof, but rejects the notion of an arbitrary subset of ω as unintelligible. One should note that there is a rather clear sense in which in the passage quoted Poincaré is trying to have his cake and eat it. He is prepared to recognize the real line and the plane as sets or collections, at least as entirely legitimate mathematical objects, but he is not prepared to think of those collections as obtained from the natural numbers in any way. Thus he

---

[1] Hallett has distingushed two versions of this thesis – first the claim that a set cannot be said to exist until there is a specification of it, and second the much stronger claim that before a set can be said to exist we must be in possession of specifications of all its potential members. The second stronger claim cannot be satisfied by set theory. Cf. [16] §2.

regards the plane and line as independent mathematical objects and our knowledge of them as given synthetic a priori independently of any knowledge of the natural numbers. This thought will not be pursued here as it would take us too far from the main point at issue but it is certainly worth noting that this sort of semi-intuitionism as it has come to be called does leave the impression of theft over honest toil. (See particularly Folina [12], Goldfarb [13].)

3. Wright's Argument Against Cantor

Now Wright has extended considerations of what I shall call the Poincaré type to show apparently that the very concept of an uncountable collection is incoherent. Wright's argument goes as follows. Since we know that the most the constructively interpretable part of the proof establishes is the non-effective enumeration of the irrational numbers then in order to obtain a *non*-denumerable collection of irrationals we will have to avail ourselves of the notion of an *arbitrary* subset of natural numbers. But before we do this we shall have to be able to give an explanation of the notion of an arbitrary subset however widely construed as opposed to that of subset specifiable by 'constructively' acceptable means. Wright argues that no such explanation is possible and presents two arguments why this is so. The first concerns the Zermelo separation schema and the second concerns possible selections. I shall concentrate on the first of Wright's negative claims.

The Separation schema specified by Zermelo in 1908 arose as a response to the contradiction embedded in Frege's comprehension principle (1893) for class existence. Zermelo's idea was to allow determinate properties not to determine extensions absolutely but to determine sub-extensions of already existing, given extensions. Motivated by this principle of 'limitation of size' the Separation schema reads:

$$(\forall x)(\exists y)(\forall z)(z \in y \leftrightarrow (z \in x \ \& \ \phi z))$$

As Zermelo puts it

> ... sets may never be independently defined by means of this axiom but must always be separated as subsets from sets already given; thus contradictory notions such as the set of all sets ... are excluded. ([25], p. 202)

Employing this schema over the natural numbers N, we can reduce the question of what subsets of the natural numbers there are to the question: what bona fide properties of the natural numbers are there? To put it another way, what specifications are we allowed to substitute in the separation schema, what does ϕ range over? The connection between the naive comprehension schema and the Zermelo principle suggests that we can substitute whatever we like, and it is just that to which Wright objects. What Wright focuses on is the link between *property* and 'possible predication'. What he suggests is that, if (as we should) we restrict ourselves to a countable language, then the link between property and predication must be severed. He regards this as unacceptable. He writes

> If, on the other hand, the Cantorian severs the connection between property and possible predication he is open to the charge that he is obscurantizing the notion of a property and thus can use it to explain nothing. ([24], p. 136)

Clearly if every property in order to be legitimately so regarded has to be expressible in some suitably chosen countable first order language, which is a requirement very close to Poincaré's, then the second component of Cantor's proof will fail. For now we shall have at most countably many predicate expressions and so at most countably many extensions corresponding to them. From a classical point of view we will have lost all those subsets of the natural numbers which are not the extensions of predicates formalizable within our language. We cannot of course use a second order language without begging the question because we shall not be able to interpret quantification in such a language without employing the notion of arbitrary property which is just the notion whose intelligibility is being brought under question.

4. Logicism, Arithmetic and Arbitrary Properties.

Now let us pause for a moment and ask where the notion of an arbitrary *property* (as opposed to *predication*) of the natural numbers occurs. It occurs most famously in the statement of the Peano axioms for (as it is now called) Second Order Arithmetic, being used to formulate of the induction principle. So taking 0 as an individual constant, *Nat x* as expressing that x is a natural number, and *Sxy* to mean that 'x immediately precedes y' we have the following postulates for the theory PA (Second Order Peano Arithmetic):

(1)   *Nat* 0

(2) $(\forall x)$ $(Nat\,x \rightarrow (\exists y)(Nat\,y\ \&\ Sxy))$

(3) $(\forall x)(\forall y)(\forall z)(\forall w)(Sxw\ \&\ Syz \rightarrow (x=y \leftrightarrow w=z))$

(4) $\neg(\exists x)(Nat\,x\ \&\ Sx0)$

and very importantly

(5) $(\forall F)[(F(0)\ \&\ (\forall x)((Nat\,x\ \&\ Fx) \rightarrow (\forall y)(Sxy \rightarrow Fy)))$
$$\rightarrow (\forall x)\,(Nat\,x \rightarrow Fx)]$$

where the second order quantifier $(\forall F)$ ranges over *all* properties of the natural numbers. As is very well known this second order system has the key property that if A is any first or second order sentence formulated in the vocabulary of Peano arithmetic then

PA ⊢ A iff A is true in N (the standard model of arithmetic).

So PA has the very important character that it captures *all and only* arithmetic truths.

Now let us ask an epistemological question about postulate (5) above, very similar to the epistemological question asked about the non-constructively interpretable component of Cantor's proof, viz. one which asks what the second order quantifier in (5) above is supposed to denote? Clearly it is crucial to the logicist programme of Frege's *Grundlagen* that such a notion is to be well–founded, for the concept of an arbitrary property of the natural numbers plays a crucial role in the formulation of the principles used in validating precisely those inferential steps which allow Frege to sketch out the derivation of the system very much like PA in §§73–88 of the *Grundlagen* and to establish key results like the infinity of the number series. Despite this, Dummett has expressed very considerable doubts about whether Frege would have accepted the notion of an arbitrary property of the natural numbers. In discussing the origin of the paradox in the explicit system of the *Grundgesetze*, Dummett remarks that it would be an error simply to see this as a failure on Frege's part to notice that the abstraction operator once introduced will automatically extend the domain by allowing new objects as well as new singular terms, for he argues that Frege is not committed to assuming that his function variables range over the entire classical totality of functions from the domain into the domain. Whatever the truth about this matter,[2] both of the formal systems employed by Frege (the implicit one in the *Grundlagen* and the explicit one in the *Grundgesetze*[3]) are essentially second order systems, a fact which of course Dummett strongly endorses.

Recent work by Boolos has established the consistency of the (second order) formal system implicit in the *Grundlagen* (at least relative to classical analysis). Both Boolos and Wright have further shown that the full formal programme sketched out by Frege therein can be carried out completely. The two systems developed by Boolos and Wright differ in

---

[2] Hintikka and Sandu [19] argue that Frege lacked the notions of an arbitrary set and arbitrary function and also that he lacked the concept of all properties of a domain of individuals i.e. of an arbitrary subset. However their argument is flawed as Demopoulos and Bell [10] have pointed out. The last section of their paper is particularly relevant to the issue under discussion here, namely the role of the notion of arbitrary property in the interpretation of the second order quantifier in proofs of the categoricity of the Peano postulates.

[3] There is a beautiful exposition of the system for arithmetic in the *Grundgesetze*, by Richard G. Heck in [17].

detail but the idea is the same in both, namely to embed within second-order logic a single principle characteristic of the concept of number and then to show that the system PA follows from that single principle together with second-order logic. Of course it is crucial that these systems avoid the inconsistency found in Frege's own formal system in the *Grundgesetze,* that contradiction which follows immediately from the notorious first level instance of the Vth axiom of the *Grundgesetze* viz:

$$(\forall F)(\forall G)([x]Fx = [x]Gx \leftrightarrow (\forall x)(Fx \leftrightarrow Gx))$$

(The non-benign component of Law V is from left to right.) The system devised by Wright in [23], which is consistent, does this by employing what has become known as Hume's Principle, namely the claim which Wright calls $N^=$:

$$N^=   (\forall F)(\forall G)(Nx:Fx = Nx:Gx \leftrightarrow (\exists R)(F \sim_R G)))$$

(where '$(\exists R)(F \sim_R G)$' expresses 'the concepts f and g are equinumerous').

Boolos's system in [1] is based on a related principle which I shall call The Mighty Numbers:

$$(\forall F)(\exists!x)(\forall G)(\Sigma(G, x) \leftrightarrow (G \sim F))$$

which essentially does no more than embody the insight that the only extensions that Frege needs in §73 of the *Grundlagen* are extensions of concepts of the form

is equinumerous with the first level concept ...

All that is really needed to effect the programme in the *Grundlagen* is that for every first level concept (F) there is a second level concept whose extension has as members exactly those concepts equinumerous with F. Explicitly defining the complex relational symbol $\Sigma(,)$ by

$$\Sigma(G, x) \equiv (\exists H)[x = \text{ext } H \text{ \& } H(G)]$$

the existence of such a (unique) extension is precisely what the Mighty Numbers asserts. It is interesting and very important to note, as Boolos does, that that second level concept whose existence the postulate asserts has falling under it just those first level concepts equinumerous with a given concept, and not all those concepts coextensive with the given concept. The equivalent postulate for coextensiveness, i.e.

$$(\forall F)(\exists!x)(\forall G)(\Sigma(G, x) \leftrightarrow (\forall x)(Gx \leftrightarrow Fx))$$

is importantly stronger than the Mighty Numbers, indeed it is too strong, for – as Boolos [2] proves – this, like Frege's original Law V, is inconsistent.

Let us return to the central issue and ask what has all this got to do with arbitrary property of the natural numbers? The answer is just this: some authors, especially Wright and Hale (in [14]), have argued that the surpris-

136   PETER CLARK

ing viability of the programme implicit in the *Grundlagen*, shows that
number–theoretic logicism in a certain sense is explanatory, in a deeply
revealing way, of our knowledge of number. Now not only is Hume's principle (the key postulate from which the explanation flows) itself second
order, but the derivation of ordinary arithmetic from this principle depends
crucially on the deployment of the full apparatus of second order logic. In
arguing for his version of number–theoretic logicism (he calls it number–
theoretic logicism III) Wright says

> it is possible, using the concepts of higher-order logic with identity, to
> explain a genuinely sortal notion of cardinal number; and hence to
> deduce appropriate statements of the fundamental truths of number-
> theory, in particular the Peano Axioms, in an appropriate system of
> higher-order logic with identity to which a statement of that explana-
> tion has been added as an axiom. ([23], p. 153)

In a footnote to this passage Wright elaborates as follows

> The most significant departure involved [here] is a relaxation of the
> demand for the third version of number-theoretic logicism is a relax-
> ation of the demand for the existence of a base class of number-theo-
> retic truths, deductively complete for a worthwhile chosen
> axiomatization of number-theory, such that *every* member of it can be
> transcribed into a truth of logic. Now, in contrast, we are free to avail
> ourselves of $N^=$ as a non-logical axiom; if the proof of an arithmetical
> theorem has recourse only to this axiom, to certain definitions, and to
> the techniques of second-order logic with identity, that is now good
> enough – there is no further demand to show that the theorem can
> ultimately be derived from statements each of which can be translated
> into a theorem of second-order logic. We also avoid begging the ques-
> tion whether $N^=$, even for 'unproblematic' choices of Fx and Gx, can
> rank as a definition. Rather, ... it is to be viewed as an explanatory
> axiom. A third effect, of course, is that ... the infinity of the series of
> natural numbers would be revealed to be a consequence not of truths
> of pure logic but of a postulate constituting a general explanation of
> the notion of cardinal number. The fact remains that the explanation
> *utilizes* – or so it is plausible – only logical notions; hence we are still
> considering something worthy of the title of 'logicism'. If number-
> theoretic logicism (III) can be realized, there will still be a route to an
> apprehension of the basic truths of number-theory to follow which
> will require knowledge only of the concepts, truths and techniques of
> second-order logic with identity.

The key point in this version of logicism is the issue of the deduction of statements expressing fundamental truths of number–theory in 'an appropriate

LOGICISM, THE CONTINUUM AND ANTI-REALISM   137

system of higher–order logic'.[4] Thus in Wright's system, and that of Boolos, the crucial notions of the ancestral relation of the *Begriffsschrift* (Part 3)

$$R^*xy \leftrightarrow$$
$$(\forall F)[((z)(Rxz \to Fz) \, \& \, (\forall v)(\forall w)((Fv \, \& \, Rvw) \to Fw)) \to Fy]$$

and the key proof, which constituted Frege's brilliant idea, of the possibility of defining the natural numbers as entities for which induction holds, all require that quantification over arbitrary properties be legitimate. This is just to be expected (of course) if we are employing full second order logic. Clearly the role played by second order logic cannot be overestimated here, especially if our knowledge of number is to be regarded as inferential. Indeed Wright commenting on the significance of the successful attempt to carry out explicitly the programme implicit in the *Grundlagen* says

> it will be demonstrable that statements, namely the Peano Axioms as formulated, whose collective effect is that the natural numbers constitute an infinite progression are logical consequences of a statement constituting the core of an explanation of the notion of cardinal number ... There are two corollaries: first, that it is quite unnecessary to regard the infinity of the natural numbers as a kind of assumption, of which knowledge, properly so regarded is impossible; and second, that it is quite unnecessary to advert to any rational capacity of ours other than that of inference in order to explain how our possession of this knowledge is possible. ([23], p. 168)

This is surely a perceptive analysis of the significance of what as a result of its explicit articulation by Wright and Boolos has come to be known as Frege's Theorem.[5] If (and this is quite a supposition) Hume's Principle expresses the essence of what it is to be a cardinal number and that from that using second order logic we have an explanation of the infinity of the natural numbers then this explanation avowedly relies upon full second order logic and consequently upon quantification over arbitrary properties.[6] Hence the explanation will only be as strong as the coherence of the notion of arbitrary property; if this latter is incoherent, then so is the explanation of the fundamental truths of number theory.

---

[4] The simplest deduction of the system PA from Hume's Principle is in the Appendix to [3], pp. 275–76.

[5] Bell has provided an interesting extension of Frege's theorem by showing how the existence of an infinite well-ordering can be established in a completely Fregean manner from a very general version of Zermelo's well-ordering theorem. This result is given in the Appendix to [10], and is a result wholly within second order logic.

[6] Heck and Boolos have given very general arguments to the effect that contextual definitions like Hume's principle cannot be possessed of a privileged logical status. See especially [18]; [3], pp. 271–75; [4], pp. 14–19; and [5].

138   PETER CLARK

5. A Dilemma

The difficulty in defending both logicism in the way advocated by Wright
and his anti-Cantorian argument is now obvious. Either Wright is correct
that no content can be given to the notion of arbitrary property of the natural numbers (i.e. arbitrary subset) employed in Cantor's proof that $P(\omega)$ is
uncountable, in which case any attempt to revive Frege's number theoretic
logicism must fail with it, or the deployment of full second order logic in
the explanation from Hume's Principle of the Peano axioms is viable and
consequently so is its deployment of the notion of arbitrary property. But
in this latter case why can not a Cantorian avail himself of the very same
notion in his informal proof of the existence of an absolutely uncountable
collection?

It might be thought that although in carrying out the programme
implicit in the *Grundlagen*, full second order notions of consequence and
validity were employed and quantification over arbitrary properties
allowed, it is unnecessary to use the full machinery of second order logic
to achieve this end. As such we need not be committed to quantification
over arbitrary properties including impredicatively specified ones. So for
example, we might take to heart the constraint employed by Wright in his
attack on the non-constructive component of Cantor's proof and insist on
a connection between property and predication.

There are a number of points to make about this putative objection, but
let me say immediately that even if we employ the vocabulary of second
order arithmetic, and allow quantification over arbitrary properties the
constraint which Wright wishes to impose on the Cantorian *cannot* be met
by the number theoretic logicist (III). If by the notion of 'possible predication' Wright means definability within the vocabulary of the formal theory
(the set ϕ is definable in second order arithmetic if there is a second order
formula $B(x)$ such that for any k, $k \in \phi$ iff $B(\underline{k})$ is true in N) then there are
lots of properties of the natural numbers which are not definable in second
order arithmetic – e.g. consider the property of being the Gödel number of
a true second order sentence of the language of arithmetic (which has only
four non-logical symbols for each of zero, successor, plus and times), or the
property of being the Gödel number of a valid second order sentence.
There is no predicate corresponding to these in second order arithmetic,
despite the enormous expressive power of that theory.

Further if one were to replace the second order induction axiom with the
first order induction schema, i.e. replace it by a countably infinite list of
axioms one for each property expressible in the vocabulary of first order
Peano arithmetic, thereby recapturing an intimate connection with property and possible predication, the resulting system would fail to be categorical as Gödel's incompleteness theorem shows. It would then hardly be

open to the logicist to say that he had characterized by logical means the notion of natural number, when what he had given as that characterization was satisfied by all sorts of 'non–standard' objects not obtainable from 0 by application of the successor operation. As Dedekind pointed out such a proof of categoricity is essential to any programme of characterizing by logical means the notion of natural number. He remarks

> I have shown in my reply (III), however that these facts [that N is simply infinite, that every number has a unique successor, etc.] are still far from being adequate for completely characterizing the nature of the number sequence N. All these facts would hold also for every system that, besides the number sequence N, contained a system *T*, or arbitrary additional elements *t*, to which the mapping φ [taking the elements of N onto their immediate successors] could always be extended while remaining similar and satisfying $\varphi(T) = T$. But such a system is obviously quite different from our number sequence N, and I could so choose it that scarcely a single theorem of arithmetic would be preserved in it. What then must we add to the facts above in order to cleanse our system *S* again of such alien intruders *t* as disturb every vestige of order and restrict it to N? This was one of the most difficult points of my analysis and its mastery required lengthy reflection ... Thus how can I, without presupposing any arithmetical knowledge, give an unambiguous conceptual foundation to the distinction between the elements n and the elements t? ([9], p. 100)

It is crucial to note that in the proof of categoricity of second order PA following Dedekind's idea we rely precisely on the fact that any subset whatsoever (including those impredicatively specified i.e. ones which can only be specified by reference to the set of natural numbers itself) can be a subject of the induction postulate.[7] Now the connection between categoricity proofs like the one provided by Dedekind and the notion of arbitrary subset crucial to Cantor's proof can be made much more explicit when we try to present the standard Dedekind proof in a first order system of set theory, for example Z or ZF. Inside that theory, we can state the Peano axioms PA, formalize the notion of being a model, show that ω is a model of these axioms, and then finally show that any other model of PA must be isomorphic to ω. This latter stems from recasting the induction axiom of PA in the form

$$(\forall x)((0 \in x \ \& \ (\forall y)(y \in x \to y' \in x)) \to \omega \subseteq x)$$

---

[7] The impredicativity of induction remains even if we regard induction as part of the explanation of the notion of natural number (See particularly Parsons discussion in [20]). This seems to me to present a very severe obstacle in the way of providing an *anti realist* construal of the concept of natural number, though not a logicist one.

140   PETER CLARK

(where y' here stands for the set theoretic successor of y). In fact what is shown is that ω is precisely the smallest inductive set, thus the intersection of *all* inductive sets. But since ω is itself an inductive set, the proof of the above result has got to depend on impredicative specification.

If we have this machinery for showing categoricity, then surely we possess the machinery for proving Cantor's theorem. For we have some principle of formation of subsets, one which most certainly must allow for impredicative specification. Thus if 'arbitrary subset of the natural numbers' is not obscure here, why is it obscure in the case of Cantor's theorem? If the Cantorian is to be found guilty of 'obscurantizing the notion of a property' by severing its link with predication then by the essential appeal to arbitrary property in the second order induction principle in order to exclude non-standard 'numbers', isn't the logicist to be found guilty too? This seems to me an essential point, for even if Frege, as Dummett argues, restricted quantification to range over properties expressible in his symbolism, how could the system then be said to capture all and only the truths of arithmetic. Wasn't that the logicist aim after all?

What I have argued here is that if the arguments as given above are the reasons for objecting to uncountable sets, that is if the notion of arbitrary subset is so objectionable then so must be the notion of arbitrary property of the natural numbers. The attempt to combine logicism (in the form discussed above) about the natural numbers with constructivism about the real numbers is incoherent.[8]

The University, St. Andrews
Fife, KY16 9AL

References

[1] G. Boolos, 'Saving Frege From Contradiction', *Proc. Aristotelian Society* 87 (1986–87) 137–151.

[2] G. Boolos, 'The Consistency of Frege's *Foundations of Arithmetic*', in *On Being and Saying: Essays for Richard Cartwright,* edited by J. J. Thomson (Cambridge: MIT Press, 1987), 3–20.

[3] G. Boolos, 'The Standard of equality of numbers' in *Meaning and Method: Essays in Honor of Hilary Putnam,* edited by G. Boolos (Cambridge: Cambridge University Press, 1990), 261–76.

[4] G. Boolos, 'Iteration Again', *Philosophical Topics* (1989) 5–21.

[5] G. Boolos, 'Whence the Contradiction?', *Proc. Aristotelian Society, Suppl. Vol.* 67 (1993) 213–33.

[8] I should like to express my thanks to Stephen Read, George Boolos and especially Micheal Hallett and Jim Page for their helpful comments.

LOGICISM, THE CONTINUUM AND ANTI-REALISM 141

[6] G. Cantor, 'Über eine Eigenschaft des Inbegriffes aller reellen algebraischen Zahlen' (1874), reprinted in *Gesammelte Abhandlungen Mathematischen und Philosophischen Inhalts*, edited by E. Zermelo (Berlin: Springer-Verlag, 1980), 115–18.

[7] G. Cantor, 'Über eine elementare Frage der Mannigfalttigkeitslehre'(1891), reprinted *ibid*, 278–80.

[8] M. Davis, *The Undecidable.Basic Papers on Undecidable Propositions,Unsolvable Problems and Computable Functions* (New York: Raven Press, 1965).

[9] R. Dedekind, Letter to Keferstein from the 27th February, 1890, English translation by H. Wang and S. Bauer-Mengelberg in *From Frege to Gödel: A Source book in Mathematical Logic, 1879–1931*, edited by Jean van Heijenoort (Cambridge: Harvard University Press, 1967), 99–103.

[10] W. Demopoulos and J. L. Bell, 'Frege's Theory of Concepts and Objects and the Interpretation of Second Order Logic', *Journal of Philosophy* (1993, forthcoming).

[11] M. Dummett, *Frege: Philosophy of Mathematics*, (London: Duckworth, 1991).

[12] J. Folina, *Poincaré and The Philosophy of Mathematics* (London: Macmillan, 1992).

[13] W. Goldfarb, 'Poincaré Against the Logicists', in *History and Philosophy of Modern Mathematics*, edited by W. Asprey and P. Kitcher, Minnesota Studies in the Philosophy of Science 11 (1988), 61–81.

[14] R. Hale, *Abstract Objects* (Oxford:Basil Blackwell, 1987).

[15] M. Hallett, *Cantorian Set Theory and Limitation of Size* (Oxford: Oxford University Press, 1984).

[16] M. Hallett, 'Putnam and the Skolem Paradox', in *Reading Putnam*, edited by P. Clark and R. Hale (Oxford:Basil Blackwell, forthcoming).

[17] R. Heck, 'The Development of Arithmetic in Frege's *Grundgesetze der Arithmetik*', *The Journal of Symbolic Logic*, (1993, forthcoming).

[18] R. Heck, 'On the Consistency of Second-Order Contextual Definitions', *Noûs* 26 (1992).

[19] J. Hintikka and G. Sandu, 'The Skeleton in Frege's Cupboard: The Standard versus Non-Standard Distinction', *The Journal of Philosophy* 89 (1992) 290–315.

[20] C. Parsons, 'The Impredicativity of Induction' in *Essays in Honor of Sydney Morgenbesser*, edited by I. Levi and C. Parsons (New York: Hackett Publishing Company, 1983), 132–153.

[21] H. Poincaré, *Mathematics and Science: Last Essays* (New York: Dover, 1963).

[22] A. Turing, 'On Computable Numbers with an Application to the *Entscheidungs-problem*', reprinted in [6], 115–153.

[23] C. Wright, *Frege's Conception of numbers as Objects* (Aberdeen: Aberdeen University Press, 1983).

[24] C. Wright, 'Skolem and the Skeptic', *Proc. Aristotelian Society, Suppl. Vol. 59* (1985) 117–137.

[25] E. Zermelo, 'Investigations in the Foundations of Set Theory I', reprinted in *From Frege to Gödel*, 200–215.

Name Index